Pattern Formation in Biology, Vision and Dynamics

Pattern Formation in Biology, Vision and Dynamics

Alessandra Carbone
Department of Mathematics and Computer Sciences,
University of Paris XII

Misha Gromov
Institut d'Hautes Etudes Scientifiques and
Courant Institute of Mathematical Sciences

Przemyslaw Prusinkiewicz
Department of Computer Science,
University of Calgary

World Scientific
Singapore • New Jersey • London • Hong Kong

Published by

World Scientific Publishing Co. Pte. Ltd.
5 Toh Tuck Link, Singapore 596224
USA office: 27 Warren Street, Suite 401-402, Hackensack, NJ 07601
UK office: 57 Shelton Street, Covent Garden, London WC2H 9HE

Library of Congress Cataloging-in-Publication Data
Carbone, Alessandra.
Pattern formation in biology, vision and dynamics / Alessandra Carbone, Misha Gromov, Przemyslaw Prusinkiewicz.
p. cm.
ISBN-13 978-981-02-3792-9 -- ISBN-10 981-02-3792-8
1. Pattern formation (Biology) I. Gromov, Mikhael, 1943– II. Prusinkiewicz, Przemyslaw, 1952– III. Title.

QH491.C37 1999
570'.1'1--dc21 99-089209

British Library Cataloguing-in-Publication Data
A catalogue record for this book is available from the British Library.

Dedicated to the memory of Paul Green (1931–1998)

PREFACE

Tyger! Tyger! burning bright
In the forests of the night,
What immortal hand or eye
Could frame thy fearful symmetry?

William Blake

The science of patterns should begin by trying to give a precise definition of pattern, but this seems neither feasible nor desirable. It is not feasible because patterns permeate human perception to such an extent that an objective definition seems unlikely. It is not desirable because any precise definition would only serve to limit ourselves unnecessarily. However, we will still try to address nontrivial questions:

Do patterns reflect the reality of the world? Patterns are inherent to our mental processes so it is not even clear whether they have an existence independent of our own perception—to what extent are we under collective hypnosis, tricked by a visual system eager to see Pegasus and Pisces in the Poisson distribution of stars?

By simply opening our eyes we are overwhelmed by the presence of patterns, so we can avoid overloading the brain by closing them and looking at the universe through the eye of a physicist. At the largest scales we see Lorentzian space–time gently curving according to 20 tensorial components and the Newtonian clockwork of stars and planets. At the smallest scales we see Hilbert spaces of quantum fields and 10 dimensions of superstrings. Here our analytical brain scores highest—mathematics, to the delight of Platonists, is able to capture the evanescent nature of matter.

Encouraged by such success we retain the framework of physics and look at patterns such as clouds in the sky, water flowing in a river, sand hills on a beach, and ice crystals on a frosty glass. We then try to describe what we see. Even in this limited context, human intellect is put to shame! We cannot precisely describe what we are perceiving, nor do we have a clear understanding of rules governing these common patterns. A satisfactory analysis based on the fundamental laws of physics and probability theory seems a long way off. These patterns can still be interpreted as physical phenomena, so their complexity might eventually be understood in terms of the deep mathematical interaction between local and global dynamics of matter. Yet the foliage of a tree, the concept of chair, or the face of a human being seem hopelessly beyond pure physical modeling.

Nevertheless, we can augment our model by using computational paradigms. In this way, pattern formation can be modeled as information flow in a computation while still retaining the local to global property of physical models [Prusinkiewicz].

"Local to global," a key geometric principle in representations of the world, is often expressed analytically in the language of partial differential equations (PDE's). Ubiquitous in physics, PDE's were imported into biology by Alan Turing. He envisioned how simultaneous diffusion of several interacting chemical substances might cause mosaic patterns on animals, from zebras and tigers to sea shells and fruit flies larvae [Meinhardt] [Argentina, Coullet].[1]

Such models possess symmetry, but what is actually observed is *broken symmetry,* in other words, partial symmetry (if any).[2] "Broken symmetry" does not evoke images of beauty, but aesthetically pleasing products of nature such as flowers can trace their allure to human sensibility to discrete symmetry. In fact, many flowers have finite but not full circular symmetry. According to Paul Green [Transduction to generate plant form and pattern: an essay on cause and effect, *Annals of Botany* **78**:269–281, 1996.] petals grow layer–by–layer with each consecutive layer taking on a symmetric arrangement determined by the lowest eigenfunction of an elasticity problem constrained by the previous stage of morphogenesis. Similar models were also used by Green to model phyllotaxis, the arrangement of leaves on a stem, of scales on a pine cone and of florets in a flower. Here an amazing regularity emerges and can lead to the appearance of Fibonacci numbers. Alternative models of phyllotaxis using finite nonlinear dimensional dynamics rather than PDE's are given by [Couder, Douady]. In general, PDE's become relevant if we go beyond average properties of a system and look at geometric properties, those which depend on shape not just size, e.g., the dynamics of coral reef formation [Kaandorp, Sloot].

Is everything a PDE? Is the development of all macroscopic patterns, from galactic formation to animal growth, driven by PDE's? Richard Feynman suggested the answer: "yes" with an eye on the Navier–Stokes equation governing the flow of viscous incompressible fluids, perhaps because the simple elementary principles underlying these equations result in a bewilderingly rich set of patterns such as vortices and Bernard cells.

A biologist knows better than to stick to PDE's and even the physicist's toolkit of nonlinear dynamics used in developmental biology, ecology, and the physiology of the brain, does not appear sufficient to him. Thus the molecular factories of cells are not run by PDE's because, though the underlying biochemistry is local, it is not locally linear. It seems hopeless to model global cellular behavior directly from its underlying biochemistry. Nevertheless, we can focus on specific phenomena: division, growth, differentiation. Imagine a geometric entity composed of billions of cells. One can then ask:

---

[1] This might look paradoxical at first since chemical reactions and diffusion drive the system down an energy well towards equilibrium, and, moreover, nothing visually interesting occurs in most chemical reactions. Yet, given two or more degrees of freedom, equilibrium (i.e., a fixed point of a vector field) is not necessarily approached along a straight line, but possibly along a spiral in configurations space and such a spiral orbit might easily generate regular oscillating behavior when projected back to space–time, a well known phenomenon in physics. This was discovered in chemistry by Belousov in 1951, a time when every self–respecting chemist would refuse even to consider such nonsense as oscillatory chemistry. Stationary solutions lead for instance to stripes on a tiger and spots on a leopard.

[2] In particular, the PDE's underlying this model (the reaction–diffusion equations) are invariant under translation, and if they are also *linear,* the singularities of their Fourier transform determine the frequencies of observable periodicities, i.e., solutions which exhibit symmetry which is less than the full symmetry of the underlying equation which, in fact, can only admit constants as solutions. The full richness of what we observe is ultimately a result of nonlinearity.

What kind of local division rules generate stable global patterns? Which rules preserve features during growth? Which rules ensure an increase in complexity during morphogenesis? As yet, there is no model for cell division comparable to PDE's in its universality. A mathematically simple model of cellular development is given by *cellular automata.* The idea, due to Ulam, is to fix the spatial structure by placing cells on a regular grid, and then prescribe a local rule by which each cell changes with time: the type of a cell at each generation is determined only by its type and those of its neighbors at the previous moment of time. This may simply seem to be a naïve discrete version of PDE's, but in fact, it produces an immense variety of examples. The most popular example is Conway's game of *Life*, where cells can be thought of as squares on an infinite chessboard and each square touches eight neighbors. The update rule is: a cell which is either *alive* or *dead* is alive only if in the previous generation it had exactly three live neighbors or was alive with two live neighbors. This model is deterministic, i.e., the initial state of a configuration determines its evolution for all time. Though the update rule is simple, anything conceivable can happen given enough time and space! Formally speaking, it was proved by Conway and Gosper that a universal Turing machine can be implemented in the game of Life. Whether random Life configurations eventually lead to universal Turing machines with significant probability is unclear. This question appears to be important in the modeling of biochemical life.

Numerous variations of the game of Life have been proposed to mimic growing colonies of bacteria and natural textures which may be hard to simulate by other mathematical means [Yaroslavsky]. Some cellular automata exhibit configurations which do not change with time. Such stationary solutions are essentially equivalent to *tilings,* partitions of a plane region by translates of a few shapes, e.g., a chessboard tiled by 64 squares. Here again the deceptively simple definition leads to unexpected beauty such as honeycombs and Penrose tilings. One also rapidly encounters complex questions:

Which sets of tiles can tile the plane? Which sets of tiles can only tile the plane aperiodically? It has been shown that for special sets of tiles large regions of the plane may be tiled without there being a tiling of the whole plane, in fact, the problem is undecidable. The point is that a universal Turing machine can be encoded into a suitable tiling and along similar lines one can construct aperiodic tilings [Kari].

Other models of cell growth and division allow the ambient space itself to grow with time. For example, we may take a finite configuration of cells connected to each other according to certain well defined rules and consider a cell division law which preserves the connection conditions. Once again, this model may be inherently complicated, though it has not yet been shown that it can emulate a universal Turing machine [Cannon, Floyd, Parry]. On the other hand, Lindenmayer discovered that there are rules which lead to orderly growth remarkably similar to the development of plants (c.f. [Prusinkiewicz]). For example, consider a root which sprouts two branches each of which simultaneously sprouts its own branches, etc. This results in the standard binary tree which can be made lifelike by randomly curving its branches. Such exponentially growing trees are difficult to fit into Euclidean space (as opposed to hyperbolic space) in which actual plants live. The dichotomy between unlimited growth and the limitation of space leads to mechanical stress in growing plants which somehow regulates the cell division process [Nakielski] [Hejnowicz]. In fact, not all cells can divide indefinitely, only *meristematic* ones, those which did not differentiate in the course of ontogeny [Zagórska–Marek]. Besides, when cells divide, they follow intricate geometric and combinatorial rules [Barlow, Lück, Lück] [Lück, Lück]. This may remind a mathematician of constructions used in the study of three dimensional manifolds.

Though this appears complicated it is simple compared to what happens in a cell. DNA, RNA, proteins, move in harmony tuned by billion years of evolution. We cannot hope to completely model this incredible factory but we can attempt to use some of its machinery. For example, bacteria produce restriction enzymes which cut invading viral DNA at prescribed locations and by using this process, we can split DNA then reassemble it back into new forms and topological shapes: lines, circles, knots, links, graphs [Seeman] [Jonoska] [Head]. The actual process is quite involved, but more and more delicate genetic engineering constructions are continually being developed—a new technological world of biochemical geometry is being created. In principle one can encode everything, once again the universal Turing machine [Paun]. The "biological computer" seems within reach and perhaps ultimately, biological nanotechnology [Seeman].

In our informal investigation of the emergence of patterns we have discussed a number of problems without touching on a fundamental question:

How does the brain perceive shapes? How do human beings perceive elementary components in visual stimuli and what are the laws governing how these components are integrated into a visual whole?

Such questions were addressed at the beginning of this century by the Gestalt school of psychology. It proposed an analysis of human perception in which the main factors favoring perceptual grouping are similarity, contrast, continuation, shapes, proximity, convexity, symmetry, orientation, and parallelism. These concepts might seem rather vague but they turn out to capture a number of important effects both in visual and auditory recognition. The main thrust of the Gestalt school was a list of criteria leading to a partition of visual stimulus into n (or even $\log n$) parts, whereas a naïve division of an image might instead lead to exponentially many components.

Are the above grouping rules still valid when motion is considered? There are indications supporting the fact that dynamic perception seems to influence the sense of spatial continuity, chromatic identity, and uniformity of speeds and trajectories as perceived in static images. It is here that a more subtle study of orientation, perceptual geometry, and stereoscopy come into play. Shapes determined during motion must agree with statically determined shapes and the brain seems to perform a process of *correction* to adapt to both perceptions, as well as a process of *synthesis* leading to the representation of a global shape starting from a small number of clues derived from the visual input [Ninio].

What can we say about the way our brain groups similar objects? For example, how does the brain visually characterize animals versus means of transportation? It seems plausible that there might be a simple mechanism as opposed to the complex linguistic division that we use [Thorpe].

What is the information content of a natural image? How much does an image tell us? From the point of view of Shannon's information theory, a fully randomized picture, where the color and intensity of each pixel in the screen is chosen randomly, carries maximal information so cannot be compressed. Yet in some sense, such an image can be described completely by simply stating that it is "fully randomized." More generally, one can introduce equivalence relations between images which are indistinguishable with respect to a certain class of observations. Examples are the different equivalence classes of pictures indistinguishable to human beings in less than one second and less than five minutes. According to these relations one can isolate essential patterns in images: those seen instantaneously, and those seen after several minutes of staring. The first type of patterns are the most pronounced and are analogous to singularities of smooth maps which correspond in some sense to the patterns classified by the Gestalt school. These local forms in conjunction with spline interpolation and texture generation mechanisms allow up to 1:15–1:50 ratios of compression of natural images. This often surpasses standard techniques such as wavelet decompositions [Briskin, Elichai, Yomdin].

This preface is not meant to be an exhaustive survey of pattern formation but rather an introduction to the topics presented in this collection. We would like to thank all the speakers who participated in the conference "La Formation des Motifs," Bures–sur–Yvette, France, on December 2–6, 1997. We are grateful to all the contributors to this volume for writing papers accessible to a wide audience. Finally, we would like to thank Jean–Michel Morel and Stephen Semmes for their encouragement and stimulating presence, and Helga Dernois, Jacqueline Keller, Marie-Claude Vergne and Maya Schirmann for their help in the production of this volume.

Alessandra Carbone
Misha Gromov
Przemyslaw Prusinkiewicz
with the participation of Ilan Vardi

Bures–sur–Yvette, December 31, 1998

CONTENTS

Preface vii

GROWTH AND FORM **1**

Paradigms of Pattern Formation: Towards a Computational Theory of Morphogenesis **3**
Przemyslaw Prusinkiewicz 3
1 Introduction 3
2 Single blind agent with finite memory 6
3 Single blind agent with infinite memory 7
4 Single sighted agent receiving cues from the environment (one-way exogenous control) 8
5 Single sighted agent receiving cues from the structure (two-way exogenous control) 10
6 Single self-controlled agent (endogenous control) 11
7 Multiple blind agents with finite memory 12
8 Multiple blind agents with infinite memory 14
9 Multiple sighted agents 14
10 Multiple endogenously-controlled agents 16
11 Combined effect of multiple mechanisms 18
12 Conclusions 19

Growth and Form of Sponges and Corals in a Moving Fluid **24**
Jaap A. Kaandorp and Peter M. A. Sloot 24
1 Introduction 24
1.1 Biological background 24
1.2 Simulation of growth in a mono directional flow 26
1.3 Simulation of growth in a bidirectional (alternating) flow 28
2 Methods 29
2.1 Modelling the nutrient distribution 29
2.2 Modelling the growth process in an alternating flow 31
3 Results 32
4 Discussion 37

From Pseudo-Random Numbers to Stochastic Growth Models and Texture Images **42**
Leonid P. Yaroslavsky 42
1 Introduction 42
2 Pseudo-random number generators 43
3 Growth models 45
3.1 Eden's type growth models 45
3.2 Conway's "Game of Life" and its modifications 48
4 Algorithmic models of texture images 54
4.1 The algorithmic approach to synthesis and analysis of texture images 55

Crystal Growth, Biological Cell Growth, and Geometry 65
James W. Cannon, William J. Floyd and Walter R. Parry 65
1 Introduction 65
2 A growing tree trunk whose shape oscillates wildly in time 66
2.1 Cell growth versus crystal growth 66
2.2 The groundrules 66
2.3 The plan 67
2.4 The coordination problem 68
2.5 The signal problem 68
2.6 The delay problem 69
2.7 The signal cascade 70
2.8 Using boundaries 70
2.9 Measuring shape 70
2.10 The example 71
3 The geometry of rat tissue 72
3.1 Discrete geometry 72
3.2 Euclidean geometry, the lens of the eye, and muscle fibers 73
3.3 Spherical geometry and the bladder 73
3.4 Negatively curved geometry and the lung, brain, digestive, and endocrine systems 73
3.5 Conflicting geometries 74
4 Crystals 75
4.1 Examples of crystals 75
4.2 Ways of viewing crystals 76
5 Negatively curved crystals are asymptotically fractal 78
5.1 The Cayley crystal 78
5.2 Thin triangles 79
5.3 Asymptotic shape 79
5.4 Fractal Theorem 80
5.5 Proof 80

Recent Results on Aperiodic Wang Tilings 83
Jarkko Kari 83
1 Introduction 83
2 Algebraic construction of aperiodic tile sets 87
3 Forced tiles and determinism 89
4 Beyond aperiodicity 93
5 Conclusions and open problems 94

REACTION-DIFFUSION AND BEYOND 97

Biological Pattern Formation as a Complex Dynamic Phenomenon 99
Hans Meinhardt 99
1 The problem of pattern formation 99
1.1 The egg cannot contain the final pattern 100

1.2 Organizing regions 100
1.3 The concept of positional information 101
2 Generation of a primary pattern by autocatalysis and lateral inhibition 101
2.1 The Activator-Inhibitor concept 102
2.2 Comparison with Turing's mechanism 105
3 The wave length problem — stabilization of a monotonic gradient by a feedback on the source density 106
3.1 How to generate structures close to each other, how at a distance: head, foot and tentacle formation in *Hydra* 108
4 Activation of genes under the influence of morphogenetic signalling 110
5 Formation of filament-like branching structures 113
6 Segmentation by mutual activation of cell states that locally exclude each other 115
7 Pattern formation in secondary embryonic fields: differentiation borders obtain organizing properties 116
8 Pigmentation pattern of shells of mollusk — a natural picture book to study dynamic systems 117
8.1 Traces of a stable pattern, oscillations and travelling waves: lines parallel, perpendicular or oblique to the direction of growth 118
9 Travelling waves with unusual properties 119
9.1 Formation of branches: the sudden formation of backwards-running waves 119
9.2 Penetration of waves without annihilation 121
9.3 Not all together now 123
9.4 Travelling waves without pace-maker 124
10 Complex patterns result from the superposition of two patterns 124
10.1 Superposition of two time-dependent patterns 126
10.2 Open problems in shell patterning 127

Andronov Bifurcations and Sea Shell Patterns **133**
M. Argentina and P. Coullet 133
1 Andronov bifurcations in extended systems 136
2 Sea shell patterns 139
2.1 Cavitation like patterns 141
2.2 Patterns of excitable waves 141

Rational and Irrational Angles in Phyllotaxis **145**
Y. Couder and S. Douady 145
1 Introduction 145
1.1 The geometry of phyllotaxis 145
1.2 The dynamics of phyllotaxis 146
2 The self organisation in model iterative systems 147
2.1 Hofmeister's rules: the spiral modes 147
2.2 The dynamical selection of the Fibonacci modes 148
2.3 The generalised model: Snow and Snow's rules 153
2.4 The selection of the actually observed patterns 155

3 The selection during the plant growth 156
4 Conclusions 156

CELLULAR PATTERNS **161**

Organogenetic Cellular Patterning in Plants **163**
P. W. Barlow, H. B. Lück and J. Lück 163
1 Introduction 163
2 Cell patterning from the perspective of the cell body 164
3 What lies behind analyses of cell patterning? 167
4 A simple autoregenerating cell pattern related to organogenesis 169
5 Simulation of cell patterns and the uncovery of division rules 172
5.1 The Simulation System for the *Psilotum* Apex 172
5.2 Identification of cell types 177
5.3 Merophyte formation by apical divisions 178
5.4 Wall-offsetting leads to refinement of the wall production system 179
5.5 The distribution of apical triangular cells and the problem of dichotomous axis branching 181
6 Possible regulatory factors underlying cell patterning and its simulation at the *psilotum* apex 182
6.1 A System's View of Division Wall Site Selection 182
6.2 Cytological View of Division Wall Site Selection 186
7 Was there a precursor of the *psilotum* pattern? 189
8 Prospects 193

A Classification of Plant Meristems based on Cellworks (3D L-systems). The Maintainance and Complexity of Their Cellular Patterns **199**
Jacqueline Lück and Hermann B. Lück 199
1 Introduction 199
2 Cellworks 201
3 A classification of plant forms 203
4 Specification of a cellwork system with double labelled walls 208
5 How to find a Cellwork system with an autoreproductive initial cell 209
6 Extended classification of plant forms 210
7 Passage from one generating system to another 212
8 Limits of the classification tables 212
9 From topology to geometry 212
10 Conclusion 214

Plant Meristems and Their Patterns **217**
Beata Zagórska-Marek 217
1 Introduction 217
2 Patterns of the shoot apical meristem 218
2.1 Developmental variation of cellular patterns 218
2.2 The geometry of a growing apex 222

2.3 Phyllotaxis 224
3 Cellular patterns of vascular cambium 228
3.1 Storied pattern 229
3.2 The pattern of growth activity 231
3.3 Structural wavy patterns 232
4 Conclusions 235

Mechanical Stress Patterns in Plant Cell Walls and Their Morphogenetical Importance **240**
Zygmunt Hejnowicz 240
1 Introduction 240
2 Stress anisotropy 240
3 Stress distribution 245
4 Conclusions 248

Tensorial Model for Growth and Cell Division in the Shoot Apex **252**
Jerzy Nakielski 252
1 Symplastic Growth and the Growth Tensor 252
2 Model for Growth and Cell Division in the Shoot Apex (2-dimensional) 254
3 The Example of Application of the Model 259
4 Discussion 263

DNA AND GENETIC CONTROL **269**

DNA Nanotechnology: From Topological Control to Structural Control **271**
Nadrian C. Seeman 271
1 DNA Nanotechnology 271
2 Double Helical DNA 272
3 Genetic Recombination and the Holliday Junction 274
4 Immobile Branched Junctions 275
5 Characterization of Unusual DNA Motifs and Structures 277
6 DNA Ligation and Cohesion 278
7 Branched DNA and Sticky Ends 280
8 Connectivity 281
9 Topological Features of DNA Constructs 281
10 Construction of DNA Polyhedra 282
11 DNA Catenanes and Knots 286
12 Construction of DNA Knots 288
13 DNA Borromean Rings 289
14 Rigid Components 292
15 1-D and 2-D Periodic Arrays 294
16 Nanomechanical Devices 297
17 Prospects 300

3D DNA Patterns and Computation **310**
Nataša Jonoska 310
1 Introduction 310
2 Constructing Graphs with k-Armed Branched Molecules 312
2.1 The Hamiltonian cycle problem 312
3 The Three Vertex Colorability Problem (3VC) 315
4 Recognition of Graph Structures 318
4.1 Covering graphs and complexity 318
4.2 Graphs made of one single stranded DNA molecule 319
5 Final Remarks 321

Circular Suggestions for DNA Computing **325**
Tom Head 325
1 Introduction 325
2 Naturally Occurring Plasmids 326
3 The Delphic Plasmid, DEL 327
4 Independent Subsets and Vertex Covers for Undirected Graphs 329
5 Partitions into Hamiltonian Subgraphs 331
6 Satisfiability of Boolean Formulas 332

DNA Computing by Matching: Sticker Systems and Watson-Crick Automata **336**
Gheorghe Păun 336
1 Introduction 336
2 DNA: Structure and Annealing Operation 339
3 Formal Language Theory Prerequisites 341
4 The Twin-shuffle Language and its Relation with DNA 342
5 The Operation of Sticking 344
6 Sticker Systems 346
7 The Power of Sticker Systems 349
8 Watson–Crick Finite Automata 352
9 The Power of Watson–Crick Finite Automata 353
10 Variants of Watson–Crick Automata 356
11 Conclusions 359

IMAGES AND PERCEPTION **363**

Aspects of Human Shape Perception **365**
Jacques Ninio 365
1 Shape from motion 365
2 From moving to static shapes 367
3 Static shapes: the gestalt approach 367
4 Static shapes: the importance of orientations 370
5 2D perceptual geometry 373
6 From 2D to 3D 374
7 Shape synthesis 378

Pattern Recognition in The Visual System and The Nature of Neural Coding **382**
Simon Thorpe 382
1 Introduction 382
2 The efficiency of pattern recognition by the human visual system 382
3 The Coding Problem 385
4 An alternative — Using patterns of spikes 386
5 Rank Order Coding 387
6 Rank Order Decoding 388
7 Conclusions and Perspectives 390

How can Singularity Theory Help in Image Processing? **392**
M. Briskin, Y. Elichai and Y. Yomdin 392
1 Introduction 392
2 Digital images 395
3 Compression problem 396
4 An example of data representation, based on normal forms of singularities 400
5 Singularities of digital images and their normal forms 402
6 Detection of singularities and their approximation by normal forms 408
7 Vectorized images 415
8 Creation of virtual worlds 418
9 Possible implications for human visual perception 423

Growth and Form

PARADIGMS OF PATTERN FORMATION: TOWARDS A COMPUTATIONAL THEORY OF MORPHOGENESIS

PRZEMYSLAW PRUSINKIEWICZ

Department of Computer Science, The University of Calgary
2500 University Dr. N.W., Calgary, Alberta T2N 1N4, Canada
E-mail: pwp@cpsc.ucalgary.ca

The variety of natural patterns makes it difficult to analyze and compare them in a systematic manner. We address this problem by focusing on the computational aspects of pattern formation processes. They are characterized in terms of the number of morphogenetic agents, the computing capability of each agent, and the forms of information transfer between the agents and their environment. This computational analysis can be applied to a wide range of patterns. It highlights the fundamental, algorithmic similarities between processes that may be implemented using different physical or physiological mechanisms in nature. It also confirms earlier observations that similar or identical patterns can be created by fundamentally different processes. The tradeoffs between computational characteristics of these processes lend themselves to a formal analysis, which could lead to the formulation of a "computational theory of morphogenesis" on the basis of the theory of algorithms.

1 Introduction

What is a pattern? In their treatise *Tilings and Patterns* [7], Grünbaum and Shephard wrote: "There seems to be not a single instance in the literature of a meaningful definition of a 'pattern' that is, in any sense, useful" (page 261). To address that situation, they introduced the notion of *discrete patterns*, which is based on repetitive occurrences of predefined figures called *motifs*. A pattern is assumed to be invariant with respect to some isometries in the plane, called its *symmetries*. The set of all symmetries forms a group under composition. [7]

The precision of this definition also makes it narrow: it does not encompass many geometric objects that, intuitively, we consider as patterns. This limitation was clearly stated in a subsequent paper by Grünbaum: [6]

> Sociologists of mathematics should have a field day investigating the mob phenomena that are apparent in symmetry analysis of repeating patterns. What started as a suggestion by Pólya [24] — that one *could* study ornaments from the point of view of the discrete symmetry groups — soon turned [...] into a dogma. [...] If just a small fraction of the energies that go into trying to force patterns and phenomena into a group-theoretic mold were devoted to the development of other approaches more suited in specific circumstances, both mathematics and its applications would profit.

What conceptual frameworks could be thus used to describe and analyze patterns that lack the strict repetitiveness dictated by the symmetry groups? Such patterns abound in the natural world and include, to name but a few examples for which mathematical models have been proposed, branching patterns of plants and rivers, venation patterns in leaves, arrangements of cells in organs of plants and animals,

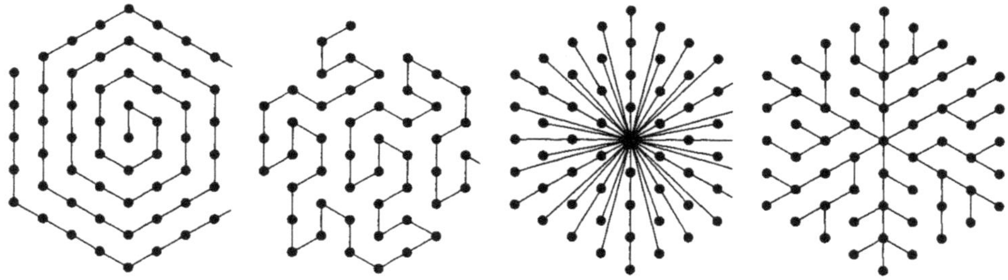

Figure 1. Prototypical patterns identified by Stevens: [38] a spiral, a meander, an explosion, and a branching pattern

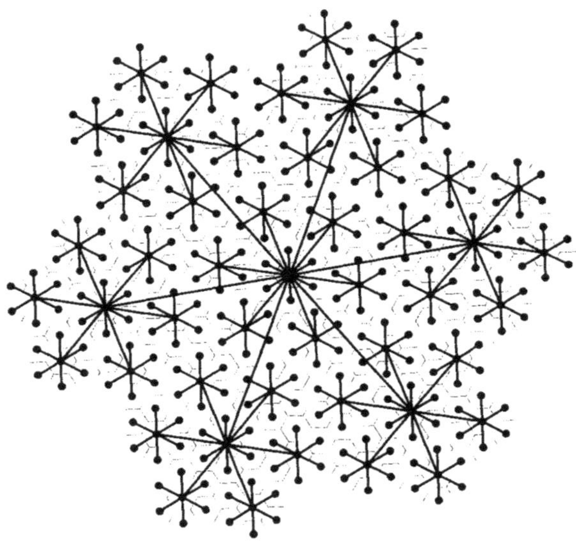

Figure 2. A variant of the branching pattern

pigmentation patterns in sea shells and butterfly wings, and animal coat patterns (see [26] for a review). Is it possible to look at these patterns in some systematic, unified manner? Can we compare different patterns, the natural mechanisms that brought them into being, and their mathematical models, using a consistent set of uniform criteria?

At some levels of abstraction, these questions have been considered in the past. Stevens [38] distinguished four prototypical classes of patterns — spirals, meanders, explosions, and branching patterns — by considering different methods for connecting a set of regularly arranged points into a graph without cycles (Figure 1). The definition of each class allows for some degree of variation, with the most significant differences occurring among branching patterns. For example, Figure 2 shows a combination of a branching pattern and an explosion.

Patterns in each class were characterized by geometric attributes: uniformity, space filling, overall length of lines, and directness of lines. This characterization proved useful when analyzing natural patterns from the viewpoint of their optimality. Such analysis has a long tradition in the description of natural patterns and forms. [9,41] Stevens extended this analysis to human-made forms, and noted that: [38]

> In the distribution of blood vessels, the route of a subway system, or the design of forked columns to carry a roof where both overall length and directness are important, the same branching pattern might turn out to be advantageous.

Unfortunately, the optimality of the result does not offer a direct insight into the mechanisms that govern pattern formation. A classification that addressed this limitation was suggested by Bell, who distinguished the following three categories of branching patterns: [3]

- 'Blind' patterns, in which branch initiation is controlled solely by the imposed programme rules, that is from without the 'organism' and its 'environment'.
- 'Sighted' patterns, in which the initiation of a new branch is influenced by factors detected by it in the immediate neighborhood, such as proximity of other organisms, or parts of the same organism.
- Self-regulatory patterns, in which branch initiation is controlled by the developing simulation itself, using communication via components of the existing framework, whether or not affected by 'environmental' factors.

This categorization is appealing for several reasons:

- It focuses on the processes of pattern formation, rather than patterns themselves. This is consistent with the biological tradition of viewing patterns and forms as derivative of developmental processes. In the words of d'Arcy Thompson, "growth deserves to be studied as a necessary preliminary to the theoretical study of form." [41]
- It sets the stage for addressing questions regarding causal relationships between the local rules and the resulting global patterns. These questions are essential to our understanding of nature, since local interactions can be implemented by a wide range of physical, chemical, physiological, and even cognitive processes.
- It focuses on the fundamental, algorithmic properties of pattern formation. This makes it possible to recognize and analyze similarities between apparently different realizations of similar patterns, such as the branching structures of plants and the branching foraging patterns of ants. [2,10]
- It does not presuppose any computational framework for model construction, such as cellular automata, [42] L-systems, [15,31] or reaction-diffusion models. [20,43] Consequently, analysis of the algorithmic nature of pattern formation processes may remain unobscured by the technical aspects of model implementations.

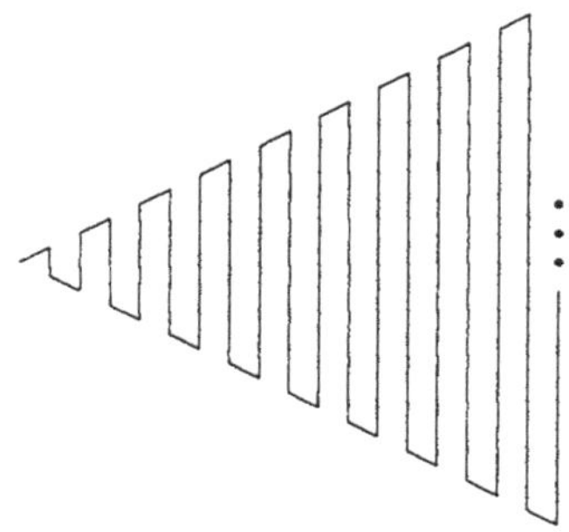

Figure 3. A zig-zag pattern

Many authors have observed that different algorithms may produce similar or identical patterns. The most extensive illustration of this point was given by Held, who devoted an entire book to the discussion of various processes that produce a pattern of periodic stripes. [8] Other case studies of different algorithms that may produce similar patterns deal with the control of branching in trees [11] and the flowering sequences in compound inflorescences. [31]

Since there is not a one-to-one correspondence between the patterns and the mechanisms that generate them, one can ask questions about the relationships between different processes that produce identical or similar patterns. In this paper we consider these questions in the context of Bell's caterization of pattern formation processes. Using a simple pattern as an example — a zig-zag with a growing amplitude (Figure 3) — we show that differences between generation mechanisms involve tradeoffs in features: the number of pattern-formation agents, the synchronization of their operation, the amount of computation that they carry out, the amount of information that they can memorize, the type of information that is exchanged between the agents, the structure and the environment, and the paths of information flow.

2 Single blind agent with finite memory

Let us consider the zig-zag pattern shown in Figure 3 as the path of an agent — a *turtle* [1] — that traces a line as it moves in the plane. The turtle operates in discrete steps: and in each step it draws a line segment of a given length or turns left or right by a given angle. The performed action is decided by a control system in the turtle's "head". Clearly, the class of patterns that can be traced by the turtle depends on the computational power of its control system. In this section we consider a "blind turtle with finite memory", in which the control is effected by an *autonomous deterministic finite automaton* (autonomous DFA). These terms mean that: (i) the turtle has finite memory, represented by a finite number of discrete states; (ii) in each step, the next state is specified in a deterministic fashion; (iii) the next state depends only on the previous state and, in particular, is not affected by any external input; and (iv) the action performed by the turtle in each step is determined by its state, thus the number of possible actions is finite as well. A

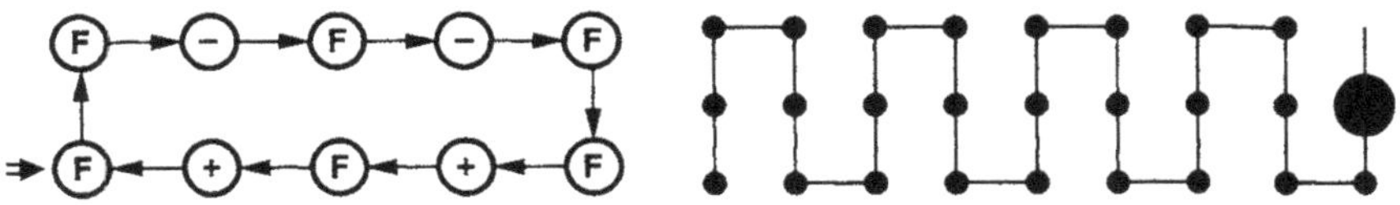

Figure 4. An autonomous finite automaton controlling the sequence of a turtle's movements (left), and the resulting pattern (right)

Figure 5. A frieze pattern generated by an autonomous finite automaton controlling a turtle

sample autonomous DFA is defined by the graph shown on the left side of Figure 4. Circles represent states, with the double arrow pointing to the initial state. Symbols in the circles indicate the output that controls the turtle: move forward and draw a line of unit length (F), turn to the left by 90° ($+$), and turn to the right by 90° ($-$). Arrows define the state transition function, which determines the next state of the automaton given its current state. As the automaton cycles through the sequence of line-drawing steps and turns, the turtle creates the zig-zag of constant amplitude shown on the right side of Figure 4. The dots in this pattern indicate its partition into individual line segments. The larger black circle represents the turtle. The short line segment associated with the turtle indicates its orientation.

A control automaton with more states may yield the frieze pattern [18] shown in Figure 5. Yet the question remains whether we can define an autonomous turtle with finite control that would trace the (infinite) zig-zag curve of growing amplitude, shown in Figure 3. This turns out to be impossible. As each state of an autonomous FDA determines the next state uniquely, the automaton must settle into a fixed cycle of state transitions after a finite number of non-periodic steps at most. Thus, any figure traced by a turtle with finite memory must have an ultimately periodic character. The zig-zag pattern of Figure 3 is not periodic, because its corners are separated by increasingly long vertical lines. Consequently, it cannot be generated by a turtle with an autonomous finite-state control.

In spite of its simplicity, the autonomous FDA pattern formation model yields many questions. For example, what are the properties of paths created by a blind finite-memory turtle on a sphere or on a torus? What would happen if the blind turtle was moving in an expanding space? What would be the consequences of allowing a turtle to create new turtles, which would contribute lines to the overall pattern on their own? Some of these questions will be addressed in Sec. 7.

3 Single blind agent with infinite memory

How can we augment the capabilities of the turtle so that it can create a zig-zag of increasing amplitude? One intuitively obvious possibility is to allow the turtle to

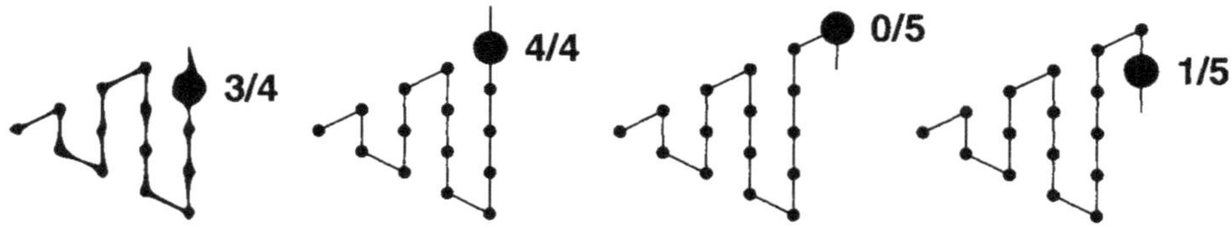

Figure 6. Four consecutive steps in the creation of the zig-zag from Figure 3 by a blind turtle with two counters. Symbols m/n indicate step m along vertical line n.

count. In the case illustrated in Figure 6, the turtle is equipped with two counters, in addition to the finite-state control. There is no upper limit on the magnitude of numbers that can be represented by either counter, thus their inclusion implies infinite memory. Counter m indicates the number of line segments that have already been created along the vertical line that is being drawn. Counter n indicates the target length of that vertical line, which is equal to its ordering number, counting from the left side of the pattern. When $m = n$, the turtle performs a U-turn to the left or right. The value of m is then set to 0 and the value of n is incremented by 1, which initializes the drawing of the next vertical line.

How realistic is this algorithm as a possible description of a real-world process? We can imagine that it could be used by a blindfolded person who was given the task of tracing the zig-zag curve of Figure 3, although the maintenance of straight lines and turning angles of proper magnitudes would not be easy. It is more difficult to imagine an agent with infinite memory in the context of a natural morphogenetic process, but counting capabilities have been postulated to explain the formation of plants that produce a constant number of leaves before they start flowering, for example in [31].

The blind agent with counters is a special case of a blind agent with infinite memory. We can expect that agents with different computational capabilities may lead to different classes of generated patterns. A study of these relationships may link, in an interesting manner, problems of pattern formation to the theory of automata and formal languages. Previous work along these lines has been reported by Maurer *et al.*, [19] and Sudborough and Welzl. [39]

In many cases, the assumption of infinite memory is unrealistically strong. In the following sections we will examine several ways in which it can be avoided.

4 Single sighted agent receiving cues from the environment (one-way exogenous control)

The computational capacities of the turtle can be reduced if the turtle is allowed to receive cues from its environment. In the example shown in Figure 7 it is assumed that the environment is a wedge limited by two intersecting guiding lines. The turtle moves vertically until it runs into any of these guiding lines, at which time it performs a U-turn.

This process can be related to the formation of zig-zag patterns by leaf miners, insect larvae that live and feed within a leaf. In the process of feeding, they create

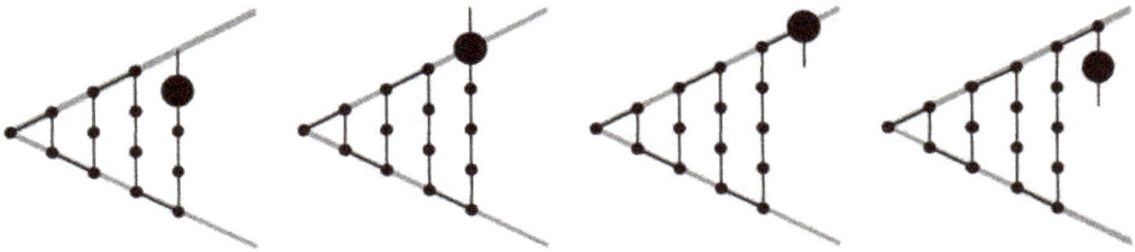

Figure 7. Zig-zag created by a turtle receiving cues from the environment

Figure 8. A leaf with a pattern of tunnels created by a leaf miner

tunnels that are visible as thin trails in the leaf blade. An example of the resulting pattern is shown in Figure 8. It has a meandering character reminiscent of a zig-zag curve, and is limited by the major leaf veins, which the larva cannot easily cross.

From a more general perspective, the guiding lines in Figure 7 provide the turtle with the information about its position within its environment. Positional and directional cues have a fundamental impact on animal behavior, ranging from the maintenance of proper distance from the shore by many littoral organisms to the orientation of insects and birds with respect to the sun, and the use of landmarks by animals that are heading towards a familiar goal. [4] Although these behaviors often do not leave visible traces, they nevertheless can be conceptualized as paths in space, and interpreted as patterns.

The use of environmental cues is also a ubiquitous factor in plant development. For example, the development of climbing plants is guided by the geometry of their support. [28]

At a lower level of organism development, the assumption that morphogenetic agents have access to positional information is the cornerstone of the pattern formation theory proposed in 1970 by Wolpert. [47] He suggested that a cell in a developing organism may "know" its position with respect to one of more reference points, and may use this information to determine the nature of its differentiation.

Since position with respect to a set of reference points is a global attribute of a cell, the use of positional information may seem inconsistent with the goal of describing pattern formation as a local-to-global process. This apparent inconsistency can be solved by considering morphogenesis as a two-level phenomenon. At the higher level, a field of values representing positional information is established by a local mechanism, for example, as a field of concentrations of a substance that diffuses from a source to a sink. At the lower level, each cell accesses the local field value associated with its position, and interprets this information as a factor guiding its development. [47]

5 Single sighted agent receiving cues from the structure (two-way exogenous control)

The mechanism of positional control considered in the previous section operated under the assumption of one-way information flow from the environment to the agent. If we allow the turtle not only to passively obtain environmental cues, but also to change its environment, the use of global positional information may become unnecessary.

To illustrate this, let us consider the generation of the zig-zag pattern shown in Figure 9. In this case, the pattern becomes a part of the environment as it is created. The turtle can perceive this environment within some finite neighborhood around itself. To form the zig-zag pattern, the turtle walks next to the previously created vertical line, sensing its presence in each step (Figure 9a). When the turtle detects that it has walked past the previous line (b), it turns around (c) and resumes the creation of the next vertical line (d).

Cues from the structure created so far may have been used already by the leaf miner to keep an approximately constant distance between adjacent parts of the tunnel, and prevent it from self-intersecting. Similar patterns, known as trace fossils, have been created by early multicellular organisms (over 500 million years ago), and have been preserved in the fossil record. As described by Prescott and Ibbotson, [25] "the earliest traces reflect simple 'scribbling' behaviors, with tracks that often cross themselves and indicate relatively crude foraging strategies. [Later],

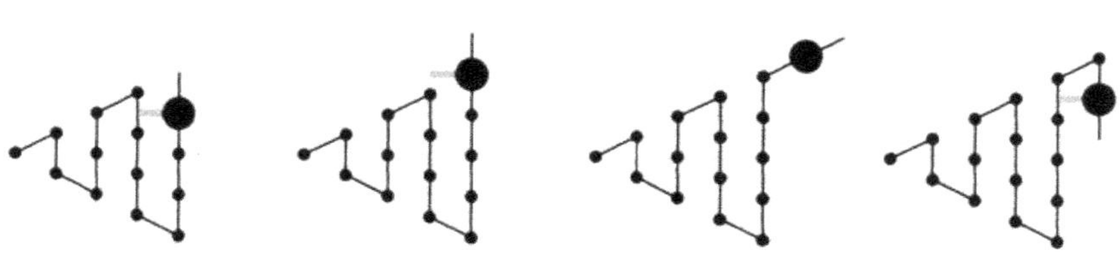

Figure 9. Zig-zag created by a turtle receiving cues from the structure (a–d)

however, more regular trails appear that form spirals or 'meanders' that loop back on themselves without crossing." Computer simulations of these patterns were presented as early as 1969 by Raup and Seilacher. [33,34] More recently, Prescott and Ibbotson recreated similar patterns using robots laying a paper trace on the floor, and sensitive to the track created so far. [25]

Construction of orb webs by spiders is another example of pattern formation in which the agent receives cues from the structure being constructed. As described by Krink and Vollrath, [12] for example, an orb spider begins web construction by creating an outer frame and a set of radii that run between the hub near the web center and the frame. It then lays typically two spirals: an auxiliary spiral (from the inside out) and the capture spiral (from the outside in). When constructing the spirals, the spider uses the previously laid coil as a reference for measuring distance to the next coil.

6 Single self-controlled agent (endogenous control)

In the case of exogenous control, the agent can receive information about the structure completed so far through the space in which this structure is embedded. As noted by Bell [3] (see Section 1), another important possibility is the transfer of information through the structure itself. We will use the term "endogenous information transfer" as a synonym for the term "self-control" to denote this type of pattern-formation mechanism. [26]

Construction of the zig-zag pattern using endogenous information flow is illustrated in Figure 10. As in the case of exogenous control (Figure 9), the turtle uses the last completed vertical line to guide the creation of the next line. Now, however, all information propagates through the zig-zag curve. Specifically, the turtle uses signals (grey triangles in Figure 10) to communicate with a pebble (grey circle), which can be placed at any node between two consecutive line segments. When the creation of a new vertical line begins, the pebble is located at the far end of the previous line (Figure 10a). The turtle then moves forward, drawing a line segment, and sends signal A to the pebble (b). Upon the receipt of this signal, the pebble

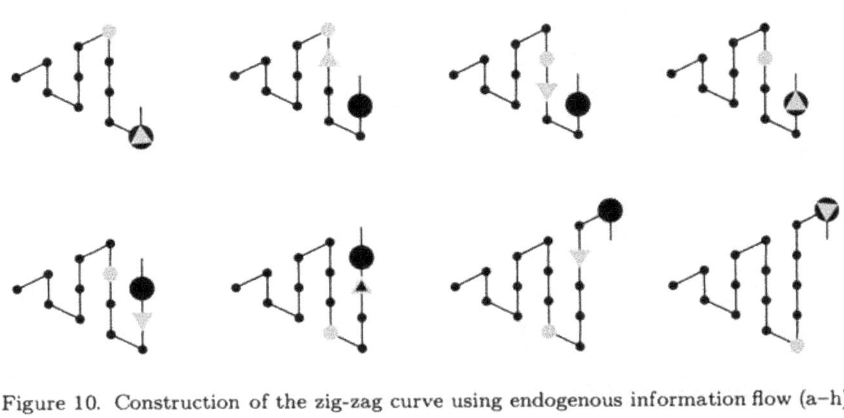

Figure 10. Construction of the zig-zag curve using endogenous information flow (a–h)

moves to the next node and sends a return signal B to the turtle (c). When this signal is received (d), the turtle moves forward, creating the next line segment, and resends signal A to the pebble (e). This initiates another sequence of steps similar to that depicted in figures (b,c,d). The back-and-forth exchange of signals between the turtle and the pebble synchronizes the addition of line segments by the turtle with the movements of the pebble from one node to the next. Thus, the previous line is used as a template for creating the next line. This cycle is modified when the pebble reaches a corner. In that case, a different signal B' is sent to the turtle (f), instructing it to create an additional vertical line segment followed by a diagonal line (g). This modified cycle is responsible for adding an extra segment to the current line, making it one segment longer than its predecessor. Afterwards, the pebble is moved to the beginning of the next (just completed) vertical line, and the turtle is ready to draw the first segment of the new vertical line (h). The state of the system is now similar to that depicted in Figure (a), and the extension of the zig-zag curve with the subsequent line segments will proceed in a manner analogous to that already described.

Taken literally, this example may seem contrived: it is hard to imagine a morphogenetic process that would use signals and a pebble in a similar way. Nevertheless, the proposed mechanism is related to self-reproduction, an attribute of life that has been extensively studied from a formal perspective. [13,14,46] Mathematically, self-reproduction is a process in which a structure embedded in a homogeneous space (originally defined in terms of cellular automata) creates a potentially infinite number of copies of itself. The described method for generating the zig-zag curve uses the same general paradigm: it creates new vertical lines by copying — with the addition of an extra line segment — the previously constructed lines.

Many physiological mechanisms, such as propagation of nutrients, water, and hormones in plants can also be conceptualized as endogenous information flow. We will consider models of processes that use endogenous information flow in Sections 10 and 11.

7 Multiple blind agents with finite memory

In Sections 2 to 6 we have shown that a finite-state turtle can generate a zig-zag curve of increasing amplitude if it receives appropriate cues from the environment, or if it can inspect the structure constructed so far. The next example illustrates another possibility. A growing set of agents, acting in concert, can create the zig-zag curve even if each agent has only finite memory and acts autonomously, without any input.

Recall that an autonomous finite-state turtle can create a zig-zag line of constant amplitude (Figure 4). A modified version of that pattern, with slanted line segments instead of horizontal lines, is shown in Figure 11 (top). Let us now assume that the turtle spawns auxiliary agents, represented as grey squares, and deposits them in obtuse-angled corners of the curve. If these agents act in synchrony with the turtle, and each agent extends its associated vertical line once per every two turtle steps, a zig-zag curve of increasing amplitude will result (Figure 11, bottom). Neither the turtle nor the auxiliary agents receive any information from the environment:

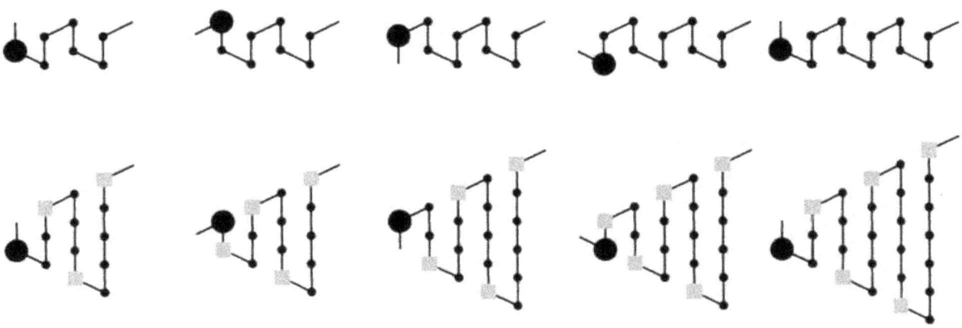

Figure 11. Zig-zag created by multiple autonomous agents with finite-state control

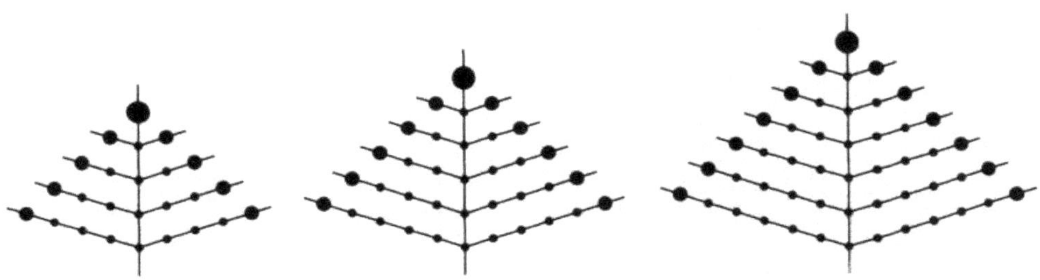

Figure 12. Development of a basitonic branching structure

each of them acts autonomously, cycling through a predefined sequence of steps and actions. The increasing zig-zag amplitude is a consequence of the sequential initiation of vertical lines by the turtle. Since the agents extend all vertical lines synchronously, the lines created earlier have more time to grow and therefore will become comparatively longer.

This mechanism can be thought of as a prototype of the development of many plants. The agents correspond to apical meristems that contribute new line segments — the internodes — to a branching structure, and create new meristems. A sample structure produced by such a process is shown in Figure 12. The main meristem creates a sequence of internodes along the vertical axis of the model. Each of these internodes is associated with a lateral meristem which, in turn, produces a sequence of internodes that become a lateral branch. The operation of all meristems is synchronized. Due to the sequential creation of the lateral apices, the branches at the bottom of the structure have more time to grow than those at the top, and the generated structure has a characteristic, *basitonic* shape: it is more developed at the base than at the top.

The development of branching structures by populations of autonomous agents with finite memory can be conveniently expressed using the formalism of L-systems without interactions (0L-systems).[16] This class of L-systems has been extensively studied [35] and applied to the synthesis of many realistic plant models. [31,32]

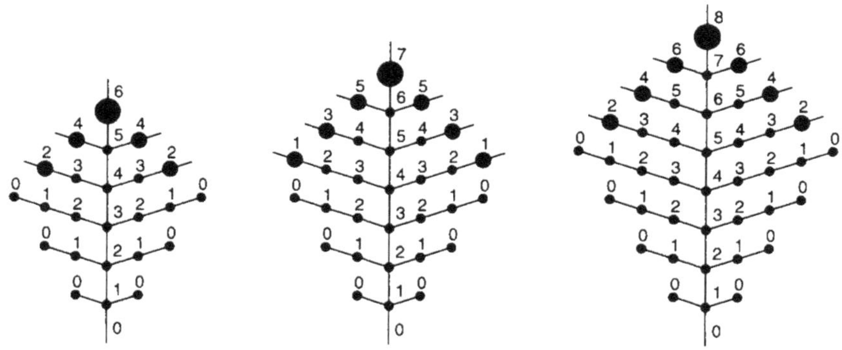

Figure 13. Construction of a mesotonic branching pattern by a growing set of turtles with counting capabilities. Numbers indicate the states of the turtles at their current and past positions.

8 Multiple blind agents with infinite memory

In nature, one observes not only basitonic, but also *mesotonic* and *acrotonic* structures, characterized by extensive development of the middle or upper parts of the plant, respectively. These structures are counterintuitive in the sense that higher branches, although produced by lateral apices that have been created more recently than the bottom ones, are nevertheless longer then the lower branches. It has been shown that (potentially infinite) mesotonic and acrotonic structures cannot be generated by a population of autonomous meristems with finite memory, located at the apices of the axes as observed in real plants. [5,30] Consequently, several more powerful mechanisms have been proposed in the literature. Specifically, Lück *et al.* [17] described a model in which the meristems can count. As shown in Figure 13, the apex of the main axis counts the segments that it has created so far, and assigns the resulting values to the counters associated with the newly created lateral apices. These apices count down as they create new segments, and stop at zero. Consequently, the length of each fully developed lateral axis is equal to its position on the main axis (measured in the number of internodes), yielding a mesotonic structure. (It is not acrotonic because of the juvenile axes of decreasing length at the top.) Biologically, the counter values might be represented by concentrations of substances, which accumulate in the apices and can be dissipated by them.

In the above example, each agent has only one counter. In contrast, the autonomous turtle with infinite memory that created the zig-zag pattern described in Section 3 (Figure 6) required the use of two counters. Apparently, the addition of one counter does not increase the computing power of a single autonomous agent with finite memory, but increases the power of a set of such agents.

9 Multiple sighted agents

In the case of curves created by a single agent, we have observed a tradeoff between memory size and the input information available to the agent. For example,

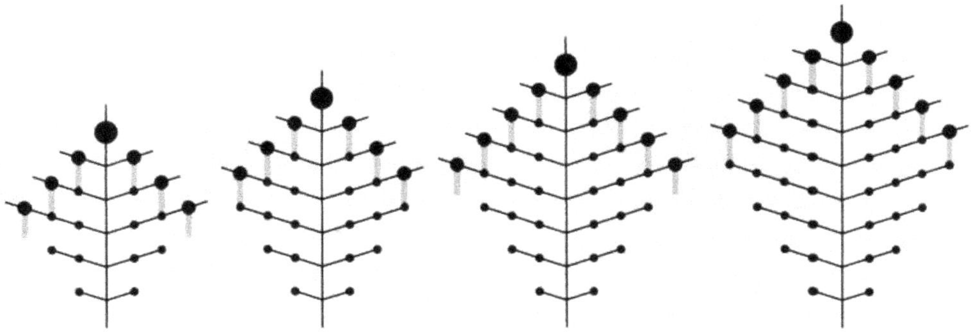

Figure 14. Construction of a mesotonic branching pattern by multiple sighted agents

Figure 15. Examples of patterns created by multiple sighted turtles that stop the creation of new branches to prevent collisions: Sierpiński gasket [36] (left) and Ulam's Maltese crosses [44,45] (right)

the exogenously-controlled turtle, which perceived its environment within some finite neighborhood, no longer needed infinite memory to create the zig-zag curve (Section 5 and Figure 9). The development of branching structures by multiple agents is subject to similar tradeoffs. Specifically, a mesotonic branching structure can be created by a collection of exogenously-controlled turtles with finite memory (Figure 14). To accomplish this, the turtle creating branch $n + 1$ follows branch n immediately below it, and adds one extra segment when the end of that branch has been reached. Branch $n + 1$ will then have one more segment than branch n.

The collective behavior of a growing set of sighted finite-memory agents is often much less obvious than in the previous example. For instance, Figure 15 shows two patterns created by sets of agents that create new branches (and new agents) only if this does not lead to collisions. In both cases, the fractal nature of the resulting patterns would be difficult to predict.

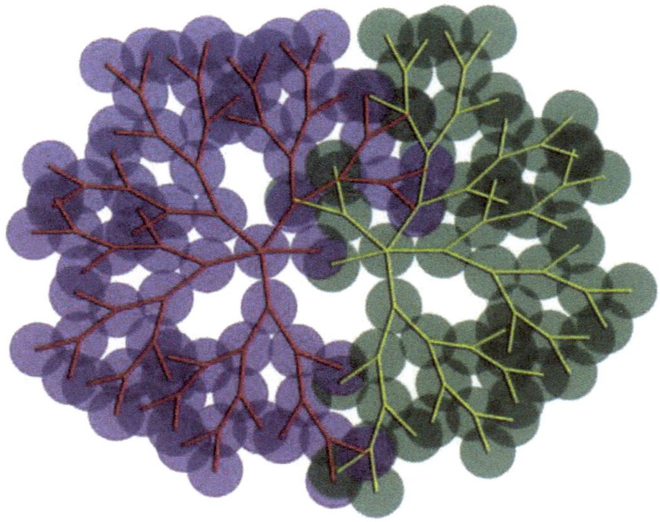

Figure 16. A top view of a branch tier created according to the model by Honda *et al.* [11]. The apices stop growing if they run into clusters of leaves located in the nodes of the structure or associated with other apices.

Plants use forms of collision-detection mechanisms to prevent excessive proliferation of branches. For example, Figure 16 shows a model of a branch tier of the tropical tree *Terminalia catappa.* This model was originally proposed by Honda *et al.* [11] The development of a branch is terminated when its apical meristem runs into a leaf cluster at any of the existing or newly formed nodes of the structure. For a detailed description of this model see [11,22].

Honda's model does not simulate the physical or physiological processes through which a collision between an apex and leaf cluster might be detected, and only captures the effect of this interaction. In nature, the proximity of other branches is oftern sensed as a change in the intensity and quality of light (red / far red ratio) reaching a meristem. The meristem may respond by slowing down the production of new internodes and lateral apices, and eventually terminating the growth of an axis altogether. For example, Figure 17 shows the developmental model of two clonal plants (*Portulaca oleracea* L.), based on a qualitative description the neighbor-sensing mechanism presented by Novoplansky *et al.* [23] Proximity of other branches is perceived as a local change in the red / far red ratio of light distributed in the canopy. Light distribution was simulated using the Monte Carlo method. [21]

10 Multiple endogenously-controlled agents

As evident in the simulations of zig-zag patterns discussed earlier, exogenous and endogenous control mechanisms may yield similar structures, although the internal mechanism of their operation is different. Correspondingly, a mesotonic branching structure can also be created by a growing population of apices that communicate

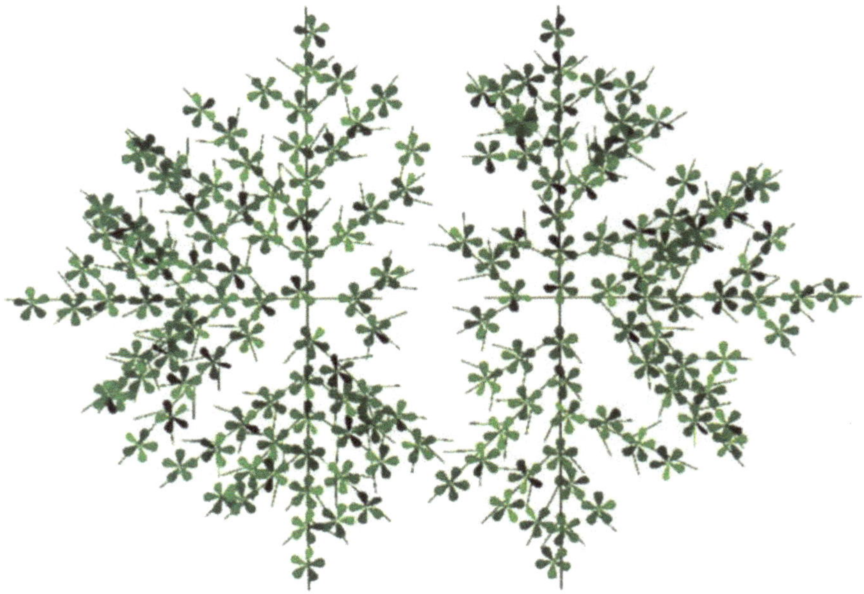

Figure 17. A model of two portulaca plants, growing close together. The apices detect the proximity of parts of the same or another plant, and use this information to avoid collisions. In reality and in the model, the neighbor-detection mechanism is based on sensing the quality of light (red / far red ratio), which is changed by reflection and transmission of light through the leaves.

endogenously. An example is given in Figure 18. The apex of the main axis produces a sequence of vertical internodes and lateral apices as in Figures 13 and 14. These apices create lateral branches by periodically adding new internodes to them. This process is controlled by signals that propagate from the base of the plant towards the apices (grey triangles in Figure 18). When a signal reaches the apex of a lateral

Figure 18. Development of a mesotonic branching pattern, with the branch lengths controlled by signals (endogenously).

branch, it terminates its growth. If proper signal propagation rates along the main axis and in lateral branches are maintained, each successive branch is allowed to grow for a longer time than its predecessor, resulting in a gradual increase of the branch length. For a detailed description of this model see [28].

11 Combined effect of multiple mechanisms

Analysis of individual pattern formation mechanisms is important from the theoretical point of view, since it illuminates the generative capacity of each mechanism, and the relationships between them. In nature, different mechanisms often operate in concert. The pruning of trees is one example. In this case, a plant receives cues from its environment (location of cut points on selected branches), and translates them into endogenous signals. The signals activate dormant buds, prompting their growth into new branches near the pruning sites. By repetitive pruning and regrowth, plants can be sculpted into artificial forms (topiary). A topiary tree model that simulates these processes is shown in Figure 19. A detailed description of this model is presented in [29].

Another multi-mechanism model, proposed by Takenaka [40] (see also [22]), explains the branching pattern and overall shape of trees that grow in isolation or in stands. Light reaching leaves (an exogenous factor) causes production of photosynthates. These photosynthates are used locally to maintain current structure and create new branches or branch segments. The remainder is transported towards branch bases (an endogenous mechanism). If the amount of photosynthates

Figure 19. Simulated trees pruned to a spiral shape. From [29].

Figure 20. Simulated relationships between tree form and its position in the stand. The model integrates exogenous and endogenous control mechanisms. From [22].

available for the production of new branches is too small, the apex will add new segments to the current axis without initiating new branches, or will terminate the growth of the current axis altogether. Furthermore, if the amount of photosynthates reaching the base of an entire branch is less than the amount needed to maintain it, the branch will become a liability and will be shed by the tree. A simulated model of a tree stand, with each tree employing the above mechanisms, is shown in Figure 20. The global shape of the crowns is an emergent property of local, exogenous and endogenous control mechanisms incorporate in the model. For details see [22].

Comprehensive models of plant structures that integrate various developmental mechanisms have been termed *functional-structural* models. [37] They are being studied with agriculture, forestry, and horticulture applications in mind. [27] This shows that studies of pattern formation are not only interesting from a theoretical perspective, but may also lead to practical applications.

12 Conclusions

In this paper we considered the generation of simple patterns — a zig-zag curve and a mesotonic branching pattern — as a vehicle for illustrating different algorithms of pattern formation. The relevance of these algorithms was supported by examples of their occurrence in nature. The description of the algorithms was guided by three independent criteria: the number of agents (one or many), the computing capability of each agent (a finite automaton or an automaton with counters), and its capability of exchanging information with the environment (one-way exogenous, two-way exogenous, or endogenous information flow).

Some obvious cases were not included in the discussion — for example, patterns created by a constant number of agents greater than one, patterns created by multiple asynchronous agents, and self-assembling patterns created by agents that move freely in space and connect themselves according to locally specified criteria. Furthermore, the presentation of different pattern formation mechanisms was carried out at the intuitive level, without formal definitions and proofs. Nevertheless, it seems possible to complete this classification, formulate the theorems, and present their proofs using the main lines of thought outlined in this paper. The resulting theory would formally characterize the tradeoffs between the information-processing capabilities of individual agents, their forms of communication, and their individual or concerted operation. It is possible that these tradeoffs would echo the relationships between different complexity measures studied in the theory of algorithms, effectively extending it to patterns and pattern formation processes. This could provide a firm basis for hypothesizing about the most plausible mechanisms responsible for the formation of specific patterns in nature using the parsimony argument: the most likely explanation is the simplest one. The creation of a computational theory of pattern formation seems a fascinating and rewarding task.

Acknowledgments

I would like to thank Radomír Měch for his work on several images included in this paper, Martin de Boer for the leaf with the miner pattern, Shane Dorosh for surveying the literature on the behavior and modeling of spiders, and Lynn Mercer for helpful editorial comments. This work has been supported in part by research and equipment grants from the Natural Sciences and Engineering Research Council of Canada.

References

1. H. Abelson and A. A. diSessa. *Turtle geometry.* M.I.T. Press, Cambridge, 1982.
2. A. Bell. Dynamic morphology: a contribution to plant population ecology. In R. Dirzo and J. Sarukhán, editors, *Perspectives on plant population ecology*, pages 48–65. Sinauer, Sunderland, MA, 1984.
3. A. D. Bell. The simulation of branching patterns in modular organisms. *Philos. Trans. Royal Society London, Ser. B*, 313:143–169, 1986.
4. F. C. Dyer. Cognitive ecology of navigation. In R. Dukas, editor, *Cognitive ecology. The evolutionary ecology of information processing and decision making*, pages 201–260. University of Chicago Press, Chicago, 1998.

5. D. Frijters and A. Lindenmayer. Developmental descriptions of branching patterns with paracladial relationships. In A. Lindenmayer and G. Rozenberg, editors, *Automata, languages, development*, pages 57–73. North-Holland, Amsterdam, 1976.
6. B. Grünbaum. Periodic ornamentation of the fabric plane: lessons from Peruvian fabrics. *Symmetry*, 1(1):45–68, 1990.
7. B. Grünbaum and G. C. Shephard. *Tilings and patterns.* W. H. Freeman and Company, New York, 1987.
8. L. I. Held. *Models for embryonic periodicity*, volume 24 of *Monographs in Developmental Biology.* Karger, Basel, 1994.
9. S. Hildebrandt and A. Tromba. *The parsimonious universe: Shape and form in the natural world.* Copernicus (Springer), New York, 1996.
10. B. Hölldobler and M. Möglich. The foraging system of *Pheidole militicida* (Hymenoptera: Formicidae). *Insectes Sociaux*, 27(3):237–264, 1980.
11. H. Honda, P. B. Tomlinson, and J. B. Fisher. Computer simulation of branch interaction and regulation by unequal flow rates in botanical trees. *American Journal of Botany*, 68:569–585, 1981.
12. T. Krink and F. Vollrath. Analysing spider web-building behaviour with rule-based simulations and genetic algorithms. *Journal of Theoretical Biology*, 185:321–331, 1997.
13. R. Laing. Automaton models of reproduction by self-inspection. *Journal of Theoretical Biology*, 66:437–457, 1977.
14. C. Langton. Self-reproduction in cellular automata. *Physica D*, 10:135–144, 1984.
15. A. Lindenmayer. Mathematical models for cellular interaction in development, Parts I and II. *Journal of Theoretical Biology*, 18:280–315, 1968.
16. A. Lindenmayer. Developmental systems without cellular interaction, their languages and grammars. *Journal of Theoretical Biology*, 30:455–484, 1971.
17. J. Lück, H. B. Lück, and M. Bakkali. A comprehensive model for acrotonic, mesotonic, and basitonic branching in plants. *Acta Biotheoretica*, 38:257–288, 1990.
18. G. E. Martin. *Transformation geometry. An introduction to symmetry.* Springer-Verlag, New York, 1982.
19. H. A. Maurer, G. Rozenberg, and E. Welzl. Using string languages to describe picture languages. *Information and Control*, 54:155–185, 1982.
20. H. Meinhardt. *The algorithmic beauty of sea shells.* Springer-Verlag, Berlin, 1995.
21. R. Měch. *Modeling and simulation of the interactions of plants with the environment using L-systems and their extensions.* PhD thesis, University of Calgary, October 1997.
22. R. Měch and P. Prusinkiewicz. Visual models of plants interacting with their environment. Proceedings of SIGGRAPH '96 (New Orleans, Louisiana, August 4–9, 1996) ACM SIGGRAPH, New York, 1996, pp. 397–410.
23. A. Novoplansky, D. Cohen, and T. Sachs. How portulaca seedlings avoid their neighbours. *Oecologia*, 82:490–493, 1990.

24. G. Pólya. Über die Analogie der Kristallsymmetrie in der Ebene. *Z. Krist.*, 60:278–282, 1924.
25. T. J. Prescott and C. Ibbotson. A robot trace maker: modeling the fossil evidence of early invertebrate behavior. *Artificial Life*, Vol 3:289–306, 1997.
26. P. Prusinkiewicz. Visual models of morphogenesis. *Artificial Life*, 1(1/2):61–74, 1994.
27. P. Prusinkiewicz. Modeling of spatial structure and development of plants: a review. *Scientia Horticulturae*, (74):113–149, 1998.
28. P. Prusinkiewicz, M. Hammel, J. Hanan, and R. Měch. Visual models of plant development. In G. Rozenberg and A. Salomaa, editors, *Handbook of formal languages,* Vol. III*: Beyond words*, pages 535–597. Springer, Berlin, 1997.
29. P. Prusinkiewicz, M. James, and R. Měch. Synthetic topiary. Proceedings of SIGGRAPH '94 (Orlando, Florida, July 24–29, 1994). ACM SIGGRAPH, New York, 1994, pp. 351–358.
30. P. Prusinkiewicz and L. Kari. Subapical bracketed L-systems. In J. Cuny, H. Ehrig, G. Engels, and G. Rozenberg, editors, *Graph grammars and their application to computer science; Fifth International Workshop*, Lecture Notes in Computer Science 1073, pages 550–564. Springer-Verlag, Berlin, 1996.
31. P. Prusinkiewicz and A. Lindenmayer. *The algorithmic beauty of plants.* Springer-Verlag, New York, 1990. With J. S. Hanan, F. D. Fracchia, D. R. Fowler, M. J. M. de Boer, and L. Mercer.
32. P. Prusinkiewicz, A. Lindenmayer, and J. Hanan. Developmental models of herbaceous plants for computer imagery purposes. Proceedings of SIGGRAPH '88 (Atlanta, Georgia, August 1–5, 1988). In *Computer Graphics* 22, 4 (August 1988), pages 141–150, ACM SIGGRAPH, New York, 1988.
33. D. M. Raup. Modeling and simulation of morphology by computer. In *Proceedings of the North American Paleontology Convention*, pages 71–83, 1969.
34. D. M. Raup and A. Silacher. Computer simulation of fossil foragin behavior. *Science*, 166:994–995, 1969.
35. G. Rozenberg and A. Salomaa. *The mathematical theory of L systems.* Academic Press, New York, 1980.
36. W. Sierpiński. Sur une courbe dont tout point est un point de ramification. *Comptes Rendus hebdomadaires des séances de l'Académie des Sciences*, 160:302–305, 1915. Reprinted in W. Sierpiński, *Oeuvres choisies*, S. Hartman et al., editors, pages 99–106, PWN – Éditions Scientifiques de Pologne, Warsaw, 1975.
37. R. Sievänen, A. Mäkelä, E. Nikinmaa, and E. Korpilahti. Preface. *Silva Fennica*, 31(3):237–238, 1997. Special issue on functional-structural tree models.
38. P. S. Stevens. *Patterns in nature.* Little, Brown and Co., Boston, 1974.
39. I. H. Sudborough and E. Welzl. Complexity and decidability for chain code picture languages. *Theoretical Computer Science*, 36:173–202, 1985.
40. A. Takenaka. A simulation model of tree architecture development based on growth response to local light environment. *Journal of Plant Research*, 107:321–330, 1994.
41. d'Arcy Thompson. *On growth and form.* University Press, Cambridge, 1952.

42. T. Toffoli and N. Margolus. *Cellular automata machines: A new environment for modeling.* MIT Press, Cambridge, Massachusetts, 1987.
43. A. Turing. The chemical basis of morphogenesis. *Philosophical Transactions of the Royal Society of London B*, 237:37–72, 1952.
44. S. Ulam. On some mathematical properties connected with patterns of growth of figures. In *Proceedings of Symposia on Applied Mathematics*, volume 14, pages 215–224. American Mathematical Society, 1962.
45. S. Ulam. Patterns of growth of figures: Mathematical aspects. In G. Kepes, editor, *Module, proportion, symmetry, rhythm*, pages 64–74. Braziller, New York, 1966.
46. J. von Neumann. *Theory of self-reproducing automata.* University of Illinois Press, Urbana, 1966. Edited by A. W. Burks.
47. L. Wolpert. Positional information and pattern formation. In C. H. Waddington, editor, *Towards a theoretical biology,* Vol. 3, pages 198–230. Edinburgh University Press, Edinburgh, 1970.

GROWTH AND FORM OF SPONGES AND CORALS IN A MOVING FLUID

JAAP A. KAANDORP AND PETER M. A. SLOOT

University of Amsterdam
Parallel Scientific Computing and Simulation Group
Faculty of Mathematics, Computer Science, Physics & Astronomy
Kruislaan 403, 1098 SJ Amsterdam
The Netherlands
E-mail: {jaapk, sloot}@wins.uva.nl
URL: http://www.wins.uva.nl/research/pscs/

In this paper the impact of hydrodynamics on a nutrient driven growth process is studied by using a simulation model. A very simple model is used to simulate the growth process, growth is modelled as an aggregation process in three dimensions, in which new sites are added with a certain probability to the aggregate. The growth probability is determined by the local amount of absorped simulated nutrient. The nutrient distribution is modelled by using the lattice Boltzmann method in combination with a tracer step, and where tracer (the simulated nutrient) is released from certain source sites and being absorbed at the boundary nodes of the aggregates. With this method it is possible to study the dispersion of nutrients caused by a combination of flow and diffusion, an increasing influence of hydrodynamics can be expressed by a relatively higher Péclet number, a Péclet number of zero indicates that nutrients are only dispersed by diffusion. Growth of the aggregates is studied under bi-directional flow conditions. In the simulations it is found that the degree of compactness gradually increases with an increasing influence of hydrodynamics. Furthermore it is observed that the aggregates (for all Péclet numbers) develop a roughly radiate symmetry under bi-directional flow conditions. Both observations correspond to qualitative observations made in marine sessile organisms, as for example sponges and corals, where growth forms tend to change gradually from compact shapes under conditions where the organism is exposed to water movement, and into thin-branching forms under sheltered conditions. The growth forms of these organisms have either a roughly radiate symmetry or a flattened morphology. Furthermore it is was found that the highest degree of nutrient absorption takes place at the upstream side of the simulated objects, which corresponds to experimental work on food particle absorption in octocorals and stony corals under low flow velocity conditions.

1 Introduction

1.1 Biological background

Many marine sessile suspension feeders from various taxonomical groups, as for example sponges, hydrocorals, and stony corals, exhibit a strong morphological plasticity, which is in many cases related to the impact of hydrodynamics. The exposure to water movement represents one of the dominant environmental parameters. The environmental parameters sedimentation and transport of food particles are closely related to the influence of hydrodynamics. Examples of studies in which the morphological plasticity, and the relation to the exposure to water movement, is investigated are: growth forms of the hydrocorals *Millepora* [1,2], the stony corals *Pocillopora damicornis* [3], *Madracis mirabilis* [4], and the sponge *Haliclona oculata* [5,6].

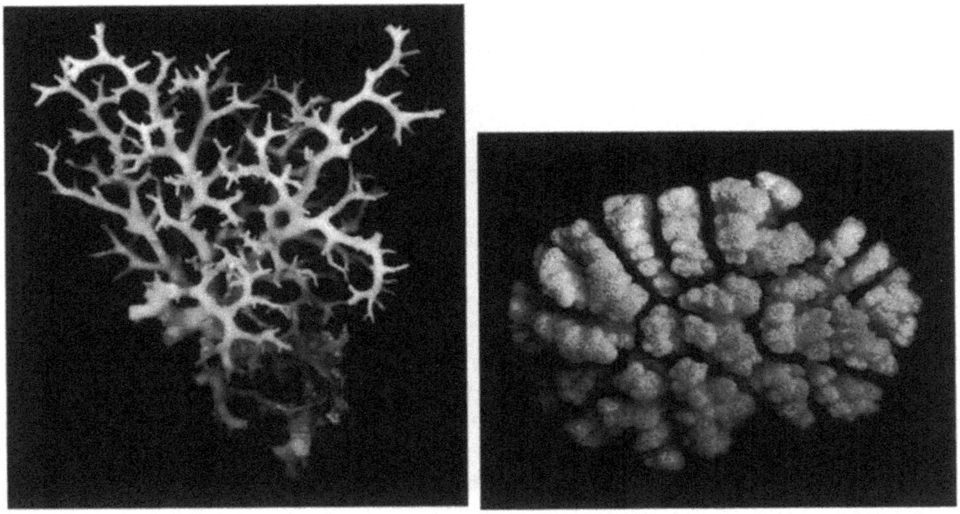

A B

Figure 1. Growth forms of the stony coral *Pocillopora damicornis*. Form A originates from a site sheltered to water water movement, form B originates from an exposed site

There is a strong impact of hydrodynamics on the growth process. In a number of cases it is possible to arrange growth forms of sponges, hydrocorals, and stony corals along a gradient of the amount of water movement [7]. In Veron and Pichon [8] several series of growth forms of stony corals (for example *Pocillopora damicornis* and *Seriatopora hysterix*) are shown, which are arranged along a gradient of the amount of water movement. In the examples of both stony corals, the growth form gradually transforms from a compact shape under exposed conditions, to a thin-branching one under sheltered conditions. In Figure 1 two extreme examples of growth forms are shown of *Pocillopora damicornis*. Form A originates from a site sheltered to water movement, while B was collected from an exposed site. Between both extremes, a range of gradually changing intermediate growth forms exist. In growth forms of the sponge *Haliclona oculata*, plate-like, more compact shapes emerge at exposed sites. This shape gradually transforms into a thin-branching form when the exposure to water movement decreases [5 9]. A similar trend can be observed in growth forms of the hydrozoan *Millepora alcicornis* [2]. In this species the shape changes from plate-like forms at shallow and exposed sites, to thin-branching forms at deeper and sheltered locations.

Water flow has a strong influence on the local supply of food particles in suspension feeders. There exists a fairly large amount of literature on the relation between the ability of species to capture food particles and the flow rate. A review about the effects of flow on suspension feeders is given by Shimeta and Jumars [10]. In several studies [11 12 13] it was demonstrated that locally around a sessile suspension feeder, exposed to various flow velocities, areas may occur which are depleted from food particles. By several authors it was found that there exists an asymmetric type of food capture. Patterson [14] demonstrated that the highest amount of food capture is found at the upstream side of the octocoral colony for low flow velocities, while the situation reverses for higher flow velocities where the highest amount of food particles is captured at the downstream side of the colony. A similar phenomenon was observed by Sebens et al [4] for the stony coral *Madracis mirabilis*, where the highest amount of food particle capture was found at the upstream side of the colony for low flow velocities (below $15cm \cdot s^{-1}$), while the situation again reverses for higher flow velocities. In studies on branching stony corals [4 15] and the influence of hydrodynamics it is demonstrated that branch spacing is a crucial morphological property which determines the microflow pattern inside the colony. The flow direction will basically reverse twice a day due to the tidal movements and in branching sessile suspension feeders growth forms emerge which have a roughly radiate symmetry or show a tendency to develop a flattened growth form [16 1 17]. In the last case most branches will develop in a plane perpendicular with respect to the governing flow direction. Local food particle absorption patterns which are determined by the direction and the velocity of the flow, local areas which surround the organism and which are depleted from food particles, as well as micro flow patterns in branching organisms have important consequences for the growth process and the resulting morphology.

To obtain insight into the influence of hydrodynamics on the growth process of sessile suspension feeders, a morphological simulation model was developed. In the absence of flow, the distribution of nutrients around the growth form can be modelled as a diffusion process in a steady state: there is a source of suspended material and the organism continuously consumes nutrients from its environment. In general in a marine environment, there will be a significant contribution of the hydrodynamics to the dispersion pattern of the suspended material around the growth form. In this case the distribution of nutrients around the organism will be determined by a combination of flow and diffusion. The basic idea of the model [18] will be briefly discussed in the next section, a further extension of this model will be discussed in Section 1.3 and in the rest of this paper.

1.2 Simulation of growth in a mono directional flow

In a previous study [18] the impact of hydrodynamics on a growth process was studied using a simulation model. In this paper the growth process was represented by a highly simplified model. The growth process was modelled as an aggregation process in a three dimensional lattice. In these simulations a system was used consisting of 144^3 lattice sites. In the aggregation process, lattice sites are added with a certain probability to the aggregate. The probability was determined by the local amount of absorbed "nutrient" at the aggregate, where higher probabilities are defined for a relatively higher amount of absorbed "nutrient". In these simulations the nutrient particles are being dispersed by the combined process of flow and diffusion. The method which we applied in this study was the lattice Boltzmann method in combination with a tracer step to study the dispersion of "nutrient" in the system. We will give a brief introduction to the method in Section 2.1, a more detailed account of this method is given in [19 20 21 18 22]. In this method, boundary conditions with a complex geometry can be easily specified. The method is very suitable for scalable implementations on Massive Parallel Computers [23 24], which makes large scale simulations of biological systems feasible.

In this study the morphology of the aggregates was compared for various Péclet numbers, where the Péclet number is defined by:

$$Pe = \frac{\overline{u}l}{D} \tag{1}$$

where $\overline{u}$ is the mean flow velocity, l a characteristic length, and D the diffusion coefficient of the nutrient. The contribution of flow to the nutrient distribution of nutrients can be quantified by the Péclet number. A low value of the Péclet number indicates that particles move mainly by diffusion (no influence of hydrodynamics, diffusion dominates) and a high value indicates that their motion is dominated by flow. In this study it was demonstrated that the aggregate gradually changes from a thin-branching morphology (diffusion dominates) into a compact shape (flow dominates), for an increasing Péclet number. Examples of a thin-branching aggregate ($Pe \approx 0$) and a compact aggregate ($Pe = 3.0$) are shown in Figure 2. To give an impression of the overall morphology of the aggregates, a triangulated surface is

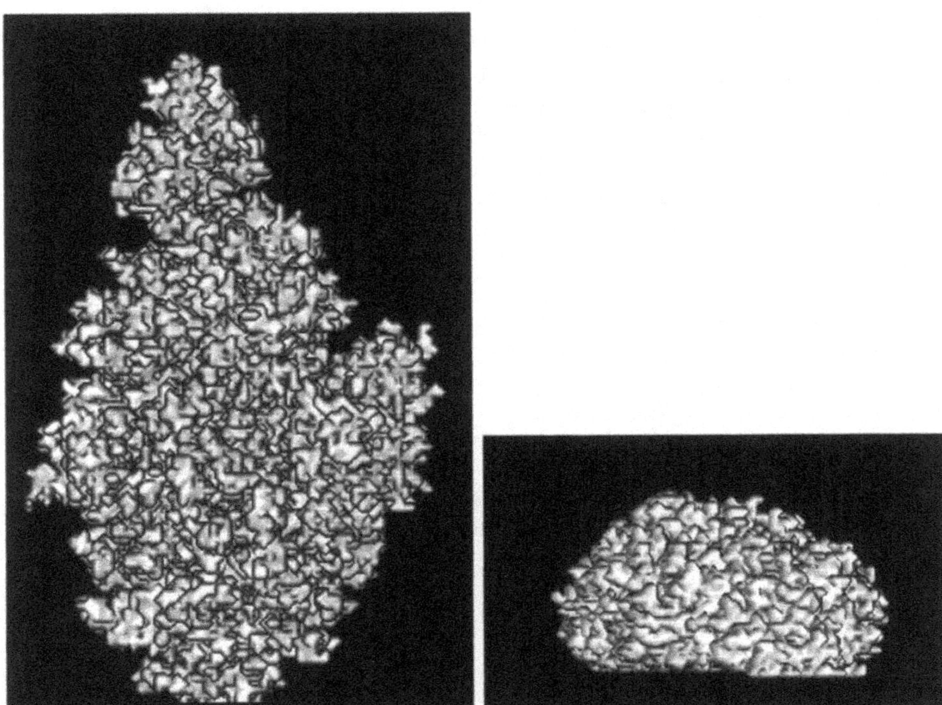

A B

Figure 2. 3D visualizations of the aggregates for $Pe = 0.0150$ (A) and $Pe = 3.0000$ (B) generated in the mono-directional flow experiment, the flow is directed from the left to the right.

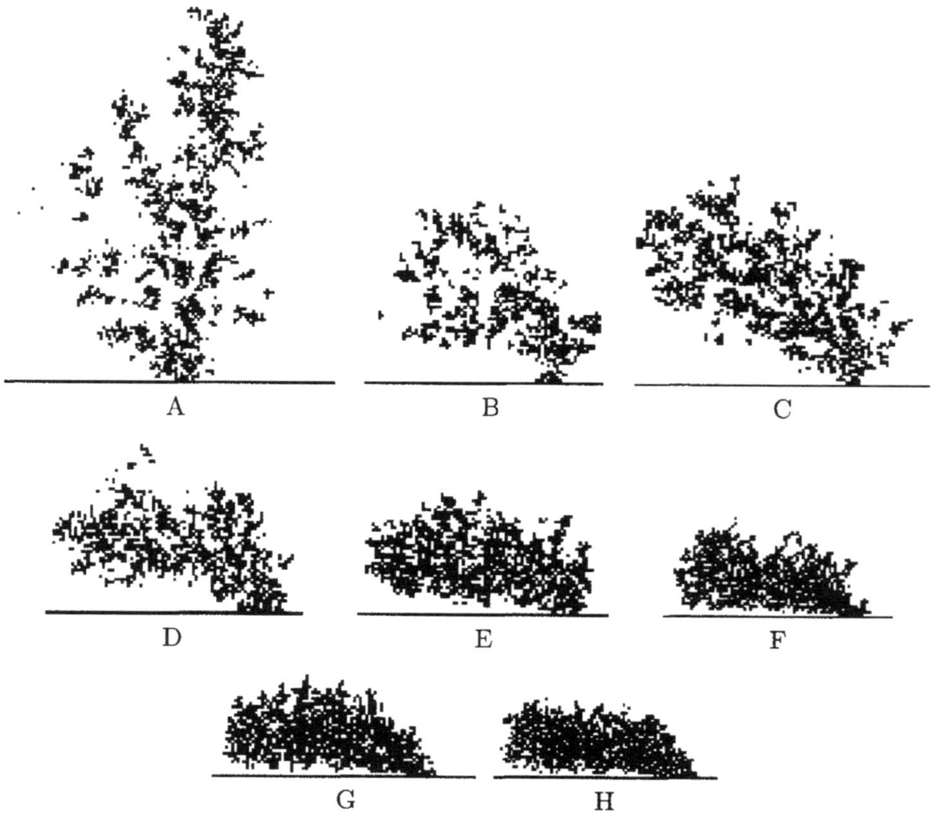

Figure 3. Slices through the middle of the aggregates in the xy-plane at $zmax/2$, one-directional flow experiment: A-H, Péclet number increases from approximately 0.0 to 3.0. The flow is directed from the left to the right.

constructed over the aggregates by using the marching cubes method [25]. In Figure 3 the results of the simulations are summarized by showing slices through the aggregate. The 144^3 system is sectioned parallel to the direction of the flow. In the sequence A-H in Figure 3, the Pe number increases. In this picture it can be observed that the degree of compactness increases for larger Pe numbers. Furthermore, the effect of the mono-directional flow can be clearly observed: the aggregates tend to grow towards the direction of the flow (the flow is directed from the left to the right).

1.3 Simulation of growth in a bidirectional (alternating) flow

In the simulation model discussed in the previous section some important simplifications have been made. One simplification is that it is assumed that the growth form develops under mono-directional flow conditions. As a consequence an asymmetric form develops, as shown in Figure 3, where the aggregates tend to grow in the upstream direction. This trend becomes stronger for higher Pe numbers. In reality, the flow direction will basically reverse twice a day due to the tidal movements. In this paper a growth process in a bidirectional flow is studied by using an aggregation model in which the flow direction is reversed after each growth step. The nutrient distribution is computed in a similar way as discussed in the previous section. The results of the simulations will be compared to the results of the simulations in a monodirectional flow [18] and qualitatively compared to the actual growth forms.

2 Methods

2.1 *Modelling the nutrient distribution*

In the simulations a cubic lattice is used consisting of 144^3 nodes. Each node is connected with 18 other nodes. There are 6 links with the length of one lattice unit and 12 links with the length of $\sqrt{2}$ units. All parameters and variables in the lattice Boltzmann method are expressed in lattice units. The mean populations of particles move simultaneously from one node to one of the 18 neighbours. The evolution of the lattice is described by the following dynamical rule:

$$n_i(r + c_i, t + 1) = n_i(r, t) + \Delta_i(r, t) \qquad i = 1, .., 18 \tag{2}$$

where $n_i(r, t)$ is the continuous particle distribution function, which describes the number of particles at a node at position r, at time t and with direction c_i. The moments of the particle distribution function $n_i(r, t)$:

$$\rho = \sum_i n_i \qquad j = \sum_i n_i c_i = \rho u \qquad \Pi = \sum_i n_i c_i c_i \tag{3}$$

correspond to the hydrodynamic fields mass density ρ, momentum density j (u is the flow velocity at node r), and momentum flux Π.

A linearized collision operator $\mathcal{L}_{ij}$ about an appropriate equilibrium n^{eq} can be constructed:

$$\Delta_i(n) = \Delta_i(n^{eq}) + \sum_j \mathcal{L}_{ij}(n_j - n_j^{eq}) \tag{4}$$

where $\Delta_i(n^{eq}) = 0$ by definition since the equilibrium is collision invariant, and where n^{eq} is the Maxwell Boltzmann distribution for a gas. The distribution function can be split into an equilibrium and a non-equilibrium part:

$$n_i = n_i^{eq} + n_i^{neq} \tag{5}$$

The moments of the equilibrium distribution function n_i^{eq} are:

$$\rho = \sum_i n_i^{eq} \tag{6}$$
$$j = \sum_i n_i^{eq} c_i$$
$$\Pi = \sum_i n_i^{eq} c_i c_i = p1 + \rho uu$$

where p is the pressure and 1 is the unit vector. The momentums of the particles are changed by adding an external force, the driving force F, to the system. The particle collisions conserve mass and momentum but change the non-equilibrium part of the momentum flux:

$$\Pi^{neq} = \Pi + \Pi^{eq} \tag{7}$$

By using properties as conservation of mass, momentum, and symmetry, it is possible to construct a number of eigenvalue equations (a more detailed account of the collision process is given by Ladd [20]) for the moments of $\mathcal{L}_{ij}$:

$$\sum_i \mathcal{L}_{ij} = 0 \tag{8}$$
$$\sum_i c_i \mathcal{L}_{ij} = 0$$
$$\sum_i \overline{c_i c_i} \mathcal{L}_{ij} = \lambda \overline{c_j c_j}$$
$$\cdots$$

where $\overline{c_i c_i}$ indicates the traceless part of $c_i c_i$ ($\overline{c_i c_i} = c_i c_i - \frac{1}{3} c_i^2 1$). The first two equations conserve mass and momentum. The eigenvalue λ is related to the shear viscosity:

$$\nu = -\frac{1}{6}\rho(2/\lambda + 1) \tag{9}$$

One update step of the lattice now involves a propagation step where n_i particles travel from node r to node $r + c_i$ and a collision step in which the post-collision distribution $n_i + \Delta_i(n)$ can be computed in the following three steps. First at each node ρ, j, and Π are calculated and the equilibrium momentum flux Π^{eq} is determined. In the second step the momentum flux, including the equilibrium and nonequilibrium part, can be updated using the moments of the linearized collision operator in Eq. (8). In the third step, after the post collision momentum is determined, the post-collision distribution $n_i + \Delta_i(n)$ can be calculated.

Two types of boundary conditions are used: at the borders of the lattice periodic boundary conditions are applied, while in the nodes adjacent to the nodes representing the obstacle, solid boundary conditions are used. Periodic boundary conditions can be implemented by exchanging the n_i's of the links at the borders of the lattice. Solid boundary conditions can be represented by exchanging the n_i's between the adjacent node and a neighbouring fluid node.

After the lattice Boltzmann iteration, a tracer step is applied where populations of tracer particles are released from source nodes, while the tracer particles are absorbed by the sink nodes: the nodes adjacent to the growth form. The tracer particles can travel from a node at site r in the lattice to one of the 18 adjacent nodes $r + c_i$, where the Péclet number Eq.(1) determines if flow or diffusion dominates. When flow dominates, most particles will move in the direction of the governing flow, while under diffusion dominated conditions the amount of particles which travels in all 18 directions will be about equal. In the simulations ν is set to the value $\frac{1}{6}$, while the diffusion coefficient D varies, and $\overline{u}$ is kept constant by adjusting the driving force F of the system. Due to the growth of the aggregate the velocity in the free fluid would gradually decrease if the driving force is not adjusted.

2.2 Modelling the growth process in an alternating flow

In the aggregation model, the growth process is represented in a similar way as done in the Diffusion Limited Aggregation model [26]. The basic construction of the aggregate in (alternating) bidirectional flow is shown in Figure 4. The cluster is initialized with a "seed" positioned at the bottom plane of the lattice. The bottom plane, "the substrate", is positioned at the xz-plane at $y = 1$, while the seed is one lattice site, located at the position $(xmax/2, 2, zmax/2)$. In both the cluster and substrate sites solid boundary conditions are applied. The flow in the lattice is directed, alternating from the left to the right and the right to the left in Figure 4 Initially the flow is directed from the left to the right. The flow velocity in the free fluid nodes is kept to a constant value $\overline{u}$ by adjusting the driving force F of the system. The flow velocity in the system is kept at a low value, all simulations are done in the laminar flow regime. The flow pattern about the obstacle (substrate

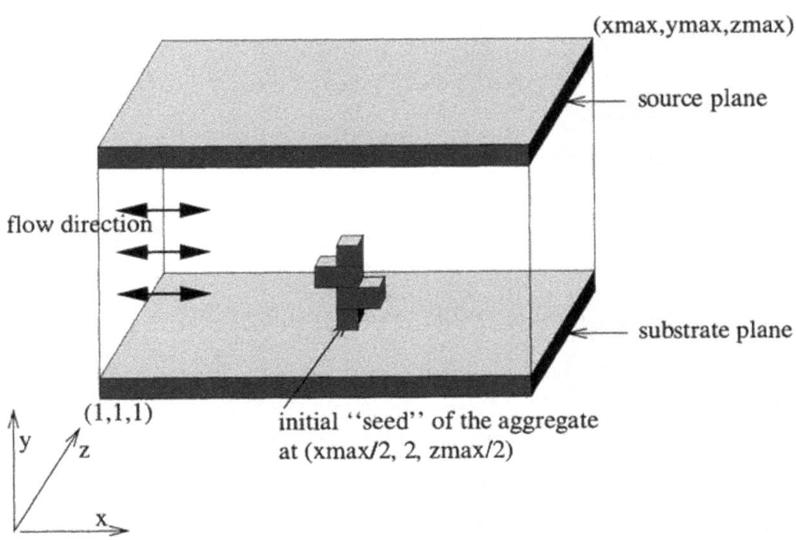

Figure 4. Basic construction of the aggregate in a bidirectional flow.

and cluster) is determined in the lattice Boltzmann iteration, using 10 iteration steps followed by a tracer step. In each growth step 10 iteration steps are used to ensure that an equilibrium is obtained in the flow pattern. In the last step tracer particles are released from the source plane, the lattice sites located at the xz-plane at $y = ymax$. The tracer particles are absorbed by the fluid nodes adjacent to obstacle nodes, which can be nodes in the substrate plane (the xz-plane at $y = 1$) and the aggregate nodes. In this growth model it is assumed that both the tracer distribution and flow velocities are in equilibrium and the growth velocity of the aggregate is much slower than the dispersion of the tracer. In the sink nodes the amount of absorbed tracer particles is determined and a new node is added to the aggregate. The probability p that k, an element from the set of open circles $\circ$ (the adjacent sink nodes) will be added to the set of black circles (the aggregate nodes) is given by

$$p(k \in \circ \to k \in \bullet) = \frac{(a_k)}{\sum_{j \in \circ}(a_j)} \tag{10}$$

where $a_k =$ absorbed amount of tracer particles at position k

After each growth step the flow direction is reversed. The aggregation model for an alternating flow is summarized in pseudo-code below:

```
initialise aggregate
initialise flow direction

do {
-compute flow velocities until equilibrium is reached;
-compute tracer distribution until equilibrium is reached;
-compute probabilities p that nodes neighbouring to the aggregate
will be added to the aggregate;
-select randomly with probability p one of the growth candidates and
add it to the aggregate;
-reverse flow direction;
} until ready
```

3 Results

The results of the simulation experiments in an alternating flow are summarized in Table 1. We have estimated error bounds on the various measurements by repeating the simulations at $Pe = 0.0150$ and $Pe = 3.0000$ four times. The ratio R, of the total sink nodes to the total cluster size indicates the compactness of the aggregates. The typical cluster size is about 6.4×10^4. The fractal dimension D_{box} of the surface of the aggregate was determined using a 3D version of the (cube)box-counting method described by Feder [27]. D_{box} measures the space-filling properties of the surface of the aggregate. In three dimensions its value varies from a minimum of 2, for a solid object with a perfectly smooth surface, to a maximum of 3 for a solid with a space-filling surface. The morphology of the aggregates for

Table 1. Ratio R , of the total sink nodes to the total cluster size, the 3D fractal dimension D_{box}, the average absorption $\overline{a}$, the measure D_{abs} of the uniformity of the nutrient distribution, and the average x, y, z coordinate of the aggregate in the alternating flow experiment for a series of increasing Pe numbers.

Pe	0.0150	0.1322	0.2521	0.4918	1.0000	2.0000	2.5000	3.0000
R	2.41 ± 0.02	2.29	2.16	1.86	1.49	1.15	1.05	1.01 ± 0.01
D_{box}	2.25 ± 0.01	2.24	2.26	2.29	2.24	2.19	2.17	2.15 ± 0.02
$\overline{a}$	0.03 ± 0.01	0.17	0.17	0.69	1.0	6.3	8.4	10.0 ± 0.1
D_{abs}	2.2 ± 0.2	1.3	1.3	0.9	1.0	0.7	0.8	0.8 ± 0.1
$\overline{x}$	0.50 ± 0.01	0.52	0.50	0.52	0.51	0.51	0.51	0.51 ± 0.01
$\overline{y}$	0.35 ± 0.03	0.23	0.21	0.17	0.14	0.12	0.11	0.10 ± 0.01
$\overline{z}$	0.57 ± 0.04	0.49	0.50	0.49	0.50	0.49	0.49	0.51 ± 0.03

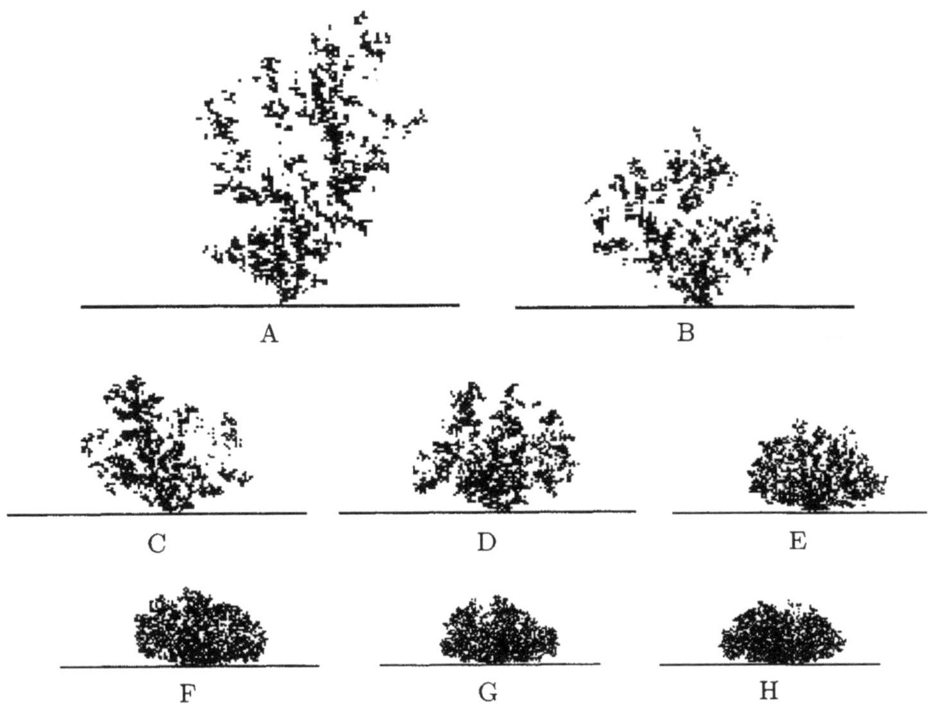

Figure 5. Slices through the middle of the aggregates in the xy-plane at $zmax/2$, alternating flow experiment: A-H, Péclet number increases from approximately 0.0 to 3.0

various Pe is demonstrated in Figure 5 by showing slices through the middle of lattice (in the xy-plane). In the simulations we have determined the number of sink nodes that have a nutrient absorption between a and $a + \Delta a$ for the aggregates obtained at various Péclet numbers. The number of sink nodes with absorption rate a, $N(a)$ can be related to a by a power law [18]:

$$N(a) \sim a^{-D_{abs}} \tag{11}$$

Table 2. Ratio R, of the total sink nodes to the total cluster size, the 3D fractal dimension D_{box}, the average absorption $\overline{a}$, the measure D_{abs} of the uniformity of the nutrient distribution, and the average x, y, z coordinate of the aggregate in the mono-directional flow experiment for a series of increasing Pe numbers

Pe	0.0150	0.1322	0.2521	0.4918	1.0000	2.0000	2.5000	3.0000
R	2.35 ± 0.10	2.21	1.88	1.64	1.29	1. 20	1.31	1.27 ± 0.06
D_{box}	2.27 ± 0.02	2.20	2.27	2.20	2.16	2 10	2.09	2.05 ± 0.04
$\overline{a}$	0.04 ± 0.01	0.85	1.75	3.2	5.5	11.0	13	15 ± 5
D_{abs}	1.7 ± 0.2	1.0	1.0	0.9	0.8	0.7	0.8	0.7 ± 0.1
$\overline{x}$	0.47 ± 0.03	0.38	0.33	0.32	0.32	0.24	0.24	0.24 ± 0.01
$\overline{y}$	0.51 ± 0.04	0.39	0.34	0.33	0.30	0.19	0.18	0.16 ± 0.01
$\overline{z}$	0.48 ± 0.02	0.53	0.50	0.49	0.50	0.49	0.5	0.49 ± 0.02

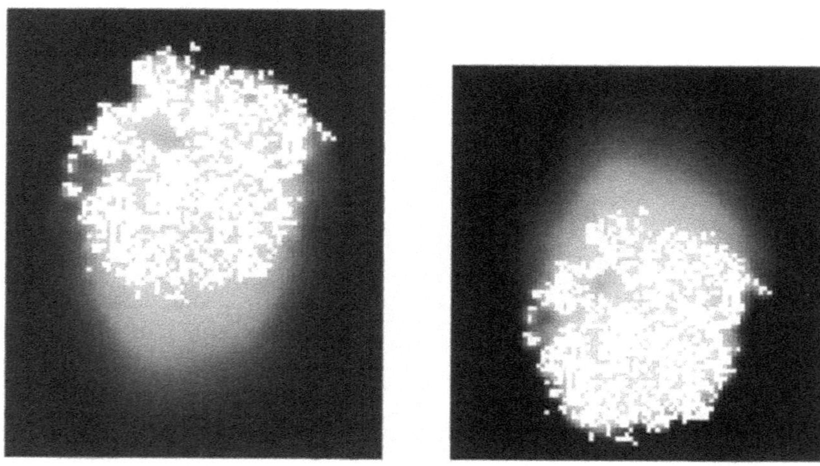

A B

Figure 6. Slice through the simulation box in the xz-plane showing the nutrient distribution in two successive growth stages in the alternating flow experiment in which Pe is set to the value 3.000 (flow dominates) in the flow is directed from top to bottom in A and directed from bottom to top in B.

The value of D_{abs} can be estimated from a log-log plot. The exponent D_{abs} can be interpreted as a measure of the uniformity of the nutrient distribution. In Table 1 the average absorption $\overline{a}$ in the boundary nodes, and the values of D_{abs}, are listed for the various Pe numbers. Furthermore, for each cluster the average x, y, z coordinate (the centre of gravity) was measured. The same measurements have been made in the aggregates obtained in a monodirectional flow experiments [18] and are listed in Table 2. The morphology of the aggregates, in the monodirectional flow experiment, is depicted in Figs. 2 and 3.

In Figs. 6 and 7 slices through the simulation box in the xz-plane are shown in which the nutrient distribution is visualized, for respectively the Pe numbers 3.000 (flow dominates) and 0.015 (diffusion dominates). The colour shift black — white in these pictures indicate a depletion of nutrient, black indicates the maximum

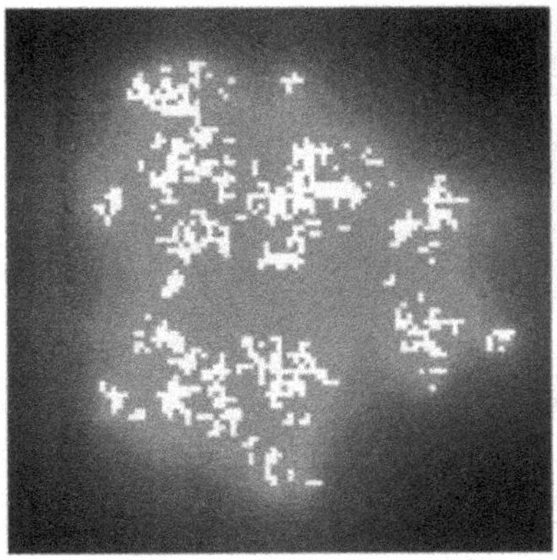

A

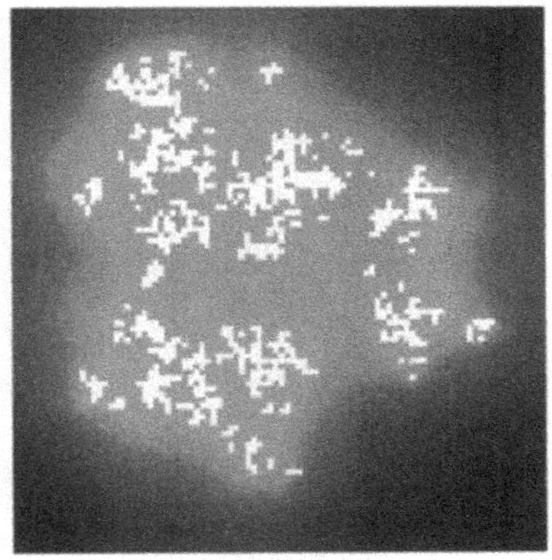

B

Figure 7. Slice through the simulation box in the xz-plane showing the nutrient distribution in two successive growth stages in the alternating flow experiment in which Pe is set to the value 0.015 (diffusion dominates) in flow is directed from top to bottom in A and directed from bottom to top in B.

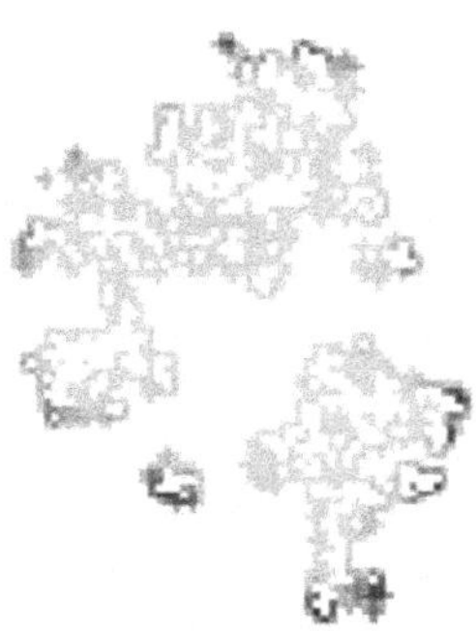

A

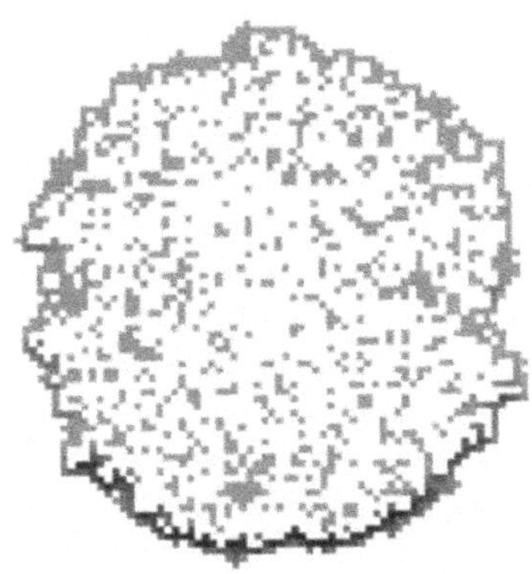

B

Figure 8. Slices through the simulation box in the xz-plane showing the absorption in the boundary nodes for respectively the Pe numbers 0.015 (A, (diffusion dominates)) and 3.000 (B, flow dominates).

concentration of nutrient, while the regions with nearly zero concentration are shown in white. In both pictures the nutrient distributions of two successive phases in the algorithm discussed in Section 2.2 are shown. In Figs. 6A and 7A the flow is directed from the top to the bottom, while in Figs. 6B and 7B the flow direction is reversed. In Figure 8 slices through the simulation box in the xz-plane are shown in which the absorption in the boundary nodes around the aggregates is visualized for respectively the Pe numbers 0.015 and 3.000, where black indicates a high absorption rate and white an absorption of about zero.

Figure 9. Aggregate and the corresponding nutrient distribution in a slice made in xy-plane in the alternating flow experiment in which Pe is set to the value 0.015 (diffusion dominates, flow is directed from the left to the right).

Figure 10. Aggregate and the corresponding nutrient distribution in a slice made in xy-plane in the alternating flow experiment in which Pe is set to the value 3.000 (flow dominates, flow is directed from the left to the right).

In the Figs. 9, 10, and 11, the three dimensional morphology of the aggregates is visualized in combination with the corresponding nutrient distribution. The colour shift white-red in the aggregates indicates which parts of the aggregate are added

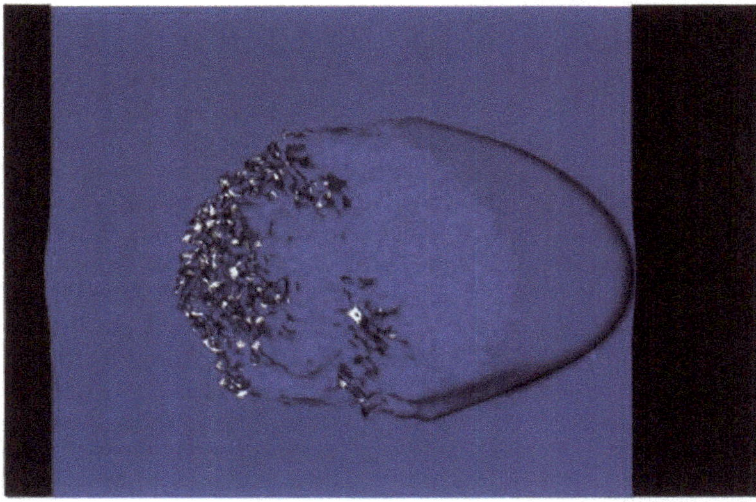

Figure 11. Top view of the aggregate from the alternating flow experiment in which Pe is set to the value 3.000 (flow dominates, flow is directed from the left to the right). The depletion zone around the aggregate is visualized by constructing an isosurface at a nutrient concentration of nearly zero.

most recently on top of the previous growth stage, red indicates the "youngest" part of the growth form in Figure 9, while (for visualization reasons) the youngest parts of the aggregate are shown in white in Figure 10. The nutrient distribution is depicted using a colour shift blue-white, where blue indicates the highest concentration, and white the lowest. In Figs. 9 and 10 the nutrient distribution is shown in one slice in the xy-plane. In Figure 11 the shape of the depletion zone is depicted by the construction of an isosurface at a level of very low (nearly zero) nutrient concentration.

4 Discussion

The nutrient distributions shown in Figs. 6, 7, 9, and 10 demonstrate the main differences between diffusion and flow dominated regimes. For low Péclet numbers the distribution of nutrient is roughly symmetric about the centre of the aggregate, where the highest concentration reside at the tips of the aggregate and where between the branches an area depleted from nutrients is found with a very low growth probability. In Figure 7 it can be observed that there is hardly any difference between the two successive phases, there is very little influence of the governing flow direction on the nutrient distribution. At higher Péclet numbers a clear asymmetry develops in the distribution with a depleted region developing downstream of the object (see Figure 6). In a top view of the aggregate and nutrient distribution (see Figure 11) it can seen that this depletion zone has more or less a parabolic shape. As a consequence, there exists little probability of growth in the depleted region.

The absorption patterns shown in Figure 8 show an asymmetry in the absorption of nutrients at the boundary nodes of the aggregates for increasing Pe numbers. For the diffusion limited case (Figure 8A, the highest degree of absorption is found at the tips of the object, while near the centre hardly any nutrient is being absorped. In Figure 8B the highest amount of absorption is found at the upstream side of the object. This corresponds to the observation done in experimental work on the absorption of food-particles by the stony-coral *Madracis mirabilis*, where the highest amount of food particle capure was found at the upstream side of the colony [4]. In Table 1 it can be seen that the average absorption $\overline{a}$ increases for increasing Pe numbers The absorption pattern characteristic of the diffusive regime changes into a more slowly decaying distribution. In other words, the flow results in a more even distribution of nutrient in boundary nodes. This is illustrated by the decrease in the measure D_{abs} of the uniformity of the nutrient distribution. As a result the probability that boundary sites are added to the aggregate becomes more equal, which accounts for the increasing degree of compactness of the aggregate. When comparing Figs. 9 and 10, it can be observed that the new added part in Figure 10 (visualized in white) "hides" the previous growth stages, all boundary sites seem to have a more or less equal probability to be added to the aggregate, while in in Figure 9 the "trunk" of the object remains visible and growth only occurs at the tips of the object. A gradual increase of compactness is demonstrated in Figure 5 and by a decrease of both D_{box} and the ratio R, of the total sink nodes to the total cluster size in Table 1 for an increasing influence of hydrodynamics. This gradual increase of compactness corresponds qualitatively to the observations made in stony corals, hydrocorals, and sponges, where growth forms gradually transform from a compact shape, under conditions exposed to water movement, into a thin-branching one under sheltered conditions [8 9 7].

The impact of the alternating flow model, presented in Section 2.2, can be demonstrated by comparing the measurements carried out for the mono-directional flow model (see Table 2) to measurements done in the alternating flow model (see Table 1). The centre of gravity (the average x, y, z coordinate of the aggregate) tends to move in the negative x direction (the upstream direction) in the mono-directional flow experiment, while the centre of gravity remains more or less in the centre of the xz-plane in the alternating flow experiment for increasing Pe numbers. When visually comparing the slices through the aggregates shown in Figs. 3 and 5 it can also be observed that the increasing degree of asymmetry in the aggregate in the mono-directional flow experiment, for increasing Pe numbers, has disappeared in the alternating flow experiment. In the last experiments, aggregates have developed with a roughly radiate symmetry, which corresponds qualitatively to the shape of branching sessile organisms, as for example *Pocillopora damicornis.* These experiments seem to indicate that an alternating flow, a reversal of the flow direction basically twice a day, leads to radiate symmetrical growth forms.

The alternating flow model is a strong simplification of the actual growth process. In many stony corals, as for example the species *Pocillopora damicornis* photosynthesis represents a major energy input. The actual growth process in many sponges, stony-corals, hydrocorals [9] consists of adding layers of material (varying in thickness) on top of the preceding growth stage, and not by the addition of particles. An accretive growth model, in which layers of material are constructed on top of the previous layers and where the local thickness is determined by the local amount of absorbed nutrients or light intensity [9], also offers possibilities for a quantitative morphological comparison of simulated and actual growth forms [28 29]. Furthermore, there is no mechanism present in the aggregation model for nutrient accumulation at some distance from the object and there are no effects of erosion included in the model. In our alternating flow simulations objects are found with a roughly radiate symmetry and where the degree of compactness increases for an increasing influence of hydrodynamics, and may be used as a simple explanation of a similar phenomenon which is observed in the morphology of various marine sessile organisms.

Acknowledgments

This work was carried out within the MPR (Massive Parallel Computing) project "The discreteness–continuity dichotomy in Individual-Based Population Dynamics using massively parallel machines" funded by the Dutch foundation for basic research. Dr. J. E. N. Veron (Australian Institute of Marine Sciences) kindly permitted us to use the photographs of *Pocillopora damicornis.* The authors would like to thank Robert Belleman and Frank Lakeman (University of Amsterdam, Parallel Scientific Computing and Simulation Group) for their assistance to create the colour plates shown in this paper.

References

1. C. W. Stearn and R. Riding. Forms of the hydrozoan *Millepora* on a recent coral reef. *Lethaia*, 6:187–200, 1973.
2. W. H. de Weerdt. Transplantation experiments with Caribbean *Millepora* species (Hydrozoa, Coelenterata), including some ecological observations on growth forms. *Bijdr. Dierk.*, 51(1):1–19, 1981.
3. M. P. Lesser, V. M. Weis, M. R. Patterson and P. L. Jokiel. Effects of morphology and water motion on carbon delivery and productivity in the reef coral, *Pocillopora damicornis* (linnaeus): diffusion barriers, inorganic carbon limitation, and biochemical plasticity. *J. Exp. Mar. Biol. Ecol.*, 178:153–179, 1994.
4. K. P. Sebens. Water flow, growth form and distribution of sceleractinian corals: Davies reef (GBR), Australia. *Proceedings of the 7th International Coral Reef Symposium*, 1:557–568, 1992.
5. J. A. Kaandorp. Modelling growth forms of the sponge *Haliclona oculata* (Porifera; Demospongiae) using fractal techniques. *Mar. Biol.*, 110:203–215, 1991.
6. J. A. Kaandorp and M. J. de Kluijver. Verification of fractal growth models of the sponge *Haliclona oculata* (Porifera; class Demospongiae) with transplantation experiments. *Mar. Biol.*, 113:133–143, 1992.
7. J. A. Kaandorp. Morphological analysis of growth forms of branching marine sessile organisms along environmental gradients. (in prep.)

8. J. E. N. Veron and M. Pichon. *Australian Institute of Marine Science Monograph series vol 1, Scleractinia of eastern Australia part 1.* Australian Government Publishing Service, Canberra, 1976.
9. J. A. Kaandorp. *Fractal modelling: growth and form in biology.* Springer-Verlag, Berlin, New York, 1994.
10. J. Shimeta and P. A. Jumars. Physical mechanisms and rates of particle capture by suspension feeders. *Oceanogr. Mar. Biol. Annu. Rev.*, 29:191–257, 1991.
11. M. Frechette, C. A. Butman and W. R. Geyer. The importance of boundary-layer flows in supplying phytoplanton to the benthic suspension feeder *mytilus edulis* L. *Limnol. Oceanogr.*, 34(1):19–36, 1989.
12. L. W. Buss and J. B. C. Jackson. Planktonic food availability and suspension-feeder abundance: evidence of in situ depletion. *J. Exp. Mar. Biol. Ecol.*, 49:151–161, 1981.
13. A. J. Pile, M. Savarese and V. I. Chernykh. Trophic effects of sponge feeding within lake baikal's littoral zone. 2. sponge abundance, diet, feeding efficiency, and carbon flux. *Limnol. Oceanogr.*, 42(1):178–184, 1997.
14. M. R. Patterson. Patterns of whole colony prey capture in the octocoral, *alcyonium siderium. Biol. Bull.*, 167:613–629, 1984.
15. J. A. Chamberlain and R. R. Graus. Water flow and hydromechanical adaptations of branched reef corals. *Bull. Mar. Sci.*, 25:112–125, 1977.
16. A. S. Johnson and K. P. Sebens. Consequences of a flattened morphology: effect of flow on feeding rates of the scleractinian coral *meandrina meandrites. Mar. Ecol. Prog. Ser.*, 99:99–114, 1993.
17. S. A. Wainwright and J. R. Dillon. On the orientation of se a fans (genus Gorgonia). *Biol. Bull.*, 136:130–139, 1969.
18. J. A. Kaandorp, C. Lowe, D. Frenkel and P. M. A. Sloot. The effect of nutrient diffusion and flow on coral morphology. *Phys. Rev. Lett.*, 77(11):2328–2331, 1996.
19. S. Chen, Z. Wang, X. Shan and G. Doolen. Lattice Boltzmann computational fluid dynamics in three dimensions. *Journal of Statistical Physics*, 68(3/4):379–400, 1992.
20. A. J. C. Ladd. Numerical simulations of particulate suspensions via a discretized Boltzmann equation Part I. theoretical foundation. *J. Fluid Mech.*, 271:285–309, 1994.
21. A. J. C. Ladd. Numerical simulations of particulate suspensions via a discretized Boltzmann equation Part II. numerical results. *J. Fluid Mech.*, 271:311–339, 1994.
22. A. Koponen, D. Kandhai, E. Hellen, M. Alava, A. Hoekstra, M. Kataja, K. Niskanen, P. M. A. Sloot and J. Timmonen. Permeability of three-dimensional random fiber webs. *Phys. Rev. Lett.*, 80(4):716–719, 1998.
23. G. Punzo, F. Massaioli and S. Succi. High-resolution lattice Boltzmann computing on the IBM SP1 scalable parallel computer. *Computers in Physics*, 8(6):705–711, 1994.
24. D. Kandhai, A. Koponen, A. Hoekstra, M. Kataja, J. Timmonen and P. M. A. Sloot. Lattice Boltzmann hydrodynamics on parallel systems. *Computer Physics Communications*, 111:14–26, 1998.

25. W. E. Lorensen and H. E. Cline. Marching cubes: a high resolution 3D surface construction algorithm. *IEEE Computer Graphics*, 21(3):163–169, 1987.
26. T. A. Witten and L. M. Sander. Diffusion-limited aggregation, a kinetic critical phenomenon. *Phys. Rev. Lett.*, 47(19):1400–1403, 1981.
27. J. Feder. *Fractals.* Plenum Press, New York, London, 1988.
28. J. A. Kaandorp. Analysis and synthesis of radiate accretive growth in three dimensions *J. Theor. Biol.*, 175:39–55, 1995.
29. R. G. Belleman, J. A. Kaandorp and P. M. A. Sloot. A virtual environment for the exploration of diffusion and flow phenomena in Complex geometries. *Future Generation Computer Systems*, 14:209–214, 1998

FROM PSEUDO-RANDOM NUMBERS TO STOCHASTIC GROWTH MODELS AND TEXTURE IMAGES

LEONID P. YAROSLAVSKY[1]
Centre Emil Borel - UMS 839 (CNRS/UPMC)-Paris
and
Department of Interdisciplinary Studies
Faculty of Engineering, Tel Aviv University,
Tel Aviv 69978, Israel

Pseudo-random number generators, stochastic evolutionary growth models and texture images are treated in a unified way in terms of dynamic systems with nonlinear feedback and their evolutionary behavior.

1 Introduction

The problem of describing and modeling formation of stochastic patterns is addressed. Two kinds of patterns are dealt with in the paper: patterns that simulate processes of growth in biology and similar processes in material sciences, and images that are studied in image processing, computer vision and computer graphics and are conventionally called *textures.* Traditionally, two completely different methodological approaches are applied to treat these two kinds of pattern formation problems. While simulating growth processes is naturally based on phenomenological modeling and imitating involved biological and similar evolution processes, in texture synthesis and analysis attempts prevail to customize techniques of the theory of random processes that originate from mathematical models in statistical physics and communication theory.

Both approaches suffer from a certain "one-sidedness". Phenomenological growth models frequently ignore stochastic nature of growth processes and growing formations. Random process theory based texture image models and texture analysis methods that have been a subject of intensive research efforts for many years ignore physical meaning of imaging and images, the fact that texture images as any other images are reflections of objects of the real physical word and therefore carry a dint of real processes that generate and govern those objects.

The paper is intended to show that, in fact, there is no substantial difference between textures and growing formations and that they can be treated in a unified way. Basic points of this unified approach are:

- Both texture images and growing patterns can be modeled in the same way in terms of *dynamic systems* with nonlinear feedback and their evolutionary behavior.

- Both texture images and growing patterns are stochastic. This means that they are controlled by a set of parameters that define only an ensemble of the

[1]ON LEAVE FROM INSTITUTE OF INFORMATION TRANSMISSION PROBLEMS, RUSSIAN ACADEMY OF SCIENCES, 101447, B.KARETNY, 19, MOSCOW, RUSSIA.

patterns of the same kind rather than each individual one. Each individual pattern is obtained as a result of a "random" choice from the ensemble.

- The process of the "random" choice can itself be regarded as a zero-dimensional growth (evolution of a single variable) and modeled by generating *pseudo-random numbers* conventionally used in Monte-Carlo computer simulations.

The exposition is arranged in the following way. In Section 2, two modifications of generators of pseudo-random numbers are represented in terms of dynamic systems with nonlinear feedback and described by schematic diagrams. In Section 3, it is shown that rather simple modifications of these schematic diagrams give rise to a family of stochastic growth models that are illustrated by Eden's type models [5–7] and by several modifications of evolutionary models that originate from Conway's "Game of Life" [8,11–13]. Section 4 is devoted to texture image modeling. It demonstrates how texture images can be generated and treated as a product of evolution of dynamic systems with feedback and illustrates several concrete examples of such systems of different complexity. In conclusion, a short glossary of terms used in the paper is provided.

2 Pseudo-random number generators

Nothing in Nature is random.
A thing appears random only through
the incompleteness of our knowledge.
B. Spinoza [1]

Anyone who considers arithmetical methods
of producing random digits is, of course, in the state of sin.
J. Von Neuman [1]

First generators of pseudo-random numbers were suggested by Von Neuman at the very beginning of the computer era. Since then many attempts have been undertaken to improve "randomness" of the generated numbers including even attempts to introduce hardware random number generators that exploit "random" nature phenomena such as radioactivity or Brownian motion. Finally software generators have become commonly accepted for generating numbers that seem "random" in particular applications. Since these numbers are generated in a complete deterministic way they are called "pseudo-random" which reflects a faith that true random numbers do exist. Usually, these generators produce pseudo-random number recursively from one initial number by using relatively simple computational rules. For instance, Knuth [2] recommends an algorithm that can be described by the following recursive relationship:

$$\xi^{(t)} = (C_1\xi^{(t-1)} + C_2) mod C_3 \quad (12)$$

Here $\xi^{(t)}$ is a pseudo-random number generated at the t-th iteration, C_1, C_2, C_3 are constants, $(\cdot) mod C_3$ is an operation finding the residual of the division of the input value by C_3.

The algorithm generates, one by one, pseudo-random numbers with uniform distribution in the range [0, 1]. The very first number is called "seed" and has to be set in advance. A schematic diagram of the algorithm is shown in Figure 1.

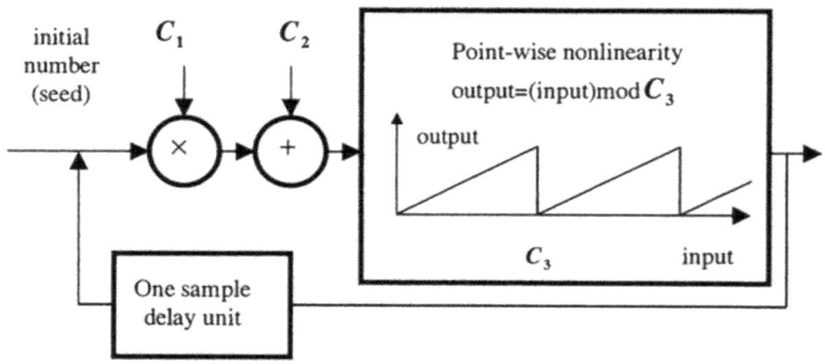

Figure 1. Schematic diagram of a pseudo-random number generator. The graph shows transfer function of the point-wise nonlinearity unit

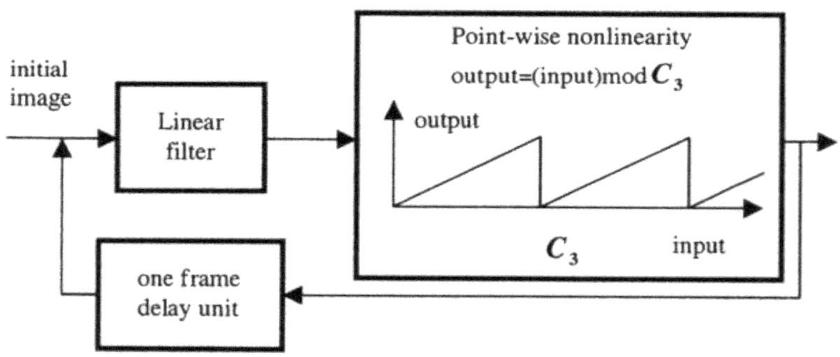

(a)

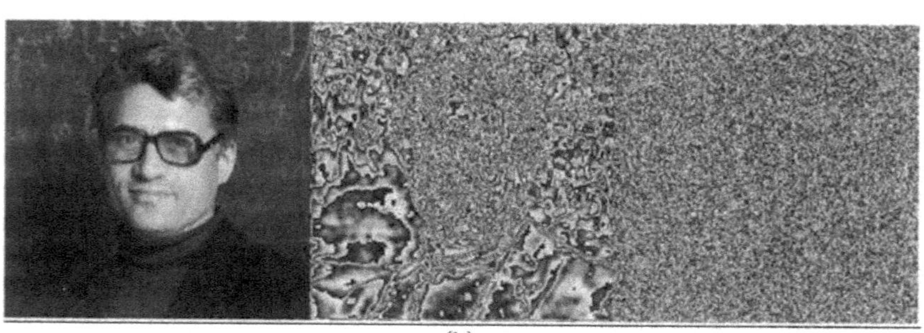

(b)

Figure 2. A modification of the pseudo-random number generator with a linear filter in the feedback (a) and examples of an initial image (b, left) and generated images after one (b, center) and 10 iterations (b, right).

Represented in this way, the algorithm assumes using the following processing units: a multiplication unit, a summation unit, a *point-wise nonlinearity* that implements the operation $(\cdot) modC$ (its transfer function is shown in Figure 1), and a one sample *delay unit*. The latter is a very important component of the scheme that makes it iterative, or, in another word, evolutionary.

This scheme represents an example of a very simple nonlinear dynamic system. It is well known now that such systems potentially may exhibit *cycles* and *fixed points*. A common requirement to the pseudo-random number generators is that they should avoid cycles and fixed points and provide numbers with nearly uniform distribution and without noticeable correlations. In practice this is achieved by an appropriate selection of the model parameters C_1, C_2 and C_3.

The above scheme is obviously a simplified version of a more general one presented in Figure 2. Multiplication and summation units in the scheme of Figure 1 are replaced here by a *linear filter*, a device that computes output signal by weighted summation of certain number of input samples, the weights being defined by the filter *impulse response* (*point spread function*), and one-sample delay unit is replaced by a one-frame delay unit where a frame is a certain group of samples. If the signal samples in this scheme are arranged in a form of a rectangular array they can be displayed as an image. Figure 2(b) illustrates an example of evolution of an image taken as a "seed". The linear filter in this example is a simple two-dimensional "box" filter with a uniform 3×3 samples impulse response. Such a filter computes, for each image sample (*pixel*), the image local mean value over the window of 3×3 pixels centered at this sample. A constant C_3 in the point-wise nonlinearity was set equal to the half of the image maximal gray level. One can see on this image how the nonlinearity and feedback destroy in only a few iterations all pixel correlations that existed in the initial image.

3 Growth models

> Turing (1952) has demonstrated by mathematical methods that certain types of dynamic system which are initially homogeneous undergo a progressive change which leads to the appearance of spatial heterogeneity.
>
> J.W.S. Pringle, F.R.S. [3]

3.1 Eden's type growth models

Stochastic growth models aimed at simulating biological grows have been studied since very first years when digital computers became available [3,4]. One of the first models was suggested by M. Eden [5–7]. In Eden's model, growth was simulated as a sequence of random "births" taking place on a rectangular lattice (*raster*) with the probability proportional to the number of already "live" cells in the nearest spatial 3×3 vicinity of the given cell (left and right neighbors at the same row, 3 and 3 neighbors on the rows from above and from below). Eden's model can be mathematically represented as an iterative equation:

$$output(k,l)^{(t)} = randb(S_8^{(t-1)}/8) \oplus output(k,l)^{(t-1)} \tag{13}$$

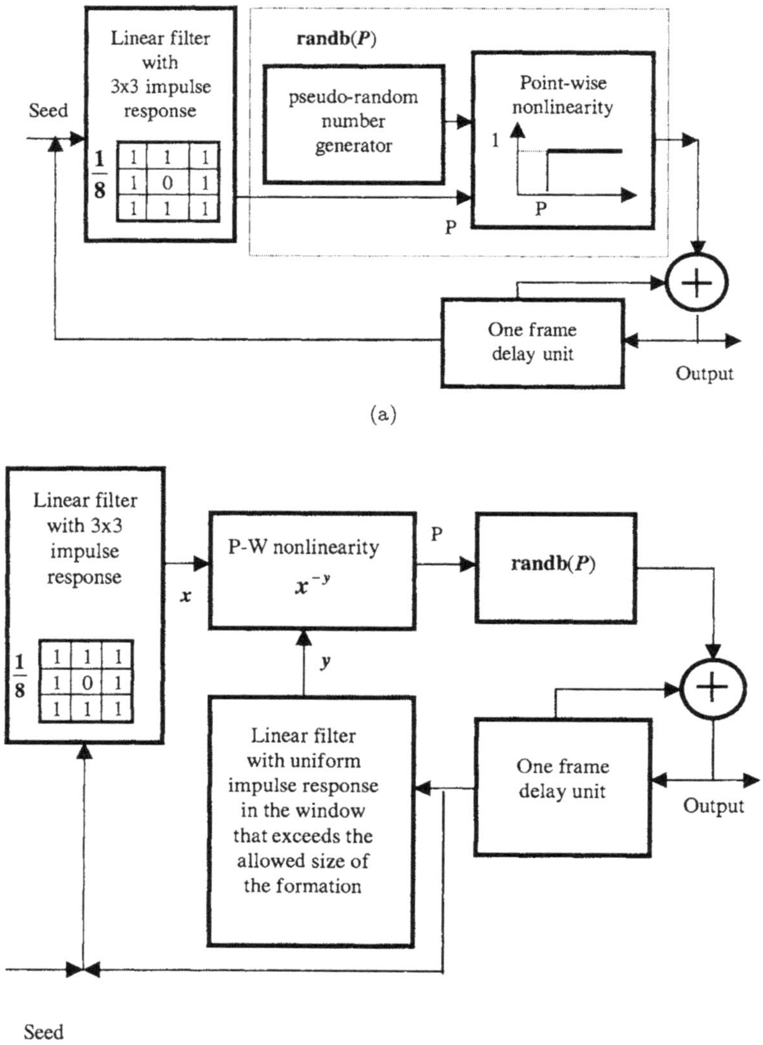

Figure 3. (a), (b) — Schematic diagram of two modifications of Eden's model. The tables in the boxes named "Linear Filter" present the linear filter impulse response. The graph in the box "Point-wise nonlinearity" in (a) shows the nonlinearity transfer function.

where (k, l) are pixel coordinates on the lattice, $S_8^{(t-1)}$ is the sum of pixel values in 8 neighbor points in the 3×3 neighborhood of the given pixel, t is iteration index, $randb(P)$ is a binary random variable that takes value of one with probability P and $\oplus$ denotes modulo 2 addition of binary numbers. Figure 3(a) shows how this model can be implemented in a system that is just a slightly modified end extended version of the system of Figure 2. This system contains, as an individual unit, a

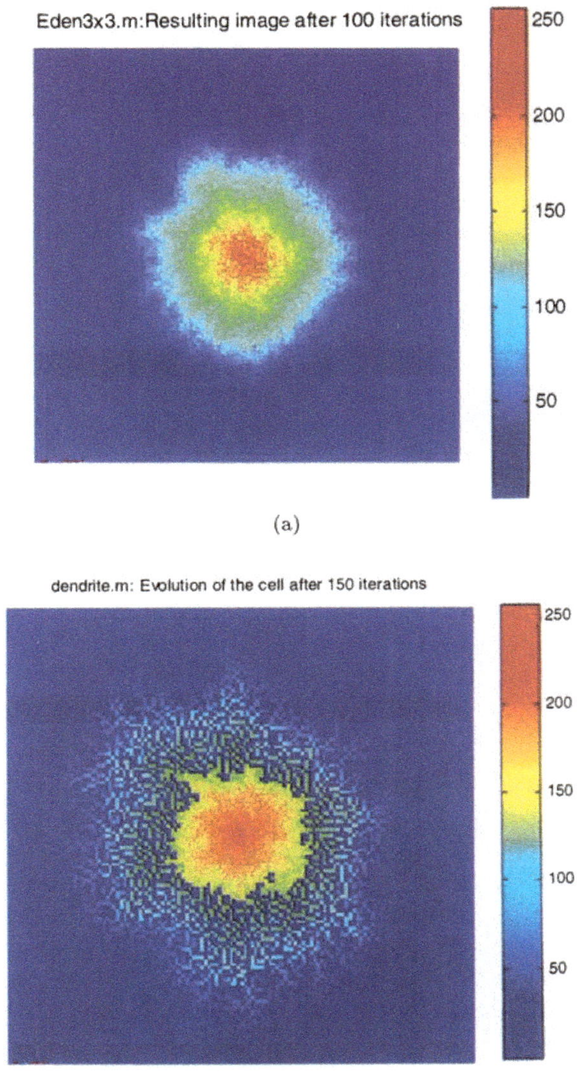

Figure 4. (a), (b) — Examples images generated by the respective Eden's model with saturation (a) and its modification that evolves into growing dendrites (b).

pseudo-random number generator of Figure 1 which is now included in the same loop of a linear filter, point-wise nonlinearity and a delay unit. The combination of the pseudo-random number generator and the point-wise nonlinearity form a unit that implements an operation $randb(P)$ of generating binary numbers 0's and 1's with a given probability of 1's. The model is binary. It assumes that each cell (pixel) has one of two possible states, say, 0 and 1. On binary images, the linear filter with impulse response as shown in Figure 3(a) computes the number of 1's in the 3×3 neighborhood (8-neighbor sum S_8) of each pixel thus defining the threshold level of the point-wise nonlinearity.

Clearly, this simple model describes unlimited growth. One can, however, easily modify this model to simulate drain of "sources of food" by measuring the size of the growing formation and introducing a saturation to the probability of "birth" as it is shown in the schematic diagram of Figure 3(b). In this scheme, the randb(P) unit of Figure 3(a) is preceded by a (x^{-y})-point-wise nonlinearity that implements the saturation. This modified model can be described by the equation:

$$output(k,l)^t = randb\left((S_8^{(t-1)})^{-cS_{gl}^{(t-1)}}/8\right) + output(k,l)^{(t-1)} \tag{14}$$

where $S_{gl}^{(t-1)}$ is a "global" sum over the entire field that defines the size of the formation on the $(t-1)$-th iteration (evolution) step.

If the saturation is introduced to all probabilities except the probability of "birth" from only one neighboring live cell, one arrives at a modification of the model which begins growing dendrites after (statistically) the cell reaches a certain size. Figure 4(a) and Figure 4(b) illustrate the work of these models. Images are displayed here in color that correspond to the "age" of each pixel (i.e. the number of evolution steps from its birth). Color bars show "age"-to-color coding scale. Other modifications of the model aimed, for instance, at imitating dependence of growth from "age" of cells are also more or less straightforward.

3.2 Conway's "Game of Life" and its modifications

A mathematical model known as Conway's "Game of Life" [8] represents yet another type of growth models where cells on a rectangle lattice (raster) can give a "births" or "die out" depending on the number of "live" and empty cells in their nearest spatial neighborhood. The rules of the original "Game of life" are very simple: if an empty cell has exactly 3 "alive" neighbor cells in its 3×3 neighborhood, "birth" takes place on the next step of the evolution; if an alive cell has less than 2 and more than 3 "alive" cells in the neighborhood it will die on the next step; otherwise nothing happens. These rules can be formally described by the equation:

$$output(k,l)^t = \lfloor output(k,l)^{t-1} \rfloor \delta(S_8^{t-1} - 2) + \delta(S_8^{t-1} - 3) \tag{15}$$

where $\delta(\cdot)$ is Kronecker delta, S_8^{t-1} is the sum of the values in 8-neighborhood of (k,l)th pixel, t is iteration number and "alive" and empty cells are represented by 1's and 0's, respectively. The original model [8] assumed a deterministic initial distribution of 0's and 1's in the field. By introducing "random" initial distribution of "alive" and empty cells the model can be made stochastic. The corresponding schematic diagram of this model is shown in Figure 5(a).

One can see that this diagram contains essentially the same units as Eden's model but in two parallel *branches* (one for "births" and one for "deaths"), and the *randb*(P) generator of the Eden's model is placed at the input of the model and is used for generating "random" initial distribution of 1's and 0's for "alive" and empty cells.

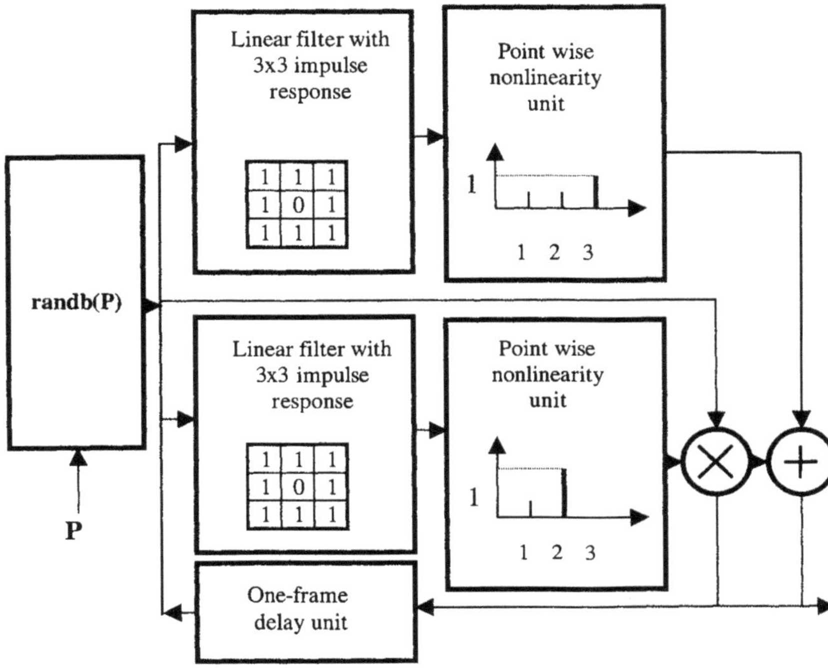

Figure 5. Schematic diagram of Conway's model. Tables and graphs in the boxes show impulse response and the transfer functions of the corresponding units.

The evolutionary behavior of the model is very interesting. The model generates different types of formations: stable formations that once appeared keep staying unchanged unless they are destroyed by other formations; growing crystal-like formations that grow until their fragments form stable formations or die out; moving formations featuring iteration-wise cycles ("gliders"); different combination of stable and moving formations. Boundary conditions of the model are also important for its evolution. Under pseudo-random boundary conditions, when pseudo-random binary numbers are permanently generated at the borders of the field the model generates patterns that do not converge to a fixed (stable) point though always contain certain number of formations that live during considerably large number of iterations (evolution steps). The pattern evolution is illustrated in Figure 6(b)–(g). One can see on these images randomly placed stable formations such as 2×2 pixel square blocks and hexagonal formations called "beehives", formations that grow like crystals, and "gliders" that move across the lattice with a period of 4 evolution steps (the right-upper corner of the south-west quarter of the images shows two gliders), etc.

An important component of the model is the direction of the spatial interaction as defined by the linear filter impulse response. In the original Conway's model, the spatial interaction is *isotropic*: all cell's 8 neighbors play the same role in the defining next state of the cell on each iteration step. In the model of Figure 5,

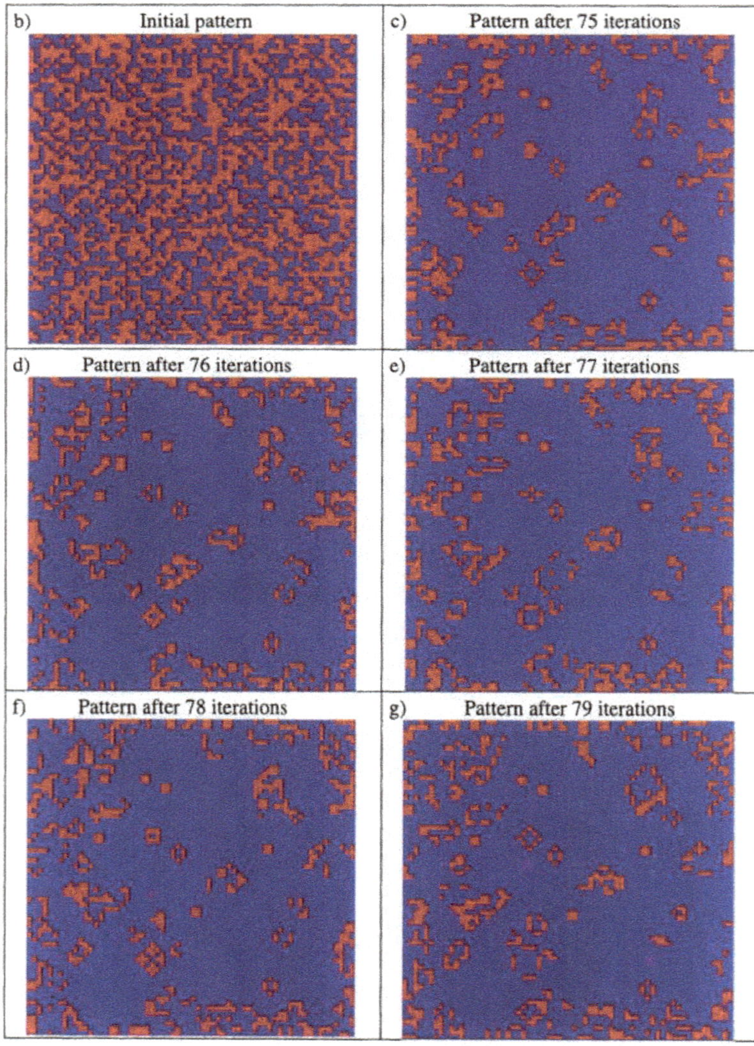

Figure 6. Evolution of pattern (b) in the Game of Life model. On images (c)–(g) one can see how different formations evolve. In particular, in right upper corner of the left-bottom quarter of each image two "gliders" are seen that move with a period of 4 iterations.

this is reflected in the linear filter isotropic impulse response equal to 1 for all 8 neighbor pixels. In general, the filter impulse response may be anisotropic rather then isotropic as in the original Conway's model. In particular, it may define only one-dimensional interaction (only left and right neighbors of each cell affect its next state) thus producing one-dimensional models. An interesting special case of such a 1-D model is one described by the equation:

$$output(k)^{(t)} = output(k)^{(t-1)}\delta(S_2) \oplus \delta(S_2 - 1) \tag{16}$$

P=0.3

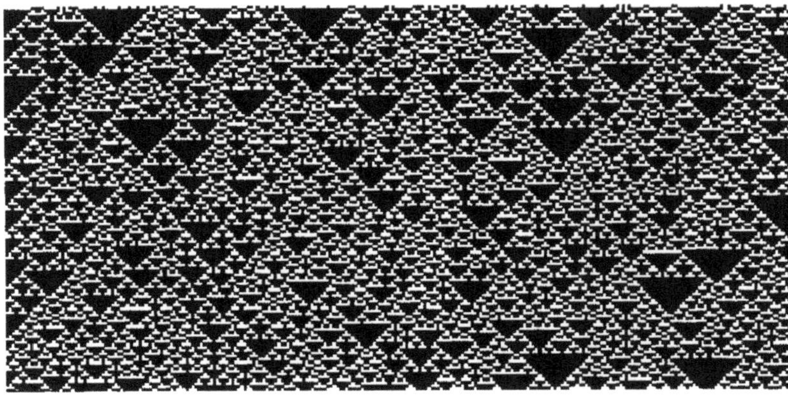

Figure 7. Evolution (downward in the vertical direction) of a one-dimensional (in horizontal direction) modification (Eq. 16) of the Game of Life (initial rate of white points in the first row was 0.3).

where S_2 is the sum over 2 neighbor cells of the k-th cell (from the left and from the right). Figure 7 shows row by row an example of the evolutionary behavior of such a one-dimensional model. It is interesting to observe that patterns that appear in the process of the evolution remind the so called Sierpinski Gasket [9] and those patterns that some sea-shells develop in their life-time.

One can further modify this model by introducing random "death" and "birth" events:

$$output(k,l)^t = randb(P_d)\lfloor output(k,l)^{t-1}\rfloor \delta(S_8^{t-1}-2) + randb(P_b)\delta(S_8^{t-1}-3) \quad (17)$$

where $randb(P_d)$ and $randb(P_b)$ are the same binary pseudo-random number generators as in Eden's model (Equation 13) that produce "ones" with probabilities P_d (probability of "death") and P_b (probability of "birth"), respectively. In the initial Conway's model, $P_d = P_b = 1$. If $P_d < 1$, the evolutionary behavior of this model radically changes. It begins to produce labyrinth-like formations with irregular dislocation whose positions depend on realization of the initial pattern. While the "body" of the patterns stabilizes after a few iterations their periphery continue growing independently until the pattern fills the entire lattice. Depending on the probability of 1's in the initial distribution it may happen that several such formations arise and start growing until they merge into one. An example of such an evolution is shown in Figure 8. One can see that most cells in the initial pattern (Figure 8(a)) die out after a few iterations. There are however some "lucky" combinations of cells that survive and give rise to growing formations (Figure 8(b)) that grow until they merge (Figure 8(c)) and finally fill the field (Figure 8(d)). The same is observed if the initial binary patterns is highly correlated as it is shown in Figure 8(e)–(f) .

Far more rich evolutionary behavior can be observed with a modification of the Conway's model in which a Kronecker delta-function $\delta(\cdot)$ in Equation 15 is replaced

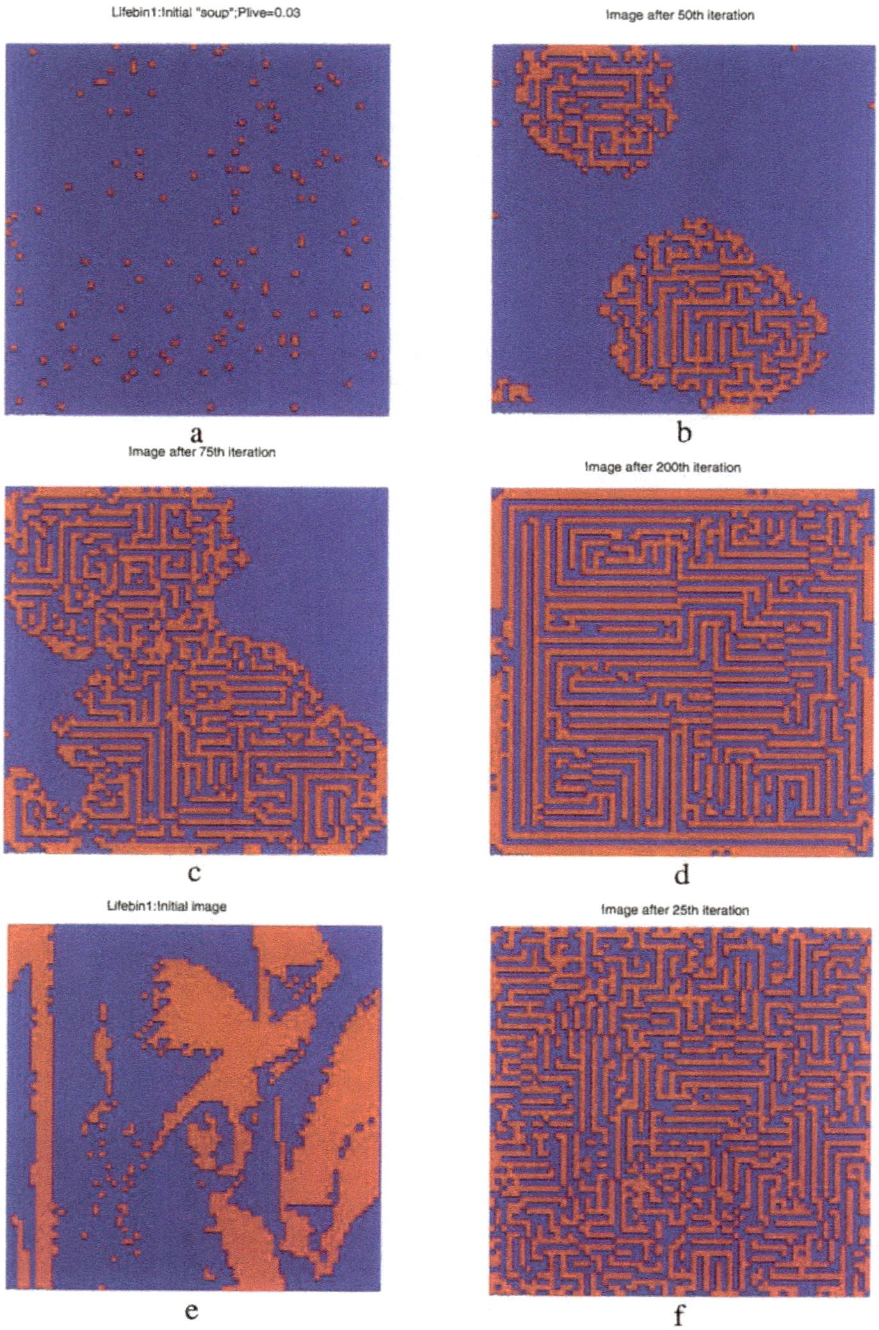

Figure 8. Example of the modified Conway's model evolution with $P_d = 0.25$ and $P_b = 1$ (a) and (e) -initial binary patterns; (b), (c), (d), (f) -corresponding evolution results.

by a "fuzzy delta" [11–13]:

$$output^{(t)}(k,l) = [output^{(t-1)}(k,l)]\Delta(L_1^{(t_1)} - C_1) + \Delta(L_2^{(t-1)} - C_2) \qquad (18)$$

where $\Delta(\cdot)$ is a unimodal function (a "fuzzy delta") that replaces the delta function in the model of Equation 15, L_1 and L_2 are outputs of linear filters that replace summations over 8 neighbors in the model of Equation 15 and C_1 and C_2 are constants that replace thresholds 2 and 3 in the model of Equation 15. In this modification, states of cells are not binary and they are modeled by numbers that can have any value in the range $[0, 1]$.

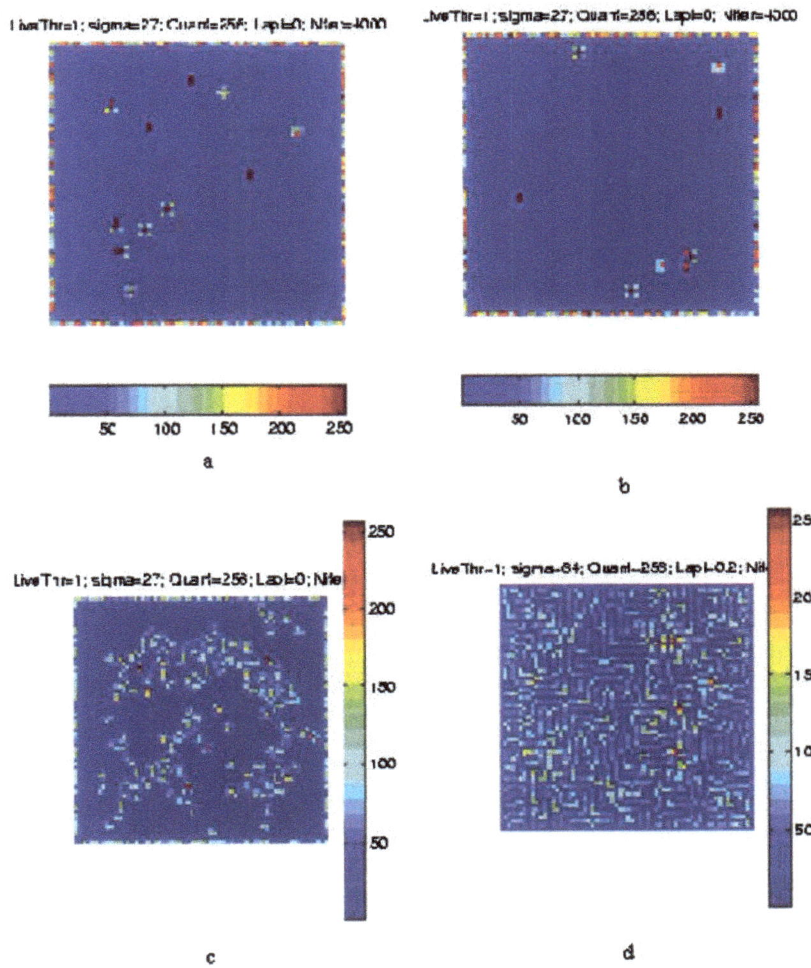

Figure 9. Examples of evolutionary behavior of the modified Conway's model of Eq. (18): (a), (b) — stable constellations that, once appeared, stay forever (model fixed points); (c) — "dying clouds"; (d) — labyrinth-alike pattern. Cell value levels in the images are varying here from 0 to 255 and are coded in color as it is represented by the color bars.

Experiments show that, with this model, the following three major types of evolutionary behavior can be observed depending on the spread of the "fuzzy delta": "stable chaos", "ordering of chaos" and "re-emerging of chaos". In the "stable chaos" mode, the initial chaotic pattern produced by the primary 2D random number generator gradually evolves into visually correlated patterns that then remain to look (visually) similarly though pixel values keep changing with iterations. In the "ordering of chaos" mode, the initial chaotic pattern degenerates in the course of iterations into spatial constellation-like or labyrinth-like patterns that remain stable spatial-wise but may exhibit "temporal" (iteration-wise) cycles. Obviously these are the model fixed points. The most complex and varying is the behavior of the "re-emerging of chaos" type. Its basic feature is a quick degeneration of the initial pseudo-random pattern into a uniform field (a trivial fixed point of the model) or into "constellations". After that a new chaotic pattern emerges through growing crystal-like formations from the constellations, left from the initial pattern, through spatial waves from the borders when they kept to be random, or through appearance of different types of "gliders" that move across and collide producing clouds of new "particles". These emerging formations gradually fill in the field with visually correlated patterns similarly looking to those characteristic of the "stable chaos" mode. Examples of the evolutionary behavior of such a model are shown in Figure 9(a)–(d). It is interesting to note that being initially non-quantized, cell values of the patterns tend, in the process of evolution, to stabilize at a certain set of quantized levels. This evidences in favor of the hypothesis that the quantization phenomena are characteristic not only for spatial behavior of the evolving patterns (spatial quasi-periodicity of labyrinth-like formations), or for their temporal behavior (temporal cycles of constellations), or for their spatial-temporal behavior (gliders as spatial-temporal cycles) but also for cell value levels.

4 Algorithmic models of texture images

> "Texture is an elusive notion which mathematicians and scientists tend to avoid because they cannot grasp it. Engineers and artists cannot avoid it, but mostly fail to handle it to their satisfaction."
>
> B. Mandelbrot [9]

Texture images represent an important class of images very frequently used, analyzed and generated in science, art and technology. Usually they are described by examples. There exist, for this purpose, quite a number of special albums of texture images for educational and applied purposes such as, for instance, Brodatz's album for artists and designers [15]. However, there is no commonly accepted formal definition of what is texture image and synthesis and analysis of textures remains to be one of the most controversial issues in image processing. This section outlines, in the same framework of dynamic systems, an algorithmic approach as a possible constructive solution of the texture image formalization, synthesis and analysis problem [11,14,17]. It demonstrates also the relation between this problem and the problems of generating pseudo-random numbers and growth modeling.

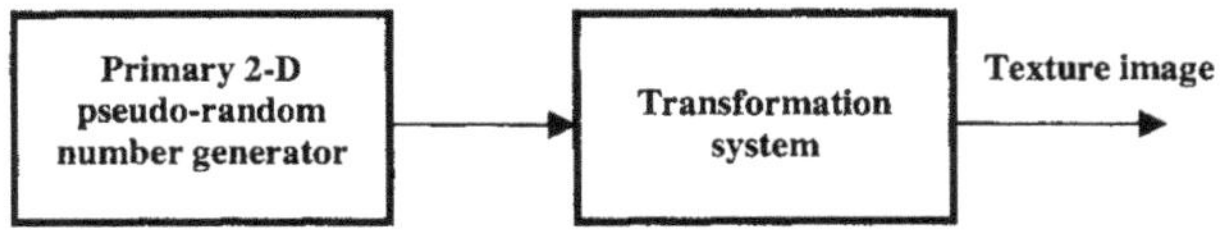

Figure 10. Algorithmic model of texture images.

4.1 The algorithmic approach to synthesis and analysis of texture images

Within the algorithmic approach, texture images are regarded as a result of a certain transformation of an initial standard pseudo-random pattern (primary texture) in a transformation system composed of a set of certain standard (elementary) units (Figure 10). Parameters of these units and the transformation system structure form the set of parameters that define texture of a certain class. Thus, with the algorithmic approach, texture synthesis is reduced to the synthesis of the transformation system, and texture image analysis is reduced to the problem of parameter identification of the transformation system.

The specific selection of the set of structural units of texture algorithmic models is governed by considerations of convenience of their parameterization and their computational complexity. It is only natural to use, for instance, units that form a basic instrumentation tool of digital image processing such as [10,11]:

- the above mentioned point-wise nonlinearity (PWN) that transforms signal samples according to the relationship:

$$output(k,l) = F(input(k,l))$$

  where $F(\cdot)$ is a, generally, nonlinear function that defines the transformation and (k,l) are sample indexes.

- the above mentioned linear filters (LF). Linear filters are defined by the equation of weighted summation:

$$output(k,l) = \sum_{m,n} h(m,n;k,l) input(k,l)$$

  where $h(m,n;k,l)$ is filter impulse response.

- Rank filters (RF) [11,16,17]. *Rank filters* operate with image *local histograms* computed over certain neighborhood of each pixel and are defined by the equation

$$output(k,l) = F_{nbh}(input(k,l))$$

  where $F_{nbh}(\cdot)$ is a function defined by the local histogram of the image over a certain neighborhood of the pixel (k,l).

- Logical filters. Logical filters assume working with binary images and are defined by a certain Boolean function of input pixels. For binary images, logical filters can implement both linear and rank filters.

- Image geometrical transformation (zoom, rotation, etc.) unit.

One is free to further extend or to modify this list to include other units proved to be useful components of image processing and computer graphics tool.

For the connection of units into a system, the following types of interconnections can be assumed:

- serial connection;
- parallel connection;
- feedback.

As a primary 2D pseudo-random number generator one can assume using the generator described in Section 2.

Some particular illustrative algorithmic models of textures are presented in Figure 11–Figure 18. Figure 11 represents the simplest PWN-model that contains only a point-wise nonlinear transformation unit. Note that a particular version of this model with a threshold nonlinearity was used in the growth models of Figure 3 to implement operation *randb*(P). In the PWN-model, one can easily control distribution density of the gray levels of generated patterns by an appropriate selection of the nonlinearity.

The next step in the hierarchy of texture algorithmic models are LF-models that consist of a linear filter as the transformation unit (Figure 12).

With LF-models, one can generate pseudo-random patterns with Gaussian distribution of pixel gray levels. Selection of the linear filter frequency response (Fourier Transform of its impulse response) controls Fourier power spectrum (spectral density) of the pattern and, therefore, LF-models can be used for generating correlated textures with a given correlation function (power spectrum). Figure 13 represents two examples of such textures obtained from initial patterns of uniformly distributed non correlated pseudo-random numbers. The top one was generated with the linear filter frequency response in a form of a ring in frequency domain, the bottom one with the isotropic filter frequency response which is inversely proportional

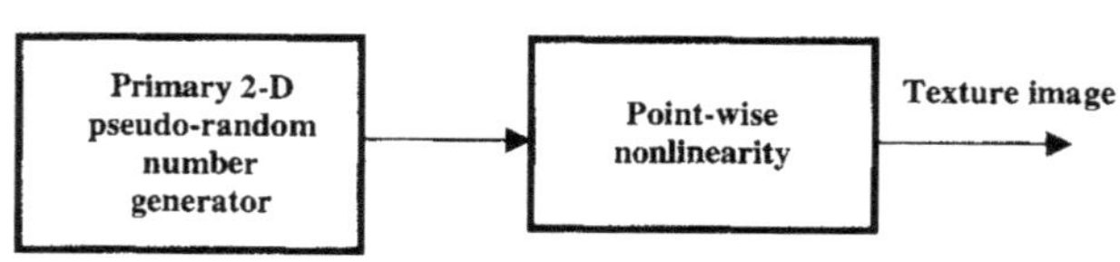

Figure 11. PWN-model of textures.

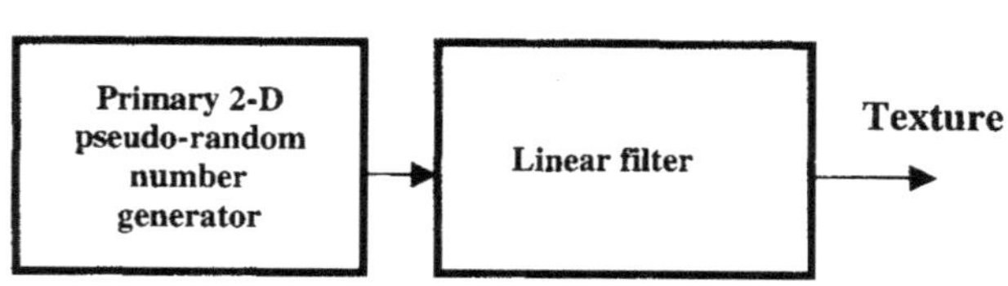

Figure 12. LF-models of texture images.

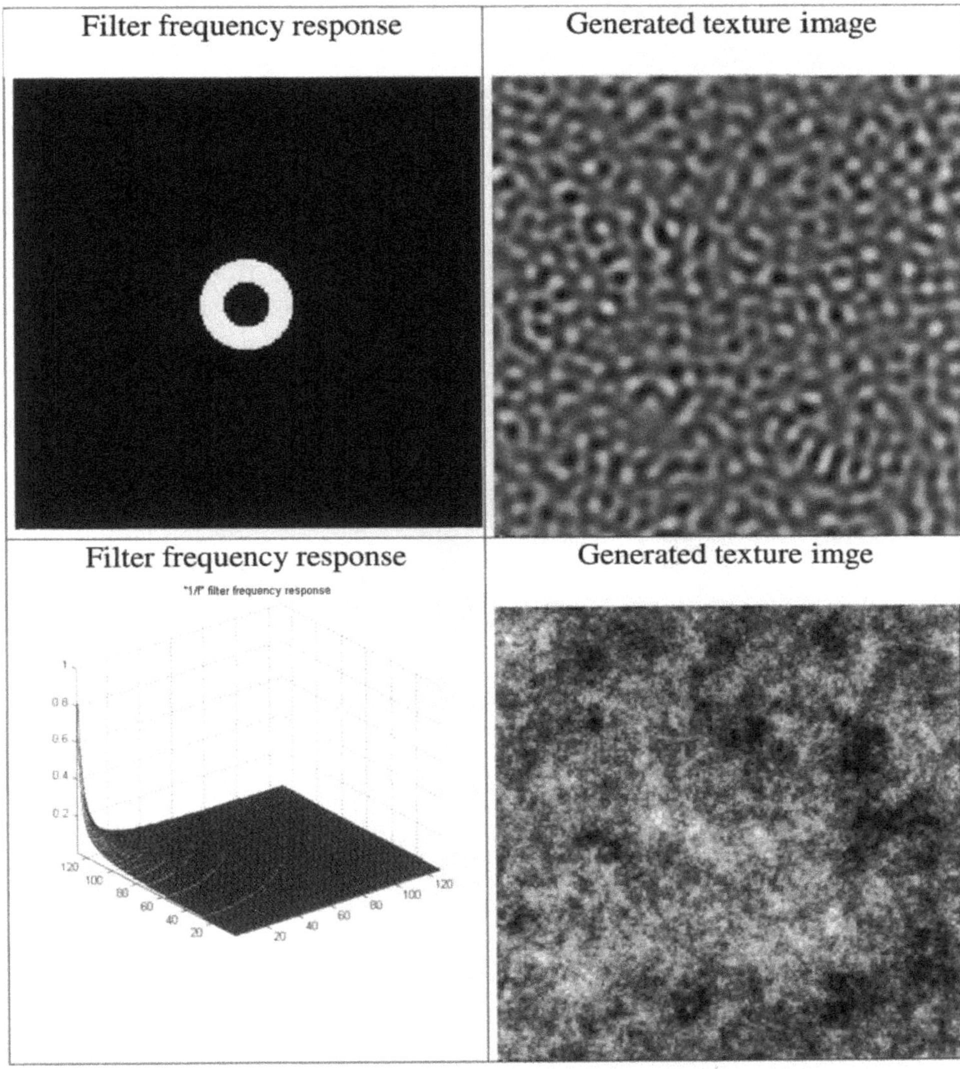

Figure 13. Examples of texture images generated by LF-models and the corresponding filter frequency responses. Images in the right column are obtained from uniformly distributed independent random numbers at the output of the primary 2-D pseudo-random number generator by multiplication of their Fourier spectra by masks shown in the left column (ring shown as an image and monotonically decaying function of spatial frequencies shown as a plot).

to the absolute value of the spatial frequency. This latter texture image illustrates what is known as $(1/f)$-fractals [9].

Even such a simple model as LF-one allows to imitate quite a number of natural texture images as it is illustrated in Figure 14.

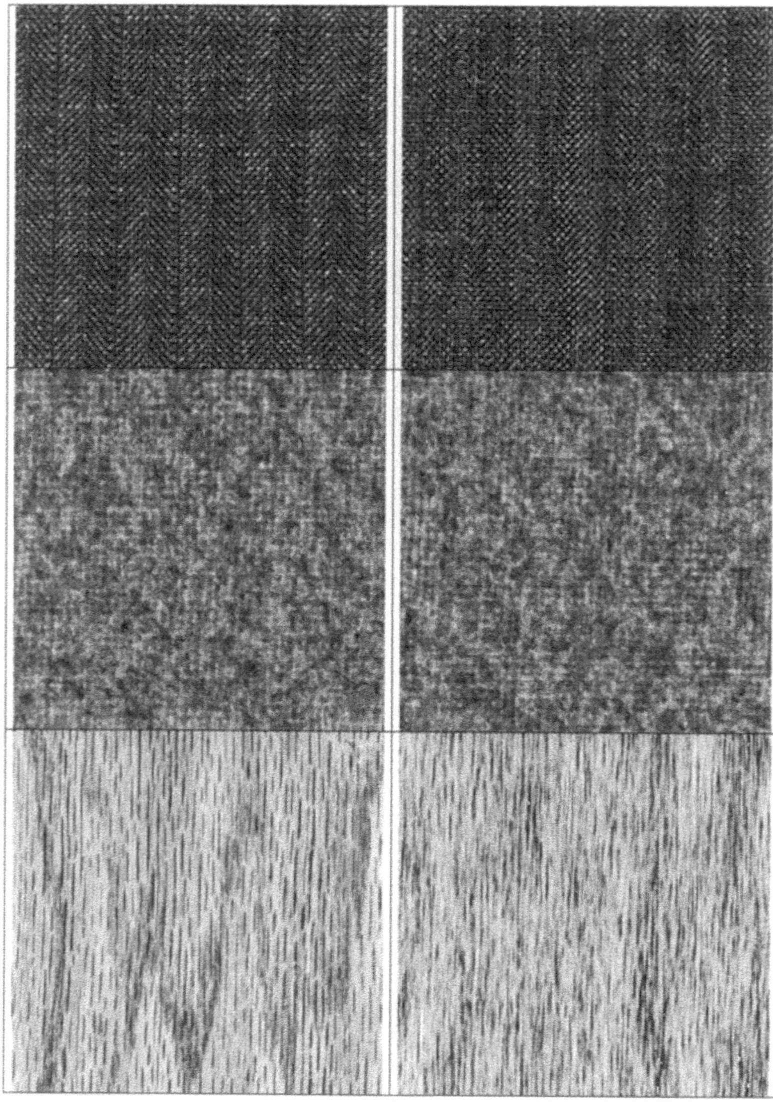

Figure 14. Natural texture images from Brodatz' album [15] (left column) and their synthetic copies (right column) generated from uniformly distributed pseudo-random numbers.

Combination of the threshold type point-wise nonlinearity and a linear filter in PWN-LF models (Figure 15(a)) allows to generate patterns of randomly distributed filter impulse responses as those shown in Figure 15(b).

Reversion of the order of point-wise nonlinearity and linear filter in PWN-LF models results in LF-PWN-models (Figure 16(a)). LF-PWN-models allow to generate textures with a given correlation function (controlled by the linear filter impulse

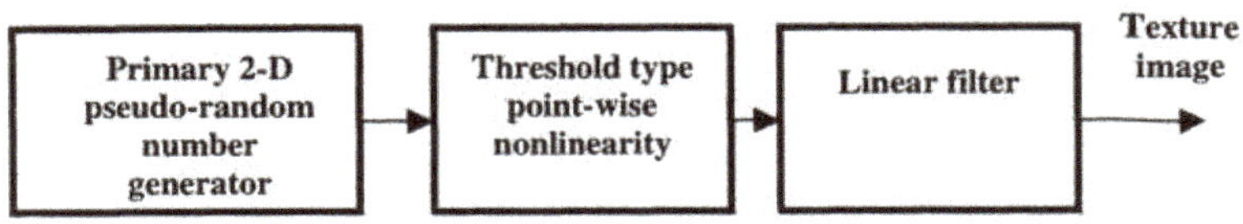

(a)

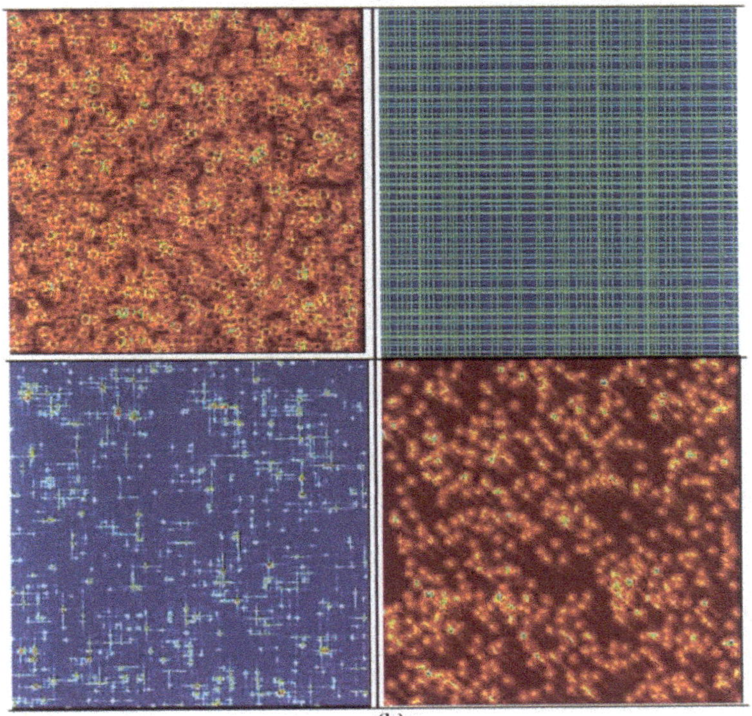

(b)

Figure 15. PWN-LF-model of texture images (a) and 4 examples of generated textures (shown in pseudo-colors).

response) and a given distribution density (controlled by the nonlinear unit). Four examples of such texture images are shown in Figure 16(b).

All above described models contain only one branch (several units in cascade). The texture model can have several branches whose outputs can be combined in different ways. For instance, outputs of branches can be multiplied, or the output of one branch can be used to switch between outputs of other branches, or the output of one branch can control parameters of the transformation units in another branch, etc. The growth models of Figure 3(b) and Figure 5(a) described above exemplify such multiple branch models. Figure 17 shows four examples of texture images generated by models with more sophisticated multiple branches.

Up to now no feedback connection was assumed in the texture models except for the one in primary pseudo-random number generators. Clearly, feedback gives

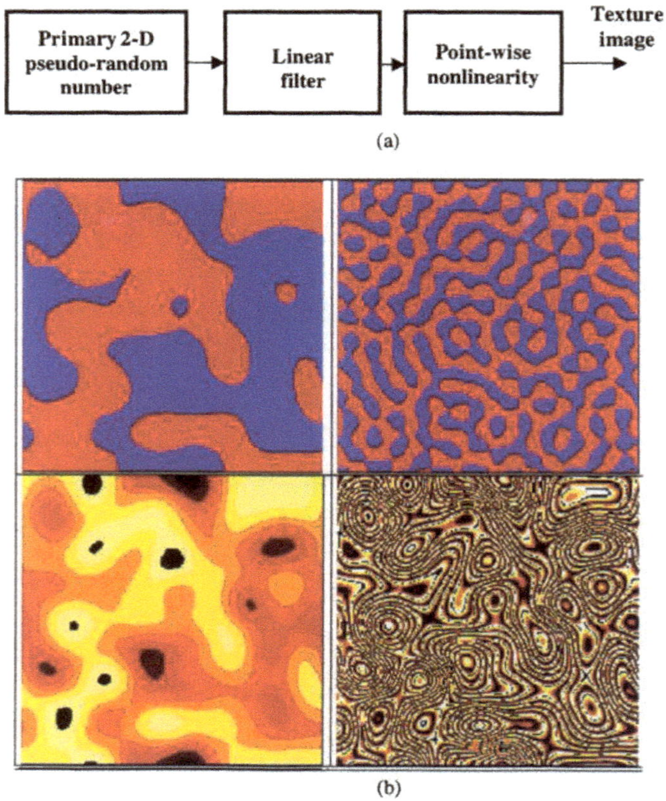

Figure 16. LF-PWN-model (a) and examples of textures images (b).

to the models evolutionary features. However, introducing feedback loop to simple PWN- or LF-models does not provide to them anything nontrivial since the result of their evolution can be obtained in appropriately modified models without feedback. Systems with non-trivial evolutionary behavior should contain both a linear filter (or any other filter with spatial interaction) and non-monotonic nonlinearity in the loop. The above random number generator of Figure 2 and growth models exemplify such systems. Inserting into the loop rank filters [11,16,17] that combine spatial interaction and substantial nonlinearity in a more sophisticated way than just by cascading linear filters and point-wise nonlinearity gives rise to a new family of evolutionary models.

An example of such an evolutionary model with a rank filter is illustrated in Figure 18. In this model, a particular rank filter is used that replaces, in each iteration, gray level of every pixel by the value that corresponds to the maximum of the image local histogram computed over certain spatial neighborhood of the pixel (this operation is called MODE(nbh)). As one can see from Figure 18(b), evolution of the texture in this model results in piece-wise constant images with

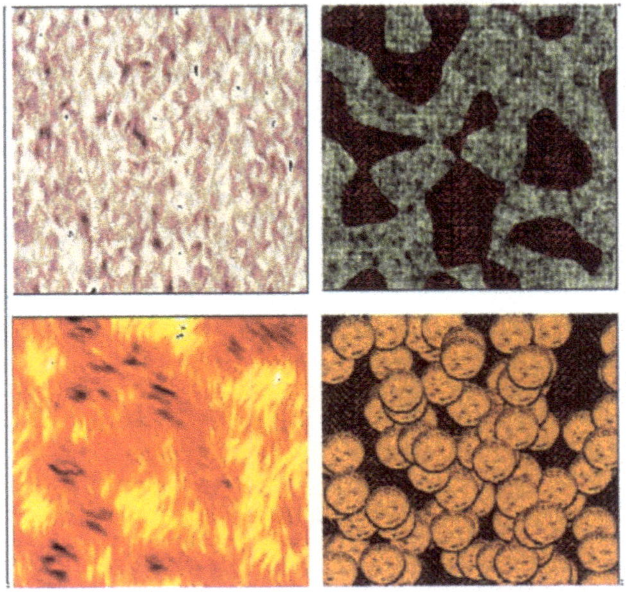

Figure 17. Texture images (represented in pseudo-colors) generated by models with multiple branches.

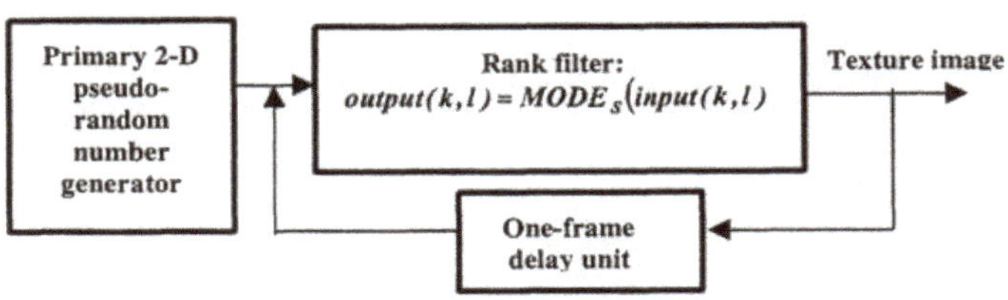

(a)

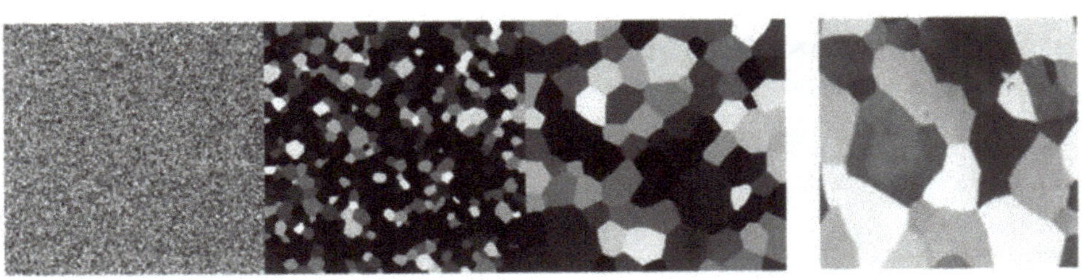

(b) (c)

Figure 18. Evolutionary texture model (a) with a rank filter and examples: (b) of primary pseudo-number pattern (left) and texture images generated by the model (center and right) and natural texture image of crystals (c).

"randomly" arranged patches whose size depends (statistically) on the size of spatial neighborhood used in the rank filter. These images are fixed points of this model. It is remarkable to see how well they resemble some natural crystal textures [Figure 18(c)].

Glossary

This section provides verbal explanations of some signal processing terminology used in the paper.

DYNAMIC SYSTEM: is a system in which the output signal is fed back to its input (system with a feedback loop). An important component of the feedback loop is the **delay unit** which delays the signal by a certain amount of time before it is put at the system input. Because of the feedback, dynamic systems exhibit evolutionary behavior in time. Signals that do not change when they pass through a dynamic system are called **fixed points** of the system. There might be system output signals that repeat themselves periodically. This is the case of system's **cycles**.

IMAGE HISTOGRAM: shows how frequently particular gray levels occur in the image. If the histogram is computed not for an entire image but for a certain neighborhood of a pixel, it is called **local histogram**.

LINEAR FILTER: is a signal processing device that computes output signal samples by the weighted summation of a certain number of input signal samples. The set of the summation weight coefficients is called **impulse response**, or **point spread function** of the linear filter. Fourier Transform of the filter impulse response is called its **frequency response**.

PIXEL: is a picture element (sample). Spatial arrangement of image samples (pixels) is called **raster**.

POINT-WISE NONLINEARITY: is a signal processing device whose output, for each input signal sample, is a certain function (in general nonlinear) of only this sample. This function is called **transfer function** of the device.

PSEUDO-RANDOM NUMBER: are computer generated numbers that appear, in a particular application, random, although they are generated by means of a fixed algorithm and are fully reproducible.

RANK FILTERS: operate with pixel values in more sophisticated way than linear filters do. They measure, for each input sample, certain parameters of image local histograms and then use them to produce the output.

STOCHASTIC PATTERNS: are patterns that exhibit (visually) elements of randomness, irregularity, chaoticity.

Acknowledgment

The work was completed during the author's stay at the Center Emil Borel, Institut Henri Poincare (Unité mixte de Paris VI-UPMC/CNRS) in the frame of the trimester "Mathematical questions in signal and image processing" and of the Arc en Ciel/Keshet Program.

References

1. H. O. Peitgen, H. Juergens and D. Saupe. *Fractals in Classroom, Pt. 1., Introduction to Fractals and Chaos.* Springer Verlag, New York, pp. 319, 1993.
2. D. E. Knuth. *The Art of Computer Programming.* Seminumerical Algorithms, vol. 2, Addison Wesley, London, 1969.
3. J. W. S. Pringle and F. R. S. . The origine of Life. Symposia of the Society for Experimental Biology, *Evolution* **VII** (1953).
4. A. Turing. The Chemical Bases for Morphogenesis. *Phil Trans. Roy. Soc. London*, **B337**, pp. 37–72, 1952.
5. M. Eden. *A Probabilistic Model for Morphogenesis.* pp. 359–370, Pergamon Press, N.Y., 1958.
6. M. Eden. A Two-dimensional Growth Process. *Proc. of the Fourth Berkeley Symposium on Mathematical Statistics and Probability*, Univ. of California Press **4** (1961) 223–239.
7. M. Eden, Ph. Thévenaz. History of a Stochastic Growth Model. In *6-th Int. Workshop on Digital Image Processing and Computer Graphics* (DIP-97), E. Wenger and L. Dimitrov, Eds., 20–22 Oct. 1997, Vienna, Austria, SPIE Proceedings Series, vol. **3346**, pp. 43–54.
8. M. Gardner. *Mathematical Puzzles and Diversions.* Penguin, 1971.
9. B. Mandelbrot. *The Fractal Geometry of Nature.* W. H. Freeman and Company, N.Y., 1983.
10. L. Yaroslavsky. *Digital Picture Processing.* Springer Verlag, Heidelberg, 1985.
11. L. Yaroslavsky. Digital Signal Processing in Optics and Holography. *Radio I Sviaz*, Moscow, 1987 (In Russian).
12. L. P. Yaroslavsky and I. F. Siverguina. Textures, Game "Life" and Non-linear Dynamic Systems. *Problems of Control and Information Theory* **19(4)** (1990) 349–372.
13. L. Yaroslavsky, I. F. Siverguina, R. Thaller and L. Dimitrov. Textures, the Game "Life" and nonlinear dynamic systems. In *Image Analysis and Synthesis*, W. Poelzleitner and E. Wenger (Eds.), R. Oldenburg, Wien, München, 1993, pp. 293–315.
14. V. V. Lyubarsky and L. P. Yaroslavsky. Linear Algorithmic Models of Texture Images. In *Tsyfrovaya Optika. Obrabotka izobrazheniy v eksperimental'nykh issledovaniyakh* (Digital Optics, Image Processing in Experimental Studies), Nauka Publ., Moscow, 1989 (In Russian).
15. Ph. Brodatz. *Textures for artists and designers.* Dover Publication Inc. N.Y., 1966.

16. L. P. Yaroslavsky. Linear and Rank Adaptive Filters for Picture Processing, In *Digital Image Processing and Computer Graphics, Theory and Applications.* L. Dimitrov and E. Wenger (Eds.), R. Oldenbourg, Wien, Muenchen, 1991, pp. 374.
17. L. Yaroslavsky and M. Eden. *Fundamentals of Digital Optics*, Birkhauser, Boston, 1996.

CRYSTAL GROWTH, BIOLOGICAL CELL GROWTH, AND GEOMETRY

JAMES W. CANNON
Department of Mathematics, Brigham Young University, Provo, Utah 84602-6539, USA
E-mail: cannon@math.byu.edu

WILLIAM J. FLOYD
Department of Mathematics, Virginia Polytechnic Institute and State University, Blacksburg, Virginia 24061, USA
E-mail: floyd@calvin.math.vt.edu

WALTER R. PARRY
Department of Mathematics, Eastern Michigan University, Ypsilanti, Michigan 48197, USA
E-mail: mth_parry@emunix.emich.edu

We use crystal growth as a simplified model for biological cell division, and we employ discrete approximations to Riemannian geometry as rough measures of the shapes of the organisms thus created. The first two sections are nontechnical, expository, and speculative. The first section creates a fantastical crystalline growth model of a tree trunk whose shape oscillates wildly over the lifetime of the tree, even though each cell of the trunk senses, responds, and reproduces according to simple local laws. The second section speculates on the underlying geometry in the beautiful microscopic view of rat tissue and suggests that the discrete geometry of tissue plays a role in evolutionary value. The remainder of the paper develops the notion of mathematical crystal growth and discrete geometry in more depth, with the main result being a mathematical demonstration of the experimental fact that, asymptotically, growth crystals generically exhibit the structure of self-similarity fractals.

1 Introduction

The shape of a biological organism is a marvel and a wonder. Beginning with the premise that, at the microscopic level, life is based on simple physical and chemical laws, we consider three questions about biological cell growth and organism shape:

(1) On the basis of very simple local laws, how can it happen that beautiful and complex organisms grow? How is it possible that local laws define growth shapes that are globally convoluted, or wildly oscillatory?

(2) Doesn't it seem more likely that, on the basis of local laws, organisms should take on regular, crystalline shapes like those encountered in Euclidean geometry? like the hexagonal cells of the honeycombs of bees?

(3) When cell biologists model cell growth by means of L-systems (Lindenmayer systems) or rewriting rules (Prusinkiewicz[11]), why is it that the growing organism so often seems to take on the convoluted shape of a self-similarity fractal?

In response to the first of these three questions in the next section we show how local laws can give rise to a growth model for a tree trunk whose shape oscillates wildly over the lifetime of the tree.

In response to the second, we consider as examples scanning electron micrographs of rat tissue; and we note the fact that there are more underlying geometries than the classical Euclidean one, that crystals in different geometries have different physical properties and different biological advantages, and that biological tissues seem to encorporate geometries appropriate to their function.

Finally, in response to the third, we refer to mathematical results which indicate that most local laws should be thought of as 'negatively curved' rules, and that negatively curved rules always form organisms which are asymptotically fractal.

2 A growing tree trunk whose shape oscillates wildly in time

A version of this example, with some additional mathematical details, appears in our earlier paper (Cannon, Floyd, and Parry[6]).

2.1 Cell growth versus crystal growth

A *crystal* is a configuration of cells (abstract, molecular, biological), having a high degree of global and local symmetry, which grows from a seed, or perhaps more accurately *accretes* further cells, on the basis of simple local (abstract, chemical) laws. A *biological organism*, on the other hand, is characterized not by a high degree of global symmetry but rather by global organization and cell differentiation, though perhaps still realized in some way by local (abstract, chemical) laws.

We wish to explore here the line between crystal growth and biological cell growth by asking the following question: how can we construct a nearly crystalline model of cell growth which exhibits global asymmetries? in particular, can we make the crystal or organism shape oscillate wildly over time?

Two biological observations suggest the possibility of such examples: (1) Mothers observe alternate and irregularly spaced periods in the lives of their children where a child first becomes plumper, then grows tall and lean. (2) The fact that certain trees have highly convoluted trunks suggests that growth occurs at different rates at different times and different positions on the trunk.

We give a nearly crystalline example of a growing tree trunk whose shape oscillates wildly over time in Sec. 2.10 below. We precede the description with a discussion of the considerations that dictated the form of the example.

2.2 The groundrules

We want our organism to be constructed of cells of finite capacity: finite memory, finite sensory apparatus, finite ability to respond. We would like those cells to be as uniform in nature and as finite in capacity as possible — that is, as close to crystalline in response as imaginable.

These finiteness conditions are in the tradition of J. von Neumann's cellular automata (Aspray[1]) and J. Conway's game of life (Berlekamp, Conway, and Guy[2]). However, we do not restrict the action to take place in some prescribed cellular framework but instead allow the local laws to determine their own growth framework. We like to call such an 'organism' a *generalized cellular automaton*: growth takes place according to local rules, but the global shape of the growing organism

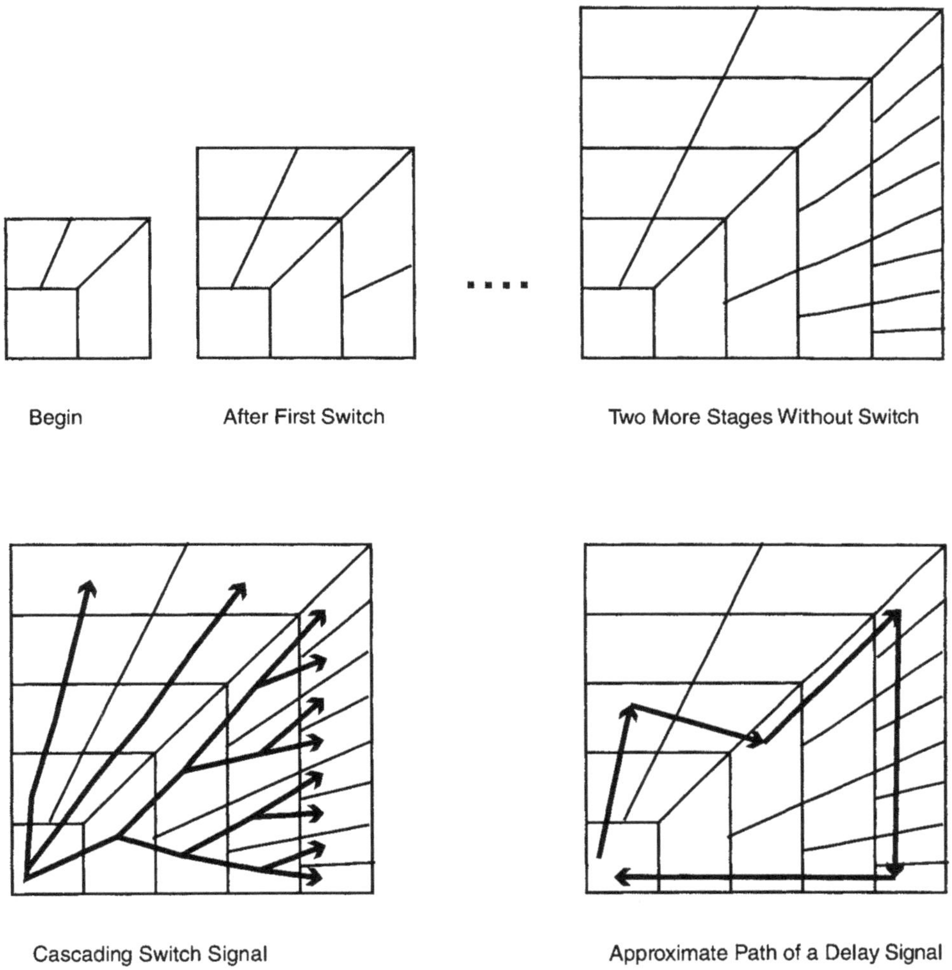

Figure 1. The Oscillating Trunk

is not determined by a presupposed geometrical framework but rather by the local rules themselves. We attempt to discover simple abstract local laws of growth by means of which an organism might exhibit complicated growth patterns.

2.3 The plan

We shall model only a portion of the trunk, namely, a quadrant in a cross section of the trunk. That is, our growing organism will look roughly like a rectangle, with one corner at the center of the trunk, two adjacent sides serving as "radii" from the center of the trunk to the outer edge of the trunk, the other two sides serving as growth layer where new growth rings are accreted by cell division. (See 1.)

The simplest way to change the shape of an organism is to have certain portions of the organism grow more rapidly than other portions. The two growth edges of the rectangle will form the two portions of the trunk where the growth rate differs. Along one of these two edges, cellular subdivision will be exponential in the sense that, at each cellular division, each cell will deposit two children to aid in forming the new growth ring. Along the other of these two edges, cellular subdivision will be linear in the sense that, at each cellular division, each cell will deposit a single child to aid in forming the new growth ring.

As these subdivisions proceed, the shape of the trunk will slowly change, with the segment of exponential subdivision bulging outward under the pressure of accumulating cells.

The only theoretical difficulty in creating the organism is, then, coordinating switches back and forth between the sides which subdivide exponentially. Once the sides have switched subdivision behavior from exponential to linear, or from linear to exponential, the side whose growth has been smaller will begin slowly to catch up with the side whose growth has been exponential. If this new pattern persists long enough, then the side which was previously slower will overtake the other side and the bulge of the tree trunk will pass to the other side.

Our plan is to build a superexponential delay between times at which the cellular subdivision behavior switches from one portion of the growth layer to the other so that the slower side, once it becomes the faster side, can persist so long in exponential growth that it not only catches up with the other side but vastly surpasses it in size.

2.4 The coordination problem

How can we coordinate the actions of an organism globally when each cell has only a fixed, finite capacity to respond? The easiest solution in our view is to have a dictator, a coordinator, a brain, a heart, a core, a patriarch or matriarch — a pacemaker that sends coordinating signals to the rest of the organism.

In our organism, this coordinator will be the original cell of the organism, which recognizes its position at the core of the trunk by the fact that it has no parent cell. Most of the time will find our pacemaker quiescent. But at widely separated times, this cell will awake and send a signal outward from the core to the growth layer which directs that layer to switch its mode of reproduction from linear to exponential or vice versa. The type of signal it sends will be called a 'switch' signal. This signal will proceed outward from the core cell toward the growth layer in a cascade which will be described below.

2.5 The signal problem

We certainly note the use of signals in biological organisms, signals mediated by electrical and chemical processes. Because of finiteness of cell capacity, exotic signal control will have to rely on natural geometric features of the growing organism. See the section on boundaries below.

But there is also another curious geometric aspect of signalling which arises from the fact that signals are not instantaneous but have finite velocity: as the organism gets larger, it takes longer and longer for signals to pass from one side of the organism to another. There is both advantage and disadvantage in the finiteness of signal speed. If communication is too slow, then control methods can founder because growth outstrips control. On the other hand, growth supplies a natural delay mechanism as signal times grow.

For the purposes of control, it seems advantageous to the organism, therefore, even when it is subdividing exponentially in certain directions, to maintain a skeleton of relatively short signal pathways. Such a skeleton seems to be supplied in our human bodies, for example, by our nervous system.

If there is to be control, the signal transmission rate must certainly exceed the growth rate in any direction in which one wishes to exercise control. A growth rate that is too rapid will escape control. And since many signals may have to pass back and forth before there is enough information to dictate a coordinating change, and since controlling signals may involve large scale chemical constructions, either signals must be much more rapid than growth or short pathways must be maintained.

In mathematically defined "organisms" there is no restriction on cell growth rate. It is commonly exponential in all directions, in which case it is impossible to send global signals, for the signals travel at fixed linear rate, and simply disappear locally into the exponential subdivision without ever reaching the desired target.

In our oscillating tree trunk, we maintain linear growth radially so that signals can be sent to the entire organism. In fact, we assume that signals travel at twice the rate of radial growth.

2.6 The delay problem

The most difficult problem in creating our example is that of having the coordinating core cell send 'switch' pulses only at extremely infrequent intervals. Each subdivision creates only a tiny change in global shape; and, since we want shapes to oscillate wildly over time, we must maintain the direction of change much much longer at each direction switch.

We have managed only one plausible idea for creating long delays, namely, sending out a signal along a long path which, by some mechanism returns to the core cell and reawakens it. Since the organism is growing, paths are getting longer and longer, so that such a signal could plausibly take a long time to return.

Using finite limits on cell response, an individual cell can actuate only constant time delays. Bouncing signals off of linearly expanding growth rings, it is possible to actuate linearly expanding delays. Sending signals through an exponentially expanded growth ring, we can design exponentially expanding delays. But even these devices are not enough to obtain the superexponentially growing delays necessary to make shapes change as wildly as we require of ourselves. We obtain superexponential delays by sending a signal first through an exponentially expanded growth ring, then sending it through another such layer that has managed to grow exponentially during the traversal of the first very large ring.

There are certainly other ways to create long delays. A second idea, which we have not yet pursued, is to study interference patterns among different signals. If these different patterns return to the core at fairly random times, then they might reinforce one another negligibly often.

One of the participants at this conference suggested that real biological systems produce delays by creating very fine filaments along which information flow is slow. Any one such structure would however seem to delay signals for only a fixed amount of time, and one would still have the problem of constructing such a structure in the first place. Nevertheless, one should never neglect nature in exploring effective possibilities.

All in all, the construction of delays seems like an interesting problem.

2.7 The signal cascade

Once the core cell has determined to send the 'switch' signal, one can carry that signal to all the cells in the growth layer in a uniform and rather plausible way so that all of the growth cells receive the 'switch' signal at the same time: namely, if we assume that signals are transmitted from cell to neighbor at each unit of time, while subdivisions take place at every second unit of time, then a signal cascade from the core cell can catch the growth layer in finite time; each cell receiving a switch signal simply has to pass that signal on to its children at the next beat of nature's clock.

2.8 Using boundaries

Boundaries of the organism and interfaces within the organism can be used to control signals. To use them as such seems a very plausible method of signal control in actual organisms. In our example, we shall use them as mirrors at which signals originate, are reflected, or are refracted. Thus we see that if we can define a very long path starting at a distinguished time from a distinguished position and if the path follows natural directions within the organisms and is reflected or refracted at natural boundaries or seams in the organism, then such a path can be used as a signal path within the organism and can be defined by the restricted finite rules that we allow ourselves.

2.9 Measuring shape

The shape pressures caused by exponential subdivision must surely cause the shape to buckle outward so that shape changes in time. There are many ways to quantify this shape change mathematically. We were led to the problem of constructing a wildly oscillating tree trunk by a mathematical problem which actually dictated a particular measure of shape, namely the *discrete conformal modulus* of a tiled quadrilateral. A *quadrilateral* is a disk with four distinguished boundary points, called *corners*, which divide the disk boundary into four arcs, called *top*, *right*, *bottom*, and *left* sides. The *Riemann mapping theorem* says that any planar quadrilateral can be mapped conformally onto a geometric planar rectangle so that the 'corners' of the quadrilateral map to the corners of the rectangle, top to top, bottom to bottom, left to left, right to right. Then the ratio of image rectangle height to width is called the conformal modulus of the quadrilateral. If the quadrilateral is tiled, then there is a discrete version of conformal modulus, which arises by mapping tile to square, quadrilateral to rectangle, all preserving the original combinatorial relationships as carefully as possible. The *discrete conformal modulus* is then the ratio of image rectangle height to width. This conformal modulus is uniquely determined by the combinatorics of the tiling of the quadrilateral. See our paper (Cannon, Floyd, and Parry[6]) for details.

Though we shall not carry out the mathematical calculations here, in our example, the discrete conformal modulus will become very large positive, then slowly come very near to zero, then even larger positive, then even nearer to zero, and so on.

2.10 The example

With all of these aspects of the example well in mind, it is easy to complete the description of the example. We use the cell at the intersection of left and bottom edges as core cell. We use the top and right edges as growth layers. One starts the process as described in the plan, with top edge dividing exponentially, right edge dividing linearly: top edge cells each have two children at even ticks of the clock, right edge cells have one child at even ticks of the clock, the lower left cell is the core or brain cell.

At some prescribed finite time after organism birth, the core cell sends its first cascading 'switch' signal. This signal cascades outward toward the two growth edges (top and right); it moves outward one layer at each tick of the clock so that, as a signal, it is moving twice as fast as the growth process. Hence it reaches the two growth edges at some point in time. This 'switch' signal is interpreted by the growth edges as a command to change mode of reproduction. A cell which had previously been designed to have two children, has, instead, one child, and its child also has one child, etc. A cell which had previously been designed to have one child, has, instead, two children, and its two children also have two children, etc.

At the same time that the core cell sends the 'switch' signal, it also begins a 'delay' signal. The core cell remembers enough to know which side is currently exponentially subdividing. The core cell sends this 'delay' signal down the outer edge of that side. The 'delay' follows edges or organism interfaces and turns at each new edge or interface according to the following rules:

(1) When a growth cell is reached, the 'delay' signal turns to follow the layer containing the cell, without crossing the center line of the organism.

(2) When, in the process of following a layer, the 'delay' signal encounters the center line, the 'delay' signal crosses the centerline, then turns outward to follow the centerline seam toward the ever moving growth edge.

(3) When the 'delay' signal follows a layer and encounters a radial edge of the organism, it turns toward the core of the trunk and follows the edge back to the core.

When the core cell receives the 'delay' signal after its long journey, the core cell awakens long enough to send a new 'switch' and 'delay' signal.

3 The geometry of rat tissue

This section consists of geometric speculation considering one of Cannon's favorite books: Richard G. Kessel and Randy H. Kardon, Tissues and Organs[8]. This marvelous book contains more than seven hundred scanning electron micrographs, mostly of rat organs and tissue. Though these complex tissues seem completely foreign to the forms of geometry we were taught in school, to the student of discrete geometry and geometric group theory these tissues seem to illustrate all of the features that have become theoretically important in these areas of mathematics.

3.1 Discrete geometry

The classical geometries are Euclidean geometry, spherical geometry, and hyperbolic noneuclidean geometry. Euclidean geometry is the geometry we learned in school. Spherical geometry is the geometry carried out on the surface of a ball where straight lines are segments of great circles; it is largely the geometry of navigation on the earth's surface. Hyperbolic noneuclidean geometry was discovered in the nineteenth century when geometers finally managed to show that the parallel postulate of Euclid is not a consequence of the other postulates. It is much less intuitive to our Euclidean eyes: the sum of the angles of a triangle is always less than 180 degrees, there are infinitely many lines through a point which fail to meet a line not containing the point (that is, there are infinitely many parallels through a point), similar triangles are in fact always congruent, etc. As unintuitive as this geometry is, it seems particularly well suited to the description of space time in the sense of Einstein's relativity theory.

The three classical geometries are distinguished by a single parameter which is called *curvature.* One of the easiest ways to distinguish curvature is to compare the radii, circumferences, and areas of circles. In Euclidean geometry, radius and circumference are proportional, while area is proportional to the square of radius and circumference. In spherical geometry, circumference and area grow more slowly than in Euclidean geometry; while in hyperbolic geometry, area and circumference grow exponentially as a function of radius. We think of these geometries as being flat (Euclidean), positively curved (spherical), and negatively curved (hyperbolic).

Discrete structures can similarly be described as having curvature, in the large. Combinatorial incidence structures, such as organisms made out of cells, can be analyzed in terms of radius, circumference, and area — or in higher dimensions in terms of radius, surface area, volume: start with a single cell; consider its neighbors as first layer, the new neighbors of the first layer as second layer, and so forth. In the *ball* consisting of n layers, we may consider the radius to be n, the *sphere* of radius n to be the nth layer, the *volume* of the ball to be the number of cells in the ball, and the *area* of the sphere to be the number of cells in the sphere. We say that the organism is *flat* if area and volume grow at the rate of a polynomial of positive degree as a function of radius. We say that the organism has *positive curvature* if the growth rate is sublinear. And finally we say that the organism has *negative curvature* if the growth rates are exponential.

These definitions of discrete curvature should be taken as only approximate and tentative. A given organism may grow exponentially in some direction or set of directions, sublinearly in other directions. The strangest growth patterns involve mixtures of curvatures. The definitions can be made more precise when the organisms considered are more symmetric and regular, as in classical mathematical structures.

Mathematically, discrete notions of curvature have proved to be very valuable (Gromov[7]). And these mathematical results suggest that it might also be instructive and helpful to look for manifestations of discrete curvature in biological cell growth, machine computations, etc. Different geometries have different structural advantages. Do they show up as appropriate in actual biological structures?

3.2 Euclidean geometry, the lens of the eye, and muscle fibers

As one studies discrete structures that are combinatorially flat mathematically, one comes to expect the standard crystalline structures of chemistry. It is precisely such structures that one sees in the Kessel and Kardon micrographs (Kessel and Kardon[8]) of the lens of the eye. Observe in particular the micrographs on pages 99 and 100. The lens of the eye looks much like neatly stacked lasagna noodles, each noodle appearing flat and Euclidean. We quote from Kessel and Kardon: "Cortical lens fibers may be 10-12 mm long (about 7 μm wide, 4.5 μm thick). Transverse sections of cortical cells, or fibers, ... illustrate their flattened hexagonal shape and the precise alignment of the cells in regular rows." Geometrically, it is precisely the flat structures that one would expect in order to minimize the distortion of light. Muscle tissue also seems to exhibit very classical crystalline structure: see in particular the rectangular lattices on page 140 and the classical triangular lattices on page 142.

3.3 Spherical geometry and the bladder

The most obvious place to seek discrete spherical geometry is in the organs of the body which acts like sacks, for example the urinary bladder. When urine is voided, the curvature changes: "As can be seen from the luminal view ... the bladder mucosa is folded; this folding is a result of contraction of the underlying layer of smooth muscle. The folding of the mucosa is a mechanism by which the lumen of the bladder can accommodate volume losses as urine is voided. ... [The] superficial epithelial cells are unique; they respond to forces of tension by changing their surface configuration. Depending upon the particular force exerted, these cells may have a smooth surface or a folded puckered surface." (Kessel and Kardon[8], page 245) The physical processes of life involve changes in underlying geometry.

3.4 Negatively curved geometry and the lung, brain, digestive, and endocrine systems

One of the most common manifestations of negative curvature is the branching tree structure. Another common manifestation of negative curvature is extreme folding. Such structures seem common in tissues where large or numerous interfaces are advantageous, such as lungs, brain, and digestive systems. One would perhaps also expect such structures in systems designed to produce large numbers of objects, such as the male reproductive system; or perhaps in the endocrine system.

Joseph LeDoux[9], in The Emotional Brain, Chapter 6, discusses the high connectivity of the brain: "Only a few neural links separate any particular set of neurons in the brain from most others." "There are billions of neurons, and each gives rise to one or more axons (nerve fibers that allow neurons to communicate with one another). The axons themselves branch, so that the number of synapses (the connection made by an axon from one neuron to another) is far greater than the number of neurons. And each neuron has multiple dendrites that receive thousands of synaptic contacts from many others."

High connectivity has the effect of reducing the combinatorial radius or diameter of the brain. The volume or cell count of combinatorial balls therefore increases rapidly as a function of radius in the presence of high connectivity. And consequently the brain becomes negatively curved. This negative curvature shows up in the convolutions of the brain.

Many of the micrographs suggest at least superficially that some tissues grow in simple Euclidean ways, that they then fold or buckle under various forces (physical constraint, gravity), and that interconnections are then superimposed which raise the connectivity of the tissue. What we mean to suggest by this observation is that constraints are not only local chemical constraints that create combinatorial structures satisfying various curvature conditions but that physical mechanical constraints probably also play an important role in creating global curvature.

3.5 Conflicting geometries

In the previous subsection, we mentioned folding as one of the common manifestations of negative curvature. In fact, if one creates a correct local model of the hyperbolic plane, when viewed in our Euclidean three dimensional space, the model folds in such a manner that it looks like one of those women's skirts often used in western square dancing, tremendously frilly and folded. The mathematical and physical fact is that negatively curved objects do not fit well into flat Euclidean space. Trees branch only a limited number of times (giving large interface with sunlight) and then end in leaves which essentially fill space. There is only a limited amount of room for branching. Similarly, the peel of an orange with its positive curvature does not easily fit flat on the table; instead, it fractures when one attempts to flatten it. Objects with one kind of geometry do not fit gracefully into objects with other geometries.

Because of this conflict of geometries, one would not expect an object that is crystalline of one geometry to appear as crystalline in another geometry. We would expect instead that there would be folding, fracturing, and distortion of other kinds. Is it possible that the fractal objects we observe in nature are manifestations of the conflict between incompatible geometries? Are some of the convoluted objects that we view in nature actually crystalline in cell structure, but combinatorial crystals in geometries other than Euclidean?

We usually expect one geometric object, when placed in another geometry to 'relax' so as to reduce energy to a minimum. If the rigidity of an object places particular constraints on this relaxation, do we expect fracturing or folding of particular kinds? How do different geometries interact with one another? Is there a good mathematical-physical description of what happens? Do we expect the fracture or folding set to have a certain Hausdorff dimension given in terms of the dimensions of the constituent objects?

4 Crystals

We define a *crystal* as a highly symmetric configuration of cells or vertices.

Symmetry can be formalized using the mathematical notion of *group*. A*group* is a set G whose elements can be multiplied, and where multiplication is assumed to satisfy four simple axioms: (1) (closure) the product of two elements of the group is an element of the group; (2) (identity) there is an identity element e of the group which is defined by the property that $e \cdot x = x \cdot e = x$; (3) (inverses) for each element $x \in G$ there is an element x^{-1} such that $x \cdot x^{-1} = x^{-1} \cdot x = e$; (4) (associativity) for every $x, y, z \in G$, $x(yz) = (xy)z$.

A subset K of a group G is said to be a *generating set* for G if each element of G can be expressed as a product of finitely many elements of K. We shall only consider groups which have *finite generating sets.*

We say that C is a *crystal with symmetry group* G if there is a function (called *action*) $G \times C \to C$, where the image of (g, c) is denoted by gc, such that $ec = c$ and $x(yc) = (xy)c$. If the elements of C satisfy an incidence relationship, then we require that the multiplication $c \mapsto gc$ preserve that incidence relationship. We think of each element of G as defining a symmetry of the crystal C.

We think of the configuration C as being *highly symmetric* if, for every two elements c_1 and c_2 of C, there is an element $g \in G$ such that g carries c_1 to c_2. More generally, we say that C is highly symmetric if there is a finite (or compact) set of elements of C which can be carried to every element of C by the elements of G.

4.1 Examples of crystals

As we noted in the previous sections, discrete objects can have underlying geometry. One of the most natural ways to form a crystal is to start with a continuous classical geometry and to consider a group G of isometries of that geometry (an *isometry* is a distance preserving mapping of the geometry to itself). The elements of the crystal C, which we call *cells*, then appear as geometric subobjects in the geometry which fit together to tile or tessellate the geometry and which are permuted by the group G. Here we describe sample tessellations from each of the classical two dimensional geometries (see 2):

(1) The soccer ball is a spherical crystal. The underlying symmetry group has 120 elements: a rotation of order five about the center of any of the twelve pentagons, twelve rotations that take the twelve pentagons to any fixed pentagon, a reflection which turns the ball inside out: $5 \times 12 \times 2 = 120$.

(2) The hexagonal tiling of the plane is a typical Euclidean crystal. Three hexagons come together at each vertex. The symmetry group contains the

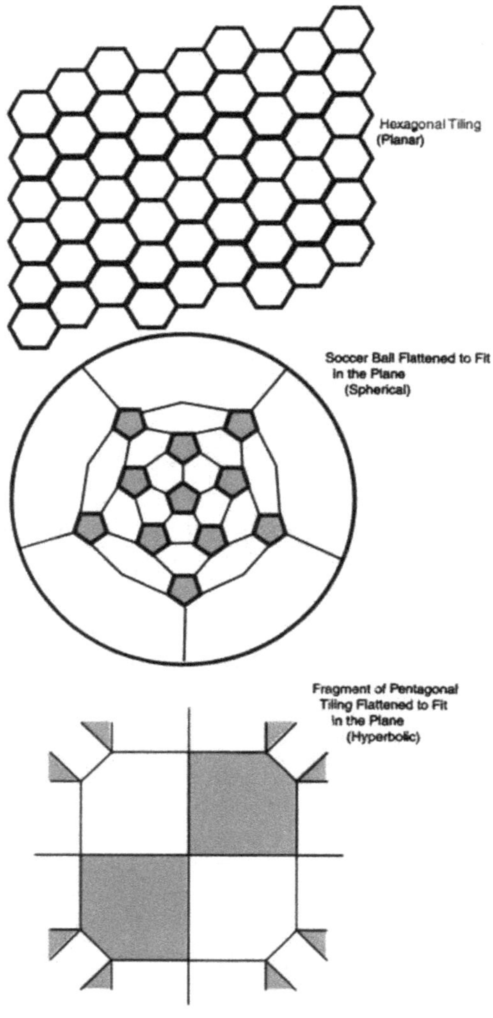

Figure 2. Crystal fragments from the three classical two-dimensional geometries

translations which take any hexagon to any other, the rotations about the center of any hexagon, reflection about a central axis of any hexagon.

(3) One can form a tiling of the noneuclidean plane by sewing together pentagons, four at each corner.

4.2 Ways of viewing crystals

There are many ways of viewing a crystal. Two of the three examples we gave in the last section were infinite. By truncating these examples we obtain objects more like what we encounter in the physical world. Here are some of the ways we can do this:

(1) The 'static' growth model. In the tessellation described above, start with one cell as seed. Define the first layer or sphere of the crystal to be the set of cells intersecting the seed. Define the sphere of radius 2 to be the set of new cells intersecting the sphere of radius 1, and so forth. Then we may think of the cell as 'growing' by accreting new layers with time. This model is termed 'static' because we think of the whole configuration C as already existing before the process starts, and one is simply organizing this already existing or 'static' model into 'growth' layers.

(2) The 'dynamic' growth model. Instead of prescribing the crystal C ahead of time, we start with an abbreviated description of the symmetry group G we hope to attain in the end. The usual procedure is to describe *generators* and *relators* for the group G. One thinks of the elements or constituent molecules of the crystel C as cells. The generators may be thought of as describing the faces of these constituent cells (one face for each generator and one for the inverse of each generator); and how they are attached to the faces of adjacent cells (the face corresponding to a generator attaching to a face corresponding to its inverse). That is, generators describe the molecules of the crystal C and the way that adjacent cells attach to one another. The relators prescribe patterns of cells which are required to form cycles or loops in the resulting crystal. In other words, relators give slightly larger scale descriptions of the structure of the crystal. The relators may be thought of as describing the local laws according to which these new fundamental molecules interact with their neighbors to form slightly higher level structures. We start with one cell as seed and allow the crystal to grow according to the prescribed local laws. The process is very much like the L-systems of Lindenmayer and his associated cell biologists using rewriting systems (Prusinkiewicz and Lindenmayer[11]). The mathematical situation has an interesting extra twist: it may happen that after a time one will encounter certain local incompatibilities in the construction that are forced by the sewing and looping conditions. If that happens, then one can, and does, force compatibility by folding the structure up until the incompatibilities disappear. The idea and the process of folding are very simple, but take us beyond the scope of this expository paper.

By analogy with what happens in formal language theory, we say that the group admits a *rewriting system* if the rules can be described in such a way that the incompatibilities never arise, so that there is never any collapsing or folding. In that case, the 'dynamic' model coincides with the 'static' model.

In the generic case, the group G does admit a rewriting system (Cannon[4]). But there are exceptional examples where the incompatibilities are so unpredictable that there is no finite algorithm describing what will happen. In such exceptional cases, the crystal grows, collapses, grows more, folds some more, etc. After *infinitely* many growth and collapse stages, one finally arises at the static model.

(3) The 'population' model. If one simply tracks the shape of the crystal or of the growth layer from time to time, then one can view the crystal as a population growth model. The 'civilization' can grow, expand, collapse, stay static, etc.

(4) The 'asymptotic' model. One can attempt to take some sort of mathematical limit of the successive populations viewed in the population model. Does the model have a limiting geometric shape? Does it have a space at infinity?

(5) 'Slice' models. One need not just examine spheres or growth layers in a crystal. We can slice the crystal in other ways. Many of those interesting objects called quasicrystals can be viewed as arising from noneuclidean crystals by slicing.

Note that all of these views of crystals give pictures which, in their finiteness, do not share all of the symmetries of the original, possibly infinite, crystal being viewed. In particular, the quasicrystals look very nearly symmetric but may have essentially no symmetries at all.

5 Negatively curved crystals are asymptotically fractal

Those who have studied rewriting systems have noted that the crystals so obtained are often, asymptotically, self-similarity fractals. This section outlines a proof that this fractal behavior, at least in our mathematical setting, is generic. The proof has been known for almost two decades now in mathematics (Cannon[3] and Gromov[7]), but, of course, we do not always communicate in a timely way between disciplines.

5.1 The Cayley crystal

Every group G acts on itself in a natural way by left multiplication. Thus we may think of the group as a crystal, and we may attempt to view it in any of the ways described above. We choose a finite generating set K, and proceed as follows to associate an abstract crystal or graph Γ with G. This abstract crystal is generally called the *Cayley graph*. It has as vertices the elements of the group G. Its interconnecting edges depend on the generating set K of G. We join two elements h_1 and h_2 of G by an edge denoted (h_1, k, h_2) if there is an element k of K such that $h_2 = h_1 \cdot k$. The group acts as a permutation group on itself and on its Cayley graph by left multiplication: $k'(h) = k' \cdot h$, and $k'(h_1, k, h_2) = (k' \cdot h_1, k, k' \cdot h_2)$. Geometric group theory treats the Cayley graph as a geometric object. One thinks of each edge as being a geometric path having length 1 and calculates the distance between two points by finding paths of shortest total length between two points. Because the group acts on Γ in the manner required by our definition of crystal, we may think of the Cayley graph as a crystal.

Every crystal C is closely related to the Cayley graph or Cayley crystal. Indeed, choose some cell $c_0 \in C$ as a base cell and place a geometric point p_0 in that cell. Think of p_0 as the *origin* and label it by the identity element of the group G which acts on the crystal C. For each $g \in G$, let $g(p_0)$ be the image of p_0 under the permutation g, and label this point by the element g of the group. One thereby obtains geometric images of the vertices of the Cayley graph, namely $g(p_0)$ for the vertex g. Now, if the crystal C is pathwise connected, for each k in the generating set K we may choose a path $\alpha(k)$ in C which joins p_0 and $p_0 \circ k$. We may label this path by the edge (id, k, k). The action of G on C carries this edge, and all others as well, to various positions in C and creates a natural image of the Cayley graph in C. Thus, one can usually replace the study of the crystal C by the study of the Cayley graph Γ.

5.2 Thin triangles

We shall call a shortest path in the Cayley graph Γ a *geodesic*.

If x, y, and z are points of Γ, and if xy, yz, and zx are particular geodesics joining these three points (geodesics are not in general unique, so there may be other geodesics joining these points as well), then we say that these three geodesics form a *triangle* with *edges* xy, yz, and zx.

We say that this triangle is *δ-thin*, where δ is a nonnegative number, if each point of each side lies at a distance no more than δ from the union of the other two sides. M. Gromov has asserted that most Cayley graphs, viewed as geometric objects, have *uniformly thin triangles* in the sense that there is a nonnegative constant δ depending on the group G and its generating set K such that every triangle in Γ is δ-thin (Gromov[7]); Olshanskii[10] has managed to write down for us a proof of Gromov's assertion.

The thin triangles condition is equivalent to a version of negative curvature: a Cayley graph satisfies the thin triangles condition if and only if the crystal has 'spherical' areas and 'ball' volumes which grow exponentially in every direction.

The first author has proved that such Cayley graphs satisfy a condition called *almost convex* (Cannon[4]) whose intuitive content is that the group admits a rewriting system in the sense described above. S. Hermiller has announced a language theoretic version of the theorem which implies that every almost convex group admits a rewriting system in the sense used by people studying formal languages.

5.3 Asymptotic shape

We shall examine the Cayley group Γ of a group G and shall assume that the triangles of the graph Γ are all δ-thin. Our aim is to show that, asymptotically, this graph looks like a self-similarity fractal.

How do we examine the local asymptotic shape of a Cayley graph? We first partially order the graph Γ by defining $x \leq y$ if there is a geodesic from the identity vertex of Γ to y which passes through x. In terms of this partial order, we may define, for each x in Γ, the *cone* $[x, \infty)$, which is the set of all points y such that $x \leq y$. We think of this cone as defining a local geometric shape or *shadow at infinity*.

When do two of these shadows or cones have the same shape? Because the group G acts on the graph Γ by left multiplication, we can translate any vertex x_1 of Γ to any other vertex x_2 of Γ by the multiplier $g = x_2 \cdot x_1^{-1}$. Although this multiplier takes x_1 to x_2, there is no guaranty that it will take the cone $[x_1, \infty)$ to the cone $[x_2, \infty)$. We say that two vertices x_1 and x_2 have the same *cone type* if the corresponding cones match under left multiplication.

In any asymptotic picture of Γ we expect the cone $[x_2, \infty)$ to look like the cone $[x_1, \infty)$ precisely when the two vertices have the same cone type. We expect the shadow of one vertex at infinity to look smaller than the shadow of the other vertex at infinity precisely when the first vertex is further from the origin or base vertex. Thus, if two vertices have the same cone type, but one is further from the origin, then there should be a 'similarity' of the pictures which takes the one to the other.

In any of the ways of viewing a crystal described above, if we intersect our view with the cones of two vertices having the same cone type, then we expect to see finite approximations to the same combinatorial shape. If there are only finitely many cone types, then we expect to see the same finite number of shadow behaviors copied over and over everywhere at all different sizes and scales. That is, we expect to have a self-similarity fractal.

With these concepts in hand, we are prepared to state our fractal theorem.

5.4 Fractal Theorem

(Cannon[3]) Let G be a finitely generated group whose Cayley graph Γ has uniformly thin triangles. Then the vertices of Γ have only finitely many cone types. Thus the shadows of these vertices at infinity fall into only finitely many similarity classes, and the shape of Γ at infinity is a self-similarity fractal.

5.5 Proof

We first describe the manner in which we shall recognize that two cones are equivalent. Since the generating set K (in terms of which Γ is defined) is finite, the graph Γ is locally finite. Hence, for fixed positive integer N, the ball or radius N centered at a vertex of Γ is a finite graph. In fact, it is the same finite graph, combinatorially, for all of the vertices of Γ. These balls, however, do admit different partial orders. But clearly, there are only finitely many possible partial orders that such balls can inherit from the partial order on the entire graph. We complete the proof by showing that, for a fixed large positive integer N, the inherited partial order on the N-ball centered at vertex x determines the cone type of x. It suffices to take $N > \delta$, where δ is the thinness constant associated with the triangles of Γ. We say that two vertices have the same *N-type* if the balls of radius N about these two points do inherit the same partial order.

Suppose therefore that two vertices x_1 and x_2 of Γ are given, each having N-ball inheriting the same partial order from the partial order on the whole graph; that is x_1 and x_2 have the same N-type. (See 3.) This means that the left multiplication which takes x_1 to x_2 (namely, multiplication by $g = x_2 \cdot x_1^{-1}$), when restricted to the N-ball about x_1 maps that ball in a partial-order preserving fashion onto the N-ball centered at x_2. We may express this relationship numerically, if we use $|\alpha\beta|$ to denote the distance between points α and β. Thus if a_1 and b_1 are points of the N-ball about x_1, if $a_2 = g \cdot a_1$ and $b_2 = g \cdot b_1$, and if 0 denotes the origin or identity vertex, then

$$|0b_2| - |0b_1| = |0a_2| - |0a_1|. \tag{19}$$

Let y_1 be a point in the cone of x_1; that means that there is a geodesic from 0 to y_1 which is the union of two geodesics $0x_1$ and x_1y_1. Then we need to show that the union of the two geodesics $0x_2$ and $x_2y_2 = g(x_1y_1)$ is a geodesic from 0 to y_2.

To that end, let α be a geodesic from 0 to y_2. Consider the union of the three geodesics $0x_2$, x_2y_2, and α. The union obviously forms a geodesic triangle, but we may actually choose an unusual vertex set for this geodesic triangle, namely, 0 and

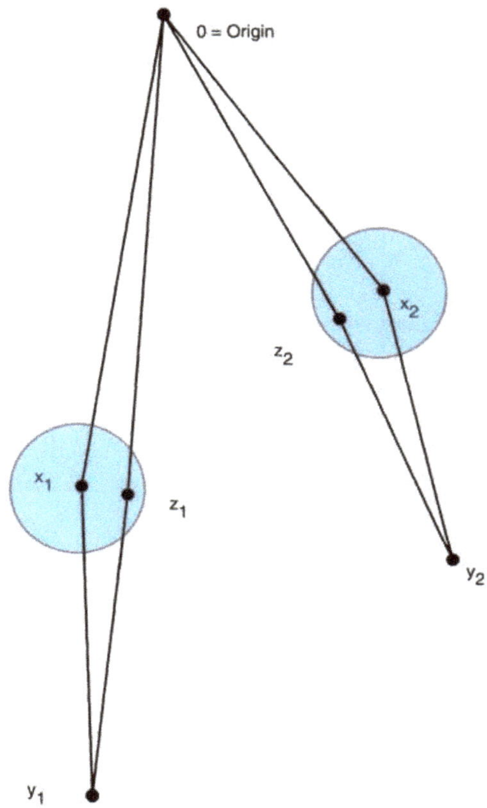

Figure 3. Comparing cone types of vertices having the same N type

y_2 (no surprises there) and the last point z of x_2y_2 which lies in the N-ball about x_2. It follows easily that the arc $0x_2 \cup x_2z$ is geodesic from Eq. (19). Consequently, x_2 must lie within δ of some point z_2 which lies in one of the other sides of the triangle. It is easy to see that z_2 must lie in the arc α since the other candidate points of the triangle lie outside the δ neighborhood of x_2. Thus we may realize α as the union of two geodesic arcs $0z_2$ and z_2y_2, with z_2 in the N-neighborhood of x_2.

Finally, we define $z_1 = g^{-1}(z_2)$ and $z_1y_1 = g^{-1}(z_2y_2)$; and we let $0z_1$ denote a geodesic from 0 to z_1. We have the following obvious relationships:

$$|0x_1| + |x_1y_1| \leq |0z_1| + |z_1y_1| \text{ and } |0z_2| + |z_2y_2| \leq |0x_2| + |x_2y_2|. \tag{20}$$

$$|x_1y_1| = |x_2y_2| \text{ and } |z_1y_1| = |z_2y_2|. \tag{21}$$

The two inequalities arise because the smaller distances are geodesic distances. The two equalities arise because multiplication by g preserves path length.

From Equations (19), (20), and (21), we easily obtain the series of equalities and inequalities,

$$\begin{aligned}|0z_2| + |z_2y_2| &= |0z_2| - |0z_1| + |0z_1| + |z_1y_1| \\ &\geq |0x_2| - |0x_1| + |0x_1| + |x_1y_1| \\ &= |0x_2| + |x_2y_2|.\end{aligned} \tag{22}$$

Consequently the union of $0x_2$ and x_2y_2 is a geodesic, as we wished to show.

Acknowledgments

The authors wish to express gratitude for financial support of the National Science Foundation, USA.

References

1. W. Aspray, *John von Neumann and the Origins of Modern Computing* (MIT Press, Cambridge, MA, 1990).
2. E. R. Berlekamp, J. H. Conway and R. K. Guy, *Winning Ways for Your Mathematical Plays* (Academic Press, London New York, 1982).
3. J. W. Cannon, Geom. Dedicata **16**, 123–148 (1984).
4. J. W. Cannon, Geom. Dedicata **22**, 197–210 (1987).
5. J. W. Cannon, W. J. Floyd, and W. R. Parry, Contemp. Math. **169**, 133–212 (1994).
6. J. W. Cannon, W. J. Floyd and W. R. Parry, in *Geometric Group Theory Down Under, Proceedings of a Special Year in Geometric Group Theory* (de Gruyter Verlag, 1998).
7. M. Gromov, in *Essays in Group Theory (S. M. Gersten, ed.)* (Springer-Verlag, New York Berlin, 1987).
8. R. G. Kessel and R. H. Kardon, *Tissues and Organs: a text-atlas of scanning electron microscopy* (Freeman, San Francisco, 1979).
9. J. LeDoux, *The Emotional Brain* (Simon and Schuster, New York 1996).
10. A. Y. Olshanskii, J. Algebra Comput. **2**, 1–17 (1992).
11. P. Prusinkiewicz and A. Lindenmayer, *The Algorithmic Beauty of Plants* (Springer-Verlag, New York Berlin Heidelberg London Paris Tokyo Hong Kong, 1990).

RECENT RESULTS ON APERIODIC WANG TILINGS

JARKKO KARI[1]
Department of Computer Science, 14 MLH
University of Iowa, Iowa City, IA 52242, USA
E-mail: jjkari@cs.uiowa.edu

Wang tiles are square tiles with colored edges. The plane is tiled with Wang tiles by placing them edge-to-edge in such a way that colors on contiguous edges are identical. A tiling is called periodic if there is a translation of the plane that keeps the tiling unchanged, otherwise the tiling is non-periodic. A set of tiles is called aperiodic if it only admits non-periodic tilings. This article discusses new results and applications of aperiodic Wang tilings. We present recently discovered small aperiodic tile sets. Instead of geometric arguments the construction is based on elementary arithmetics on the tiles. Then we discuss special types of tile sets that have the property that each tile is completely determined by two of its neighbors. Such tiles provide examples of one-dimensional cellular automata that exhibit unexpected behaviour.

1 Introduction

The art of decorative plane tessellations with simple geometric shapes is very old and can be found in all ancient cultures. Usually ceramic or stone tiles were fitted together in regular fashion to cover a surface without gaps. The artistic appearance of the tessellation was important, but also the geometric aspects of different shapes were investigated. Almost 400 years ago Kepler systematically studied periodic tessellations with regular polygons, and since then tiling properties of more and more complex shapes have been established. The book "Tilings and Patterns" by B.Grünbaum and G.C.Shephard is a comprehensive treatment of the history and the mathematics of tilings and tessellations [9].

During this century mathematicians have made remarkable new discoveries in the theory of tilings. In the 1960's it was found out that certain shapes force the tiling to be non-periodic. Such tile sets are called *aperiodic*. There are many shapes that allow non-periodic as well as periodic tilings, but the essential property of aperiodic tile sets is that every tiling they admit is necessarily non-periodic.

The purely mathematical construction of aperiodic tile sets was soon followed by the discovery of their material world counterpart, quasicrystals. In 1984 material was produced that showed icosahedral symmetry, previosly believed to be impossible in crystalline state [22]. Because no periodic pattern can have rotational symmetry of order five, the atoms in the new material had to be arranged non-periodically. Soon other examples of non-periodic crystals were found, and the term "quasicrystal" was introduced to stand for crystals that exhibit symmetries that are not compatible with periodicity. See "Quasicrystals and geometry" by M.Senechal for more discussion on the role of aperiodic tilings in crystallography [23].

The most widely known example of an aperiodic tile set is due to R.Penrose [19], and it consists of just two shapes called *Kites* and *Darts*. This elegant solution

[1]WORK SUPPORTED BY NSF GRANT CCR 97-33101

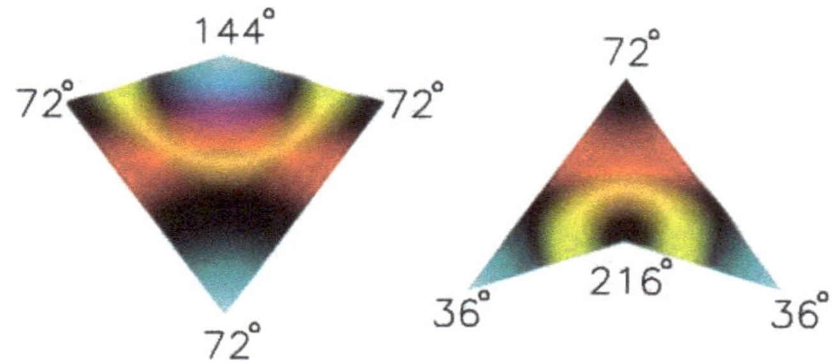

Figure 1. Penrose Kite and Dart

Figure 2. Part of a non-periodic plane tiling using Penrose Kites and Darts

is shown in Figure 1. The long and short edges of Kites and Darts have lengths in the golden ratio $\frac{1+\sqrt{5}}{2}$. Penrose tiles form a "jigsaw puzzle" where only two different pieces are present. In valid tilings not only the shapes must fit without gaps and overlaps, but also the coloring must continue uninterrupted between adjacent tiles. Equivalent tiling condition using shapes only can be obtained by introducing suitable dents to the shapes that simulate the coloring.

Part of a valid plane tiling with Penrose tiles is depicted in Figure 2. In fact, the depicted pattern appears in every valid Penrose tiling, as does every finite pattern that appears in some valid tiling. The coloring makes the five-fold symmetry of the tiling noticable. The red and yellow curves are called the Conway-lines of the tiling.

Penrose's aperiodic set of two tiles is the smallest known aperiodic tile set: no-one has been able to come up with a single shape that would force non-periodicity on the plane.

Historically, the first aperiodic tile sets were defined in terms of *Wang tiles.* Wang tiles are unit square tiles with colored edges. We use compass headings N, E, S and W to refer to the sides of a square. A *tile set* T is a finite set of Wang tiles. In the rest of this article, unless otherwise indicated, the word "tile" always refers to a Wang tile.

A *tiling* using a given tile set T consists of copies of the elements of T placed on the integer lattice points of the plane with their edges oriented horizontally and vertically, just like squares on a checker-board. On valid tilings contiguous edges must have the same color. Also the tiles must be used in the given orientation — they may not be rotated.

The local matching rule of Wang tiles is defined in terms of edge coloring but it can easily be converted into shape based matching condition by introducing suitable "holes" and "dents" along the edges. However, square tiles with edge coloring provide useful analogies to computation theory which gives new insight and simplifies their treatment. Applications in other areas of mathematics such as mathematical logic, algebraic topology and automata theory indicate that Wang tiles are a useful abstraction.

A tiling f using T can be formalized as a function

$$f : \mathbb{Z}^2 \longrightarrow T$$

that specifies the tiles at all positions of the plane: Location $(x, y) \in \mathbb{Z}^2$ contains tile $f(x, y)$. A tiling f is called *valid* if in each position $(x, y) \in \mathbb{Z}^2$ the N-edge and and the E-edge of the tile $f(x, y)$ have the same color as the S-edge and the W-edge of the tiles $f(x, y + 1)$ and $f(x + 1, y)$ above and to the right, respectively. In other words, the colors on contiguous edges have to match everywhere.

A tiling f is *periodic* if it is invariant under some translation of the plane. More precisely, f is periodic with period $(x_0, y_0) \neq (0, 0)$ if

$$f(x, y) = f(x + x_0, y + y_0)$$

for every $(x, y) \in \mathbb{Z}^2$. A tiling is *non-periodic* if no such translation exists. A tiling is called *doubly periodic* if it is periodic in both horizontal and vertical direction, that is, if it is periodic with some periods $(a, 0)$ and $(0, b)$ where $a, b > 0$. One can view a doubly periodic tiling as a horizontally and vertically repeating rectangular pattern of size $a \times b$. It is straightforward to prove that any tile set that admits a periodic tiling necessarily admits also a doubly periodic tiling. A tile set T is called *aperiodic* if (i) it admits a valid tiling, but (ii) it does not admit any valid periodic tilings.

Basic questions concerning any given tile set T are the following: Is it possible to tile the entire infinite plane ? What kinds of tilings are possible ? Especially: does the given tile set allow periodic tilings of the plane ? In general these questions are very difficult to answer, as indicated by the following classic theorem by R.Berger:

Theorem 1 *(Berger 1966* [1]*) There is no algorithm able to determine for all tile sets T whether T admits a valid tiling of the infinite plane.*

Berger's theorem states that no mechanical decision procedure (that is, an algorithm) is able to determine for every tile set T if a tiling is possible or not. Each algorithm is going to fail on some input T: it either gives a wrong answer or enters an infinite computation that never stops.

Interestingly, the theorem implies the existence of an aperiodic tile set. To see this, assume no aperiodic tile sets would exist. Then every tile set T either would not admit a valid tiling of the plane, or would admit some doubly periodic tiling. In the first case, some sufficiently large rectangular region can not be tiled. In the second case, the period of a doubly periodic tiling provides a periodic tiling of some rectangular region. We would have the following algorithm for determining if a given tile set admits a valid tiling: The algorithm tries one-by-one all tilings on rectangles of all sizes, starting with smallest rectangles and moving towards larger and larger rectangles. Sooner or later it either finds a valid period of a doubly periodic tiling, or it finds a rectangle that can not be tiled with the given tiles. In the first case the tile set admits a plane tiling, in the second case it does not. Berger's theorem states that such an algorithm does not work, and consequently there must exist aperiodic tile sets that make the algorithm fail.

In fact Berger's proof of the theorem includes a construction of the first known aperiodic set. The number of tiles in the set is very large, 20426, as the tiles are composed of several components. Berger himself shorty afterwards constructed an aperiodic tile set that contained only 104 tiles. A number of researchers, among them some well known people in discrete mathematics, logic and computer science, found smaller aperiodic sets of Wang tiles. Between 1966 and 1978 progressively smaller aperiodic sets were found by D. E. Knuth [17] (92 tiles, 1966), H. Läuchli [9]

Figure 3. Amman's 16 tile, 6 color, aperiodic NW-deterministic tile set

(40 tiles, 1966), R. M. Robinson (56 tiles, 1967, 35 tiles, 1971), R. Penrose (34 tiles, 1973), R. M. Robinson [9] (32 tiles, 1973, and 24 tiles, 1977) and finally R. Amman [21,9] (16 tiles, 1978). Amman's aperiodic set is shown in Figure 3. A discussion of these and related results, and in most cases the first published version of them is included in Chapters 10 and 11 of "Tilings and Patterns" by B. Grünbaum and G. C. Shephard [9].

Recently, a new technique based on elementary arithmetic has been developed that provides even smaller aperiodic tile sets [15,3]. Section 2 of this article describes this technique. Currently, the smallest aperiodic set of Wang tiles is based on our technique and is due to K.Culik. It contains 13 tiles but it is conjectured that one tile in this set is not present in all valid tilings. If the conjecture is true, an aperiodic set of only 12 tiles results.

Consider Amman's aperiodic set of 16 tiles in Figure 3. It is based on a clever encoding of the Penrose Kite and Dart tiles (Figure 1). Amman's tile set has the interesting property that each tile is determined by the colors on its north and west edges. In other words, any two tiles differ in their north or west edges, or both. We call such a set NW-deterministic. On a valid tiling each tile is forced by its north and west neighbors.

Coincidentally, Amman's tile set is also SE-deterministic. Deterministic tile sets and forced tiles are discussed is Section 3. It turns out that there are aperiodic tile sets that are even 4-way deterministic, that is, simultaneously deterministic in NW-, NE-, SW- and SE-directions.

2 Algebraic construction of aperiodic tile sets

Traditionally, aperiodicity arguments have been based on some geometric properties of the tiles. Most known examples of aperiodic tile sets allow self-similar, hierarchical tilings of the plane [10]. In such cases it is relatively simple to demonstrate that the tiles admit a valid tiling. On the other hand, it is more difficult to prove that aperiodicity is forced, that is, that no periodic tilings are admitted.

However, in some cases forced aperiodicity can be established easily if the tiles have certain algebraic properties. Let us start by interpreting colors as numbers. We say that Wang tile

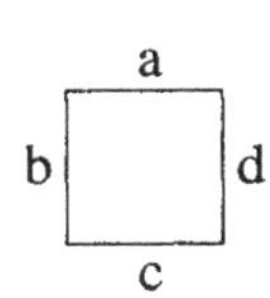

multiplies by number q if $qa + b = c + d$. In other words, the tile multiplies the input a from its N-edge by q, adds the carry b from the W-edge, and splits the result into the output c on the S-edge and the new carry d on the E-edge.

Consider, for example, the 14 tiles of Figure 4. The Figure shows the colors and their interpretation as numbers. Because the tiles are not to be rotated, we may use different interpretation on horizontal and vertical edges.

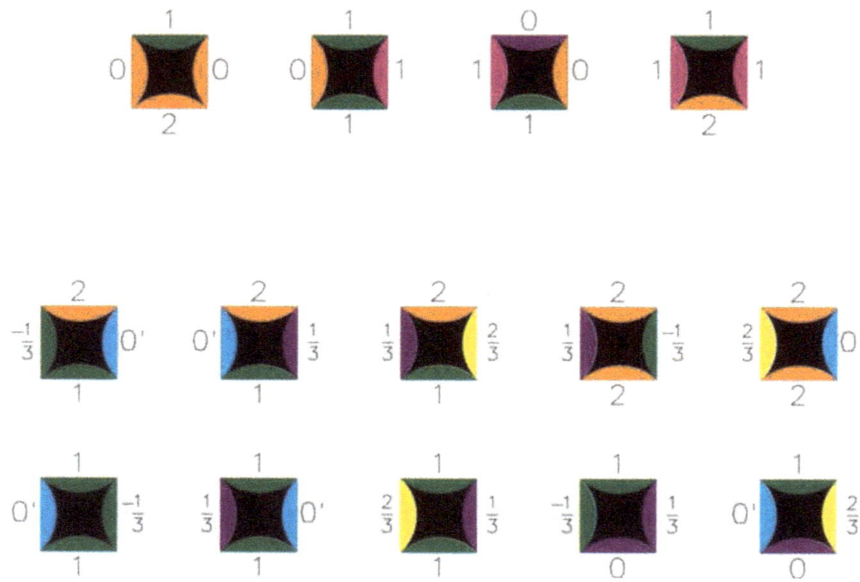

Figure 4. Aperiodic set of 14 Wang tiles

The first four tiles of Figure 4 multiply by number 2, and the remaining 10 tiles multiply by 2/3. Let us denote these sets of tiles by T_2 and $T_{2/3}$, respectively. The vertical edges on T_2 and $T_{2/3}$ are colored using disjoint color sets, which means that on any valid tiling each horizontal row must come from the same set T_q.

Matching tiles in the horizontal direction means transmitting the carry numbers correctly from tile to tile. Any horizontal row of matching tiles that all multiply by the same number q performs a multiplication by q of the entire input: On validly tiled horizontal segment

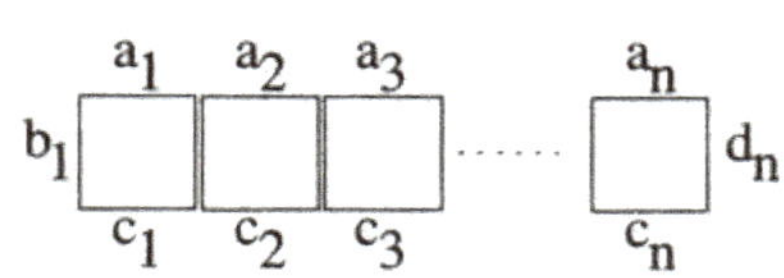

we have

$$q \cdot (a_1 + a_2 + \ldots + a_n) + b_1 = (c_1 + c_2 + \ldots + c_n) + d_n \,.$$

If the segment is a horizontal period of a periodic tiling then $b_1 = d_n$ which means that

$$q \cdot (a_1 + a_2 + \ldots + a_n) = c_1 + c_2 + \ldots + c_n \,.$$

Matching tiles in the vertical direction means transmitting the output of the tile above to the input of the tile below. Consider m valid horizontal periods of length n tiled vertically on top of each other:

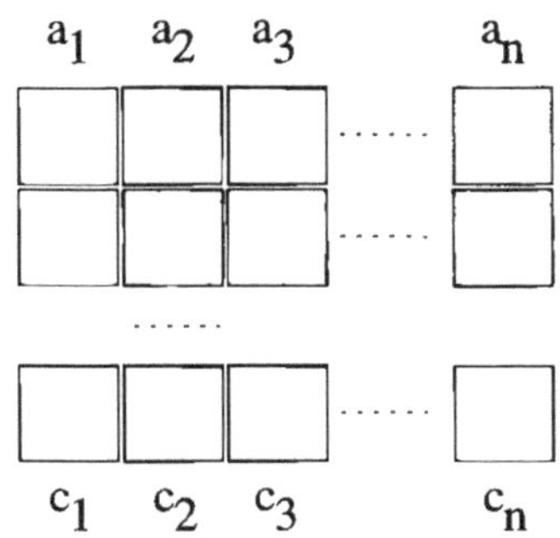

If $A = a_1 + a_2 + \ldots + a_n$ is the total input to the first row, $C = c_1 + c_2 + \ldots + c_n$ is the total output from the last row, and if tiles on row i multiply by q_i then the entire rectangle performs the multiplication

$$(q_1 q_2 \ldots q_m) \cdot A = C\,.$$

If the rectangle is a repeating pattern of a doubly periodic tiling then $A = C$. It is obvious from the 14 tiles of Figure 4 that a horizontal row of 0's is impossible. Therefore A must be greater than 0, and it can be divided from the equality leaving

$$q_1 q_2 \ldots q_m = 1\,.$$

No product of numbers 2 and 2/3 is equal to 1 because 2 and 3 are relatively primes. This means that there does not exist a choice of numbers q_i that would satisfy the equality. Consequently, no doubly periodic tiling using the tiles of Figure 4 is possible. This means that no periodic tiling exists either, and we have proved that the tiles force aperiodicity.

In order to prove that the tile set of Figure 4 is aperiodic we still have to demonstrate that a valid tiling of the plane is possible. Although the construction of a valid tiling is not complicated, we skip it and refer an interested reader to reference [15]. Valid tilings are based on implementing the dynamical system

$$x \mapsto \begin{cases} 2x, & \text{if } x \in \left[\frac{1}{2}, 1\right), \text{ and} \\ \frac{2}{3}x, & \text{if } x \in [1, 2]. \end{cases}$$

Figure 5 shows part of a valid tiling using the tiles of Figure 4.

Based on a similar argument K.Culik constructed an aperiodic set that contains only 13 tiles [3]. It is the smallest known aperiodic set of Wang tiles. It is likely that even smaller sets exist, though, but they remain to be discovered. In fact, we suspect that one tile of the Culik set is unnecessary and can be removed. This would provide an aperiodic set with 12 tiles. Naturally the trimmed set does not admit periodic tilings, because the original Culik set does not permit periodic tilings. The problem is to demonstrate the existence of at least one valid tiling using only the remaining 12 tiles.

3 Forced tiles and determinism

Some patches of tiles have the property that there is only one way to extend the tiling at some adjacent positions. We say that the patch forces the adjacent tiles.

Figure 5. 20 × 20 pattern from a valid plane tiling with our 14 Wang tiles

For example, consider the Penrose tiles of Figure 1. There is only one way to extend the tail of a Dart, namely two Kites must lie inside the fork as shown in Figure 6. Larger patches containing more Penrose tiles can force wider areas, and even tiles in non-adjacent positions may be forced. The set of tiles forced by a given patch is called its empire. Chapter 10.6 of Grünbaum–Shephard [9] discusses forced tiles and empires on Penrose tilings.

Let us consider Wang tiles and define precisely patches, forced tiles and empires. Let T be a set of Wang tiles. A *patch* is a partial tiling on the plane. It can be formalized as a pair (A, p) where $A \subseteq \mathbb{Z}^2$ is a subset of tile positions, called the *support* of the patch, and p is a function

$$p : A \longrightarrow T$$

that assignes tiles from tile set T to all positions in the support A. The patch is called finite if its support A is finite.

A tiling $f : \mathbb{Z}^2 \longrightarrow T$ contains patch (A, p) if it coincides with p on the support A, that is, if $f_{|A} = p$. The patch *forces* tile t in position $(x, y) \in \mathbb{Z}^2$ if $f(x, y) = t$ for every valid tiling f that contains the patch. The *empire* of a patch is the set of all positions (x, y) where some tile is forced by the patch.

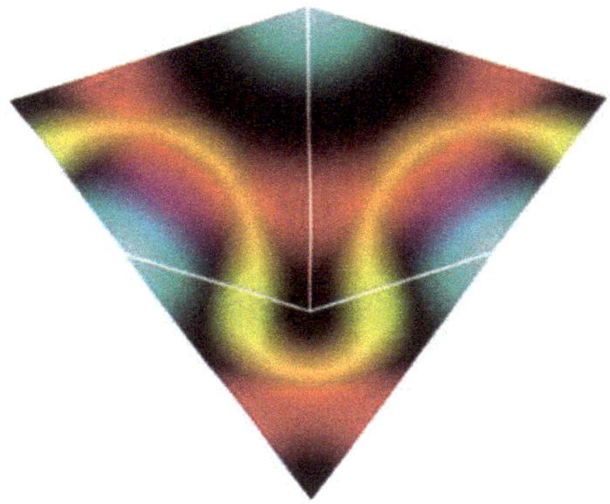

Figure 6. Two Kites forced by a Dart

One can also define a stronger version of forced tiles: We say that tile t is *immediately forced* in position $(x, y) \in \mathbb{Z}^2$ by a patch (A, p) if every validly tiled extension of p into $A \cup \{(x, y)\}$ assigns tile t to position (x, y). Tiles can be immediately forced only in positions that are adjacent to the support of the patch. Any tile that is immediately forced is of course forced in the weaker sense as well, but the converse is not necessarily true: some forced tiles may not be immediately forced. The reason is that not all local extensions of a patch can be extended further into a tiling of the whole plane.

Consider Amman's aperiodic set of 16 Wang tiles, shown in Figure 3. It has the interesting property that no two tiles are identical on both N- and W-edges. We say that the set is *NW-deterministic.* Two tiles aligned diagonally (immediately) force a tile below and to the right: For any tiles A and B in

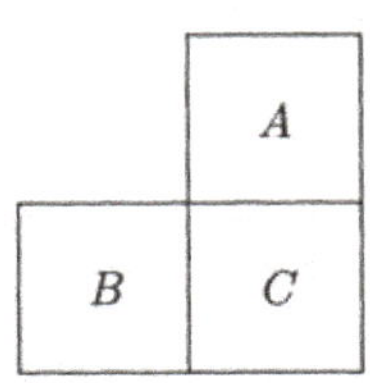

there is at most one matching tile C. Amman's set is also *SE-deterministic*, meaning that the tile above B and to the left of A is forced as well. This means that the entire tiling of the plane is forced by the tiles along the diagonal

$$D_0 = \{(x, x) \mid x \in \mathbb{Z}\},$$

or along any parallel diagonal

$$D_t = \{(x, x - t) \mid x \in \mathbb{Z}\}.$$

Moreover, the tiles on the diagonal D_t locally determine the tiles on the diagonals D_{t-1} and D_{t+1} above and below.

Let us interprete t as time, and take tiles along diagonal D_t as a configuration of a discrete dynamical system at time t. The configuration at time $t+1$ is then locally determined by the configuration at time t. In order to have well defined dynamics on every possible sequence of tiles on the diagonals, we have to decide what happens when tiles A and B do not admit any matching tile C. There are different alternatives, depending on what type of dynamics is preferred.

One solution is to introduce a special killer symbol k that is assigned to cell C if no matching tile exists. Once a killer symbol is introduced it does not disappear, instead it starts spreading: if A or B is k then C is k as well. Now every possible choice of A and B from $T \cup \{k\}$ determines one $C \in T \cup \{k\}$, i.e. we have specified a discrete dynamical system called (one-dimensional) *cellular automaton* (CA) over the 17 element state set $T \cup \{k\}$.

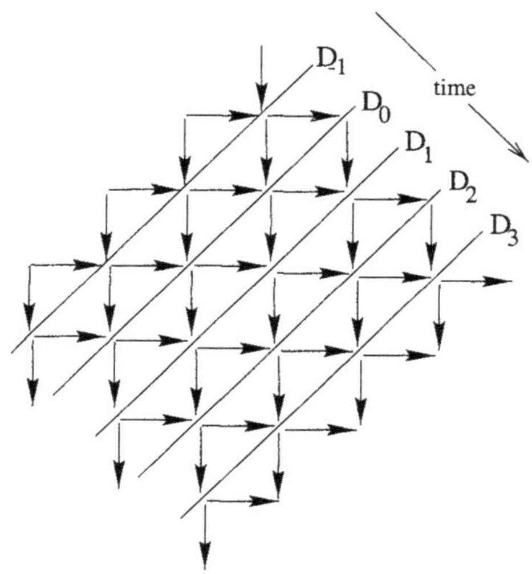

Let us investigate some dynamical properties of this CA. The configuration K consisting entirely of killer states k is a fixed point. Each spatially periodic configuration evolves into this fixed point in finite amount of time. Otherwise its time evolution would generate a periodic tiling where each diagonal D_t is tiled as the configuration of the CA at time t. But T does not admit periodic tilings, so the dynamics on periodic configurations is trivial.

On the other hand, there exist time evolutions that do not involve the killer state at all. They correspond to (non-periodic) valid tilings. In this CA periodic and non-periodic configurations exhibit different dynamics, and simulations of the CA using finite configurations with periodic boundary conditions give incomplete picture of its dynamics.

Analogously, any NW-deterministic tile set T gives rise to a one-dimensional cellular automata. Tiling properties of T are reflected in the dynamics of the corresponding CA as follows: If T does not admit any valid tiling of the plane then the CA has trivial dynamics on all configurations: they all evolve in bounded time to the fixed point K. Such CA are sometimes called *nilpotent.* If T admits periodic tilings then the corresponding CA has more than one periodic orbit. And if T is aperiodic, as in the case of Amman tiles, then the CA has only one periodic point, namely the fixed point K. All periodic configurations evolve into K, but there exist also (non-periodic) configurations whose time evolutions do not involve state k.

It turns out that Berger's celebrated undecidability result, Theorem 1, is true even if one only allows NW-deterministic tile sets. There does not exist any algorithm that could determine for all NW-deterministic tile sets if they admit a tiling of the plane [12]. In terms of cellular automata, this result means that no algorithm can determine if a given one-dimensional cellular automaton is nilpotent or not. The same is true for any other question concerning the long term dynamics of CA: all non-trivial questions are undecidable [14].

Even more restrictive notions of determinism have been studied. A tile set is called *4-way deterministic* if it is simultaneously deterministic in NW-, NE-, SW- and SE-directions. If T is 4-way deterministic then a tiling is forced not only by the tiles along any diagonal D_t but also by the tiles along any continuous curve that intersects every horizontal and vertical line of the plane. Any connected finite patch forces all tiles on the smallest rectangle enclosing the patch. There exist aperiodic sets that are 4-way deterministic [16].

4 Beyond aperiodicity

In periodic tilings patterns are repeated regularly throughout the entire tiling. This means that they appear densely everywhere: Every pattern that is present can be found in close vicinity of each point. More precisely, there exists a constant $r > 0$ such that any finite pattern that exists in the tiling is found withing distance r from any given point of the plane. In fact it can be shown that this condition is equivalent to periodicity [7]: Aperiodic tilings do not have such constant r. Note that the constant r is uniform to all patterns.

If we allow the maximum distance r vary from pattern to pattern we get a weaker regularity condition: A tiling is called *recurrent* if for every finite pattern appearing in the tiling there is a constant r (that may be different for different patterns) such that there is a copy of the pattern within distance r from any given point of the plane. So also in recurrent tilings patterns appear densely, but the density depends on the pattern. In the literature recurrent tilings have been called also repetetive tilings [23] and quasiperidic tilings [7]. The term recurrence is adopted from the dynamical systems theory.

Unlike periodic tilings, recurrent tilings can not be avoided. There do not exist any "arecurrent" tile sets: Every tile set T that admits a valid tiling of the plane must admit a recurrent tiling. This surprising result can be proved using basic notions of topological dynamics [8] as follows: Consider an arbitrary valid tiling f of the plane. The topological closure of the set of all tilings obtained by translating f contains a minimal invariant subset A. It is a (i) non-empty, (ii) topologically closed set of valid tilings such that (iii) all translations of all tilings in A are also in A, and A has no proper subset that would have properties (i)–(iii). It is easy to see that tilings in A are recurrent. Namely, let $g \in A$ be a tiling, and let P be any finite patch appearing in g. Let U denote the set of all tilings containing patch P. Because A is minimal, the set U and all its translates form an open cover of A and therefore contain a finite subcover of A. This means that all translates of g contain a translated copy of patch P in one of finitely many positions, i.e. P recurs in g.

It remains an interesting open problem to find the borderline between avoidable and unavoidable regularity properties. Which other, weaker conditions than periodicity one can prevent using local matching rules, and which stronger conditions than recurrency can not be forbidden?

It has been known for a long time that a weaker condition than periodicity that a tile set can be forced to avoid is recursiveness. Tiling f is *recursive* if it can be constructed using a mechanical procedure, i.e. an algorithm. More precisely, there must exist an algorithm that, given any location (x, y) as input, determines the tile $f(x, y)$ in that location. Any periodic configuration is naturally recursive: a tiling algorithm only needs to determine the relative position of given (x, y) inside the period. Surprisingly, there exist sets of Wang tiles that can tile the plane, but none of the tilings is recursive [18].

5 Conclusions and open problems

We have discussed recent developments on Wang tilings. Our algebraic technique to enforce aperiodicity produces the smallest known aperiodic sets [15,3]. Applications in cellular automata theory prompted investigation of deterministic tile sets, which has resulted in several new undecidability results concerning cellular automata [11,12,13,14]. The refered articles contain detailed proofs of these results, all of which are based on variants of Berger's theorem [1].

Several open questions remain. A challenging problem is to come up with even smaller aperiodic sets of Wang tiles. The author has been working in co-operation with K.Culik to trying to prove that one tile in the 13 tile Culik set [15] is not needed. It would be surprising if even that would be the smallest aperiodic set. The set includes an "unnatural" partitioning of the tile set into two parts that cannot be mixed on same horizontal rows.

Small and simple 4-way deterministic aperiodic sets remain to be discovered as well. Our construction is quite complicated and results in a very large tile set [16]. It is also not known if Berger's theorem remains valid in the restricted case of 4-way deterministic tiles. Finally, it would be very interesting to know other regularity conditions than periodicity (and recursiveness) that a tile set can be forced to avoid.

References

1. R. Berger, The Undecidability of the Domino Problem. *Mem. Amer. Math. Soc.* 66 (1966).
2. J. H. Conway and J. C. Lagarius, Tiling with polyminoes and combinatorial group theory. *J. of Comb. Theory*, Ser. A 53, 183–208 (1990).
3. K. Culik II, An aperiodic set of 13 Wang tiles. *Discrete Mathematics* 160, 245–251 (1996).

4. K. Culik II, L. P. Hurd and S. Yu, computation theoretic aspects of cellular automata. *Physica D* 45, 357–378 (1990).
5. K. Culik II and J. Kari, An aperiodic set of Wang cubes. *Journal of Universal Computer Science* 1, 675–686 (1995).
6. B. Durand, Self-similarity viewed as a local property via tile sets. Proc. MFCS'96, Lecture Notes in Computer Science, vol. 1113, 312–323 (1996).
7. B. Durand, Tilings and quasiperiodicity. Proc. ICALP'97, Lecture Notes in Computer Science, vol. 1256, 65–75 (1997).
8. H. Furstenberg, *Recurrence in ergodic theory and combinatorial number theory.* Princeton University Press, 1981.
9. B. Grünbaum and G. C. Shephard, *Tilings and Patterns.* W. H. Freeman and Company, New York, 1987.
10. C. Goodman-Strauss, Matching rules and substitution tilings. *Annals of Mathematics* 147, 181–223 (1998).
11. L. Hurd, J. Kari and K. Culik II, The Topological Entropy of Cellular Automata is Uncomputable. *Ergodic Theory and Dynamical Systems* 12, 255–265 (1992).
12. J. Kari, The nilpotency problem of one-dimensional cellular automata. *SIAM Journal on Computing* 21, 571–586 (1992).
13. J. Kari, Reversibility and surjectivity problems of Cellular automata. *Journal of Computer and System Sciences* 48, 149–182 (1994).
14. J. Kari, Rice's theorem for the Limit Sets of Cellular Automata. *Theoretical Computer Science* 127, 229–254 (1994).
15. J. Kari, A small aperiodic set of Wang tiles, *Discrete Mathematics* 160, 259–264 (1996).
16. J. Kari and P. Papasoglu, Deterministic aperiodic tile sets. *Geometric and Functional Analysis*, submitted.
17. D. E. Knuth, *The Art of Computer Programming*, Vol.1, p.384. Addison-Wesley, Reading, MA, 1968.
18. D. Myers, Nonrecursive tilings of the plane, II. *J. Symbolic Logic* 39, 286–294 (1974).
19. R. Penrose, Pentaplexity. *Eureka* 39, pp. 16–22, 1978.
20. R. M. Robinson, Undecidability and Nonperiodicity for Tilings of the Plane. *Inventiones Mathematicae* 12, 177–209 (1971).
21. R. M. Robinson, Undecidable tiling problems in the hyperbolic plane. *Inventiones Mathematicae* 44, 259–264 (1978).
22. D. Shechtman, I. Blech, D. Gratias and J. W. Cahn, Metallic phase with long-range orientational order and no translational symmetry. *Physical Review Letters* 53, 1951–1953 (1984).
23. M. Senechal, *Quasicrystals and geometry.* Cambridge University Press, 1995.
24. H. Wang, Proving theorems by pattern recognition II. *Bell Systems Tech. J.* 40, 1–41 (1961).

Reaction-Diffusion and Beyond

BIOLOGICAL PATTERN FORMATION AS A COMPLEX DYNAMIC PHENOMENON[1]

HANS MEINHARDT

Max-Planck-Institut für Entwicklungsbiologie,
Spemannstr. 35, D-72076 Tübingen, Germany

Self-enhancement coupled with one or more antagonistic reactions is the crucial element in pattern forming reactions. Depending on the parameter, this can lead to patterns in space and/or in time, they can be either extremely robust and reproducible or highly variable. Complex patterns result from a linkage of many pattern forming reactions, one pattern generates the prerequisites for the next. The support these models have obtained recently by molecular-genetic observations give rise to the hope that in the future an interplay between theory and experiment will lead to a still better understanding of this central issue.

Free of functional constraints, the diversity of patterns of on shells of mollusk provide a rich source to study the properties of dynamic systems in general. Everyday we are confronted with systems that have an inherent tendency to change. The weather, the stock market, or the economic situation are examples in which self-enhancing and antagonistic processes play also a decisive role. The shell patterns are sufficiently complex to be a challenge but also sufficiently simple to be accessible to modelling. Their one-dimensional character and the preservation of the history of their formation provide unusual help for deciphering these patterns. They illustrate the range of behavior that can be generated by modifications of a basic mechanism. They can be regarded as a natural exercise book to study dynamic systems.

1 The problem of pattern formation

The generation of the complex structure of a higher organism within each life cycle is one of the most fascinating aspects of biology. The similarity of identical twins underlines of how reproducible this process can be when the genetic information is the same. Reference to the genes, however, does not provide on its own an explanation for the generation of spatial structures since with each cell division both daughter cells obtain, as the rule, the same genetic information. This leads to the question of how different parts of the developing organism can become different from each other. How can patterns emerge in an initially more or less structure-less system? How can different genetic information be activated at different positions?

Basic concepts, such as *Positional Information* [Wolpert, 1969] or the *embryonic organizer* [Spemann and Mangold, 1924] have been derived from experiments involving perturbations of normal development. From the observed regulatory phenomena one cannot directly deduce the molecular mechanism on which development is based. We have used such observations to develop specific models for different developmental situations. By computer simulation we have shown that the regulatory features of the models correspond closely to the experimental observations. More recently, the new tools of molecular biology have opened a second inroad. In this way, several of these models have obtained direct support.

[1]This is a modified version of an article that appeared in the *International Journal of Bifurcation and Chaos*, 7, 1-26 (1997)

In the present article I would like provide first a brief description of the basic mechanisms we envisage for pattern formation and space-dependent gene activation. By using a small freshwater polyp as example it will be shown how complex structures can be generated in a reliable way by an appropriate coupling of several such reactions. The resulting patterns are very stable. They can become restored even after severe perturbations. The simulations resemble closely the biological observations. In a third part I would like to show that the same type of reactions in a different range of parameters can also lead to highly variable patterns. The richness of the pigment patterns on tropical sea shells will serve as an example.

1.1 The egg cannot contain the final pattern

Although many eggs have pronounced asymmetries, the final pattern cannot already be present in the egg in a hidden form. If it where so, the egg must also contain the pattern of the eggs of the next generation and so on, a process that comes soon to an end due to the atomic nature of the matter.

At an early stage, many embryos can be fragmented into two part and each part forms a complete organism. The sea urchin embryo at the 16 cell stage is an example. In some vertebrates, this can also occur spontaneously, giving rise to identical twins. This indicates that in these cases the early embryo is not a mosaic-like arrangement of differently determined cells that have a fixed fate. Instead, a communication must exist between different parts, such that the removal of some parts become detected and the missing parts replaced.

1.2 Organizing regions

Some small specialized regions obviously play a decisive role for the overall organization of the developing organism. Such organizing regions direct pattern formation in the surrounding tissue. One of such regions is the dorsal lip of the amphibian embryo [Spemann and Mangold, 1924]. The embryo forms first a blastula, a hollow sphere. The embryo proper becomes shaped in the process of gastrulation, an invagination of cells that starts at a particular position. This process leads to a tube-like structure within the cavity touching at one side the outer surface. In this way the embryo obtains its basic cell sheets, the ecto-, meso, and endoderm. The first invaginating cells have organizing properties as mentioned above. A transplantation to the opposite (ventral) side causes the formation of a second embryo. By using tissue for transplantation that differ in their pigmentation from that of the host it has been shown that the secondary embryo is formed mostly from host tissue, not from the transplant. Obviously, the transplanted cells have an instructive influence on the surrounding tissue. The resulting pattern has no discontinuity at the border between the transplanted and the host cells. Meanwhile, many genes involved in the formation of the organizing regions are known [see DeRobertis, 1995].

Small regions with an organizing influence on the surrounding tissue are also found in the generation of substructures. For instance, at the posterior margin of a chick wing bud a small nest of cells exists that organizes the anteroposterior pattern of the limb. Transplantation of these cells into a more anterior position causes mirror image duplications of the bones of the wing with a "little finger" at the anterior *and* at the posterior wing margin and two "thumbs" in the centre [Tickle et al., 1975].

1.3 The concept of positional information

To account for the long range effect of small specialized regions and for the spatial continuity observed after many experimental interferences, the "positional information" scheme has been proposed by Wolpert [1969]. Organizing regions are assumed to be the source of morphogenetic substances. By diffusion and decay graded distributions of such substances are generated. The cells are assumed to interpret this positional information by a concentration-dependent gene activation. The local concentration is a measure for the distance from the organizing region. After transplantation of such a source into an ectopic region the surrounding tissue becomes exposed to the graded concentration profile of the morphogen and, when competent, they follows the same pathway as the tissue that normally surrounds the organizing region. Meanwhile clear cases of morphogenetic gradients are available. In the *Drosophila* egg, a localized deposition of messenger RNA of the *bicoid* gene leads to a graded distribution of the corresponding protein [Driever and Nüsslein-Volhard, 1988, 1989]. The position at which particular structures such as the head fold appear and at which the downstream gene *hunchback* becomes activated depends on the local *bicoid* concentration.

However, the assumption of a localized source shifts the problem. The question remains of how a localized source can emerge in an initially more or less homogeneous tissue or at a particular position within the developing organism. Further, a local source has no handedness, but, for instance, the limb has a handedness. Models will be discussed that suggest possible mechanisms.

2 Generation of a primary pattern by autocatalysis and lateral inhibition

The driving force of all biological processes is the interaction of molecules. In his pioneering paper, Turing [1952] has shown that pattern formation requires the interaction of two substances with different diffusion rates.

Pattern formation is not a privilege of biology. High sand dunes are formed although the sand is permanently redistributed by the blowing wind. Sharply contoured rivers are formed by erosion although the rain is more or less homogeneously distributed over a particular region. The formation of lightnings from diffuse clouds is another example. Together with Alfred Gierer I have shown that the crucial condition for pattern formation is local self-enhancement and long range inhibition [Gierer and Meinhardt, 1972; Gierer, 1981; Meinhardt, 1982]. It is easy to see that the examples for inorganic pattern formation mentioned above, sand dunes, rivers and lightnings, are based on the same mechanism. A small deviation from a homogeneous distribution has a strong feedback such that the deviation grows further. This increase becomes eventually restricted by an antagonistic effect, resulting either from a depletion of a prerequisite in the surrounding or by an inhibition that spreads out from the self-activating centre. The same idea has been proposed almost simultaneously by Segel and Jackson [1972] for interacting predator — prey systems with different mobilities, a problem in population dynamics.

The paper of Turing marks the beginning of many theoretical investigations, applying the concepts of chemical reactions and diffusion to biological pattern formation [Lefever, 1968; Lacalli, 1981; Murray, 1989; Harrison, 1993; Borckmans et al., 1994; Koch and Meinhardt, 1994]. Recently, pattern formation has been observed in chemically completely defined systems [Ouyang and Swinney, 1991; Castets et al., 1990]. Pattern formation based on local self-enhancement and long ranging inhibition has been also shown to take place in appropriate electronic circuits [Berkemeier et al., 1986, Chua et al., 1995] and in gas discharges [Astrov et al., 1995]

2.1 The Activator-Inhibitor concept

Since all biological processes are assumed to be accomplished by the interaction of molecules, a theory of biological pattern formation has to describe the change of substance concentration in space and time as function of the concentration of the other substances involved. According to our theory, a simple molecular interaction with pattern-forming capability would consist of an "activator" whose autocatalysis is slowed down by a long ranging "inhibitor". The following coupled differential equations provide an example for an interaction between an activator $a(x)$ and an inhibitor $h(x)$ that can generate patterns [Gierer and Meinhardt, 1972]:

$$\frac{\partial a}{\partial t} = \rho \frac{a^2}{h} - \mu_a a + D_a \frac{\partial^2 a}{\partial x^2} + \sigma_a \tag{1a}$$

$$\frac{\partial h}{\partial t} = \rho a^2 - \mu_h h + D_h \frac{\partial^2 h}{\partial x^2} + \sigma_h \,. \tag{1b}$$

(1a) states that a concentration change of the activator per unit time ($\partial a/\partial t$) is proportional to an autocatalytic production term (a^2), and that the autocatalysis is slowed down by the action of the inhibitor ($1/h$). As any other biological substance, the activator molecules become degraded. It is natural to assume that the number of disappearing activator molecules is proportional to the number of activator molecules present. This is expressed by the term $-\mu_a a$. The autocatalysis must be non-linear since it must overcome the disappearance by the linear decay. Non-linearity of the reaction will result if the active component is not the activator itself but a dimer of two activator molecules.

The concentration of a in a given cell may also vary due to an exchange of molecules with neighboring cells. Diffusion is simplest way to take this exchange into account. Although diffusion is a good approximation, the real process is often much more complicated and requires a chain of several molecules: molecules that are exposed on the cell surface, molecules that act as receptors on neighboring cells and molecules that transmit the signal from the cell surface to the nucleus. Other modes of redistribution of molecules are conceivable as well. In plants, for instance, active transport plays a major role.

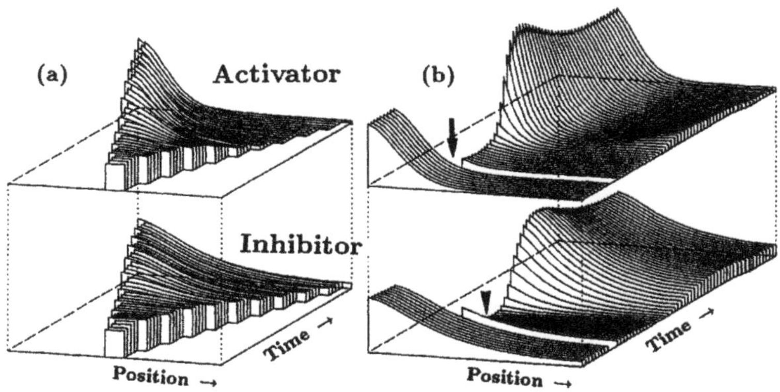

Figure 1. Pattern formation by interaction of an autocatalytic activator (top) and its long ranging antagonist, the inhibitor (Eq. 1). Assumed is a linear array of cells that grows at both margins. Both concentrations are plotted as functions of position and time. (a) If a critical size (the range of the activator) is exceeded, random fluctuations are sufficient to initiate pattern formation. A high concentration appears at one marginal position of the field since a central maximum would require space for two activator slopes for which no space is available at the critical size. A graded concentration profile results that can be maintained upon further growth. (b) Regeneration: after removal of the activated region (arrow), a new activator maximum regenerates after the decay of the remnant inhibitor (arrow head).

The last term in (1a), σ_a, describes a small basic (activator-independent) activator production. This term insures that the concentration of the activator never sinks to zero. This is important for the initiation of the autocatalytic reaction at low activator concentrations. The term ρ, the *source density*, describes the general ability of the cells to perform the autocatalytic reaction. Slight asymmetries in the source density may have a strong influence on the *orientation* of the emergent pattern. A necessary condition for the formation of a stable pattern is that the inhibitor diffuses much faster than the activator and has a shorter half life, i.e., $D_a \ll D_h$ and $\mu_a < \mu_h$ must be satisfied. In other ranges of parameters oscillations and travelling waves can occur. These modes will be discussed further below in connection with the patterns on the shells of mollusks.

Figure 1 shows numerical integrations of this set of equations. The simulations demonstrate that the interaction Eq. (1) has properties basic for the explanation of biological pattern formation. A pattern emerges whenever the size of the field exceeds the range of the activator. In fields with a size comparable to the activator range, the high activator concentration is formed at one end of the field. This is a very important aspect for the understanding of pattern formation in an early embryo. One side of the embryo becomes different from the other. The graded activator and/or inhibitor distribution can activate different genetic information in different parts of the tissue in an ordered sequence (see below). Thus, the system described by Eq. (1) is able to generate "positional information" [Wolpert, 1969].

It is only recently that pattern forming systems have been uncovered that follow this type of regulation. In the *Drosophila* fly the precursor cells of the peripheral nervous system (neuroblasts) and the ectoderm forms from two lateral bands of ectodermal cells. Crucial for neuroblast formation are the genes of the *achaete-scute* complex. Both genes have a direct autocatalytic feedback on their own synthesis [Van Doren et al., 1992]. Cells that express these genes activate the gene *Delta* that codes for a secreted molecule. This message is received by *Notch*, a receptor molecule ubiquitously present on all cells in that region and that acts on the *achaete-scute* complex in an inhibitory way. In this way, particular cells are singled out to participate in the formation of the peripheral nervous system while the remaining cells form the ectoderm.

A similar mechanism has been recently described for the ventral midline cells in Drosophila. These cells express the gene *orthodenticle (odt)*. Several genes are involved to form an autocatalytic loop. A small secreted molecule, *argos* interferes with the molecule *Spitz* that is involved in the positive loop, i.e., *argos* acts as an inhibitor [Golembo et al., 1996]. If the *argos* gene is mutated, the restriction does not work and all cells competent to express *odt* will activate that gene.

As mentioned, there are many genes involved in the generation of the primary embryonic organizer [see DeRobertis, 1995]. Recently, a secreted protein has been uncovered with anti-organizing properties [Moos et al, 1995]. Although it has its highest concentration in the organizing region, its overexpression leeds to a repression of all genes normally expressed in the organizing region. This corresponds to the expected property of the inhibitor in such a system.

An important feature of many developing systems is their ability to regenerate. For instance, in the freshwater polyp *Hydra* a removal of the head leads to the formation of a new one. This is a property of the activator-inhibitor system. After removal of the activated region, the remnant inhibitor decays until a new maximum is formed via autocatalysis (Figure 1b). However, the actual pattern formation in *Hydra* is more complex (see Figure 3 and Figure 4).

In most inorganic reactions the antagonistic effect results from a depletion of ingredients that are required for the self-enhancing process. Eq. (2) describes a possible interaction between an activator $a(x)$ and a substrate $s(x)$ that is consumed during the autocatalysis Gierer and Meinhardt, 1972].

$$\frac{\partial a}{\partial t} = \rho s a^2 - \mu_a a + D_a \frac{\partial^2 a}{\partial x^2} + \sigma_a \tag{2a}$$

$$\frac{\partial s}{\partial t} = \sigma_b - \rho s a^2 \ \ (-\mu_b s) \ + D_s \frac{\partial^2 s}{\partial x^2} \tag{2b}$$

This reaction has similarities with the so-called Brusselator reaction [Prigogine and Lefever, 1968; Lefever, 1968, Lefever et al., 1988] but is somewhat simpler. It has some properties distinctly different of the activator-inhibitor mechanism. For instance, in growing activator-substrate systems new maxima are formed preferentially by a split and movement of existing maxima, in contrast to the generation of new maxima at a distance to existing maxima in the activator-inhibitor system (Figure 2). I will come back to this reaction in the discussion of the patterning of sea shells.

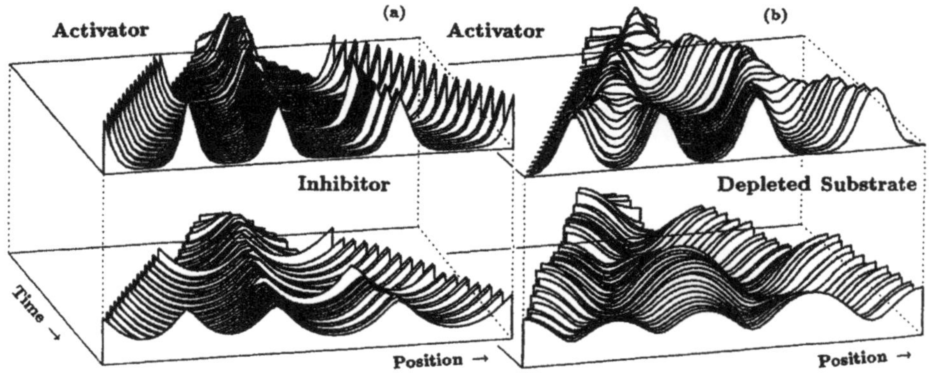

Figure 2. Different behavior upon growth. (a) If the antagonistic reaction results from an inhibitor h (Eq. 1), in a growing field new maxima are inserted whenever the inhibitor drops below a certain threshold. Thus, new maxima appear at a distance to the existing one. (b) If the antagonistic reaction results from the depletion of a substrate s (Eq. 2), upon growth the substrate concentration increase in the non-activated regions. Therefore, activator production may become higher at the side than in the centre of a peak. This leads to a split of a maximum followed by a shift towards a region of higher substrate concentrations.

2.2 Comparison with Turing's mechanism

As mentioned, the possibility to generate a pattern by the interaction of two substances with different diffusion rates was discovered by Alan Turing. However, in his paper he did not mention anything like 'lateral inhibition'. It seems, however, that he has recognized this crucial point later. In unpublished notes found after his death the following sentence occurred "The amplitude of the waves is largely controlled by the concentration V of 'poison' " (Hodges [1983], p. 494). In his paper, he exemplified the mechanism he envisaged by the following set of equations [Turing, 1952, p.42].

$$\frac{dx}{dt} = 5x - 6y + 1 \tag{3a}$$

$$\frac{dy}{dt} = 6x - 7y + 1 \quad (\,+\text{ diffusion}) \tag{3b}$$

Both Eqs. (3a) and (3b), look very similar. It is not immediately obvious why such interaction leads to a pattern. However, it is easy to see that this interaction satisfies our condition of local autocatalysis (of x) and long range inhibition via a molecule (y) who's production is under x control and that acts antagonistically, in this case, by destruction the x molecules. Therefore, Turing's mechanism can generate basically the same types of pattern as the lateral inhibition mechanism, i.e., graded concentration profiles and isolated maxima [Bard and Lauder, 1974; Lacalli and Harrison, 1978]. But knowing that self-enhancement must be balanced by a long-ranging inhibitory reaction facilitates substantially the design of appropriate reactions.

The mechanism proposed by Turing has an essential drawback: its molecular basis is not reasonable. According to Eq. (3a), the number of x molecules disappearing per time unit is assumed to be proportional to the number of y molecules but independent of the number of x molecules. In other words, x molecules can disappear although no x molecules are present. This can lead to negative concentrations. Turing has seen this problem and proposed to ignore negative concentrations. One could repair this deficit by assuming a degradation also proportional to x and a non-linear autocatalytic activator production. In the activator-inhibitor notation given above this would read

$$\frac{\partial a}{\partial t} = \rho a^2 - \mu_a a h + D_a \frac{\partial^2 a}{\partial x^2} \tag{4a}$$

$$\frac{\partial h}{\partial t} = \rho a^2 - \mu_h h + D_h \frac{\partial^2 h}{\partial x^2} \tag{4b}$$

Although this works, this interaction is also not reasonable for pattern formation. First, the lifetime and therefore the range of the activator changes during pattern formation. The shortening of the half life can lead to a transition from a stable pattern to oscillations. Secondly, in an embryo that depends on the use of stored energy it appears more reasonable to assume that the inhibition works by a slowing down of the production rate instead by an increase in the destruction rate combined with a production at a very high rate.

3 The wave length problem — stabilization of a monotonic gradient by a feedback on the source density

Usually the size of a morphogenetic field increases during the growth of the embryo. In a usual reaction-diffusion system a graded concentration profile can be maintained only over a range of about a factor two. With increasing field size, a tendency exists to change from a monotonic distribution into a symmetric and ultimately into a periodic distribution either by insertion of new or by splitting of existing maxima (Figure 2). This is inappropriate if the graded concentration should be used in the growing embryo as positional information for the determination of the primary body axes since, for instance, several heads would appear instead of one.

Some basic biological observations provide important hints of how this wave length problem has been solved by nature. *Hydra*, a small fresh water polyp maintains its polar structure over substantial growth but, nevertheless, a fragment 1/10 of the normal body size is still able to regenerate. In such a fragment the regeneration of the head occurs always at the side pointing towards the original head. Morgan [1904] interpreted similar observations with *Tubularia* by assuming that a systematic change in the ability for head regeneration exists and that a competition takes place. The tissue originally closer to the head has a head start and wins the competition. The *relative* position is decisive.

In terms of our model, the range of dominance of the activated over the non-activated region (the so-called apical dominance) can be increased by an order of magnitude if a feedback of the activator or inhibitor on the source density (ρ in

Eq. 1) exists [Meinhardt and Gierer, 1974]. Due to this feedback, the source density becomes graded too. In a region of low source density, the initiation of secondary maxima is unlikely. In contrast, the rising source density at an existing maximum stabilizes this maximum. Thus, the polar distribution becomes stabilized. The graded source density provides the long-lasting information about the polarity of the system. A small fragment regenerates a pattern according to the original polarity since the graded source density provides a systematic headstart for some cells to outcompete the others. Since the source density has a long time constant, it remains essentially unchanged during pattern regeneration (Figure 3). Although secondary maxima are successfully suppressed, the regeneration is not impaired. Regeneration can be fast since no symmetry breaking and no communication by diffusion over the total field is required (Figure 3).

The source density must have a more shallow distribution than the activator. This may be achieved by a small diffusible molecule. Alternatively, it may not the activator but the inhibitor that controls the source density. Due to its higher

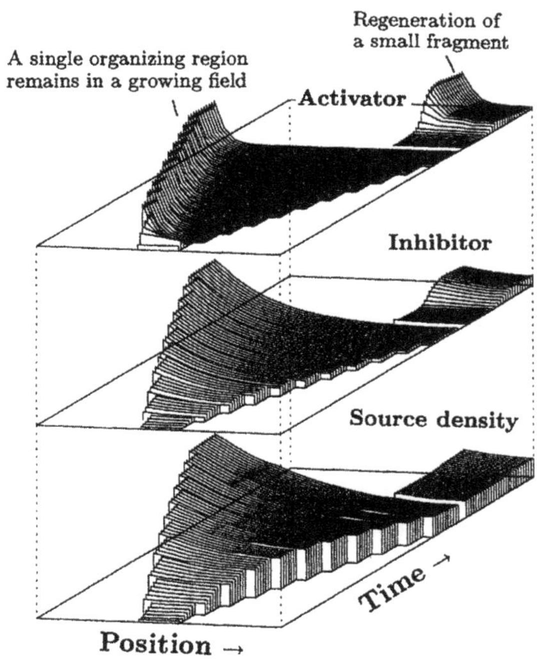

Figure 3. A solution of the wave length problem: stabilization of polar distributions by a feedback of the pattern onto the efficiency of the pattern forming reaction (source density *rho* in Eq. 1). It is assumed that the inhibitor (middle) has a feedback on the source density (bottom) as given in Eq. (3). The tissue becomes polar. Since the source density changes only slowly, it remains nearly unchanged during regeneration. A new activator maximum (top) forms reproducible at the highest source density, i.e., at the site pointing towards the original maximum. The tendency to form periodic structures upon growth (see Figure 2) is reduced. A monotonic gradient is maintained although the range of the activator is very small. This has the consequence that very small fragments can regenerate.

diffusion rate, the latter has anyway a more shallow distribution. No further diffusion of the sources would be required. In Eq. (1) the activator a is nearly independent of the absolute ρ concentration. Therefore a feedback of a on ρ does not create stability problems. This is, however, not true if the inhibitor accomplishes this feedback. Eq. (3) provides an example in which the increase of the source density slows down at higher levels of the source density. This avoids the stability problems.

$$\frac{\partial \rho}{\partial t} = \mu_\rho(h/\rho - \rho)$$

In an interaction according to Eqs. (1) and (3) the inhibitor has a dual role (Figure 3). Firstly, it localizes the activator maximum. Secondly, it increases the local source density. In this way it is also responsible for the generation of polarity within the tissue. Although inhibitor and source density would have in the steady state similar distributions, they have otherwise very different properties. The inhibitor must have a high diffusion rate and a short time constant to enable a stable pattern. In contrast, the source density must be non-diffusible and must have a long time constant. Since the wavelength of the system is no longer determined by the activator diffusion, the latter can be very small. Therefore, small fragments can regenerate a pattern. Since after regeneration the ρ-gradient will be restored in the course of time (Figure 3), this gradient will not be diluted out in repeating rounds of regenerations. The ρ-gradient may also be realized by the stable space-dependent activation of different genes (see below).

Since the source density concept is very important to explain many biological observations, let me explain the mechanism again with a somewhat anthropomorphic analogy. A king (or any other local hero) has usually a strong tendency to suppress other's from becoming a king too - a long range inhibition. On the other hand, he promotes individuals in his surrounding to obtain different levels in a hierarchy, to become ministers etc. In this way, the centre of power generates a hierarchy. If the centre of power would become void, due to this non-uniformity is usually clear who will win the subsequent competition.

3.1 How to generate structures close to each other, how at a distance: head, foot and tentacle formation in Hydra

The complexity of the patterns in higher organisms requires a hierarchical linkage of many pattern forming reactions. One or more patterns generate the precondition for a subsequent pattern. For instance, by an appropriate coupling it can be achieved that two pattern forming systems (anteroposterior, dorsoventral) emerge perpendicular to each other [Meinhardt, 1989]. The combinatorial possibilities are very large, making modelling very difficult. That nevertheless the modelling of complex patterning processes is possible should be illustrated with a model for the freshwater polyp *Hydra* (Figure 4; [Meinhardt, 1993]).

Hydra is under control of two organizing regions located at opposite ends of the tissue, the head and the foot. This is common of many morphogenetic fields. How can it be achieved that two organizing centres reliably appear at opposite positions of an extended field? For *Hydra* a simple cross-inhibition is not appropriate since

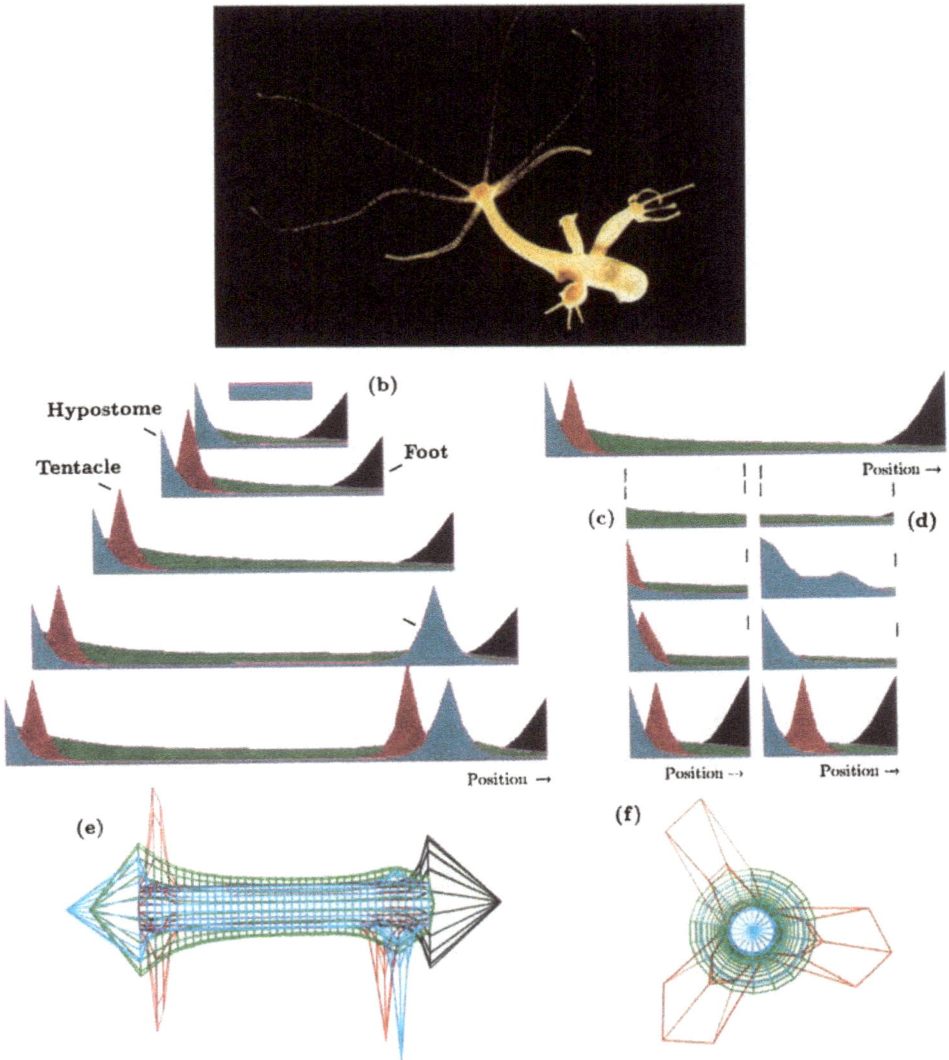

Figure 4. Generation of complex patterns by linkage of several pattern forming reaction. (a) The fresh water polyp *Hydra*. (b-f) Simulation of hypostome, tentacle, bud and foot formation in Hydra. (b) Primary head (blue) and foot activation (black) appear at opposite end of the field due a coupling via the source density (green). Tentacle activation (red) appears close to the hypostome since it requires a high source density but it is locally suppressed by head activation. Budding results from a second head activation. Due to the long-ranging head inhibitor, this can occur only at a large distance from the original head. (c) In regenerating near-head fragments tentacle activation precedes head activation. It occurs first at the tip and shifts later on, in agreement with the experimental observations. (d) In more basal fragments head activation appears first. Tentacle activation takes place later at the final position after the source density has obtained a threshold value. (e, f) Lateral and top view of a Hydra simulated as a cylinder: the periodic arrangement of the tentacles (red) around the hypostome and the lateral localization of the bud (blue) is correctly described. (The simulations describe only pattern of the signalling molecules on a cylinder, but not the subsequent shape changes of the tissue; after Meinhardt [1993]).

in small (young) animals head and foot must appear very close together. If at such short distances a mutual inhibition between the head and the foot system would be at work, this would lead to a suppression of the foot by the nearby head or vice versa. This problems disappears when the spacing between the head and foot system is achieved by an interaction via the source density. As mentioned above, the head activation appears at the position of the highest source density. If the foot system has the opposite behavior, i.e., it appears at the lowest source density, the foot is formed at the maximum distance from the head (Figure 4). Nevertheless, head and foot system can coexist at a close neighborhood in small animals since no direct inhibition is involved. The graded source density only generates a preference. Experimental evidence indicates that the foot also lowers the source density, contributing in this way to the maintenance of the source density gradient.

Many structures emerge during development close to each other in a precise arrangement. We have shown that a controlled neighborhood of structures is enforced if one structure activates the other on long range but excludes it locally [Meinhardt and Gierer, 1980]. In *Hydra*, the tentacles appear around the hypostome, the opening of the gastric column. Many experiments can be accounted for by the assumption that tentacles are under control of a separate activator-inhibitor system that also depends on the source density. Since the source density increases under the influence of the primary head system, the latter generates the precondition for tentacle initiation. Locally, however, the head signal suppresses tentacle formation. Thus, tentacle formation is possible only at a sub-hypostomal position (Figure 4b,e).

The model accounts for a strange-appearing observation. With tentacle-specific antibodies Bode et al. [1988] have shown that after head removal, tentacle activation first reappears at the very tip of the gastric column. It is only later that this activation becomes shifted to the position where the tentacles eventually appear. Since the tentacles are formed close together, the tentacle inhibitor need not to diffuse very far. In terms of the model, the tentacle inhibitor can have a short half life. Thus, after removal of the head and the tentacles, the tentacle inhibitor fades away more rapidly than the head inhibitor. Therefore, the tentacle activator can reappear sooner than the head activator. Since no suppressing head activator is present, this happens at the highest possible source density, at the front end of the remaining gastric column (Figure 4c). After the trigger of the primary head activation at the same position, tentacle activation becomes shifted to the final location. The prediction that the sequence of events is the reverse in more basal fragments (Figure 4d) or in buds (Figure 4b) has found meanwhile direct experimental support [Technau and Holstein, 1995].

4 Activation of genes under the influence of morphogenetic signalling

The generation of signals by the exchange of molecules via diffusion works only in small fields. In larger fields the time required to exchange information by randomly moving molecules would be much too long. Indeed, the sizes of embryonic fields in which developmental decisions take place are always small, smaller 1 mm and less than 100 cells across [Wolpert, 1969]. Therefore, the signals generated at small scales have to be translated into more permanent states that can be maintained upon further growth. The obvious means is a stable concentration- (and thus space-) dependent activation of genes. The choice of a particular pathway under the influence of a morphogenetic signal requires the activation of particular genes and the suppression of alternative genes. This situation has formal similarities with the formation of a pattern: Pattern formation requires activation at a particular position and the inhibition in the remaining part. The selection of a particular pathway requires the activation of a particular gene(s) and the suppression of the alternative genes. Based on this similarity, I have proposed that gene activation requires a direct or indirect feedback of genes on their own activation and their mutual competition such that only one of the alternative genes can remain active in a particular cell. (Figure 5; Meinhardt, [1978]). Meanwhile many such genes have been found. The genes *deformed* [Regulski et al., 1991], *hunchback* [Simpson-Brose et al., 1994] or *twist* [Leptin, 1991] are examples.

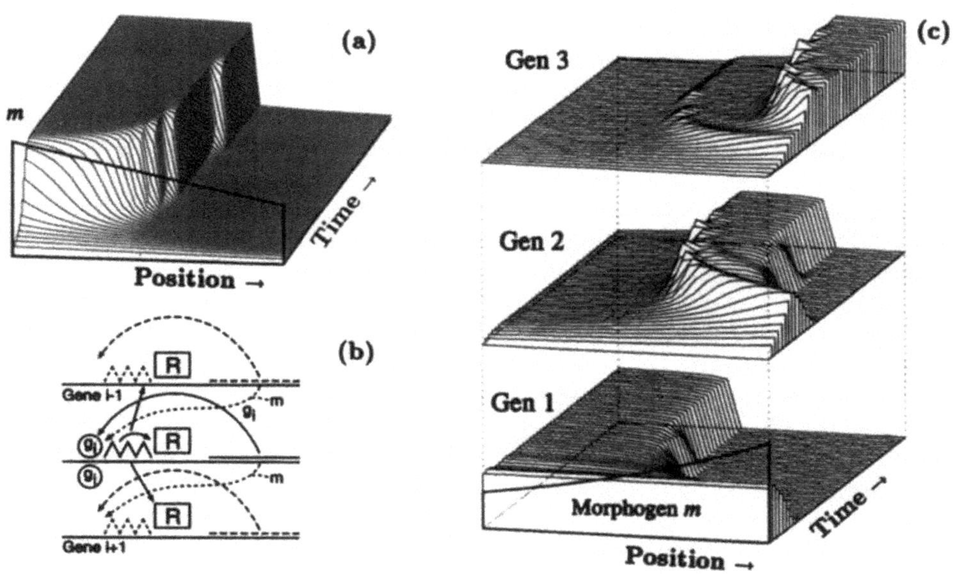

Figure 5. Space-dependent gene activation by a morphogenetic signal. (a) Switching behaviour of a single gene with non-linear, saturating feedback of the gene product on its own activation. At higher concentration of the morphogen m , the low state becomes unstable. Cells exposed to a certain m concentration switch irreversible into the high state. Regions with high and low gene activation appear. No zone of transition exist although the controlling signal has a smoothly graded distribution (calculated with Eq. (5d). (b, c) Activation of several genes. (b) Scheme: alternative genes are supposed to have an autocatalytic feedback on their own activity but compete with each other, for instance, by a common repressor R. The morphogen promotes the transition to a subsequent gene. (c) Starting with a default gene (Gene 1), a 'promotion' from one gene to the next under control of a morphogenetic gradient leads to sharply confined regions in which particular genes are active.

Such a dynamic regulation of gene activation has many properties that are essential for the developing organism. Firstly, such a system has a threshold. The activation of a gene becomes an all-or-nothing event (Fig 5a). If a threshold is surpassed, due to the autoregulation, the response of the cell becomes independent of the exact level of the morphogen concentration. Therefore, small deviation from a desired value are not propagated into the hierarchically next level of gene activation. Secondly, the cells obtain a long-term memory in respect to the signals they have seen. The signal is required only for the initiation of gene activity. Due to the feedback the activity is self-maintaining.

The translation of a graded concentration profile of a morphogenetic (signalling) substance into the position-dependent activation of genes requires an analog-to-digital conversion. Based on ligation experiments with early insect embryos [Sander, 1976] I have proposed that the cells do not measure a particular morphogen-concentration all at ones but that, starting from a default gene activity, other genes become activated in a step-wise manner. Each further step requires a higher morphogen concentration. This process comes to rest if the actual gene activation corresponds to the local morphogen concentration (Figure 5c, Meinhardt, 1978].

Due to the autocatalysis of the gene activation and the unidirectional promotion, the process is essentially irreversible. A reduction of the signalling molecules is without effect since the morphogen is not required for the maintenance of the gene activity. This mode of step-wise promotion has a very essential function during development. As mentioned, the the range of the signalling molecules is usually small. The initial activation of genes takes place in small fields that grow afterwards. During growth, the distance between a cell and the morphogen source necessarily increases and thus the morphogen concentration decreases. Due to the irreversible character of the step-wise promotion the cells maintain their once obtained state of determination even if they escape from the influence of a morphogen source. This stability is possible without that the determination is absolutely fixed. If, for instance, during regeneration, a new morphogen source is generated, the cells can still be "distally transformed". In the regeneration of a cockroach leg, this process plays a decisive role. The step-wise and irreversible promotion enables on the one hand the required stability to cope with the fading influence of the morphogen during growth and provides, on the other, still some (uni-directional) flexibility if regulation is required. This avoids the necessity of a narrow time window to which the measurement is confined.

Meanwhile several systems with this behavior have been described. Gurdon et al. [1995] found that a low concentration of Activin (a member of a family of growth factors) causes in cells from the blastula stage of *Xenopus* (an amphibian) an activation of the *Xbra* gene while for the gene *Xgsc* a high concentration is required. These genes remain activated after removal of Activin. However, applying Activin first in low and later in high concentrations leads to a reprogramming of cells from *Xbra* to *Xgsc*. Gurdon et al called this a ratchet-like switching, in full agreement with the mechanism I have proposed.

As mentioned, in *Drosophila*, neuroblast are formed in two lateral bands. Ventral midline cells act as organizing region to determine their fate. When grafting neuroblasts towards a more dorsal position (i.e., away from the organizer and thus presumably into a region of lower morphogen concentration), the neuroblast behave according to their origin. In contrast, if more dorsal neuroblasts are grafted into a more ventral position, they adapt to their new position [Udolf et al., 1995], in full agreement with the model.

5 Formation of filament-like branching structures

A feedback of a position-dependent gene activation on the pattern that has caused its activation can lead to very complex patterns. As an example the formation of filament-like branching structures will be discussed. This pattern is very common in almost all higher organisms. The venation of leaves, the tracheae of insects, the blood or lymph vessels as well as neurons are examples. How can such complex patterns emerge?

We have seen how a local high activator concentration can be generated and how it can be used as the signal to cause stable activation of a gene when a threshold is exceeded. The exposed cells differentiate and become, for instance, a part of a vascular system. I will assume further that differentiated cells repel the signal. The signal will be shifted into a neighboring cell which will differentiate and become thus a part of the vascular system, too. A repetition of this process — differentiation, shift of the signal, differentiation — leads to a strand of differentiated cells behind a wandering activator maximum. For the simulation shown in Figure 6, the following set of equations have been used [Meinhardt, 1976]

$$\frac{\partial a}{\partial t} = \frac{\rho a^2 s}{h} - \mu_a a + D_a \Delta a + \sigma_a y \tag{5a}$$

$$\frac{\partial h}{\partial t} = \rho a^2 s - \mu_h h + D_h \Delta h + \sigma_h y \tag{5b}$$

$$\frac{\partial s}{\partial t} = \sigma_s - \mu_s s - \nu s y + D_s \Delta s \tag{5c}$$

$$\frac{\partial y}{\partial t} = \frac{y^2}{1 + \kappa y_2} - \mu_y y + \sigma_y a \tag{5d}$$

The concentration of y within a cell is an indicator of whether a cell is differentiated or not. Due to the saturating autocatalysis, only two stable states are possible. The concentration of y is either high or low. At a certain concentration of the activator sufficient y is produced to turn on its autocatalysis and the cell switches irreversibly from low to high y production. To achieve a shift of the activator maximum away from the differentiated cells, it is assumed that all differentiated cells remove a substance s $(-\nu s y)$ that is produced by all cells. Therefore, the s concentration is lowest at the differentiated cells. If the activator autocatalysis depends on s, after differentiation of the cell and subsequent lowering of the s concentration, the maximum becomes shifted to those neighboring cells that contain the highest s concentration. Usually, these are the cells in front of the tip of the incipient filament. This cell will differentiate too. Long filaments of differentiated cells results from the repetition of signal formation, differentiation of the exposed

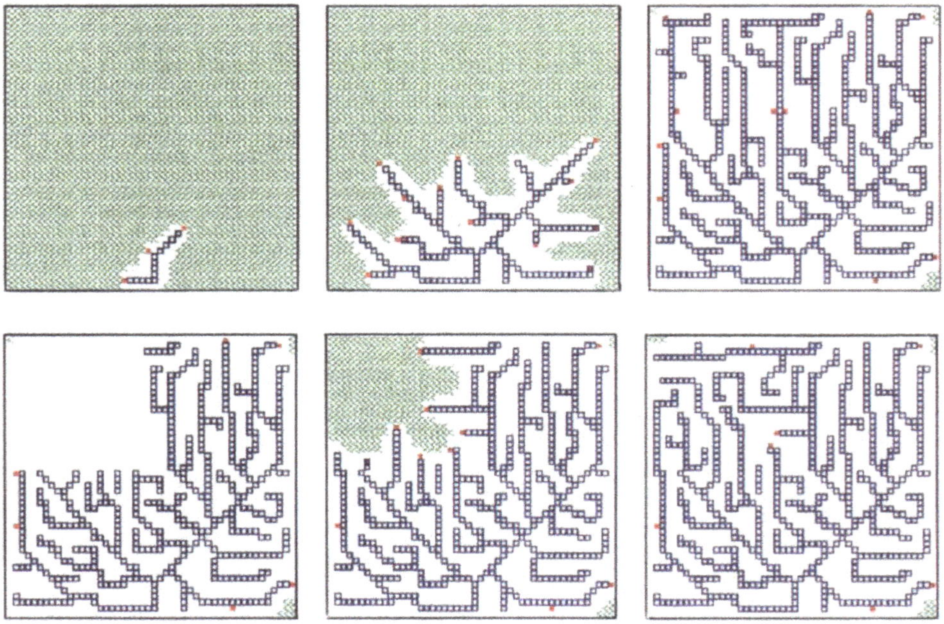

Figure 6. Formation of a net-like structure. (a-c) The interaction of four substances (Eq. 5) is sufficient to generate a structure with branching filaments. A signal for the local elongation of the filament is generated by an activator (red dots) / inhibitor system. This signal is used to differentiate cells (blue squares). Differentiated cells remove a substrate (green wavy lines). Since the activator/inhibitor system depends on this substrate, the activator maxima are shifted to that neighboring cells that have the largest distance from other differentiated cells. This is usually the cell at the tip of the filament. The patterning process comes to rest if a certain density of the filaments is reached. (d-f) After removal of some filaments, the substrate increases in that area and the defect becomes repaired. Calculated with (5) and $D_a = 0.02$, $\rho = 0.12 \pm 3\%$ fluctuations, $\mu_a = 0.12$, $\sigma_a = 0.035$, $D_h = 0.2$, $\mu_h = 0.04$, $\sigma_h = 0.015$, $D_s = 0.06$, $\mu_s = 0.02$, $\nu = .2$, $\sigma_s = 0.02$, $\kappa = 10$, $\mu_y = 0.1$, $\sigma_y = 0.0015$.

cell and shift of the signal into neighboring regions. Deflection of a growing filament can result from other nearby filaments or from the margins of the field.

Branches are formed whenever activator maxima become sufficiently remote from each other during elongation of filaments. Then, the inhibitor concentration can become locally so low that a new activator maxima is triggered along an existing vein due to the term $\sigma_a y$. Whether a branch is formed towards the one side or the other depends on minute fluctuations. However, if a branch has been formed, let us say, to the right, the next branch will probably point to the left due to the s removal by the first branch (such alternations are the rule in second order branches in many leaves). If the details depend on fluctuations but each step has a strong influence on forthcoming decisions, the actual pattern will be unpredictable. Indeed, the venation of two leaves on a tree are never identical although they are certainly formed under control of the same positional information. The term $\sigma_b y$ leads to an increasing background inhibition with increasing density of the filaments. This leads to a switching off of further filament elongation whenever a certain density is obtained.

After removal of some filaments, the system is able to regenerate the missing veins (or whatever it is) since in these regions, s is no longer removed and the rising s concentrations attract activator maxima from the non-injured region (Figure 6d-f). The regenerated pattern is similar but not identical.

6 Segmentation by mutual activation of cell states that locally exclude each other

A widespread pattern in biology is segmentation, the reiteration of similar structures. The body segments of insects or the segments of insect legs are examples. Segments have an internal pattern, frequently visible by overt structures such as bristles. Perturbation experiments have revealed that in intrasegmental organization not the natural pattern but a correct neighborhood of elements is controlled. Assume that the natural pattern can be described by a sequence of the structures 12345....9. A graft of the type 1234567+3456789 leads to a mismatch 7/3 at the graft/host border. This triggers an intercalation of structures *654* such that the discontinuity 7/3 is removed: 1234567*654*3456789.

As mentioned above, stable cell states result from self-activation and mutual competition of genes. If two (or more) such states not only exclude each other locally but activate each other over longer ranges, these cell states need each other in a close neighborhood. The local exclusiveness assures that the two states do not merge [Meinhardt and Gierer, 1980]. According to this model, segmentation requires the following molecular ingredients:

- Genes (or more general feedback loops) must exist that have a positive feedback on their own activation.
- These activities are locally exclusive; only one of the alternative genes can be active in a given cell.
- Long ranging molecules provide a mutual activation of those cell states that eventually become neighbors. Each cell state in a given cell depends on the help from different cell states in neighboring regions.

By computer simulations we have shown that the resulting structures are self-regulating. Missing structures become intercalated, if necessary with polarity reversal, in full agreement with the observation.

The mutual long ranging activation of cell states that locally exclude each other is in its core an alternative version of the local autocatalysis - lateral inhibition scheme [Gierer, 1981]. In this case the inhibition works by an long ranging activation of a competing cell state. However, a distinction between activated and non-activated regions is no longer possible.

The theoretical prediction found meanwhile direct support. In Drosophila *engrailed (en)* is a key gene for segmentation. It has a direct autocatalytic feedback on its own activation. Further, it activates a neighboring cell state via a diffusible molecule *hedgehog*. This cell state is characterized by the activation of the gene *wingless (wg)*. The *wg* protein can also diffuse into neighboring cells and stabilizes *en*. As expected from the theory, the *en* gene activity requires a functional *wg* gene in its neighborhood and vice versa although both genes are transcribed in non-overlapping regions. The prediction of such a complex molecular interaction by a theory could hardly be more precise. From the theory we would expect that the wingless gene is under transcriptional control of a second (directly or indirectly) autocatalytic gene. The gene *cuD* on which wingless-expression depends and that is repressed by *en* is perhaps a part of the missing feedback system [for references see Meinhardt, 1994].

7 Pattern formation in secondary embryonic fields: differentiation borders obtain organizing properties

Of course, the structure of a higher organism is more complex than can be achieved by the interpretation two orthogonal gradients. The generation of sub-structures like legs, wings or eyes, requires secondary pattern forming reactions. They must be initiated at precise position and the resulting pattern must obtain the correct orientation with the established axes of the embryo.

The notion of how secondary embryonic fields become established during development where highly influenced by the field concept of Harrison [1918]. Based on his experiments with amphibian limbs he proposed that first cells are set aside that will form the future the limb. At later stages, the originally homogeneous cell population was assumed to become different, first along the anteroposterior and subsequently the dorsoventral axis. However, a realization of such a scheme would require as a prerequisite that a boundary exist that separates this field from the remaining tissue. This boundary must have very peculiar properties. It must be impermeable for molecules that should generate this pattern. Especially the inhibitor should not leak out since otherwise the whole margin will become activated. On the other hand, the border must be permeable for those molecules that accomplish the orientation of the new pattern according to the established axes of the embryo. Therefore, such a model contains a hidden circular argument since complex differentiated structures must be assumed to explain the emergence of new structures. To circumvent this problems, I have proposed a different mechanism: secondary fields are formed in regions that surround the intersection of two determination borders [Meinhardt, 1983a,b]. Let us first regard two neighboring cell types, A and P and assume that both cooperate to produce a new substance m. For instance, the P cells may produce a necessary co-factor while only the A cells are able to generate the final product. In this way, the m production is restricted to the common border. If diffusion is involved, its concentration provides a measure for the distance from the border. It is therefore appropriate for the internal organization of the A and the P region. Although the pattern is symmetric in respect to the border, the resulting pattern can be asymmetric since the A and P cells can respond differently. In the vertebrate limb an extreme case is realized in that only the A cells respond at all.

A border that separates two cell types along the anteroposterior axis surrounds an embryo in a belt-like fashion. To determine the position of a limb along this line a cooperation between a second pair of cell types is required. Their border must be oriented perpendicular to the first. This leads to secondary fields around the intersection of two borders. Therefore, secondary fields consist from the beginning of four different quadrants (or, as in the insect leg, of at least three sectors). Simplifying an early embryo as a cylinder, for any reasonable subdivision along the anteroposterior and the dorsoventral axis these intersections occur in pairs, one at the right and the other at the left side of the embryo. They have opposite handedness, a feature crucial for the formation of legs, wings, eyes, etc. In contrast to the classical model, the secondary field is assumed to be never without an internal structure since the formation of the borders is the primary event. Many classical observations of insect and vertebrate appendages become explicable [Meinhardt, 1983a, 1983b, 1991]. This model has found much support by recent investigation on the molecular-genetic level [Vincent and Lawrence, 1994; Martin, 1995, Meinhardt, 1994]. The *engrailed/wingless* border mentioned above involved in the generation of segmentation is also a the primary border for the generation of insect legs and wings.

8 Pigmentation pattern of shells of mollusk — a natural picture book to study dynamic systems

A special case of biological pattern formation is the emergence of the pigment patterns on the shells of mollusks. These patterns are of great diversity and frequently of great beauty. The shells consist of calcified material. The animals can increase the size of their shells only by accretion of new material along a marginal zone, the growing edge of the shell. In most species, pigment becomes incorporated during growth at the edge. In these cases, the formation of these patterns proceeds in a strictly linear manner. The second dimension is a protocol of what happens as function of time. The shell is, so to say, a space-time plot. The shells provide a unique situation in that the complete history of a highly dynamic process is preserved.

In normal development, a strong evolutionary pressure exists to reproduce faithfully a given structure. In contrast, the functional significance of the pigment patterns on shells is not clear. Many mollusks live buried in the ground and some are covered with an opaque layer, the *periostracum.* Thus, there is presumably no strong selective pressure on the shell patterns. In Figure 7 a shell is shown with an additional pattern on the inner side of the shell, a pattern that is definitely invisible during the life time of the animal. The observed diversity suggests that the patterns have been drastically modified without endangering a species. Independent of functional constrains, nature was allowed to play. Moreover, in normal development a structure, once formed, usually should be stable at least for a certain amount of time. In the pigmentation patterns, oscillating modes play an important role as well. This contributes to the diversity.

Several attempts to model patterns on mollusk shells have been made. Formal models have been proposed by Waddington and Cowe [1969] for the tent-like patterns of Oliva shells and by Lindsday [1982] for bivalved mollusks. Cellular automata models have been discussed by Herman and Liu [1973], by Wolfram [1984], by Plath and Schwietering [1992] and by Kusch and Markus [1996]. The idea of using reaction-diffusion mechanisms to model shell patterns dates back to 1984

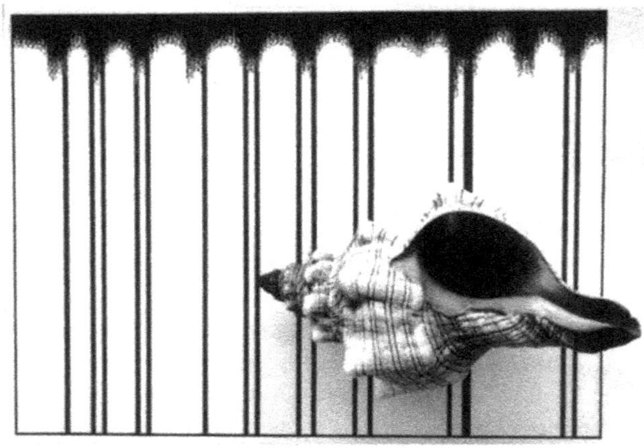

Figure 7. Stripes parallel to the direction of growth in the shell of *Pleuroploca trapezium*. Such stripes result from a stable pigment production at particular positions. On this shell, stripes occur in pairs. According to the model, this results from an early switching off of diffusion. If in an activator-inhibitor scheme (Eq. 1) the basic inhibitor production σ_b is high enough, the system is bistable. In an initially broad stripe, after switching diffusion off, only in the two marginal zones the pigment reaction remain in the ON-state. In the centre, the high inhibitor concentration causes there the transition into the OFF-state too. Note that there is a second pattern with a more uniform spacing on the inner side of the shell. This pattern is definitely not visible as long as the mollusk is alive. This supports the view that the pigmentation patterns have no signalling function.

[Meinhardt, 1984]. A model based on similar interactions but emphasizing the role of the nervous system was proposed by Ermentrout, Campbell and Oster [1986; see also Murray 1989].

Together with Martin Klingler I have shown that the same types of interactions we have proposed for pattern formation during embryonic development are also able to account for shell patterning [Meinhardt and Klingler, 1987]. Very different appearing shell patterns can be reproduced by small variation of the parameters or minor changes in the underlying mechanism. However, a large class of shell patterns was not included in the earlier work: patterns that depend on the mutual interaction of at least two time-dependent patterns. An extensive treatment of models for the elementary and the more complex patterns is provided in a recent book [Meinhardt, 1998]. The book is accompanied by a floppy disk that allows a reproduction of the simulations on a PC and includes the source code. It contains also an integration of these models into a three-dimensional description of the shell shape by Fowler and Prusinkiewicz.

8.1 *Traces of a stable pattern, oscillations and travelling waves: lines parallel, perpendicular or oblique to the direction of growth*

Basic elements of the shell patterns are lines of different orientations. Keeping their space-time character in mind, lines parallel to the direction of growth (i.e., usually perpendicular to the growing edge) indicate that pigment production occurs at particular positions that are separated by regions without pigment production

(Figure 7). This is the usual situation as discussed above for normal development. The formation of a particular structure is restricted to a particular position. As shown above, it requires that the self-enhancing reaction is more or less locally confined while the antagonistic effect has a long range. In contrast, pigmented lines perpendicular to the direction of growth (usually parallel to the growing edge) are traces of a more or less synchronous oscillation in pigment production. In terms of the model, if the antagonistic reaction follows the self-enhancing reaction too slowly, the activation increases in an avalanche-like manner. It is only somewhat later that the accumulating inhibitor (or the removal of all the substrate) causes a collapse in the activator production. A refractory phase follows. A new activation can be triggered only after the decay of the remaining inhibitor. In other words, a longer time constant of the antagonistic reaction can lead to oscillations. A rapid diffusion of either the activator or the inhibitor can lead to their synchronization. Lines oblique to the growing edge result from travelling waves of pigment production. In the model, they result if the activator has a small diffusion range while the antagonistic substance is nearly non-diffusive. An activated region can "infect" its neighboring region such that, after a certain lag phase, it becomes fully activated too. Such a cell will infect a neighboring region, and so on. The situation is very similar to the wave-like spread of an epidemic. One person can infect its neighbors. The full development of a sickness is also based on a self-enhancing effect, the replication of the virus. But after some time the immune system acts antagonistically. It captures the virus and the person will become healthy again. For the spread of the epidemic it is crucial that only the virus, but not the immune response is transmitted from one individual to the next.

9 Travelling waves with unusual properties

Modelling of travelling waves is in a very advanced state, especially due to the study of the Belusov-Zhabotinsky reaction [see Zhabotinsky, 1991, Tyson, 1979, Murray, 1989]. However, the travelling waves as seen on the shells display features that are otherwise rarely observed in waves in excitable media. As shown below, these wave may penetrate each other without annihilation or they can split and form additional waves that run in a backwards direction. These features make the shell patterns an interesting object of study.

9.1 Formation of branches: the sudden formation of backwards-running waves

Characteristic for many shell patterns with oblique lines are V-like pattern elements (Figure 8). At a the point of collision all neighboring cells are in a refractory period. Therefore, non of the waves can propagate further. Both waves terminate. This is a standard behavior for travelling waves.

The oblique lines of many shell patterns show branches. The shell of *Oliva porphyria* (Figure 8) provides an example. Keeping in mind the space-time character of the pattern, a branching is a record of the sudden formation of a backwards-running wave. This feature is very unusual for waves in excitable media since the region into which the backwards wave should run is expected to be in the refractory phase.

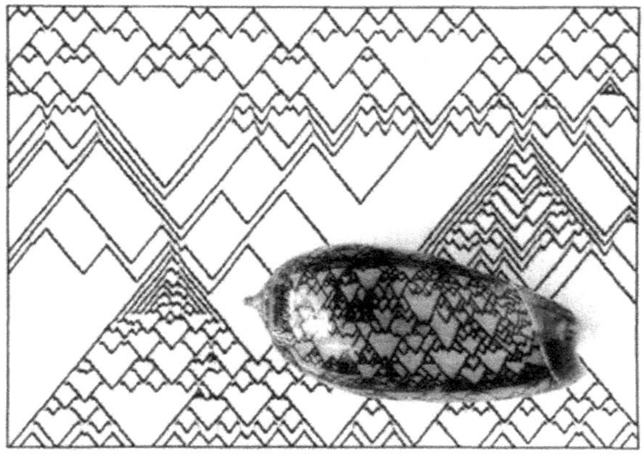

Figure 8. Branching by global control. A shell of *Oliva porphyria* in front of a computer simulation. A branch of an oblique line indicates the sudden formation of a backwards wave. According to the model, branching occurs whenever the number of travelling waves drops below a threshold value. The controlling agent is a hormone-like substance that stabilizes the inhibitor. If too few waves are left, groups of cells remain activated until backwards waves are initiated. Since the backwards waves start with almost the same speed as the original wave, the refractory period must be very short in this species.

The initiation of such a backwards wave requires a modification of the normal serial trigger mechanism that leads to the oblique lines. From a particular moment onwards and for a short time interval, the activated cells must remain activated until the refractory period of the neighboring, still oscillating cells is over. Then, this cell can be re-infected and the initiation of the backwards wave is completed. This pattern is possible only if the refractory period is very short.

What can be the signal to initiate branching? As mentioned, as the rule, travelling waves annihilate each other during a collision. Thus, in the course of time, more and more waves would disappear. Branching is a possible mechanism to maintain the number of travelling waves. In the simulation Figure 8 this is achieved by occasional shifts of the cells from a bursting mode into a longer lasting activation. This is achieved by a hormone-like substance that is produced by all activated cells. It is homogeneously distributed in the organism, for instance by the lymph or blood system. The hormone concentration is thus a measure for the fraction of activated cells. It is assumed to have a stabilizing influence on the inhibitor. Whenever the number of activated cells becomes too small, the inhibitor life time becomes so short that the just activated cells obtain a steady state mode of activation instead of producing a short burst. The just activated cells remain activated. When the refractory period of neighboring (still oscillating) cells is over, these cells will become re-activated. In this way backwards waves are initiated, causing an increase of the number of travelling waves and thus an increase in the hormone concentration. All cells return to the oscillating mode and the formation of branches is completed.

The simulation reproduces many details of the natural pattern. Branches are simultaneously initiated at many positions. In the model, this results from the global control due to the homogeneous distribution of the hormone. In the final pattern, this leads to the V -like elements with the same distance between the tip and the points of branch initiation. While the original wave goes on in an unperturbed way, a branch appears frequently only loosely connected to the original line. In the simulation, this effect results from the overshoot before obtaining the temporary steady state activation. Sometimes, a little hook appears close to the point of branch initiation. This results from an incipient initiation of a wave parallel to the primary wave. Usually this wave does not survive since the medium is not yet excitable enough. If this wave survives, however, a line close to the original line is formed in addition to the branch. Since the total number of travelling waves is maintained, large gaps and dense pigmentation may appear simultaneously at different positions along the growing edge.

The to enable backwards running wave that have a constant speed almost from the point of wave splitting, the refractory period must be very short. Moreover, since the oblique lines are straight, is must be assumed that a the speed of the waves do not increase with increasing time after the last activation. This can be achieved by adding in the inhibitory therm $1/h$ of Eq. (1a) a Michaelis-Menten term ϵ in the form of $1/(\epsilon+h)$. Then, the nominator in (1a) remains finite although the inhibitor concentration drops to very low values.

9.2 Penetration of waves without annihilation

The shell of *Tapes literatus* (Figure 9) shows oblique lines that regularly cross each other. Crossings are records of waves that do not annihilate upon a collision, but

Figure 9. Crossings of pigmentation lines on *Tapes literatus*. Crossings suggest that two waves do not annihilate but penetrate each other upon a collision. Model: assumed is a three-component system; an activator-substrate mechanism is tuned to produce a steady state activation. Travelling waves result from an additional diffusing inhibitor. Each newly activated cells extinguishes the activation in its predecessor. At the point of collision, no new neighbor is available which could accomplish this extinguishing. The cells remain in the activated until two backwards waves are initiated. The global perturbation on the shell is reproduced by the assumption of a temporary decrease of the activator concentration. Calculated with (6) and $D_a = 0.01$, $\mu_a = 0.08$, $\sigma_a = 0.0001$, $\kappa = 0.5$, $D_s = 0$, $\mu_s = 0.005$, $\sigma_s = 0.1$, $D_h = 0.4$, $\mu_h = 0.02$, $\sigma_h = 0$, $\epsilon = 1$, $\gamma = 3$

penetrate each other. This is again a very unusual behavior of waves in excitable media. Penetration can be regarded as the initiation of two new diverging waves at the point of collision. This requires that upon a collision cells remain activated for a prolonged period until the refractory period of the neighboring cells is over. We have shown that an interaction involving a second antagonist with a long range can cause such a pattern [Meinhardt and Klingler, 1987, Meinhardt, 1998]. In the simulation Figure 9, an activator-substrate mechanism has been assumed with parameters such that a cell, once activated, would remain in a steady state. An additional diffusive inhibitor is produced by the activated cells. Each newly activated cell extinguishes the activation of the preceding cell. The result are travelling waves similar to those discussed above. However, if two waves collide, the situation is very different. No newly activated cells are available to extinguish the activation of the preceding cells. Therefore, at the point of collision, cells remain activated until the neighboring cells can be re-infected again. Then, the newly activated cells extinguish via the diffusive inhibitor the activation of that cell by which the activation escaped from annihilation. Eq. (6) provides an example for a possible interaction.

$$\frac{\partial a}{\partial t} = \frac{\rho\, s}{\epsilon + \gamma\, h}\, a^{*2} - \mu_a a + D_a \frac{\partial^2 a}{\partial x^2} \tag{6a}$$

$$\text{with} \quad a^{*2} = \frac{a^2 + \sigma_a}{1 + \kappa a^2}$$

$$\frac{\partial s}{\partial t} = \sigma_s - \frac{\rho\, s}{\epsilon + \gamma\, h}\, a^{*2} - \mu_s s + D_s \frac{\partial^2 s}{\partial x^2} \tag{6b}$$

$$\frac{\partial h}{\partial t} = \mu_h a - \mu_h h + D_h \frac{\partial^2 h}{\partial x^2} \tag{6c}$$

If $\epsilon > 0$ the additional inhibitor h plays a role only at high h concentrations, while at low h concentration the system behaves the same as a standard activator-substrate system; a^{*2} subsumes activator production by autocatalysis, the baseline activator production σ_a and its limitation by saturation. In this interaction, the activator concentration (but not substrate concentration) is to a large extend independent of h, since a reduction of h also leads to a compensating decrease in substrate removal.

The specimen shown in Figure 9 shows an interesting perturbation. Many lines terminate at a particular growth line. Most remarkable, other lines bifurcate at same instance. This perturbation is a record of a simultaneous occurrence of wave termination and wave doubling, presumably caused by an external event such as dryness or lack of food. This phenomenon has a straightforward explanation in terms of the model and is reproduced in the simulation shown in Figure 9. A sudden lowering of the activator concentration cannot only cause termination of a line. The reduced activator concentration causes also a decrease of the diffusible inhibitor. This, in turn, stabilizes the activator. Thus, the situation during the perturbation is very similar as during a crossing. Two new diverging lines can emerge, in agreement with the natural pattern.

Penetration of waves and wave splitting has been also observed in a catalytic reaction on platinum surfaces [Bär et al., 1992]. This phenomenon has been interpreted to result from small imperfect regions on the crystal. The simulations provided above suggest a different possibility: the involvement of a second, long ranging antagonist with a short time constant.

9.3 Not all together now

There is a large body of literature stating that coupled non-linear oscillators have a strong tendency to synchronize each other [see Stewart, 1991 for an overview, Hopfield and Herz, 1995]. An example is the synchronization of the pulsing light emission by fireflies. Synchronization happens in the elementary pattern forming systems discussed above when either the activator or antagonist diffuses. However, the patterns on many shells show clear evidence for oscillation with opposite phases in neighboring regions. An example is the pattern reminiscent to a chess board on in *Bankivia fasciata* (Figure 10). In the following it should be shown that the second and long-ranging antagonist introduced above enforces a such a de-synchronization.

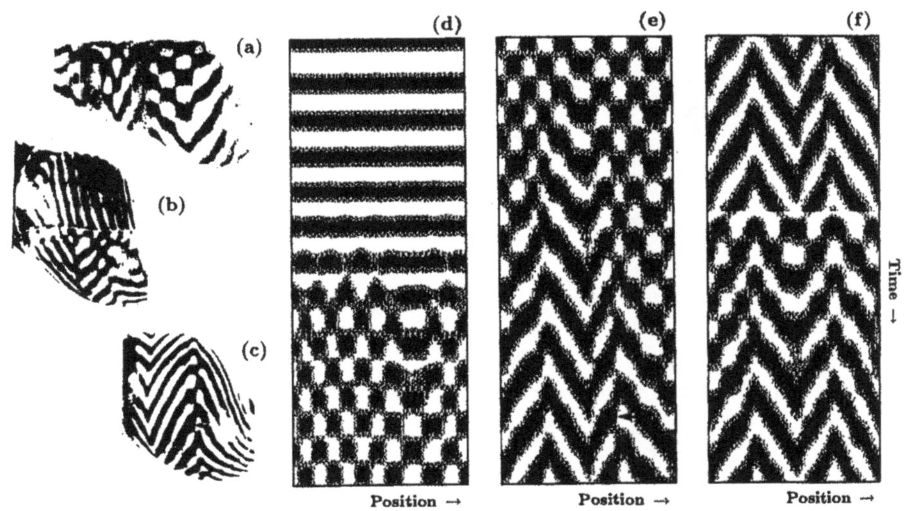

Figure 10. De-synchronization of oscillations and formation of travelling waves. (a-c) The chess-board pattern on *Bankivia fasciata* results from oscillations out of phase in neighboring regions. It can easily change into oblique lines (travelling waves). (d-f) Model: saturation of the activator autocatalysis leads to activation periods of the same length as the intervals in between. The highly diffusible antagonist (in this case, the substrate s, Eq. 6) causes the phase shift between parts of the resulting thick lines. Low inhibitor diffusion maintains the maximum phase difference and the coherence of the patches. A slight increase in substrate production (e) leads to the tendency to form oblique lines. (f) A global perturbation (drop of activator concentration to 20%) may bring back pattern elements reminiscent of the chess-board pattern. Initiation and termination of travelling waves may occur out of phase due to the long ranging antagonist, in agreement with the natural pattern (arrows) (Photographs kindly supplied by J. Campbell [see also Ermentrout et al, 1986]. Calculated with (6) and $D_a = 0.0015$, $\mu_a = 0.02$, $\sigma_a = 0.003$, $\kappa = 3$, $D_s = 0.4$, $\mu_s = 0$, $\sigma_s = 0.003$ and 0.004 (e), $D_h = 0.002$, $\mu_h = 0.004$, $\sigma_h = 0$, $\epsilon = 0$, $\gamma = 1$

Imagine a system that is able to generate travelling waves (non-diffusible antagonist). As long as no faster oscillating pace-maker region is present, all cells may oscillate in synchrony. However, a second long ranging antagonist enforces a de-synchronization. If some cells becomes activated somewhat later than their neighbors, this phase difference will increase during subsequent oscillations. The inhibitory influence that spreads from the advanced cells delays the retarded neighbors even more. As shown in Figure 10, minute difference in the parameter can lead either to travelling waves (i.e., to oblique lines) or to oscillations in cell groups with opposite phases (chess board). The common basis of both these patterns is easy to understand. The long ranging antagonist has the tendency to generate pattern in space, the short ranging antagonist (with a long time constant) leads to alternating pigmentation in time. Superimposed, both antagonist enforce an alternation of pigmentation along the space and along the time coordinate. Both the chess board and the oblique line pattern are based on a regular alternation of the pigmentation pattern along the space and the along the time coordinate.

An analogous phenomenon has been observed by Middya et al. [1993] with temperature patterns on catalytic ribbons. Due to the reaction, the temperature of an electrically heated wire increases further which has a feedback on the actual reaction rate. With a thermo-sensitive camera, the temperature has been monitored as function of time. The observed space-time plots show travelling waves and oscillations with opposite phases in neighboring regions that resembles closely the shell pattern shown in Figure 10.

9.4 Travelling waves without pace-maker

The generation of travelling waves usually requires a pace-maker, a group of cells that oscillates somewhat faster then the rest. The faster oscillating cells can trigger their neighbors before they would trigger spontaneously. If the position of the pace-maker is given, the resulting patterns are very reproducible since the waves have a defined origin and direction. In contrast, travelling waves resulting from the enforced de-synchronization by a second antagonist shows a much higher degree of variability, accounting for the diversity of pattern observed even within the same species. It should be noted that Turing [1952] discovered already that the generation of travelling waves in a homogeneous field of cells requires the interaction of three substances.

10 Complex patterns result from the superposition of two patterns

Many shell patterns exist that are far more complex than that discussed so far. The shell of *Palmadusta diluculum* Figure 11 provides an example where a stable pattern exists that controls the parameter of the oscillation. The controlling pattern can only be deduced from the position-dependent behavior of the pigmentation reaction. In some regions, broad stripes nearly parallel to the growing edge exists. In other regions, fine but very steep stripes are formed. According to the model, in the region of steep stripes, the oscillation frequency is a bit slower. Therefore, the faster oscillating cells in the remaining area act as pace-maker. Thick lines are

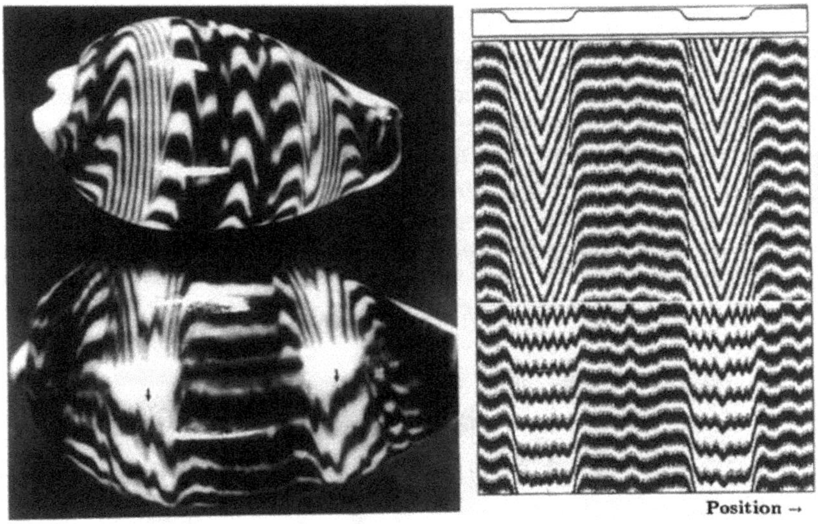

Figure 11. Superposition of stable and periodic patterns. (a) Shells of *Palmadusta diluculum*. (b) Model: a stable pattern exists that influences the parameter of the oscillating reaction (production rate of substrate σ_s, upper curve). Due to the lower substrate supply, these cells have a lower oscillation frequency and are entrained from the more rapidly oscillating cells (thick broad bands). The steep lines result from a large phase shift between neighboring cells. They are an indication for a small triggering influence between neighboring cells. After the perturbation documented in the lower shell, the slower oscillating cells trigger spontaneously (arrows). In agreement, after a temporary reduction of the activator concentration (arrow heads) the steep lines are gone. It requires time to restore the large phase difference.

formed if, due to a saturation of the activator production, an upper border of the activator concentration exists. This elongates the activation since at the reduced level the available substrate can fuel the autocatalysis for a longer period. The steep lines indicate that the infection of one cell by its neighbor requires a long fraction of the cycle time. If the period of activation is very long, the time window in which the infection can take place is long too. A very low exchange of activator with its neighbors is already sufficient to form a wave. Its speed will be slow and the resulting lines are steep. Therefore, the formation of the thick parallel lines on the one hand and thin steep lines on the other is intimately connected.

The involved of a second invisible pattern that modifies the parameter of the pigmentation reaction suggest the basis for the more complex shell patterning. It requires a change of only a single time constant and the reaction modifying the pigmentation system will become time-dependent too. However, deciphering such a pattern is very difficult. We don't know whether the invisible pattern has an enhancing or an inhibitory influence, whether the production-, decay- or diffusion rates are modified. Further we don't know what is the normal and what is the modified phase of the pattern. By a careful inspection of different patterns I have tried to deduce the mode of interference between the two pattern forming systems [Meinhardt, 1998]. Some examples will be given below.

10.1 Superposition of two time-dependent patterns

The pattern on the shell of *Conus marmoreus* (Figure 12) is characterized by a pigmented background with drop-like unpigmented regions. The long straight black-to-white border almost perpendicular to the direction of growth at the beginnings of these white drops indicates a nearly simultaneous cessation of the activation in large regions. Thus, switching pigment production on or off must be under control of two different processes. The corresponding simulation has been made under the following assumptions. The pigmentation reaction is tuned to obtain a stable steady state. Due to the activator diffusion, a small activated region will enlarge. A second reaction extinguishes the first whenever the pigment production was sufficiently

Figure 12. Complex patterning by two superimposed patterning reactions. (top) Shell of *Conus marmoreus*. In the simulation a bistable pigmentation system is assumed that becomes extinguished by a second reaction (red) whenever the pigment reaction was active for a certain period. Pigmentation survives at the margins and starts spreading again. A stronger extinguishing reaction (bottom) leads to a preferential termination at those marginal regions that was recently active. The white triangle pattern shows transitions into chains of connected black triangles.

long in the ON-state. The second reaction produces a short pulse-like activation (red in the simulation) that shortens the half life of the pigment activator. The substrate required for the extinguishing reaction is produced only when pigment synthesized, i.e., when the first reaction is in its activated state. After pigment has been produced for a certain time interval, the burst-like extinguishing reaction is triggered. Since the components of the extinguishing reaction spread rapidly by diffusion, this burst-like activation occurs almost simultaneously in a large region. In the real pattern this reaction is only indirectly visible by the sudden black-white transition at the beginning of a white drop. The pigmentation will survive at the margins of the spreading activation since there the time for the accumulation of the substrate of the extinguishing reaction was short. This has the consequence that the extinguishing reaction is less strong there. Thus, the pattern of white drops on a black background results from a permanent change between a slow spread of pigmentation, a large-scale breakdown, survival of the pigment reaction at the margin, and so on.

Other patterns indicate the opposite influence. On *Conus pennaceus* (Figure 13) many branches are formed. As mentioned above, branching requires a local temporal transition from a bursting into a steady state mode. Therefore, the second pattern may, for instance, cause an increased substrate production or an elongation of the activator half life. The model obtains support from the fact that it can account for very different appearing patterns. *Conus textile* (Figure 13) displays dark parallel lines on a brownish background. In other regions, oblique branching lines appear on a white background. According to the model, in the region where the pigmentation reaction is in a steady state (brownish background), the oscillating enhancing reaction leads to the dark stripes. In contrast, if due to a somewhat reduced substrate supply only travelling waves are possible, the same periodic enhancing reaction keeps the pigment reaction sufficiently long active that backwards-running waves are initiated. Thus, according to the model, branch formation and the thick lines have ultimately a common basis.

10.2 Open problems in shell patterning

Although many shell patterns have been successfully simulated [Meinhardt, 1998], there are still many open problems. In the last examples given above (Figure 12 and 13), it may appear unsatisfactory to have two opposite actions of the modifying pattern, an enhancing and an extinguishing of the pigmentation reaction. A close inspection of many shell pattern show that this is a constant theme. On the upper shell shown in Figure 13 the fine white spots are located along darker lines, suggesting that it is the enhancement of the pigmentation that causes also its breakdown. Also in *Conus textile* (lower shell in Figure 13), the bands with background pigmentation and dark parallel stripes terminate after a certain time. This suggest that there is in addition to the short-term enhancing effect a long-term destabilization process superimposed. The crucial point for the modelling is that the destabilization is occurs suddenly. It is not announced by any visible effect such as a slowly progressing poisoning of the system. This correctly reproduced by the extinguishing reaction (Figure 12), but so far, I was not able to integrate

Figure 13. Complex patterning by an enhancing reaction: a second reaction (green in the upper part of the figures) is assumed to elongate pigment production. Either branches as in *Conus pennaceus* (top) or dark lines on a brownish background combined with oblique lines on an unpigmented background as in *Conus textile* are possible (from Meinhardt, 1998).

enhancement and destabilization in a common scheme that reproduces all these patterns in a satisfactory way. Starting from known properties of the interactions we have proposed for pattern formation in general, the interaction used for the simulation of shell patterns have been deduced from observed complex behavior of the pigment reaction. Alternative approaches using more formal methods to deduce the appropriate interactions from the visible pattern as well as the study of their chaotic nature would be highly desirable. Many shell patterns still wait for an explanation. I hope, that this work will trigger a more widespread interest in these natural pieces of jewellery.

Acknowledgments

I wish to express my sincere thanks to Prof. Alfred Gierer. Much of the basic work described in this review emerged from a fruitful collaboration over many years.

References

1. Astrov, Y. U., Ammelt, S., Teperick, S. and Purvin, H. G. Hexagon and stripe Turing structures in a gas discharge system. *Phys. Let. A* **211**, 184–190 (1995).
2. Bard, J. B. and Lauder, I. How well does Turings theory of morphogenesis work. *J. theor. Biol.* **45**, 501–531, (1974).
3. Berkemeier, J., Dirksmeyer, T., Klempt, G. and Purwins, H. G. Pattern formation on a non-linear periodic electrical network. *Z. Phys. B — Condensed Matter* **65**, 255–258 (1986).
4. Bode, P. M., Awad, T. A., Koizumi, O., Nakashima, Y., Grimmelikhuijzen, C. J. P. and Bode, H. R. Development of the two-part pattern during regeneration of the head in hydra. *Development* **102**, 223–235 (1988).
5. Borckmans, P., Jensen, O., Pannbacker, V. O., Moosekilde, E., Devel, G. and De Wit, A. Localized Turing and Turing-Hopf patterns in *Pattern and Complex Behavior in Living Systems*, (Eds. E. Mosekilde, O. G. Mouritsen) Synergetics Series, Springer Berlin, pp 47–68 (1994).
6. Castets, V., Dulos, E., Boissonade, J. and De Kepper, P. Experimental evidence of a sustained standing Turing-type nonequilibrium chemical pattern. *Phys. Rev. Lett.* **64**, 2953–2956 (1990).
7. Chua, O. L., Hasler, M., Moschytz, G. S. and Neirynck, J. Autonomous cellular neural networks: A unified paradigm for pattern formation and active wave propagation. *IEEE Tansactions on Circuits and Systems* **42**, 559–577, (1995).
8. DeRobertis, E. M. Dismanteling the organizer. *Nature* **374**, 407–408, (1995).
9. Driever, W. and Nüsslein-Volhard, C. A gradient of *bicoid* protein in *Drosophila* embryos. *Cell,* **54**, 83–93, (1988).
10. Driever, W. and Nüsslein-Volhard, C. The *bicoid* protein is a positive regulator of *hunchback* transcription in the early *Drosophila* embryo. *Nature* **337**, 138–143 (1989).
11. Ermentrout, B., Campbell, J. and Oster, G. A model for shell patterns based on neural activity. *The Veliger* **28**, 369–338 (1986).
12. Gierer, A. Generation of biological patterns and form: Some physical, mathematical, and logical aspects. *Prog. Biophys. molec. Biol.* **37**, 1–47 (1981).
13. Gierer, A. and Meinhardt, H. A theory of biological pattern formation. *Kybernetik* **12**, 30–39 (1972).
14. Golembo, M., Schweitzer, R., Freeman, M. and Shilo, B. Z. *argos* transcription is induced by the *Drosophila* EGF receptor pathway to form an inhibitory feedback loop. *Development* **122**, 223–230 (1996).
15. Gurdon, J. B., Mitchell, A. and Mahony, D. Direct and continuous assessment by cells of their position in a morphogen gradient. *Nature* **376**, 520–521 (1995).
16. Harrison, L. G. *Kinetic theory of living pattern.* Cambridge Univerity Press, Cambridge, 1993.

17. Harrison, R. G. Experiments on the development of the fore-limb of Amblystoma, a self-differentiating equipotential system. *J. exp. Zool.* **25**, 413–46 (1918).
18. Herman, G. T. and Liu, W. H. The dauther of Celia, the frensch flag and the firing quad: Progress report on a cellular linear iterative-array simulator. *Simulation* **21**, 33–41 (1973).
19. Hodges, A. *Alan Turing: the enigma.* Simon and Schuster, New York, 1983.
20. Hopfield, J. J. and Herz, A. V. M. Rapid local synchronization of action potentials: Toward computation with coupled integrate-and-fire neurons. *Proc. Natl. Acad. Sci. USA,* **92**, 6655–6662 (1995).
21. Koch, A. J. and Meinhardt, H. Biological pattern formation: from basic mechanisms to complex structures. *Rev. Mod. Phys.* **66**, 1481–1507 (1994).
22. Kusch, I. and Markus, M. Mollusc shell pigmentation: Cellular automaton simulations and evidence for undecidability. *J. theor. Biol.* **178**, 333–340 (1996).
23. Lacalli, T. C. Dissipative structures and morphogenetic pattern in unicellular algae. *Phil. Trans. R. Soc. Lond. B* **294**, 547–588 (1981).
24. Lacalli, T. C. and Harrison, L. G. The regulatory capacity of Turing's model for morphogenesis, with application to slime moulds. *J. theor. Biol.* **70**, 273–295 (1978).
25. Lefever, R. Dissipative structures in chemical systems. *J. Chem. Phys.* **49**, 4977–4978 (1968).
26. Lefever, R., Nicolis, G., Borckmans, P. The Brusselator: It does oscillate all the same *J. Chem. Soc., Farady Trans.* **84**, 1013–1023 (1988).
27. Leptin, M. *Twist* and *snail* as positive and negative regulators during *Drosophila* mesoderm development. *Genes Development* **5**, 1568–1576 (1991).
28. Lindsday, D. T. A new programmatic basis for shell pigment patterns in the bivalue mollusc *Lioconcha castrensis* L. *Differentiation* **21**, 32–36 (1982).
29. Martin, G. R. Why thumbs are up. *Nature* **374**, 410–411 (1995).
30. Meinhardt, H. Morphogenesis of lines and nets. *Differentiation* **6**, 117–123 (1976).
31. Meinhardt, H. Space-dependent cell determination under the control of a morphogen gradient. *J. theor. Biol.* **74**, 307–321 (1978).
32. Meinhardt, H. Models of biological pattern formation. *Academic Press*, London, 1982.
33. Meinhardt, H. A boundary model for pattern formation in vertebrate limbs. *J. Embryol exp. Morphol.* **76**, 115–137 (1983a).
34. Meinhardt, H. Cell determination boundaries as organizing regions for secondary embryonic fields. *Devl. Biol* **96**, 375–385, (1983b).
35. Meinhardt, H. Models for positional signalling, the threefold subdivision of segments and the pigmentation pattern of molluscs. *J. Embryol. exp. Morph.* **83**, (Supplement) 289–311, (1984).
36. Meinhardt, H. Models for positional signalling with application to the dorsoventral patterning of insects and segregation into different cell types. *Development* (Supplement), 169–180, (1989).

37. Meinhardt, H. Determination borders as organizing regions in the generation of secondary embryonic fields: the initiation of legs and wings. *Seminars Devel. Biol.* **2**, 129–138 (1991).
38. Meinhardt, H. A model for pattern-formation of hypostome, tentacles, and foot in hydra: how to form structures close to each other, how to form them at a distance. *Dev. Biol.* **157**, 321–333 (1993).
39. Meinhardt, H. Biological pattern-formation — new observations provide support for theoretical predictions. *Bioessays* **16**, 627–632 (1994).
40. Meinhardt, H. The Algorithmic Beauty of Sea Shells. *Springer, Heidelberg, New York* (2nd edition, with PC-software), 1998.
41. Meinhardt, H. and Gierer, A. Applications of a theory of biological pattern formation based on lateral inhibition. *J. Cell Sci.* **15**, 321–346 (1974).
42. Meinhardt, H. and Gierer, A. Generation and regeneration of sequences of structures during morphogenesis. *J. theor. Biol.* **85**, 429–450 (1980).
43. Meinhardt, H. and Klingler, M. A model for pattern formation on the shells of molluscs. *J. theor. Biol* **126**, 63–69 (1987).
44. Middya, U., Sheintuch, M., Graham, M. D., Luss, D. Patterns of temperature pulses on electrically heated catalytic ribbons *Physica D* **63**, 393–400 (1993).
45. Moos, M., Wang, S. W. and Krinks, M. Anti-dorsalizing morphogenetic protein is a novel TGF-beta homolog expressed in the spemann organizer. *Development* **121**, 4293–4301 (1995).
46. Morgan, T. H. An attempt to analyse the phenomena of polarity in tubularia. *J. exp. Zool.* **1**, 587–591 (1904).
47. Murray, J. D. Mathematical biology. *Springer*, Heidelberg, New York, 1989.
48. Ouyang, Q. and Swinney, H. L. Transition from a uniform state to hexagonal and striped Turing patterns. *Nature* **352**, 610–612 (1991).
49. Plath, P. and Schwietering, J. Improbable events in deterministically growing patterns. In: *Fractal Geometry and Computer Graphics* (J. L. Encarnucao, H. O. Peitgen, Skas, G. Englert, Eds. pp. 162–172 (1992).
50. Prigogine, I. and Lefever, R. Symmetry breaking instabilities in dissipative systems. *J. chem. Phys.* **48**, 1695–1700 (1968).
51. Regulski, M., Dessain, S., Mcginnis, N. and Mcginnis, W. High-affinity binding-sites for the *deformed* protein are required for the function of an autoregulatory enhancer of the deformed gene. *Genes Dev.* **5**, 278–286 (1991).
52. Sander, K. Formation of the basic body pattern in insect embryogenesis. *Adv. Insect Physiol.* **12**, 125–238 (1976).
53. Segel, L. A. and Jackson, J. L. Dissipative structure: an explanation and an ecological example. *J. theor. Biol.* **37**, 545–549 (1972).
54. Simpson-Brose, M., Treisman, J. and Desplan, C. Synergy between the *hunchback* and *bicoid* morphogens is required for anterior patterning in *Drosophila. Cell* **78**, 855–865 (1994).
55. Spemann, H. and Mangold, H. Über Induktion von Embryonalanlagen durch Implantation artfremder Organisatoren. *Wilhelm Roux' Arch. Entw. mech. Org.* **100**, 599–638 (1924).
56. Stewart, I. All together now. *Nature* **350**, 557–557 (1991).

57. Technau, U. and Holstein, T. W. Head formation in hydra is different at apical and basal levels. *Development* **121**, 1273–1282 (1995).
58. Tickle, C., Summerbell, D. and Wolpert, L. Positional signalling and specification of digits in chick limb morphogenesis. *Nature* **254**, 199–202 (1975).
59. Turing, A. The chemical basis of morphogenesis. *Phil. Trans. B.* **237**, 37–72 (1952).
60. Tyson, J. J. Oscillations, bistability and echo waves in models of the Belusov-Zhabotinski reaction. *Annals New York Acad. Sciences* **316**, 279–295 (1979).
61. Udolf, G., Lüer, K., Bossing, T. and Technau, G. M. Commitment of the CNS Progenitors along the dosroventral axis of *Drosophila* neuroectoderm. *Science* **269**, 1278–1281 (1995).
62. Van Doren, M., Powel, P. A., Pasternak, D., Singson, A. and Posakony, J. W. Spatial regulation of proneural gene activity: auto- and cross-achtivation of *achaete* is antagonized by *extramacrochaetae*. *Genes Dev.* **6**, 2592–2605 (1992).
63. Vincent, J. P. and Lawrence, P. A. It takes three to distalize. *Nature* **372**, 132–133 (1994).
64. Waddington, C. H. and Cowe, R. J. Computer simulation of a mulluscan pigmentation pattern. *J. theor. Biol.* **25**, 219–225 (1969).
65. Wolfram, S. Cellular automata as models of complexity. *Nature* **341**, 419–424 (1984).
66. Wolpert, L. Positional information and the spatial pattern of cellular differentiation. *J. theor. Biol.* **25**, 1–47 (1969).
67. Zhabotinsky, A. M. A history of chemical oscillations and waves. *Chaos* **1**, 379–386 (1991).

ANDRONOV BIFURCATIONS AND SEA SHELL PATTERNS

M. ARGENTINA AND P. COULLET[1]

Institut Non-Linaire de Nice, route des Lucioles, 06560 Valbonne, France

E-mail: argentin@inln.cnrs.fr

Tropical molluscs exhibit very complex pigmentation patterns on their shells. In this paper we intend to show that some of those patterns can be understood as the instability of limit cycles in spatially extended dynamical systems.

Reaction diffusion models exhibit a wide variety of patterns. The reactions are described by first order differential equations (ODE) which rule the time evolution of the various chemical concentrations $C = (C_1, C_2, \dots, C_n)$

$$\partial_t C_i = f_i(C, \mu) \tag{1}$$

$\mu = (\mu_1, \mu_2, \dots, \mu_p)$ are p parameters which represent the temperature, the pressure, the reaction rates, etc... If one takes into account the diffusion of the chemicals, Eq. (1) becomes a set of partial differential equation (PDE)

$$\partial_t C_i = f_i(C) + D_i \nabla^2 C_i \tag{2}$$

where the diffusion coefficients D_i are positive constants. A. Turing [1], in a famous work, introduces the idea that morphogenesis could be described by solutions of reaction-diffusion equations.

Although an ODE can possess complex solutions, the Poincaré geometrical theory tell us [3], in principle, how to classify the robust dynamical behaviors. The question of the robustness of the solutions of partial differential equations is completely open, mainly because of the lack of qualitative understanding of such dynamical systems. For example, there are no mathematical evidences that the solutions of reaction diffusion equations are more special than the solutions of any other PDE. This situation contrasts with the common observation that many of the solutions of those equations describe universal phenomena as for example fronts, patterns, defects, growth,... In the 60's R. Feynman, in his famous lectures [2] on Physics wrote: "The next great area of awakening of human intellect may well produce a method of understanding the qualitative content of equations".

One way to generalize the idea of reaction-diffusion equations consists to study the continuous limit of coupled identical bifurcating elements. To each bifurcation corresponds a partial differential equation. These PDE generalize the normal forms of the bifurcation, when spatial variation are taken into account; they describe so to speak the spatial unfolding of the singular vector field.

The spatial unfolding of the pitchfork bifurcation, for example, is given by

$$\partial_t u = u - u^3 + \nabla^2 u \tag{3}$$

where $u(\vec{r}, t)$ is a real valued function of space and time and ∇^2 is the Laplacian operator. The spatial unfolding of the pitchfork bifurcation is a reaction-diffusion

[1]PROFESSEUR A L'INSTITUT UNIVERSITAIRE DE FRANCE

equation. This property is not true in general. Let us consider the bifurcation which leads to "small" limit cycles (Andronov-Hopf bifurcation). When the bifurcation is supercritical, its normal form is given by

$$\partial_t u = u - (1 + i\alpha)|u|^2 u \tag{4}$$

where $u(t)$ is a complex valued function which describes the amplitude of the oscillation, $u(t)\exp(i\omega_0 t)$, and α is a constant which characterizes the oscillator. In spatially extended systems u also varies in space. It is a solution of the PDE

$$\partial_t u = u - (1 + i\alpha)|u|^2 u + (1 + i\beta)\nabla^2 u \tag{5}$$

The parameter β is the dispersion constant, it measures the wave-vector dependance of the group velocity of the linear waves. The conservative limit of the Eq. (5) $(t \to -t, u \to \bar{u})$ is the well known nonlinear Schrödinger equation [4]

$$\partial_t u = -i\alpha|u|^2 u + i\beta\nabla^2 u \tag{6}$$

The spatial unfolding of singularities leads to a large class of PDE which have been studied in many different contexts [5]. From a general view point, the spatial unfolding of a singular vector field leads to a PDE of the form

$$\partial_t u = f(u, \vec{\nabla}) \tag{7}$$

which possesses the property that the differential equation

$$\partial_t u = f(u, 0)$$

is the normal form of the unfolding of the singular vector field. The symmetry arguments restricts the form of the coupling terms. Space is assumed to be isotropic. Normal forms are useful when their finite order truncation describes universally the behavior of a nearly singular system. The criterion used to perform the truncation both combine asymptotic and robustness arguments. Let us consider as an example the spatial unfolding of the pitchfork bifurcation. First, u, $\partial_t u, \vec{\nabla} u$ are assumed to be small. Isotropy of space implies that the leading coupling term is $\nabla^2 u$. Thus $|\vec{\nabla} u|^2 u$ is smaller than $\nabla^2 u$. The question of robustness is much more subtle, since it is related with the structural stability of solutions of partial differential equations. Let $u_T(\vec{r}, t)$ be a well identify solution of Eq. (7) truncated to a given order. What happens if higher order terms are taken into account? Does this solution change just a little? No mathematical tool exists today to answer this question. In many cases a great deal of intuition is necessary to decide what terms are relevant and what terms are not. Two important questions arise naturally, when a bifurcation occurs in spatially extended systems. The first one concerns the stability of the spatially homogeneous solutions by respect to space dependent perturbations. The second question concerns the solutions which spontaneously break the space translations. When the bifurcations spontaneously break some symmetry, the topological defects are precisely solutions of this kind. In the case of the pitchfork bifurcation, the two bifurcated homogeneous solutions $u = \pm 1$ are stable by respect to small space dependant perturbations. The pitchfork bifurcation breaks the Z_2 parity symmetry $(u \to -u)$. Kink like stationary solutions exist in one space dimension:

$$u_\pm(x) = \pm \tanh\left(\frac{\sqrt{2}}{2}x\right)$$

They are codimension one topological defects [5], whose existence is a direct consequence of the broken Z_2 symmetry. At the core of the defect solution u vanishes. In the case of the Andronov-Hopf bifurcation leading to a "small" limit cycle, the homogeneous solutions $u = \sqrt{\mu}\exp(-i\alpha t + i\phi)$ are stable by respect to small space dependant perturbations only if $(1+\alpha\beta) > 0$ [6]. The instability which occurs when $(1+\alpha\beta) < 0$ leads to very complicated spatio-temporal behaviors which are far from being understood. In the reversible limit this instability is known as the self-focusing instability [4]. The codimension two topological defects exist in two space dimensions. When they are stable they take the form of spiral waves. Moving fronts are other spatio-temporal solutions which are common to many PDE [5]. In the simple case fronts are moving interfaces which separate homogeneous stationary solutions. This fronts separate stable and unstable domains, stable and stable domains or identical domains. Fronts between identical domains are actually spatially localized solutions which are known as solitary excitable waves in the context of dissipative dynamical systems. The fronts which separate stable and unstable domains have been studied mathematically in ref. [7]. The mathematical model considered corresponds actually to the spatial unfolding of the saddle node bifurcation which reads

$$\partial_t u = \pm 1 - u^2 + u_{xx} \tag{8}$$

Using a simple change of variable, in the parameter range where the two solutions exist (case $+1$), this equation can be put in the simplest form

$$\partial_t v = v(1-v) + v_{xx} \tag{9}$$

Any positive initial condition which is different from zero on a compact set only, leads to the propagation of two contaminating fronts which connect the unstable solution $v = 0$ to the stable solution $v = 1$. Eq. (9) admits fronts with arbitrary velocity. The compactness of the initial condition leads to a strict selection mechanism. The selected front is the one which possesses the steepest profile without spatial oscillations [8]. In the presence of a small symmetry breaking term (imperfection), stationary kinks of the pitchfork bifurcation are transformed into fronts which connect two stable domains. The motion of the fronts is such that the "most stable" domain wins over the other [9]. The equation of the imperfect pitchfork bifurcation in one dimension reads

$$\partial_t u = u - u^3 + \eta + u_{xx} \tag{10}$$

where the parameter η measures the breaking of the Z_2 symmetry. For η small Eq. (10) possesses two stable homogeneous stationary solutions. Fronts which separate domains corresponding to these solutions propagate with a well define velocity. This velocity vanishes as $\eta \to 0$ where they becomes stationary kinks. The question of fronts between homogeneous solutions reduces itself to the problem of the existence of a heteroclinic connection for the dynamical system obtained by looking for solutions of the form $u(\zeta)$, where $\zeta = x - ct$. For example in the case of Eq. (15) the fronts obey to the differential equation

$$u_{\zeta\zeta} + u - u^3 + \eta + cu_\zeta = 0 \tag{11}$$

The two homogeneous solutions of Eq. (10) becomes hyperbolic fixed point of the flow associated with Eq. (11). A front appears as a heteroclinic connection between these two fixed points. The bifurcation leading to such a heteroclinic connection is a codimension one bifurcation which can be reached for a given value of the velocity parameter c. The simplest example of excitable front is found in the overdamped forced sine-Gordon equation which reads

$$\partial_t u + \sin(u) = \Omega + u_{xx} \tag{12}$$

In the absence of the forcing Ω, this equation possesses kink solutions which connects homogeneous solutions $2n\pi$ and $2(n \pm 1)\pi$. They are given by

$$u_K = 4\arctan\left(e^{\pm x}\right) + 2n\pi \,.$$

For Ω small these stationary solutions becomes propagating fronts which separate domains $u_s + 2n\pi$ and $u_s + 2(n \pm 1)\pi$, where $\sin(u_s) = \Omega$ and move with a velocity which vanishes as $\Omega \to 0$. The mathematical nature of these fronts is the same as the fronts connecting stable domains. They correspond to a heteroclinic or a homoclinic bifurcation. Nevertheless the physical interpretation of the motion is quite different. In the general case of fronts separating stable domains, the difference in stability of the two domains leads to mechanism of propagation. It is not the case for excitable fronts since there the two domains are identical or symmetrical. For the over-damped forced sine-Gordon equation, the two domains are symmetrical since the equation is invariant under the transformation $u \to u + 2p\pi$.

1 Andronov bifurcations in extended systems

In a famous work, Andronov and Pontryagin in 1937 [10] have described the codimension one bifurcations leading to the appearance of limit cycles in planar dynamical systems (see Figure 1). Thanks to the central manifold theorem [3], their results extend to arbitrary dimensions. In particular the simplest Andronov bifurcation is the well known Hopf bifurcation [12]. To some extend, this result was also known by Poincaré. The second Andronov bifurcation is the homoclinic bifurcation in which the limit cycle disappears as "it collides" with a fixed point. In the third Andronov bifurcation, the limit cycle disappears with the appearance of a pair of fixed points which appear through a saddle node bifurcation. The purpose of this paper is to discuss some spatio-temporal phenomena associated with the Andronov bifurcations when they occur in spatially extended systems [11].

Andronov-Poincaré-Hopf bifurcation in spatially extended system is described by Eq. (5). The stability of the bifurcated limit cycle depends upon the parameter $D_\phi = 1+\alpha\beta$. A general perturbation of the limit cycle includes phase and amplitude perturbations. For homogeneous perturbations the amplitude perturbations have finite relaxation time, while the phase perturbations are marginal. For weakly homogenous perturbation the phase obeys to a diffusion equation given by [13]:

$$\partial_t \phi = D_\phi \phi_{xx} \tag{13}$$

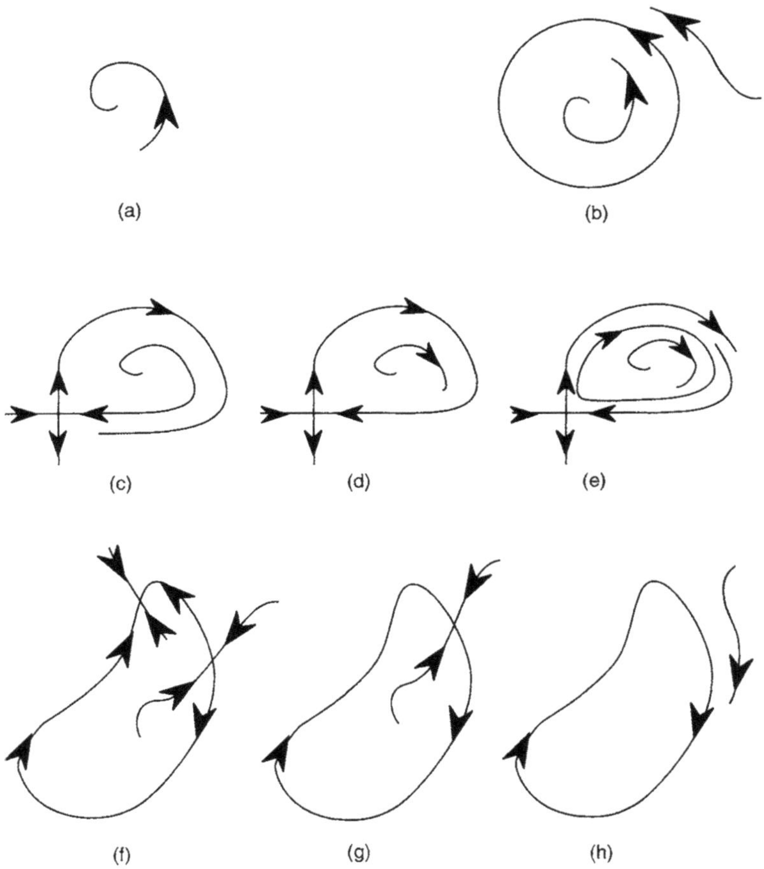

Figure 1. Sketch of the three Andronov bifurcations

For positive D_ϕ the limit cycle is stable by respect to non-homogeneous perturbation and develops a "phase instability" for negative D_ϕ. This instability leads to a "frequency modulation" which eventually couples with the amplitude perturbations leading to a complex amplitude-phase dynamics which involves both phase singularities and large localized amplitude excursion [15]. This complicate state is far from being understood. It changes from "phase turbulence" to "amplitude turbulence" as the parameters α and β are varied [14]. A typical $x - t$ diagram of Eq. (5) are shown on Figure 2.

The spatio-temporal phenomena associated to the two other Andronov bifurcations are more difficult to study since the global bifurcations cannot be described by normal forms. Nevertheless higher codimension local bifurcations contain the nonlocal bifurcations in their unfolding. Such normal forms thus appear as simple models to study nonlocal bifurcation. For example the unfolding of the codimension two Bogdanov-Takens singularity contains naturally the Andronov homoclinic

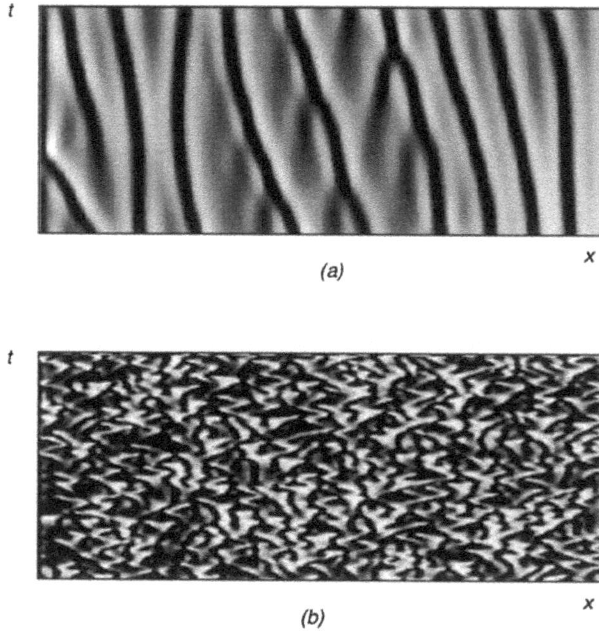

Figure 2. Spatiotemporal diagrams of the modulus of the complex amplitude of Eq. (5). (a) Phase turbulence ($\alpha = 1$, $\beta = -1.5$. $0.9 < |A| < 1.2$) (b) Defect turbulence, ($\alpha = 2$, $\beta = -1.5$. $0 < |A| < 1.5$.)

bifurcation. It corresponds to the nonlinear unfolding of a linear vector field with a non semi-simple double zero eigenvalue. Its normal form is given by:

$$u_{tt} + (\mu + \alpha u)u_t + \nu - u^2 = u_{xx} + \kappa u_{xxt}. \tag{14}$$

In the parameter regime where the stationary solutions exist ($\nu > 0$), this equation can be reduced to the complex Ginzburg Landau equation:

$$\partial_T A = -\mu_2 A - (1 + i\alpha')|A|^2 A + (1 + i\beta')\nabla^2 A \tag{15}$$

where $\alpha' = (10/(3\alpha w) + \alpha w/3)$, and $\beta' = -1/w/\kappa$. The complex number A measures the complex amplitude of the oscillation: $u = -\sqrt{\nu} + \epsilon\sqrt{\alpha}/w(Ae^{iwt} + c.c.) + \epsilon^2\alpha/(A^2Be^{iwt} + c.c. + C|A|^2)$, with $w^2 = 2\sqrt{\nu}$, $B = (-1 + i\alpha w)/(3w^2)$, and $C = 2/w^2$. The small parameter ϵ is defined as $\mu = \alpha\sqrt{\nu} + \epsilon^2\mu_2$. The bifurcation is supercritical if $\alpha > 0$. The homogeneous oscillation is stable when:

$$D_\phi = 1 - \frac{\alpha}{3\kappa}\left(1 + \frac{5}{\alpha^2\sqrt{\nu}}\right) > 0\,.$$

This criterion of stability is valid only for limit cycle close to their onset. We notice, that for α large enough or ν small enough, the limit cycle is always unstable by respect to inhomogeneous phase perturbation. In fact, we have conjectured, that close to the homoclinic bifurcation, this limit cycle will be always unstable. This conjecture have been recently proven [16].

A very interesting spatio-temporal behavior is naturally associated with the spatial unfolding of the Andronov saddle-node bifurcation. Close to the bifurcation, in the parameter regime where stationary solution exists, the corresponding dynamical system exhibits "excitable" properties. Although the stable stationary solution is a global attractor (at least in the simple models of the Andronov bifurcation), small but finite perturbations above a given threshold can lead to very large amplitude excursion in the phase portrait. The stable manifold of the unstable fixed point acts as a separatrix for the initial conditions which converge in a simple manner to the stable fixed point and the initial conditions which converge towards the stable fixed point with a large amplitude transient. For obvious reasons this large amplitude transient is called the "excitation". When the bifurcation occurs in spatially extended systems, the excitation can propagate under the form of excitable waves [17].

2 Sea shell patterns

Pigmentation patterns observed on the shells of tropical molluscs exhibit very interesting features. A great deal of modelisation was achieved by H. Meinhardt using reaction diffusion equations [18]. These equations model various chemical reactions involved in the patterning formation processes. Our approach to this problem is somewhat different. We intend to give some mathematical hints about the robustness of those patterns, using very naive non-chemical models. The seashells of molluscs are a conical shell and are composed of calcified material. The growth of the shell is made sequently by adding a new whorl to the shell. Before to add the whorl, the molluscs makes its mantle "taste " the growing edge of the shell and give the pigmentation P to the added accretion material by pigmentation cells placed at the extremity of the mantle. From this point of view the seashells just appears as a spatiotemporal diagram. In the continuous approximation $P = P(x,t)$, the first attempt to model the dynamics of the pigmentation consists in writing the following equation for the pigmentation:

$$\partial_t P = f(P) + DP_{xx} \tag{16}$$

where D measures the interaction between cells. The function f can be always written as $f = -\partial V/\partial P$. A further simplification consists to assume the existence of two stable equilibria, noted A (pigmented cell) and C (non-pigmented cell or albinos state). They corresponds to different spatially homogeneous pigmentation patterns. A simple analytic expression for V may be given by

$$V(P) = \frac{1}{2}P^2 + \frac{\mu}{3}P^3 + \frac{1}{4}P^4$$

where μ is a positive parameter. Depending on the values of this parameter the dynamical system possess one (A which corresponds to $P = 0$, for $\mu < 2$) or three fixed points (A, an unstable fixed point B, and C, for $\mu > 2$). The solutions of Eq. (16) converge to stationary solutions since this equation possesses the Lyapunov functional

$$L = \int \left(V(P) + \frac{1}{2}P_x^2 \right) dx$$

which always decreases in the temporal evolution. Nevertheless transients are obviously dynamical and could be observed on sea shells which only live a finite time. For example, in the parameter regime $2.12\cdots > \mu > 2$ an initial state which consists in patches of pigmented cells and albinos cells evolves to the homogeneously pigmented pattern through the propagation of fronts. In the $x-t$ diagram, this takes the form of a transitory triangular shaped pattern. The existence of the Lyapunov functional ruled out the possibility of temporal oscillations, which is a common feature of many sea shell textures [18]. As a matter of fact, the pigmentation can oscillate at a given position in x as a function of time, i.e. as a function of the coordinate which measure the growth of the shell. In "mechanical" terms, inertia is missing in Eq. (16). The simplest generalization of this equation which allows the existence of oscillations is given by

$$\frac{\partial^2 P}{\partial t^2} + g(P)\frac{\partial P}{\partial t} = f(P) + DP_{xx} + \kappa\frac{\partial P_{xx}}{\partial t} \tag{17}$$

where a "dissipation function" $g(P)$ and a new parameter κ which represents a diffusion constant have been introduced. The term DP_{xx} is now associated with propagation, it is responsible for the existence of inertial waves. Sustained oscillations are possible only if $g(P)$ is negative for some interval of the pigmentation variable P. In order to model the dissipation, we choose the simple function

$$g(P) = -\nu + P^2$$

To summarize, our model depends now on three parameters. The relative stability as defined by the potential V is associated with the parameter μ. For μ larger than $\mu = 2$ the only stationary is $P = 0$. The fixed points B and C appear through a saddle-node bifurcation when $\mu = 2$. The parameter ν is associated with the "negative" dissipation. For positive ν, the fixed point $P = 0$ looses its stability through the Andronov-Hopf bifurcation. Slightly above the bifurcation threshold, this bifurcation leads to a small stable limit cycle for the dynamical system obtained by dropping the spatial derivatives. For a critical value ν_h, when $\mu > 2$, this limit cycle disappears through a homoclinic bifurcation. For this parameter value the stable and the unstable manifold of the fixed point B, merge. For a larger value of ν_c the fixed point C itself looses its stability though a Andronov-Hopf bifurcation. Between ν_h and ν_c another homoclinic Andronov bifurcation occurs for $\nu = \nu_{h'}$. This bifurcation leads to a stable "large" limit cycle which encloses the three fixed points. For $\mu = 2$, the fixed points C and B disappear through a saddle node bifurcation. When this happens in the parameter range $\nu_{h'} > \nu > \nu_h$, a limit cycle appear through a saddle-node Andronov bifurcation. Our model thus contains the three Andronov bifurcations. It can be seen as a model to study spatio-temporal phenomena associated with the spatial unfolding of these bifurcations. The last parameter κ is associated with the diffusion. In the frame of this model, which can also be interpreted as the continuous coupling of Van der Pol like electrical oscillators, we now describes few phenomena which are relevant to understand some of the sea-shell patterns.

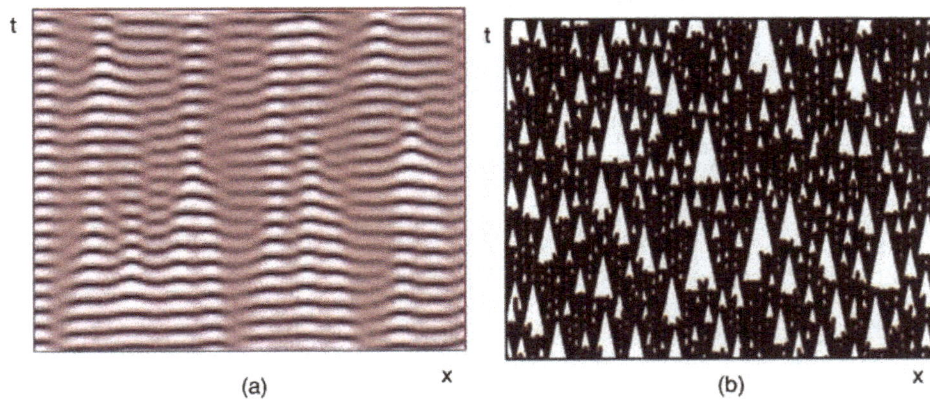

Figure 3. (a) Phase like patterns ($\mu = 2.08$ $\nu = 0.05$, $-0.4 < u < 0.2$). (b) Cavitation like pattern ($\mu = 2.08$, $\nu = 0.2$, $-1.4 < u < 0.5$). The white color corresponds to small values of u, and time is running up.

2.1 Cavitation like patterns

Near the onset of oscillations ($\nu \gtrsim 0$) Eq. (17) reduces to the complex Ginzburg Landau equation 5. The stability of the homogeneous oscillatory solutions depends on the sign of $1 + \alpha\beta$. As expected large κ stabilizes the homogeneous oscillatory solution. For large values of μ there is only one stationary solution ($P = 0$) and the homogeneous oscillation is stable. The existence of an Andronov homoclinic bifurcation is a sufficient condition of the instability of the limit cycle. This bifurcation only occurs in the parameter range $\mu > 2$. As ν increases, two types of spatio-temporal patterns are observed

- "Phase patterns". For ν close to zero, the phase instability leads to wavy patterns which are easily interpreted as the manifestation of the phase instability of the oscillatory homogeneous pattern [see Fig. 3(a)].
- "Cavitation patterns"[19]. For larger value of ν, strong self-focalization of the chaotic oscillation leads to the nucleation of the homogeneous stationary solution C. Depending upon the values of μ, the nucleated domains either retract or expand. The case where the domains disappear through the propagation of two fronts is particularly interesting. It leads to a sustained dynamical state in which new domains are permanently created. On the $x - t$ representation of such a state, one notices white triangles which corresponds to the retraction of the Albinos domains [see Fig. 3(b)].

2.2 Patterns of excitable waves

When $\mu = 2$, the fixed points B and C disappear through a saddle-node bifurcation. For $\mu \gtrsim 2$, depending upon the value of ν, the fixed point C is a global attractor of the planar vector field. Excitable solitary waves are then observed in the simulation

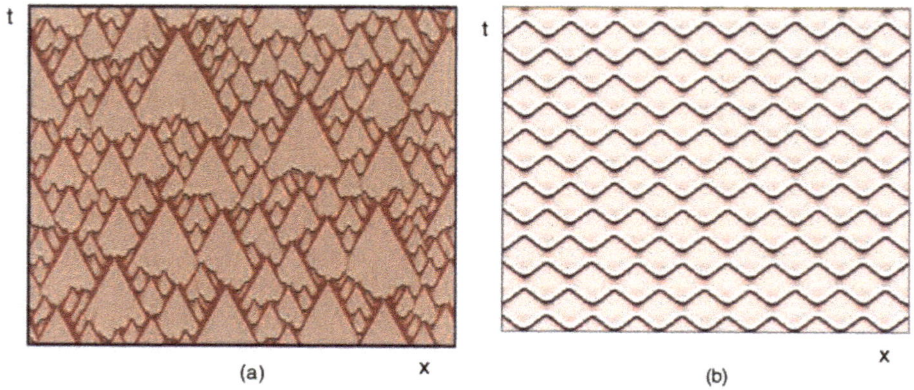

Figure 4. Patterns of excitable waves. (a) Back-Fire pattern ($\mu = 2.01$ and $\nu = 0.5$, $-1.4 < u < 0.5$). (b) Crossing pattern ($\mu = 2.01$ and $\nu = 0.83$, $-1.4 < u < 0.5$). The white color corresponds to small values of u, and time is running up.

of the partial differential equation. For μ given, as ν increases, three types of spatio-temporal behavior are observed

- "Back fire patterns" [21]. This pattern is associated with the instability of the front which connects the fixed points A and C. As the front moves, the A domains left behind is unstable and leads to the nucleation of domains of the C state [see Figure 4(a)].
- "Normal" excitable wave patterns. This is the domain of parameter where excitable exists and are stable. Because of the finite spatial domain, excitable solitary waves generally disappear at the border, where they are absorbed. The nature of the interaction between a wave and the boundary depends upon the nature of the boundary conditions. In this paper, the free boundary conditions which have been considered are such that they lead to the absorption of "normal" excitable waves.
- "Crossing patterns" [20]. The frontal collision of two excitable pulses generally leads to their disappearance. It has been recently shown that such a collision could actually leads to the reflection of pulses. The transition from non-crossing to crossing has been described in the context of the geometrical theory of dynamical system. In this parameter range, the partial differential equation, when it is solved with periodic of free boundary condition, exhibits sustained dynamical behaviors. For slightly larger value of ν, the crossing patterns takes the form of a standing wave [see Figure 4(b)].

Acknowledgments

This work has been supported by the CNRS (France) and the EU through a TMR grant FMRX-CT96-0010. One of us (P.C.) thanks the support of the "Institut Universitaire de France".

References

1. A. M. Turing, The chemical basis of morphogenesis, Phil. Trans. R. Soc. London B **237**, 37 (1952).
2. R. P. Feynman, R. B. Leighton and M. Sands, *The Feynman Lectures on Physics*, II, Addison-Wesley, Reading, Massachusetts, 1964.
3. V. Arnold, *Geometrical Methods in the Theory of Ordinary Differential Equations*, Springer, New-York, 1983. See also, J. Guckenheimer, P. Holmes, *Nonlinear Oscillations, Dynamical Systems and Bifurcation of Vector Fields.* Appl. Math. Sci. 41, Springer, New-York, 1983. G. Iooss and M. Aldelmeyer, *Topics In Bifurcation theory and Applications*, World Scientific, Singapore, 1992.
4. A. C. Newell, Envelop Equations, Lectures in Applied Mathematics 15, 157, Am. Math. Soc. Society, 1974.
5. M. C. Cross and P. C. Hohenberg, *Pattern formation outside of equilibrium*, Rev. Mod. Phys. **65**, number 3, part II, 1993.
6. A. C. Newell and J. A. Whitehead, Finite bandwidth, finite amplitude equation, J. Fluid. Mech. **38**, 279 (1969).
7. A. N. Kolmogorov, I. G. Petroskii and N. S. Piskunov, A study of the diffusion equation with increase in the amount of substance, and its application to a biological problem, Bull Moscow Univ., Math. Mech **1**, 6, 1–26 (1937). Reprinted in *Selected Works of A. N. Kolmogorov*, Vol I; *Mathematics and Mechanics*, V. M. Tikhomirov (ed.), Kluwer Academic Publisher. See also R. A. Fisher, *The Genetical Theory of Natural Selection*, Oxford Univ. Press, 1930.
8. For a review see P. Collet and J. P. Eckmann, *Instabilities and Fronts in Extended Systems*, Princeton University Press, Princeton, 1990.
9. Y. Pomeau, Front motion, metastability and subcritical bifurcations in hydrodynamics, Physica D **23**, 3 (1986).
10. A. A Andronov and L. Pontyagin. Dokl. Akad. Nauk, SSSR **14**, 247 (1937). See also A. A. Andronov and al, *Theory of Bifurcation of Dynamical Systems on the Plane*, Israel Program For Sci. Trans., 1971.
11. M. Argentina, Ph. D. thesis, University of Nice (1999).
12. E. Hopf, Ber. Mat.-Phys. Kl. Sachs Acad, Wiss, Leipzig **94**, 1 (1942).
13. Y. Kuramoto, *Chemical Oscillations, Waves, and Turbulence*, Springer-Verlag, Berlin, 1984.
14. B. I. Shraiman *et al.*, Physica D **57**, 3 (1992).
15. B. I. Shraiman, Phys. Rev. Lett. **57**, 325 (1986). See also P. Coullet, L. Gil, and J. Lega, Phys. Rev. Lett. **62**, 1619 (1989).
16. Risler, Preprint INLN 1999.
17. R. A. FitzHugh, Impulses and physiological states in theoretical models and propagation in nerve membrane, Biophys. J. **1**, 445–466 (1961). See also, J. S. Nagumo, S Arimoto and S. Yoshizawa, An active pulse transmission line simulating nerve axon, Proc. IREE aust, **50**, 2061 (1962).
18. H. Meinhardt, *The Algorithmic Beauty of Seashells*, Springer, Heidelberg, New York, 1995.
19. M. Argentina and P. Coullet: Chaotic Nucleation of metastable domains. *Phys. Rev. E* **63**, R2359 (1997).

20. M. Argentina P. Coullet and L. Mahadevan: Colliding waves in a model excitable medium: preservation, annihilation and bifurcation. *Phys. Rev. Lett.* **9**, 2803 (1997).
21. V. Petrov, S. K. Scott and K. Showalter, Excitability, wave reflection, and wave splitting in a cubic autocatylis reaction-diffusion system, Phil. Trans. Soc. Lond. **347**, 631 (1994).

RATIONAL AND IRRATIONAL ANGLES IN PHYLLOTAXIS

Y. COUDER AND S. DOUADY
Loboratoire de Physique Statistique
24 rue Lhomond
75231 Paris Cedex 05, France

Sunflowers do not inspire only painters; for botanists, mathematicians, physicists. there is a specific beauty in the spiral order of their inflorescences. They are the most complex manifestation of a quasi crystalline order which is ubiquitous in plants. The study of these structures has developed into a whole domain called phyllotaxis. It is historically remarkable that one of the pioneering works in this field is due to A. and L. Bravais (1835, 1837). This investigation preceded A. Bravais's famous work on crystallography where he gave the classification of the 14 possible periodic lattices in three dimensional space. In this latter work he used the same type of geometrical arguments for lattices of point-like elements that he had established in his botanical work. Very probably, for Bravais the order of plants served as a starting point to imagine the order in crystals.

1 Introduction

1.1 The geometry of phyllotaxis

The works of the Bravais brothers and that of Van Iterson (1907). were concerned with the order of grown up elements and were essentially geometrical. It was shown that the dispositions observed for leaves could be generalised to all the other elements, bracts, sepals, petals etc.. which grow around the stems of vascular plants. Furthermore, all the observed dispositions of these elements around the stem can be classified in two categories only.

The first mode of organisation, the most frequently observed, is spiral. Two different descriptions of this organisation can be given. If the elements are linked in their order of appearance, the resulting curve forms, around the stem, a single spiral (or a helix) called the generative spiral. The direction of winding of this spiral is with equal probability left or right. Projected in the plane perpendicular to the stem, the divergence angle j between the base of two successive elements is constant and has generally a value close to $\Phi = 2\pi(1-\tau) = 137.5^o$ where τ is the golden mean $\tau = (-1+\sqrt{5})/2$. In the case where the elements are tightly packed (as observed on e.g. pine cones or sunflower heads) it is easier to describe the pattern by linking each element to its nearest (or contact) neighbours. Two sets of intersecting spirals with opposite directions of winding are thus defined (Figure 1). They are called the parastichies by the botanists. The numbers i and j of these spirals necessary to cover the whole pattern are almost always two successive terms of the main Fibonacci series: $\{1, 1, 2, 3, 5, 8, 13, 21, 34, \ldots\}$ in which each term is the sum of the two previous ones (the first two terms being 1 and 1). The main exceptions are cases where the parastichies number are part of the Lucas series $\{1, 3, 4, 7, 11, \ldots\}$ which has the same construction rule but applied to different initial elements.

In the second type of organisation, the whorled modes, several leaves appear simultaneously at the same height on the stem. There is the same number n of

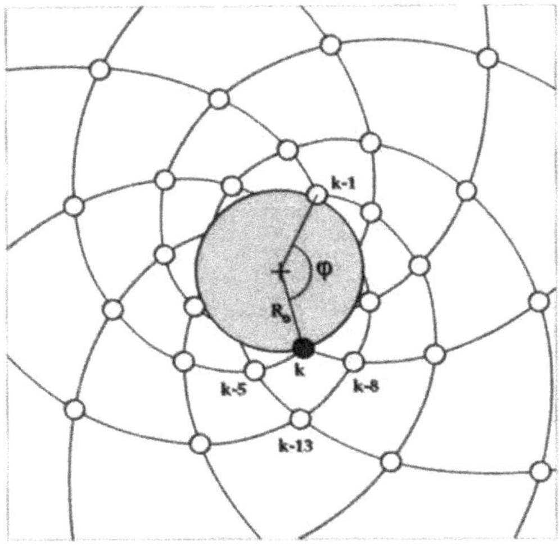

Figure 1. Sketch of the apical region in the vascular plants projected in the plane perpendicular to the axis. On the present sketch the Fibonacci mode is (5, 8). As shown here, a geometrical property of a parastichy system (i, j) is that the neighbours of an element having the order of formation k are respectively of order of formation $k - i$ and $k - j$ (Bravais 1837).

leaves at each node and they are regularly disposed at an angle $\phi = 2\pi/n$ from each other. The leaves of a given node are shifted by $\phi = \pi/n$ relatively to those of the previous and following ones. In the case of tightly packed elements (e.g. maize), when each element is linked to its contact neighbours two sets of straight radial rows of order n are observed, often called the "orthostichies".

In a contemporary perspective a remarkable asset of these orders is that while the whorled dispositions are related to finite size crystalline arrangements, the spiral organisations are related to the approximants of quasi crystals (Couder and Douady 1996).

1.2 The dynamics of phyllotaxis

The stems of the vascular plants are always terminated by a bud. Inside this bud is the apical region, sketched on Figure 1, where the phyllotactic structures are formed. A smooth region of undifferentiated tissues, the apex, forms the extremity of the shoot. Around it, a specific mitotic activity creates small protrusions, the "primordia", which are already organised in parastichies and will evolve into grown-up botanical elements: leaves, bracts, florets... In the reference frame of the apex, because of the growth of the shoot, the existing primordia drift away from the tip with a velocity V so that they leave room for the formation of new elements. The microscopic observation of the apices initiated by W. Hofmeister (1868) led him to state a general dynamical principle. He proposed that a new primordium appears with an imposed periodicity in the largest available space left by the previous ones. Some of the experiments performed by M. and R. Snow (1935)) had shown that there was a connection between the whorled phyllotactic modes and the spiral ones.

For this reason M. and R. Snow (1952) tried to generalise Hofmeister's dynamical rules by removing the imposed periodicity of the process and by giving the primordia a fixed finite size. With these hypotheses they suggested that a new primordium simply forms where and when there is enough space for its formation.

2 The self organisation in model iterative systems

Both these hypotheses form the rules of iterative systems that we tried to formalise. More complete reports on the various aspects of this work can be found in Douady and Couder (1996 a, b, c,).

2.1 Hofmeister's rules: the spiral modes

We first examined Hofmeister's set of hypotheses. The characteristics we retained for the iterative system were the following: identical elements are generated in a plane surface on a circle of radius R_0 at constant time intervals T (the plastochron of the botanists). These elements are radially advected at velocity V. We assume that there is a repulsive interaction between the elements and that the new element appears as far as possible from the preceding ones, i.e. in the largest available place. We managed (Douady and Couder (1992 and 1996 a) to implement a physics laboratory experiment in which all these characteristics existed. As a result we obtained, in this analogic experiment, patterns of ferrofluid drops organised in Fibonacci arrangements.

Our first numerical simulations were based on the same hypotheses. Also performed in a plane radial configuration, the locus of appearance of new elements is a circle C of radius R_0 centred at the origin (Figure 1). All particles after their appearance are given the same radial motion with a velocity V_0. The particles are formed at time intervals T on the circle. The whole organisation of the system derives from the choice of the position of formation of a new particle on the circle, which occurs in the largest space left available by the previous particles. In order to implement this rule we assume in the simulation that each particle is point-like but generates a repulsive energy $E(d)$ where d is the distance to this particle. To decide the place of formation of a new particle, we compute the value of the total energy due to all the previous particles in all points of the circle C. The new element is created at the point of minimum energy.

The parameter which controls the system is $G = V_0T/R_0$. Figure 2 shows a steady pattern obtained at $G = 0.01$ where the divergence ($\phi = 137.47^o$) is very close to Φ, and parastichies numbers are successive elements of the Fibonacci series: $(i, j) = (13, 21)$.

With all the results obtained using various types of initial conditions we can build a diagram (Figure 3(a)) showing the divergence angles of all the steady regimes observed at various values of G. In this diagram there is one main curve along which ϕ converges oscillatingly from 180° towards 137.508^o. Each change of sign of the slope of the curve $\phi(G)$ occurs at a value G_{ij}. Comparison of the patterns obtained just above and just below G_{ij} shows that a transition from parastichies numbers (i, j) to $(j, i + j)$ occurs at this value. A remarkable asset of the diagram

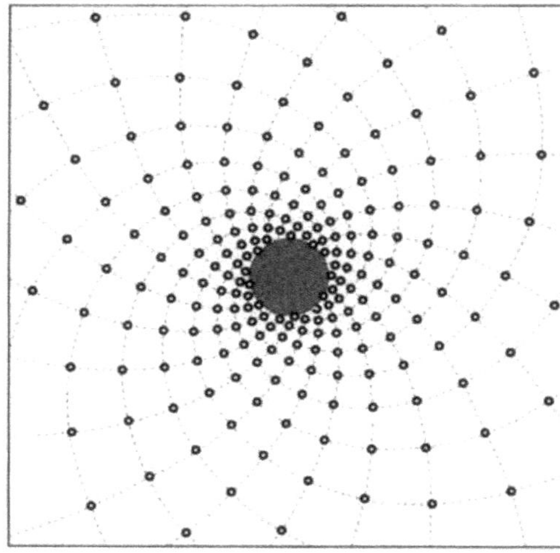

Figure 2. A pattern obtained in the numerical simulation (the parastichies and the apex radius were drawn for an easier visualisation of the structure). The steady divergence angle is $\phi \approx 137.47^o$ and the parastichies numbers $(13, 21)$.

(Figure 3(a)) is that near each G_{ij}, a new possible value of ϕ also appears. It corresponds to a regular spiral pattern with parastichies numbers $(i, i+j)$.

With a further decrease of G this new value of ϕ will evolve continuously and thus form in the $\phi(G)$ diagram a new curve disconnected from the main one. Following this new curve towards $G = 0$, ϕ tends oscillatingly towards an irrational angle as a Fibonacci type of series builds up with the same transition rule $(i, j) \rightarrow (j, i+j)$. Each of the curves $\phi(G)$ thus corresponds to a different pair of initial terms for the series and converges for $G = 0$, towards the related irrational angle (99.502^o, 77.955^o, 151.135^o, etc...). This structure of the diagram $\phi(G)$ is practically independent of the energy profile $E(d)$ and of the chosen geometry (plane, cone, cylinder, etc.).

2.2 The dynamical selection of the Fibonacci modes

It is possible in this system to reach an intuitive understanding of the convergence of the different branches towards irrational angles by the simple arguments presented in Douady and Couder (1996 b) and in Douady (1998). Let us consider the system at a value G corresponding to the organisation (i, j). It can be shown that this disposition is obtained when the position of a new particle k is determined by the repulsion of j previous elements. More specifically the element of order k is located between the elements of order $k-i$ and $k-j$ and its precise position is fixed by their repulsion (see Figure 1 and Figure 4).

We can draw another diagram $\phi(G)$ using the condition that the new particle of order k appears exactly equidistant from two older ones of order $k-i$ and $k-j$, disregarding the effect of any other element. This geometrical diagram (of a type

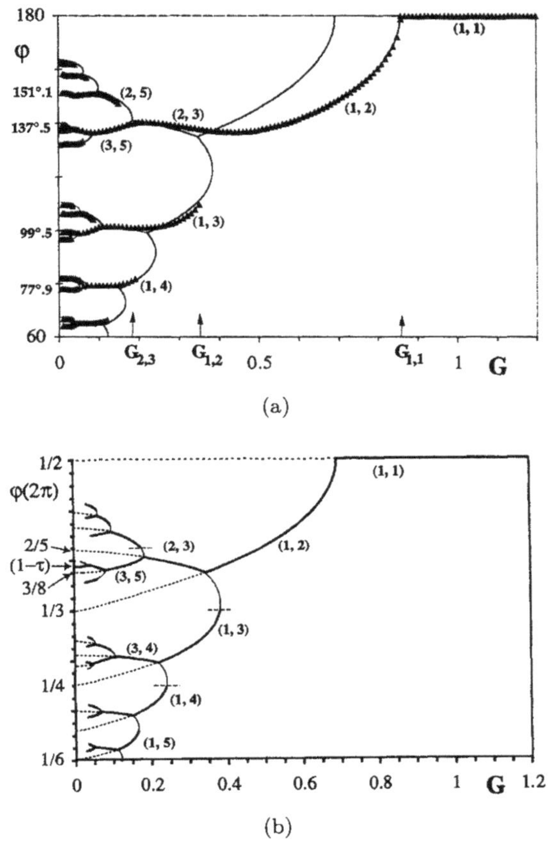

Figure 3. (a) Diagram of the values of the steady divergences ϕ obtained in the simulation (dark triangles) as a function of G. Only the values of ϕ smaller than 180° have been drawn (the diagram being symmetrical). (b) A geometrical diagram of the type introduced by Van Iterson, drawn using the condition that the new particle appears exactly equidistant from two older ones. The comparison of the two diagram is the key to the dynamical selection of the Fibonacci modes. Each geometrical branch tends at $G = 0$ towards one of the rational angles $(180, 120, 144, etc\ldots)$. At the crossing of two branches there is a bifurcation. This bifurcation is imperfect and the part of the branches drawn in thin lines cannot be obtained dynamically. The dotted lines show the rationnal divergences that would be reached if only j particles were repulsive.

introduced by Van Iterson (1907)) is shown on Figure 3(b). It is formed of many curves each representing an organisation (i, j). We can first notice that each of the curves tends at $G = 0$ towards a rational angles (180°, 120°, 144°, etc...). To understand this, let us recall that without any other repulsion, the new particle of order k is placed exactly equidistant from the two older particles of order $k - i$ and $k - j$. Then decreasing the parameter G reduces their drift away from the apex border, and their difference in radial position (see Figure 4). In particular, the new particle become closer and closer to the element of order $k - i - j$. Near $G = 0$, all the particles remain very close to the apex border, and the new particle is angularly

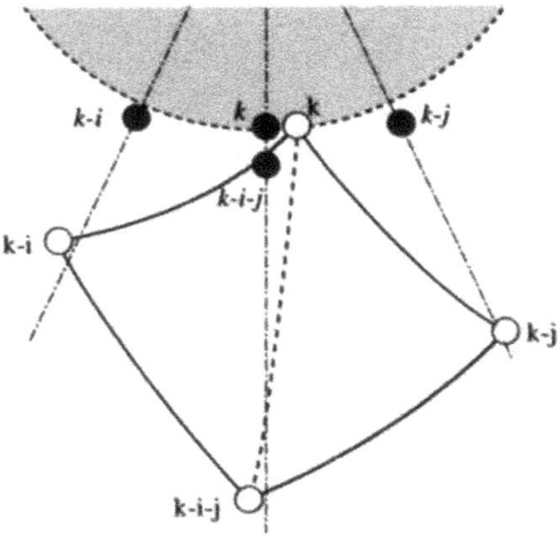

Figure 4. Scheme of the relative position of the new element k and of its nearest neighbours of order $k-i$ and $k-j$ (white disks). The element diagonally opposite to k is of order $k-i-j$. The positions drawn in dark disks correspond to what would be obtained at very low G if the position of element k was still determined by elements $k-i$ and $k-j$ only. In this case k and $k-i-j$ would become radially aligned and the divergence angle would become rational.

located halfway between the two others so that ϕ becomes rational (Figure 4). Its value can be found by the position of the older particle $k-i-j$, which is now very close and radially aligned with the new particle. Thus, knowing that the generative spiral has done p turns from particle k to $k-(i+j)$ (called the encyclic number associated to $i+j$ following Bravais (1837)), we find that the angle ϕ has converged toward the rational angle $2\pi p/(i+j)$. For instance, as shown on Figure 3(b), an organisation ($i=1, j=1$) would lead to $\phi = (1/2)(2\pi) = 180^o$; ($i=1, j=2$) to $\phi = (1/3)(2\pi) = 120^o$; ($i=2, j=3$) to $\phi = (2/5)(2\pi) = 144^o$; ($i=3, j=5$) to $\phi = (3/8)(2\pi) = 135^o$ etc... In reality this is not possible because the potentials are not switched off and the repulsion forbids the radial proximity of the elements that would result from such alignments.

The geometrical solution (i, j) thus stops being possible at a value G_{ij} when the element of order $k-i-j$ which is placed diagonally to k (Figure 4), comes to be at the same distance as $k-i$ and $k-j$. Geometrically, this occurs in the $\phi(G)$ diagram at the crossing of the curves representing the three possible arrangements: (i, j), $(i, i+j)$ and $(j, i+j)$ (Figure 3(b)). At this crossing the disposition of the elements is locally hexagonal. If only geometrical considerations are considered it is not possible to predict which solution the system will follow if the control parameter is slowly reduced below G_{ij}. Dynamically we have already shown why (i, j) becomes unstable. We can now demonstrate that dynamically, only one evolution is possible leading from organisation (i, j) to $(j, i+j)$ and thus building up the Fibonacci series.

In order to understand it, one has to consider not only the final regular result, but also the way it has grown. In the simulation, we place a new element on the apex border at the minimum of the repulsion due to all the previous elements $E(\theta) = \sum E_n$, (θ being the angular position of the new element on the apex circle).

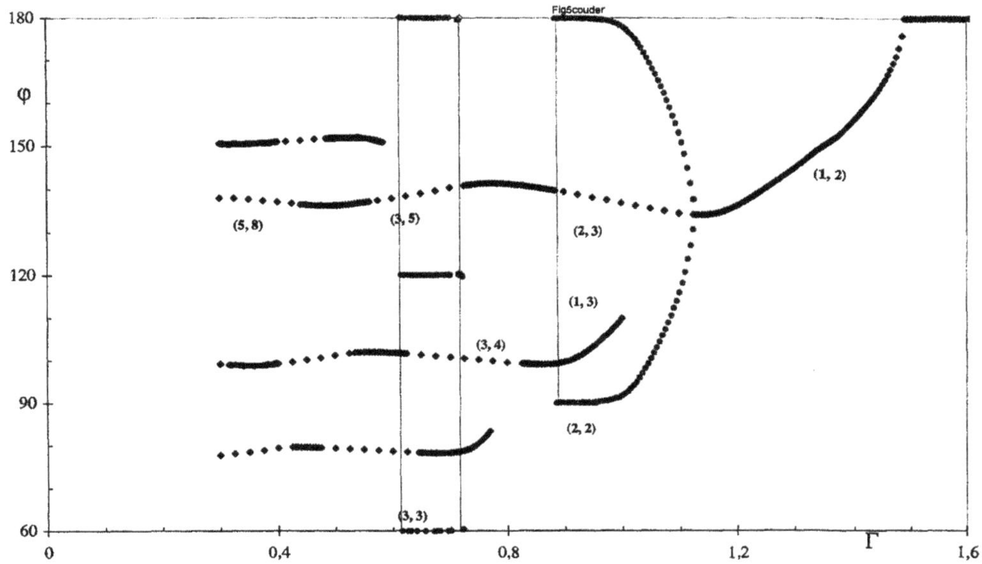

Figure 5. Diagram of the values of the steady divergences ϕ obtained in the simulations of iterative system using Snow & Snow's hypotheses. The black diamonds correspond to steady spiral organisations, the grey disks to the whorled modes ((2, 2) and (3, 3) for which ϕ is multivalued).

To have a minimum requires at least that the repulsive energy has an extrema at this place: $\partial E/\partial\theta = 0$. If the repulsion due to one particle, labelled $k-i$ say, $E_i(d)$, decreases rapidly with the distance $d(\theta)$ to it, then this repulsion is mainly due to the closest neighbour, except when two closest neighbours are at the same distance (say $k-i$ and $k-j$). A maximum is obtained when the new element is closest to one element, and a minimum when it is the further from the previous ones, i.e. equidistant from two preceding ones. In this case the extrema condition becomes: $\partial E_i/\partial\theta + \partial E_j/\partial\theta = 0$. A consequence is that the two derivatives are of opposite signs. Recalling that the repulsive energy is a decreasing function of the distance, this means that when θ is increased, one distance increases while the other decreases. In other words, having the new element at the minimum of the repulsion implies that its two nearest neighbours are azimuthally on opposite sides.

We can now return to the situation at a given bifurcation. The pattern is locally hexagonal, but this hexagonal disposition is not radially symmetrical: the two previous elements $k-i$ and $k-j$ are not of the same order (except when $i=j=1$). Then the older ($k-j$ if $j>i$) is located further from the apex border, and the rhomb consisting of (k,$k-i$,$k-i-j$,$k-j$) is tilted (see Figure 4 and Figure 6(b)). Then two parastichies (of order i and $i+j$) wind in one direction while the third (of order j) winds in the opposite direction. The bifurcation consist in the transition from (i, j) to either $(i, i+j)$ or $(j, i+j)$. Out of these two modes the former would have parastichies winding in the same direction, meaning that the two closest neighbours would be azimuthally on the same side of element k. This directly shows that this pattern cannot be obtained at the bifurcation in our

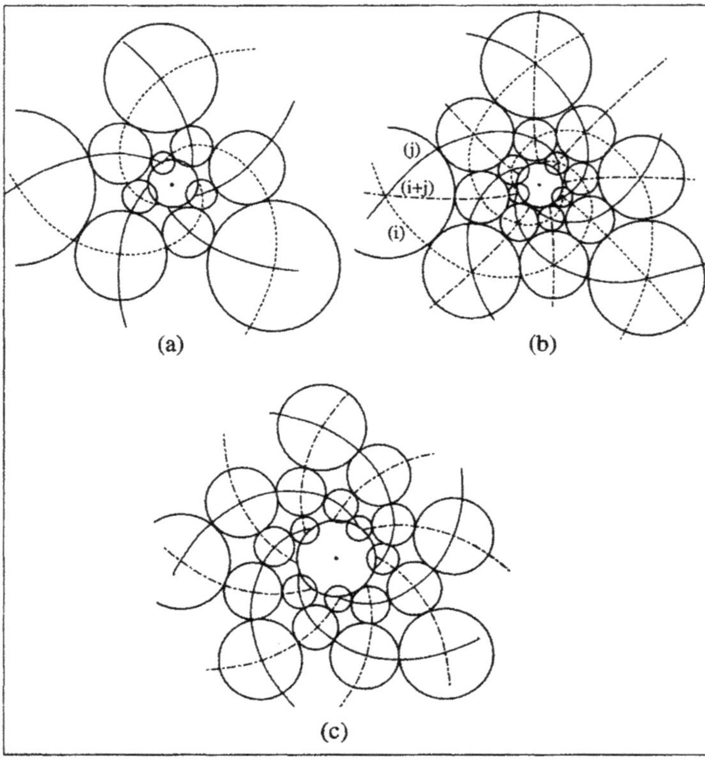

Figure 6. Reproduction of three drawings by van Iterson (1907) showing three spiral dispositions (3,5), (3,5,8) and (5,8) obtained by the radial piling of disks of increasing sizes. The difference in the density of piling β between the locally hexagonal patterns obtained at the transition and the locally square ones can be noticed.

simulation: the new element which would continue the regular packing would not be placed at a minimum of E. Such solutions are not dynamically possible *near the bifurcation*. However these organisations are not altogether impossible: as observed by a comparison of the geometrical and dynamical diagrams the mode $(i, i+j)$ can be obtained for lower values of G (Figure 3(a)). We know already that for the angle $\phi = (p/(i+j))(2\pi)$ the elements k and $k-i-j$ are aligned and the corresponding parastichy is radial. Beyond this angle the two parastichies of order i and $i+j$ will become of opposite sign of winding and the corresponding solution $(i, i+j)$ will become possible. However, it is not only needed that the new position should be a minimum but also that it should be the absolute minimum on the apex border. This brings an even stronger limitation to the dynamically significant part of the $(i, i+j)$ branch (see Douady (1998)).

As a partial conclusion we have reached an understanding of a very unusual situation. As demonstrated above the structure of the various branches of the $\phi(G)$ diagram and of their successive bifurcations can be seen as a sequence of "repulsions" by rational organisations. The convergence towards an irrational angle then results naturally from the avoidance of successive rational angles. It is worth noting that this situation is the opposite of a situation usually met in non-linear physics. Generally when a system is submitted to two frequencies the effect of a non-linearity is to lock the two frequencies onto each other so that their ratio be a simple rational number. The width of the basin of attraction of the different rational numbers increases with the strength of the non-linear terms giving rise to a structure known as Arnold's tongues (Arnold, 1965). At a given level of non-linearity the width of these tongues corresponds to the hierarchy of rationals: the simpler the rational the wider the tongue. In the diagram shown on Figure 3(a) a hierarchy of rationals is also exhibited: the simpler the rational the larger the value of G at which its starts being avoided.

A diagram very similar to that shown Figure 3(a) was obtained independantly by Levitov (1992) investigating a minimum energy condition on compressed helical lattices. The comparison of his results with ours shows a relation between them similar to that which exists in classical crystallography between the energy of a periodic structures and its growth mechanism.

2.3 The generalised model: Snow and Snow's rules

In the above simulation of Hofmeister's system we demonstrated experimentally and numerically that the formation of the spiral phyllotactic patterns could be obtained by self organisation. This system is controlled by only one parameter: G which reflects basically the rhythm of formation of the elements: the botanists' plastochron. However there is no realistic way within this model to consider the whorled and multijugate modes of organisation.

We thus turned to a new type of simulation using the rules of Snow and Snow (1952) where the periodicity is no longer imposed. The new rule is that a new primordium appears when and where an empty space of diameter d_0 has formed at the periphery of the apex. In the simulation this is implemented using the same particles with a repulsive potential. We now compute the potential in all the points of the circle all the time. This potential decreases as the particles move away. When it becomes smaller than a fixed threshold E_s a new particle is created. The value of E_s fixes the size d_0 of the necessary space (i.e. the size of the particles). In order to make the model more realistic we have the possibility to take into account the fact that many apices are conical. In this case the calculation are still performed in the plane but considering it as the plane perpendicular to the cone's axis. On the surface of the cone, constant values of the repulsive potential of a particle are circles "wrapped" around the cone. Once projected onto the plane, the equipotential circles take an azimuthally elongated folioid shape (see Figure 7). The control parameter of this system is now the ratio of the size of the primordium over that of the apex $\Gamma = d_0/R_0$. This means that the dominant factor is here, even more than above, geometrical. The first result of our simulations is that, for all values of Γ, steady regimes of the iterative system can be obtained. These are all time periodic. If only one element forms at each period the same spiral modes obtained in Hofmeister's simulations are recovered. From the periodicity of their formation T_s it is possible to deduce a value of the parameter $G = VT_s/R_0$. The diagram

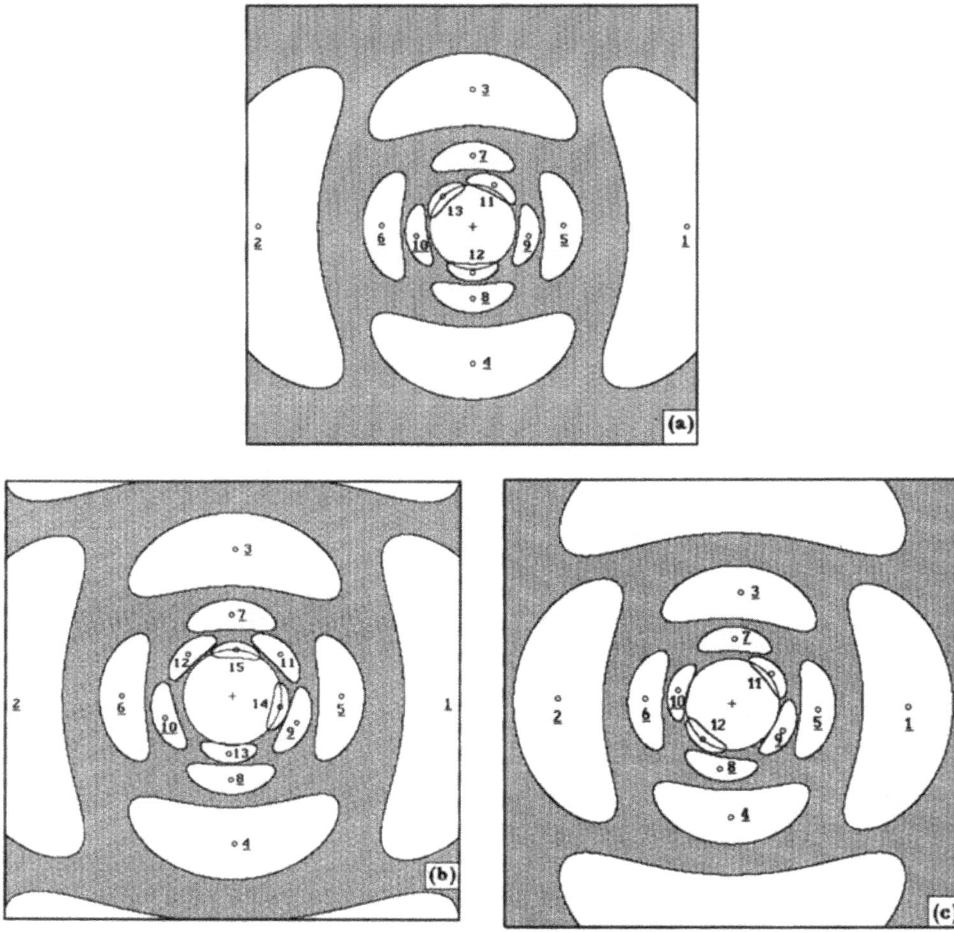

Figure 7. Results of simulations of the early stages of the growth of sunflowers. The simulations are done on the surface of a cone and represented in the projective plane perpendicular to the cone's axis. (a) The common transient leading to a Fibonacci order. (b) A rare transient, shown here, leads to a Lucas series spiral. (c) The transient leading to a bijugate mode (i.e. a spiral mode with parastichy order twice the Fibonacci numbers). The resulting spiral mode is $(2, 4)$.

$\phi(G)$ for the spiral modes is identical to that obtained in the previous case with the only difference that now G is not imposed. But there are also regimes where several elements are formed at each period. The complete diagram of the possible ϕ as a function of the new control parameter Γ (Figure 5) is thus more complex. The branches corresponding to the spiral modes have the same structure as obtained previously. But competing with them are the whorled modes. A characteristic of the whorled modes is that each of them is stable in a finite range of value of Γ. When Γ decreases they become unstable when the influence of the elements of the second elder whorl becomes non negligible. But, unlike in the spiral modes there is no possible continuous transition from a whorled mode to another. In the centre of their range of existence the whorled modes are more stable than the spiral ones so that for these values of Γ spontaneous transitions from a spiral to a whorled mode are observed if long iterations are done. In reverse if Γ is decreased below the stability limit of a whorled mode a transition to a spiral mode will be observed. Generally these transitions will be of the type $(n, n) \rightarrow (n, n+1)$.

2.4 The selection of the actually observed patterns

The coexistence of different organisations poses the problem of the selection: which amongst all the possible patterns is actually obtained at a given value of Γ? This is a complex problem because it depends on the history of the system. We will discuss separately the physical problem and the botanical one.

We can first note that there is a difference in the possibility of evolution of a given pattern in the two types of simulations. In the Hofmeister situation all the spiral solutions of Figure 3(a) are strongly stable (and the stability can be easily demonstrated (C. Gole, private communication). The reason is that any perturbation in the position of a previous particle will only create a smaller perturbation in the azimuthal position of the new one but will not affect its radial position which is fixed (by the time of formation). However, in the Snow & Snow simulation, quasi stationnary modes can slowly evolve and undergo a slow transition to another more stable mode. A perturbation in the position of a given particle can propagate more easily as it will affect both the azimuthal and the radial positions of the new elements (because of the possibility of fluctuations of the time of appearance).

In the Snow & Snow simulation, at a given small value of Γ, many different modes can be dynamically obtained. If the simulation is run during a long time some of these modes turn out to be completely stable, while others are slightly unstable; there is then a slow transition to another mode. In order to compare the patterns it is convenient to compare the average time of formation of a new element. For the spiral modes it is simply T_s. For the whorled modes (n, n) it is the period T_n of formation of the whorls divided by the order of the whorl. $\langle T \rangle = T_n/n$. In first approximation the typical time of formation of a new element varies as the square of the elements size: $\langle T \rangle = \beta\Gamma^2$. However a close inspection shows that there are fluctuations, the coefficient of proportionality β being linked to the density of the packing. For instance following one of the spiral branch the patterns exhibits minimum values of β at each bifurcation because the arrangement is locally hexagonal and therefore the most dense. Conversely β is much larger in-between two bifurcations where the piling is square and therefore the least dense. This difference in the density of piling is well observed on Figure 6 (a, b and c) which is due to Van Iterson (1907). The study of the stability of the mode shows that at a given value of Γ the most stable mode is that with the largest density. However, the slow transition corresponds simply to the appearance of a dislocation. Thus only the modes which can transform into one another through a dislocation are subject to these transitions. Thus two states of different densities are still possible for the same parameter value.

3 The selection during the plant growth

We can now return to the botanical problem. The normal growth of sunflowers (*Helianthus annuus L*), usually leads to highly organised inflorescences with parastichies corresponding to large Fibonacci numbers. However some rarer specimens exhibit a different order with Lucas series spirals, bijugate spirals or trimerous whorls. It is possible to measure the positions of the successive elements of various specimens to understand what determined these deviations. The main asset is that there is continuity in the growth. The specimens with Fibonacci inflorescence have a Fibonacci organisation of the leaves of their stems, the specimens with a Lucas series inflorescence have the same anomaly in the organisation of their leaves. This demonstrates that the transition to one or the other of these organisation occurs very early in the growth.

In fact the initial growth of the first pairs of leaves of the sunflowers is always organised in the opposite decussate mode (i.e. a whorled mode with two elements at each whorl). During the later growth of the plant a transition occurs to one of the vegetative modes. These transitions, in which the phyllotactic mode of the whole plant is selected, can be characterised. They are shown to correspond to only a few archetypes.

The possible transients from a decussate to a spiral growth were investigated in our model (Couder 1998). They coincide exactly with the observed transitions. Furthermore the various types of transitions have, in the model, a very different probability of occurrence which corresponds to the difference of their frequency of observation in plants. We show on Figure 7 three transients. In all cases the initial value of the parameter is chosen in the range of stability of the decussate mode. Then G is decreased to a smaller value located below this range. The decussate pattern thus becomes unstable and a transition is observed to a spiral mode. The most common transition, shown on Figure 7(a) leads to a mode $(2,3)$ located on the main Fibonacci branch. If this transition occurs the later growth will lead to the formation of a normal Fibonacci inflorescence. For transients very similar to the previous one but with simply slightly different rates of change of G during the transition two other types of transitions can be observed (Figure 7(b and c), leading respectively to the formation of a $(1,3)$ mode located on the Lucas branch or to a bijugate mode $(2,4)$. Figure 7 shows that this results simply from a different initial breaking of the decussate symmetry.

4 Conclusions

On the physiology

We did not consider here the physiology of the plant. From this point of view several hypotheses have been put forward. They are aimed at accounting for the formation of well defined primordia and at giving them a short range repulsive or inhibiting interaction. The most investigated model is the reaction diffusion processes of the type initially proposed by Turing (1952). The first simulations showing the spontaneous formation of a Fibonacci order were done by Veen and Lindenmayer (1977), Mitchison (1977), Young (1978), Meinhardt (1982 and this conference) and

Friedman (1992) in systems controlled by the diffusion of an inhibitor. Mitchison (1977), was the first to give an interpretation for the transition from a Fibonacci mode (i, j) to the next $(i, i + j)$ in the case of a model using a diffusive inhibition on a cylindrical surface. More recently P. Green (1992 and this conference) suggested that a buckling instability would lead to the formation of a wavy surface having finite size bumps. These bumps would interact with their first neighbours by an elastic interaction which would also be short range. Our result show that the mode of organisation is a result of very general self organisation principles so that the phyllotactic studies will presumably be of little help to decide of the process actually at work in plants.

On the relation between local and global order

In very general terms we believe that the difficulty in understanding the genesis of the Fibonacci order comes from the relation between local order and global order. The existence of high order Fibonacci modes ((55, 89) on Figure 8) seems, at first sight, to imply a process of organisation in which, at a given time, an element interacts with all the other elements present around the apex. This is wrong: our results, after many others, demonstrate that at a given time the position of formation of an element is determined by its nearest neighbours and thus by the parastichies.

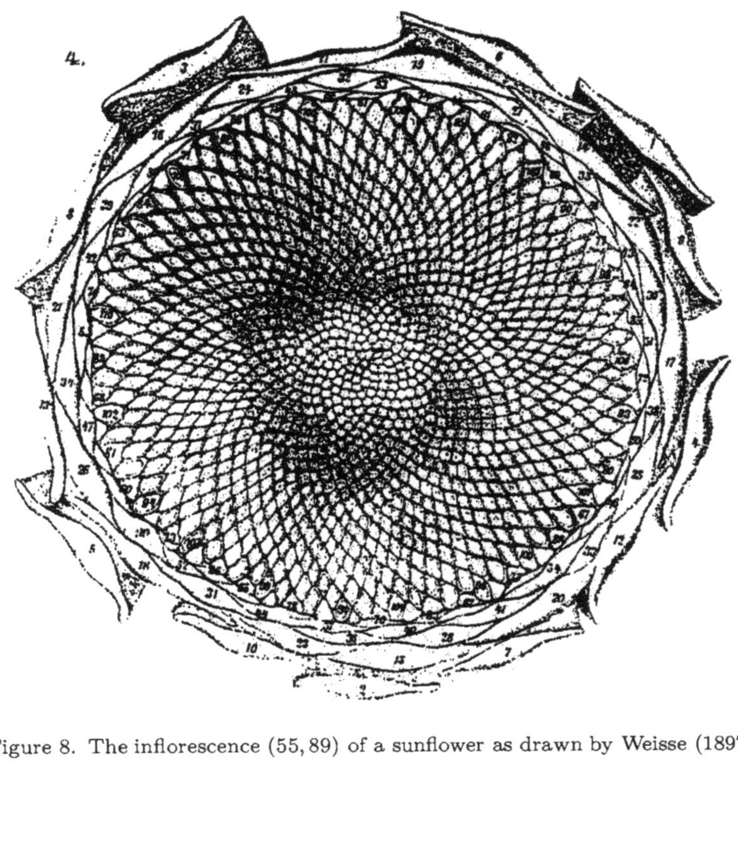

Figure 8. The inflorescence (55, 89) of a sunflower as drawn by Weisse (1897)

We have shown that, due to these growth rules, the system tends to optimise the density of its packing. However, at a small value of Γ many different systems of parastichies have similar packing efficiency and can thus keep growing. The rules of appearance of a new element create a local order but they do not lead by themselves to an interpretation of the global order of the system. In particular the selection of the Fibonacci sequence does not result from them only: it is to be sought in the ontogeny of the plant. This is demonstrated *a contrario* in the cases where the continuity of the spiral organisation lacks. Zagorska Marek (1994 and this conference) observe in the inflorescence of Magnolia a large variety of organisations different from the expected Fibonacci modes. This is presumably due a growth of the inflorescence which starts almost directly at a very small value of G. Couder (1998) observed the same absence of Fibonacci organisation in an anomalous sunflowers. This breaking of the continuity of the growth can also be obtained artificially; Hernandez and Palmer (1988) performed an experiment in which they did a circular cut in the head around the apex of a flowering sunflower. They observed that the following regeneration of the capitulum led to a disordered pattern instead of the Fibonacci one. As shown above, the build-up of the Fibonacci patterns starts at the initial stages of the vegetative growth and occurs through the successive transitions during a continuous decrease of Γ. During this decrease the parastichy orders change with a sequence of transitions from (i, j) to $(i, i+j)$. This evolution occurs locally: it is linked with the change of the nearest neighbours of an element. As the same process occurs all around the apex, the system globally undergoes a transition (i, j) to $(i, i+j)$. It thus follows the main branch of solutions of the $\phi(\Gamma)$ diagram so that at small Γ a high order global Fibonacci pattern is obtained.

References

1. V. I. Arnold (1965) *Trans. Am. Math. Soc.* **46**:213.
2. L. Bravais and A. Bravais (1837(a)) *Ann. Sci. Nat.* second series **7**:42–110.
3. L. Bravais and A. Bravais (1837(b)) 193–221 and 291–348; (1837) *Ann. Sci. Nat.* second series **8**:11–42.
4. L. Bravais and A. Bravais (1839) *Ann. Sci. Nat.* second series **12**:5–14 and 65–77.
5. Y. Couder (1998) *Acta Soc. Bot. Poloniae.* **67**:129–150.
6. Y.Couder and S. Douady (1996). *Statphys 19*, Edited by Hao Bailin, World Scientific.
7. S. Douady (1998) pp. 335–358, in *Symmetry in Plants*, R. V. Jean and D. Barab, Editors, World Scientific.
8. S. Douady and Y. Couder (1992). *Phys. Rev. Lett.* **68**:2098–2101.
9. S. Douady and Y. Couder (1996 a) *J. of Theor. Biol.* **178**:255–274 (1996b) *J. of Theor. Biol.* **178**:275–294 (1996c) *J. of Theor. Biol.* **178**:295–312.
10. D. Friedman (1992) *In Spiral Symmetry* eds. I. Hargittai & C. A. Pickover. Singapore: World Scientific, 280.
11. P. B. Green (1992) *Int. J. Plant. Sci.* **153** (3):S59–S75.
12. L. F. Hernandez and J. H. Palmer (1988) *Amer. J. Bot.* **75**:1253–1261.

13. W. Hofmeister (1868). *Allgemeine Morphologie der Gewachse, Handbuch der Physiologischen Botanik*, 1. Engelmann, Leipzig, 405–664.
14. L. S. Levitov (1991) *Europhys. Lett.* **14**:533–539.
15. H. Meinhardt(1982). *Models of biological pattern formation.* Academic Press London.
16. G. H. Mitchison (1977) *Science* **196**:270–275.
17. M. Snow and R. Snow (1935) *Phil. Trans. Roy. Soc. London Ser.* **B225**:63–94.
18. M. Snow and R. Snow (1952) *Proc. Roy. Soc. London.* **B139**:545–566.
19. M. Snow and R. Snow (1962) *Phil. Trans. R. Soc. of London* **237B**:37–72.
20. A. M. Turing (1952) *Phil. Trans. R. Soc. of London* **237B**:37–72.
21. G. Van Iterson (1907) *Mathematische und microscopisch-anatomische Studien über Blattstellungen, nebst Betraschungen ûber den Schalenbau der Miliolinen*, Gustav-Fischer-Verlag, Iena.
22. A. H. Veen and A. Lindenmayer (1977). *Plant Physiol.* **60**:127–139.
23. A. Weisse (1897) *Jahrb. Wiss. Bot.* **30**:453–483.
24. D. A. Young(1978) *J. Theor. Biol.* **71**:421–432.
25. B. Zagorska-Marek (1994) *Acta Soc. Bot. Poloniae* **63**:117–137.

Cellular Patterns

ORGANOGENETIC CELLULAR PATTERNING IN PLANTS

P. W. BARLOW[1], H. B. LÜCK[2] AND J. LÜCK[3]

[1] *Cell Biology, IACR-Long Ashton Research Station, Department of Agricultural Sciences, University of Bristol, Long Ashton, Bristol BS41 9AF, UK*

[2] *Val d'Arenc, 83330 Le Beausset, France*

[3] *Laboratoire de Botanique analytique et Structuralisme végétal, Faculté des Sciences et Techniques de St-Jérôme, CNRS ERS 6100, 13397 Marseille cedex 13, France*

1 Introduction

I with the same sharp Penknife, cut off from the former smooth surface an exceeding thin piece of it, and... casting the light on it with a deep plano-convex glass, I could exceeding plainly perceive it to be all perforated and porous, much like a Honey-comb, but that the pores of it were not so regular... (R. Hooke (1665), Micrographia, Observation XVIII)

Cellular patterning in biological organisms is perceived by the eye as regularity in the disposition of cellular boundaries. This is particularly true of plant tissues where these boundaries-the cell walls-are refractile constructions of cellulose and other polysaccharides. The arrangement of wall surfaces around a plant cell is reminiscent of the facets of a geometrical solid, and much early work in the 1940s and '50s, seemingly conducted in that same spirit of enquiry which Robert Hooke showed in preparing his *Micrographia* 300 years earlier, was concerned with determining the shape of cells and counting the number of these wall facets (Marvin, 1939), analysing how these numbers varied during a cell division cycle (Matzke, 1949), and discovering the way in which the facets, and hence the cells, were packed within different tissues (Matzke, 1950; Duffy, 1951). Apparently, interest still persists in these aspects of plant cellular geometry and patterning (Korn, 1980; Mombach *et al.*, 1990; Rivier *et al.*, 1997), although the investigations have now moved away from purely geometrical considerations to analysis of surface energy distributions. In the realm of animal cells there are also similar interests (Honda, 1983) though here, in the last few years, cell shape has begun to be investigated in connection with the development of cell function, with the concept of 'tensional integrity' (tensegrity) providing the link between these two aspects of the cellular state (Ingber, 1997; Ingber and Folkman, 1989).

From another view point, the cell walls of plant organs may be regarded as an extensive system of extracellular matrix, this matrix material being secreted by the protoplasts onto their external surfaces where it 'solidifies' as a cell wall. It is probable that, when a new wall is interpolated into this wall/matrix system, its orientation has already been specified beforehand in accordance with both the prevailing local pattern of biophysical forces imparted by the already-existing walls to the dividing cell and the more global, transcellular forces due to the external boundary of the organ. The cell wall system grows and expands within meristems, which are zones, often at the apices of roots and shoots (primary meristems), where new cells are produced. It is in the meristems, during the process of cell division, that new walls are interpolated.

Cellular patterns within any given plant organ, and especially within the meristems, are generally highly conserved and specific to a given species. The water-fern, *Azolla filliculoides*, presents an example of invariant cell patterning during the formation of its shoot-borne roots (Gunning *et al.*, 1978). However, roots of many species, especially those of higher plants with their often larger dimensions, seem to show small, but presumably insignificant, irregularities (developmental noise) in the positioning of new division walls. It is reasonable to conclude, nevertheless, that the characteristic cellular patterns of plant organs do have some vital significance for organisms as a whole. What this could be may be answered with reference to:

(1) the levels of plant organization

(2) the sub-system structure of each level, in accordance with a 'Living Systems' concept of plant organization (Miller, 1978; Barlow, 1998).

Thus, at the level of the **cell**, the wall constitutes the 'Boundary' sub-system and serves as a protective layer around the sensitive and potentially vulnerable unit of cytoplasm. At the level of the **organ**, the complete set of walls constitutes the 'Supporter' sub-system for the organ itself. Like an internal scaffold supporting the organ, the walls reinforce, and sometimes substitute for, the turgor pressure of the wall-enclosed protoplasts (Niklas, 1992, pp. 268–274). Also, as is now becoming increasingly clear, the walls serve as part of the plant's set of Information-Processing sub-systems: not only do they provide channels for the communication of biophysical information in the form of stresses and strains, but they are also repositories of informational molecules. Representatives of one class of such molecules, the oligosaccharins, which are resident within the extracellular matrix, and are thought to form part of a signal transduction system for the regulation of plant tissue growth and differentiation (Aldington and Fry, 1993). In this respect, the extracellular material of plant cells performs similar functions to analogous molecules of animal cells (see Ingber and Folkman, 1989).

Certain cell walls within an organ may be selected by 'positional information' to serve as the boundary which encloses specific sets of metabolic processes. Such walls therefore define the tissues which are familiar from conventional anatomical studies of plant tissues. Even regional differences within a tissue can be identified by techniques which discriminate between the activities of single genes (Fleming *et al.*, 1993). Such methods thus define a tissue's microstructure.

These introductory statements relate to cell patterns seen from a conventional cellular viewpoint. But another, perhaps more profound, perspective is possible. It is that the patterning of cells and cell walls reflect the activity and mutual disposition of 'cell bodies' within the cytoplasmic space of the organ in question.

2 Cell patterning from the perspective of the cell body

And could we so easily and certainly discover the Schematism and Texture of these films... we might as readily render the true reason of all their Phenomena...
(R. Hooke, ibid.)

When viewed from either the cellular or the organ level of organization, the plant cell body is seen to be a basic instrument of growth and development. The cell body is capable of replicating itself and propagating information for the development of form as the organism progresses from zygote to adult plant. Some of these features of the cell body have been briefly discussed in connection with animal cells by Mazia (1993), although most animal cytologists use the term 'cell body' in a more mundane way to indicate only where the bulk of the cytoplasm is located (see Chapter 21 on "Nerve Cells" in Lodish *et al.*, 1995). Our conception of the plant-and animal-cell body is that it fundamentally consists of a nucleus (*i.e.*, chromatin, nuclear matrix, nuclear envelope) and a perinuclear population of microtubules (MTs) (Baluška *et al.*, 1998a, 1998b). These two basic components of the plant cell body are indissolubly wedded because the interior of the nucleus is apparently the site of manufacture of MT-organizing material which oozes onto the nuclear surface where it immediately directs polymerization of MTs from free tubulin molecules contained within the surrounding cytoplasm (Baluška *et al.*, 1997). The plant cell body also contacts, by means of its MTs, the cell periphery which it can cause to differentiate, and in this way directs further aspects of cellular and organ morphogenesis.

That the cell body is a powerful engine of cellular morphogenesis and ultimately of organogenesis is related to the fact that the MTs which radiate from the nuclear surface are highly dynamic. These 'endoplasmic' MTs continually and rapidly grow and shorten, this being due, respectively, to phases of assembly and disassembly of their tubulin sub-units. Thus, MTs are continually able to probe the cytoplasmic volume (Holy and Leibler, 1994) and presumably transmit information back to the nucleus for further processing. It is even possible that cell size is determined by the length-of-reach of the MTs from the nuclear surface into the cytoplasmic space, thence defining the volume of cytoplasm which a given cell body (and especially its nucleus) can control. When a meristematic cell nucleus enters the later stages of interphase, the dynamic MTs of the cell body, in ways yet unknown, define the position at which a preprophase band (PPB) of MTs forms in the cytoplasmic cortex. This cortical PPB, in turn, seems infallibly to define the orientation of the new cell plate (incipient division wall) which will begin to form a little later, at the telophase of mitosis. It is assumed that the PPB prepares a site on the mother cell wall which is recognized by the centrifugally expanding cell plate.

The form of the cell body is much modified upon entry into mitosis. At this stage in the cell division cycle, all MTs, especially those in the cell cortex, are disassembled and all the newly available tubulin molecules are reassembled into the mitotic spindle. The spindle MTs and condensed mitotic chromosomes comprise the 'mitotic cell body'; it represents one extreme of a number of topological forms which the cell body assumes during the course of cellular development. The dispersion of the nuclear envelope at late prophase seems inevitably to result in chromosome separation at anaphase. In telophase, at the close of mitosis, a new nuclear envelope reforms around each of the two daughter nuclei and two new minimal cell bodies are recreated. The cell bodies then become active and assist in directing the growth of the phragmoplast, a process which brings the cell plate to the site formerly occupied by the PPB. When the cell plate consolidates as a new division wall, cytokinesis is complete. The remaining cytoplasm is thus divided into two portions as two new daughter cells. Cell patterning is, therefore, a consequence of two 'directed' phases of cell body activity, one which is responsible for the positioning of the PPB, the other for the formation and positioning of the phragmoplast and cell plate (Baluška *et al.*, 1998b).

The processes achieved by the directed cell body require the cooperation of actin microfibrils. With the help of actin-associated motor proteins, the cell body, by means of cytoplasmic streaming, is able to extend its control over much larger volumes of cytoplasm (*e.g.*, the cytoplasm of elongated cells of root and shoot axes) than can be commanded by the simpler, minimal cell body of meristematic cells composed of nucleus and dynamic MTs alone.

In cases where a cell within an organism is multinucleate (em i.e., the cell is a coenocyte), the nuclei adopt approximately hexagonal packing within the parietal cytoplasmic space (Goff and Coleman, 1987). The nuclei become arranged in this way because the MTs associated with them (in conformity with the structure of a minimal cell body) radiate symmetrically from the nuclear surface, making contact with the plasma membrane and with a number of other similar cell-body-associated MT clusters (La Claire, 1987; Brown *et al.*, 1996). The **coenocytic state** of a cell or tissue (or even organism) comes about because the usual activity of the post-telophase cell bodies has been disabled by the prevailing pattern of gene activity. Another possibility is that no cell plate precursors are transported into the phragmoplast which, as a consequence, fails to develop. Division of nuclei does not disturb their mutual spacing arrangement, which adjusts to the incoming pairs of new-born cell bodies at the completion of telophase. However, in the more usual **multicellular state** of an organ, an active cell body resides within a cytoplasm capable of supporting cell wall synthesis. It assists in both adding new precursor material to pre-existing walls and creating a new cell wall by constructing a phragmoplast and cell plate. There is, therefore, no fundamental difference between coenocytic and multicellular tissues, organs or organisms since all are comprised of the all-important cell bodies. They differ because of the presence or absence of cell walls around the cell bodies. Although walls are often important from a structural point of view, their absence (as in the coenocytic organism, *Caulerpa*, for example) can be compensated in other ways (see Niklas, 1992, p. 284; Jacobs, 1994).

Control over the orientation of the mitotic cell bodies and of the resulting division walls opens the possibility of non-hexagonal packing of cells within tissues. Such a modulation of cell packing can be demonstrated in leaf tissue from analyses of R values using the equation $R = l\sqrt{d}$, where l is the average distance to the nearest neighbour cell centre, and d is the cell density (number of cells per unit area). Estimates of R for arrays of epidermal cells showed that values of 1.6–1.7 were more frequently encountered than were R values of about 2.19, which would have been obtained in cases where there was perfect ordering of hexagonal cellular arrays (Korn, 1993, 1994). In addition, analyses of the relative positions of mitotic cells in a region of the leaf meristem of *Tropaeolum peregrinum*, which was assumed to be growing isotropically, showed statistically unexpected alignments of these cells (Fuchs *et al.*, 1985). These alignments might have been a consequence of dislocations, whose development was discussed by Rivier *et al.* (1997), but they could also possibly indicate the workings of a supracellular regulation of organogenesis.

Cell bodies are always divided equally at telophase because this is a property of the symmetrical bipolar mitotic apparatus. A cell, however, may be divided unequally because of the asymmetric placement of its cell body and the PPB which it fabricates. This size difference leads to unequal interdivisional periods of cell body reduplication in the two sister cells which, in turn, leads to cell size distributions and cell patterns within the cell population that are more complex than they would be if all interdivisional periods were similar (Lück *et al.*, 1997). Moreover, as mentioned earlier, the cell walls are channels (or repositories) of information, some form of which can influence the placement of new cross-walls, perhaps as a result of being able to inform the cell body about where to locate the PPB. Thus, the presence of pre-existing walls can influence where any future new wall will be inserted. Exactly how this information for wall pattern formation propagates within an organogenetic field so that developmentally relevant networks of walls are created is a fundamental question in biology, especially since the growth of the organogenetic field itself is probably regulated by the controlled expansion of the organ boundary, rather than by the growth of the individual component cells (Kutschera, 1995). This question will be addressed later in our chapter. But in this regard, it is important to appreciate that, very probably, there is an informational loop connecting the cell body and the division wall such that the orientation of the cell body influences the orientation of the division wall; the division wall can then, in its turn, influence the orientation of the cell body, and so on.

3 What lies behind analyses of cell patterning?

Our Microscope will easily inform us, that the whole mass consists of an infinite company of small Boxes... (ibid.)

Certain multicellular organisms (*e.g.*, algae) are aggregates of hexagonally packed cells (*e.g.*, coenobia of *Pediastrum boryanum* — Ingold, 1973) or, at another extreme, they are able to position their cells so that they form the edges of hexagonal meshes (*e.g.*, the cells of *Hydrodictyon reticulatum* — Marchant and Pickett-Heaps, 1972). All these diverse forms have the capacity to differentiate a limited number of cell types. But in the context of green plant development (especially within the Tracheophyta), more scope for cell differentiation comes with the ability to regulate cell shape. This can be achieved by the implementation of local cell division controls (division rules) and by regulation of local cell growth rates (differential cell growth). It is possible that both of these controls reside in modifiers of the inherent spatio-temporal organization of cellular structure and molecular processes, although the latter may also be the result of the growth-regulating chemical gradients which attend organogenesis.

Generally speaking, the patterns of cells seen in the apices of Angiosperms and Gymnosperms pose complex problems for their analysis. Irrespective of whether the three-dimensional pattern of all cells or the two-dimensional pattern of just the superficial epidermal cells are considered, the interpretation of the cellular relationships within the patterns is often too ambiguous to be useful in achieving the main objective of such a study, which is to make explicit the rules by which a given cellular pattern is generated. By contrast, the apices of many ferns (Pteridophytes) show cell patterns which are more amenable to analysis and interpretation. One great help in this respect is that groups of cells of common descent-merophytes-can be identified on account of their distinctive boundary walls. The sequence and orientation of the cell divisions which have occurred within these boundary walls can be inferred by identifying the walls that were formed at successively earlier rounds of cell division (Lück *et al.*, 1988; Barlow, 1991). The pattern of cell walls is thus a reflection of the pattern of cell body divisions. A necessary assumption in analyses of merophyte cell patterns is that the attachment points of walls to other walls remain fixed throughout merophyte development. Nevertheless, as the merophytes grow in conformity with the overall organogenetic program, so the relative lengths of walls and the angles of their attachment may change.

A comparison can be made between the cellular pattern contained in each successive merophyte. If the patterns in each merophyte are identical, then the division pattern is stereotypical. If the patterns are different, then a source of developmental perturbation must exist which influences what otherwise might be a regular division process. The question then is whether deviations from a presumed regular pattern have a biological basis or not. If they do, then they may be non-trivial and part of a larger pattern of development: it is, after all, not immediately apparent what is the normal pattern and what is a variant because some variant types of cell division might, in fact, be part of a normal, more global pattern. Alternatively, deviations could reflect origins in irrational, non-deterministic events.

Further comprehension of the development of cellular patterns is often possible because the sequence of merophyte production can be traced to a single prominent cell at the apex. In one of the simpler situations found in the shoot and root apices of some types of ferns, this 'apical cell' is pyramidal (tetrahedral) and shows a triangular outline when seen in surface view. All cells of the apex are descendents of this cell, which can hence be considered an organogenetic centre, especially since apical cells appear at the onset of organogenesis during embryogeny. Where the pattern is simple, the more likely it is to reveal the minimal number of rules that govern the behaviour of the cellular complex. If it were also possible to discover the simplest of the tracheophyte apical cellular patterns, it might be possible to infer from it the prototypic rules for organogenesis.

Evolution of land plants has been continuing for about 450M years. Some living taxa (*e.g.*, the Anthocerotopsida) persist as relicts of this earlier epoch though, naturally, details of their structure have changed since the time when they were at the 'leading edge' of evolving plant life millions of years ago. Another interesting group of plants in the early land flora, and known only from the fossil record, is the Rhyniopsida, which includes *Rhynia gwynne-vaughnii.* Sporophytes of this plant were rootless, but did have horizontal rhizome-like axes as well as upward-growing aerial axes which bore sporangia at their apices. In these respects, and also with regard to the dichotomous branching of its shoots, this Rhyniophyte is remarkably similar to present-day species of *Psilotum* (Pteridophyte). Because of these features, the Psilophytes were at one time considered to be living relatives of the Rhyniophytes. One problem with this interpretation, however, is that, after the mid-Devonian period, the fossil record lacks any trace of the Rhyniophytes. Although absence from the fossil record is not proof that Psilophytes are not relics of the Rhyniophytes, evidence from molecular biology and cladistics now suggests that the Psilophytes are a morphologically reduced type of eusporangiate fern (Kenrick and Crane, 1997). It is possible, nevertheless, that the morphological reduction of a hypothetical fern ancestor could have brought the Psilophytes back to a primitive state as a temporally retrograde homoplasy, or atavism, recapitulating the form once shown by the now extinct Rhyniophytes.

In view of all these interesting features concerning the Psilophytes, it is important to examine the cellular patterns which contribute to the growth and branching of their seemingly prototypic axes. It may be that what is revealed from analyses of these patterns also gives an indication of how one of the first land-plants developed at a cellular level of organization. Even if the cellular patterning of the *Psilotum* apex does not resemble that which prevailed in antiquity, some 400M years ago-and here there is evidence (Edwards, 1993) suggesting that the cells in the superficial cell layer of the *Rhynia* shoot apex might have been arranged differently to those of *Psilotum* (see later in this chapter)-it is still certainly one of the simplest patterns that involves a tetrahedral apical cell (tetrahedral in shape, but usually with more than 4 walls). Moreover, the cellular pattern of the *Psilotum* apex possesses the potential for its own autoregeneration: that is, new apical cells and their derivatives can be detected in each merophytic descendent of the principal apical cell. However, as described in the next section, not all these second-order apical cells function as organogenetic centres. Interestingly, non-growing buds also occurred along the stems of *Rhynia*, whereas other buds were clearly able to extend as lateral branches (Edwards, 1993).

4 A simple autoregenerating cell pattern related to organogenesis

A curious piece of Clock-work, that had had several motions and contrivances in it, which, when in order, would all have mov'd in their design'd methods and Periods. (idem., Observation XXI)

Plant axes should be considered at two levels of organization-the cell and organ levels. In their detailed study of both the shoot and the rhizoid-like subterranean axes of *Psilotum nudum*, Takiguchi *et al.* (1997) found that the principal features at the organ level were:

- Anisotomous dichotomy of axes (the most frequent type of branching) and isotomous dichotomy (the least frequent) [1]. Monopodial branching was of intermediate frequency.

[1] *Dichotomy.* This term is useful at the organ level to describe the terminal branching of an organ to give two new organ axes. Dichotomy is usually accompanied by a splitting of the apex and the continued growth of the two resulting portions. The branches derived from the split apices may be either equal in size/or growth rate (isotomous) or unequal (anisotomous). At the cellular level, however, where the degree of observational resolution is greater, it is often difficult to be sure whether dichotomy is the correct term to describe such a branching event. The term would apply if, at the organ apex, a group of organ-initiating stem-cells became separated into two daughter groups, both of which then proceeded to direct organogenesis. But in the case of *Psilotum*, evidence for true dichotomy is lacking because one branch axis is derived from a first-order apical cell, whereas the other branche is derived from a second-order apical cell originating, not directly from the first-order apical cell, but from a descendent multicellular merophyte. From

- Enations (determinate axes) were present on all shoots.
- Subterranean axes were slow-growing (no more than 0.3 cm in length per month).
- Branches bore buds. Their subsequent development was not uniform: some buds continued growth after about 1 month of dormancy, others remained dormant for up to 1 year.

The principal cellular features were:

- There was a tetrahedral primary apical cell which was triangular in surface view or in transverse section.
- Division of the triangular apical cell was regular, the production of merophytes involving each of the three interior facets in turn, in either a dextrorse (winding into the apical cell in a clockwise helix) or sinistrorse (counter-clockwise helix) sequence.
- Three different cell division patterns could be found in the superficial cells of the merophytes (Figure 1). Two of these patterns gave rise to new, second-order triangular apical cells (Figure 1, left and centre) which probably had the potential to serve as apical cells of new axes. The third division pattern (Figure 1, right) did not generate new triangular cells.

Although more research and observations are required from the *Psilotum* apices, two likely hypotheses which relate cell patterns with organogenesis are that:

(1) the formation of a second-order, triangular apical cell is dependent upon the preceding division sequence within the relevant merophyte

(2) the type of organ branching and outgrowth is dependent upon the degree of activity of the cells.

Further possibilities are that:

- Monopodial branching is related to the slower rate of development of the derivative cells surrounding a second-order triangular cell compared to those around first-order triangular cells. The former group of cells can remain dormant for up to one year before becoming active again.
- Anisotomous branching occurs when there is a lesser, though still a significant, growth differential between first- and second-order triangular apical cells and their respective derivatives.

---

the cellular perspective, this is not evidence of dichotomy, even though at the level of the organ it would seem probable that dichotomy had occurred. The cellular resolution of the branching event in *Psilotum* impels us, therefore, to consider apical bifurcation, where it occurs by means of a first-order and a derivative second-order apical cell, as a 'pseudo-dichotomy'. A true dichotomy could be considered only if one triangular apical cell gave birth directly to another triangular cell, and the two sister triangular cells then behaved identically; but such a procedure does not occur.

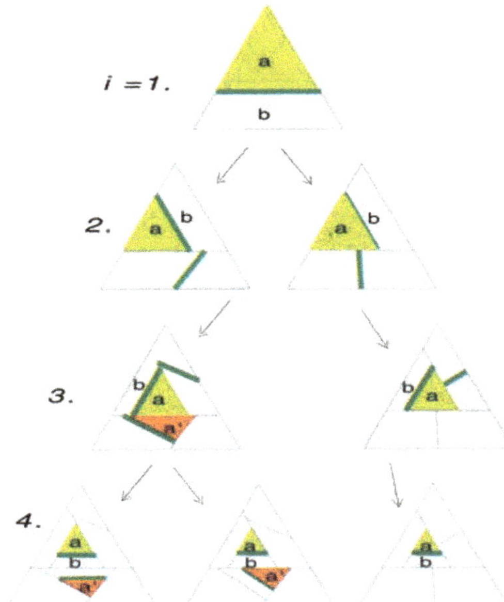

Figure 1. Schemes of three basic types of cellular segmentation found in apices of *Psilotum nudum*. The starting point is a triangular apical cell from which a daughter cell is derived at each successive time-step. Division of these daughter cells results in groups of cells of common descent which are termed merophytes. The three schemes are developed over four time-steps, i = 1 to i = 4, the triangular apical cell dividing four times in a counter-clockwise direction in each scheme (*i.e.*, divisions wind into the apical cell in a sinistrorse fashion). At $i = 1$ the self-maintaining triangular cell has already divided (the division wall is shown as a dark-green line) into a triangular daughter cell, a (light-green) and a four-sided daughter cell, b (uncoloured). Left-The most-frequent pathway of development of the b cell is traced by arrows at the *left-hand side* of the diagram. The outcome of two oblique (relative to the attachment walls for the new cross-wall) divisions, at $i = 2$ and $i = 3$, is a second-order triangular cell, a' (orange). Note that the oblique walls are off-set with respect to the division walls in the triangular cell and attach external to the perimeter of the triangular cell (*cf.* Fig. 3a). The offsetting leads to an 'S' junction forming between the three walls involved; hence, the four cousin cells form an 'S' tetrad. The a' cell then continues its division in a clockwise direction (*i.e.*, divisions wind into the a' cell in a dextrorse fashion). Centre- At bottom *centre* is an infrequent variant of the previous *left-hand* pathway. Here, a third oblique division is inserted into the merophyte at $i = 4$ (see also Fig. 4a, apex 22) and the usual clockwise spiral of merophyte production from this a' cell appears to be interrupted. Right- The pathway on the *right-hand side* traces another less-frequent sequence of b cell divisions where the new cross-wall is perpendicular to the cross-wall that had earlier divided the triangular cell (see also Fig. 4a, apex 24). No second-order triangular a' cell is derived in this pathway.

- Isotomous branching involves similar rates of development in two closely related merophytes.
- The formation of an enation may be related to non-autoregenerating cell patterns around a second-order triangular apical cell (*e.g.*, Figure 1, centre) or to an absence of autoregenerating cells (*e.g.*, Figure 1, right). The cell pattern shown in Figure 1, left, is possibly the only one of the three which invariably supports development of new axes.

It should be remarked that the terms 'first-order' and 'second-order' triangular cells are relative. A first-order cell is simply one which is the principal triangular cell in a field of observation (usually at the organ's apex), and the second-order cell is one of its derivative. However, it is possible that some difference could exist between these cells on account of the different states of their walls (this point will become evident later, see Figure 5) and the pattern of division in the merophytes which surround them.

5 Simulation of cell patterns and the uncovery of division rules

We must grope, as it were, in the dark, and onely ghess at the true reasons of things by similitudes and comparisons... (idem., Observation XVIII)

5.1 The Simulation System for the Psilotum *Apex*

Simulation of the cell patterns displayed by the outer cell layer of the *Psilotum* apex validates the rules which have been devised for this patterning and also promotes a commentary upon the cell-biological properties which may correlate with these rules. What is remarkable about the cellular patterning of the apex is that, whereas the first-order triangular apical cell is a relatively large cell, each second-order triangular cell is at first smaller and, importantly, is but one of a number of other cells which occupy not only the outer layer but also the multicellular depth of the enclosing merophyte (see Figure 2). Their continued growth, however, would no doubt bring these second-order cells to the same size as the first-order cells. In the present analysis, however, the outer layer is considered to be autonomous and its pattern of cell division independent of any cellular activities in the subjacent cell-tiers. The pattern of division of cells in the outer layer can therefore be adequately analysed with reference only to their two-dimensional, and not to their three-dimensional, disposition. Consideration of all the other cells would be needed for a complete analysis of three-dimensional cell patterning within the apical merophyte system and for this purpose 3D-L-systems (cellworks) have been proposed (Lück and Lück, 1992, and in this volume).

To simulate the development of cellular patterns in two dimensions, we have used double wall Map OL-systems, dwMOL-systems (Lück and Lück, 1981). Cell walls are represented by line segments. A segment corresponds to either an entire wall (edge to edge) or a portion of wall. Segments are numbered on each side because each wall (or portion of wall) is shared by two neighbouring cells. Within a cell, the numbering of the line segments proceeds clockwise, as is the convention for our map-systems, always starting with the youngest (which is also the last division wall to have been inserted)[2].

---

[2]Other map L-systems have been defined which use simple wall labelling (Lindenmayer and Rozenberg, 1979; De Boer and Lindenmayer, 1987). However, double walls are more accurate for the definition and analysis of categories of systems, opening the possibility for their classification (Lück and Lück, 1981) and the derivation of integrated alphabets for complete sets of cell walls, cell types and even sub-maps. Also, they can be easily translated into simple wall labelling systems which are less cumbersome for automated graphical treatment of map expansion (Fracchia, 1996).

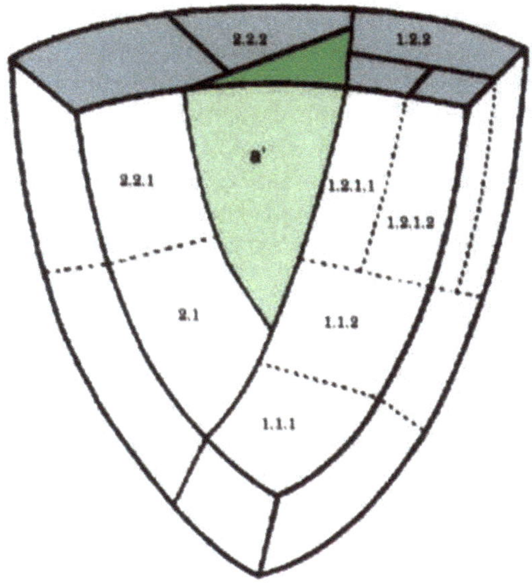

Figure 2. Three-dimensional scheme of a tetrahedral (triangular) second-order apical cell, a', and two of its associated merophytes (merophyte production is counter-clockwise, so the oldest merophyte is to the right) within which one or two further rounds of cell division are currently taking place (new internal cell walls are represented by broken lines). Cells in the two merophytes are numbered according to the order of their production. The stippled area is that which would be seen in surface view of the apex. The next division of the triangular cell would be parallel to the front wall. (From an apex of a subterranean axis of *Psilotum triquetrum*. Redrawn from Siegert, 1965).

In the present case, the starting-point for the simulation is a three-sided cell. The simulation system (Map L-System) which we apply is one which is designated S_{5-5}. This system uses wall parity: a label on one side of a wall segment has a corresponding label on the other side. The initial set of wall segment productions (label substitutions rules) is the following:

$$\begin{array}{ll} 1 \to 4.5/_6 7 & 6 \to 5/_6 7.8 \\ 2 \to 8 & 7 \to 9 \\ 3 \to 9 & 8 \to 0/_1 2 \\ 4 \to 0/_1 2 & 9 \to 3 \\ 5 \to 3 & 0 \to 4 \end{array} \qquad \text{(A1)}$$

Because the maps derivated by the system are topological objects, the exact relationship of non-intersecting regions is specified. But there is no specification of map size. For convenience of representation, the successive maps derived by the system are drawn with an increasing size. Neither angles nor segment lenghts are specified (as in geometry)? In fact, the maps can be drawn in any shape, provided that the neighbourhood of regions is respected. Nevertheless, for biological relevance, we have always introduced in our map systems an additionnal geometrical interpretation of maps: thus, segments become more or less the same lenght; and they can also be represented as stereomaps to show the three-dimensional deformation of the maps (Lück and Lück, 1985.

A point (.) separates two consecutive wall segments. A slash (/) indicates a division wall attachment node; the subscript of the slash (*e.g.*, $/_1$ and $/_6$) indicates the label associated with the division wall. In order to trace their development, the successive maps of the cell walls need a starting point. Accordingly, in its initial state, at map derivation step $i = 0$, there is a map represented by a cell with three walls (Figure 3 a.0), the segments of which are numbered 1. 2. 3. 4. 5. In this sequence, the segment labelled 1 corresponds to the last-added division wall (*dw*). Segment 1 is an entire wall. The other two walls are each composed of two segments, 2.3 and 4.5, respectively.

Because of the double labelling of walls, the set of complete walls for a given cell can be represented by an antiparallel double string of labels (such as is used to denote strings of base pairs along a DNA double helix). The triangular initial cell has three neighbouring cells, so the corresponding double-wall notation which incorporates the segments of these walls is $\frac{1}{6}$ (for the *dw*), $\frac{3.2}{9.0}$ (or $\frac{0.9}{2.3}$, which is the same wall), and $\frac{5.4}{7.8}$ for the two other walls.

At each derivation step i,

(1) the map is redrawn

(2) the labels are rewritten in parallel according to the substitution rules

(3) new *dw*s are added to the map by joining, within a cell, the two specified attachment sites with a line.

So, in the wall map at state $i = 1$ (Figure 3a.1), and using substitution rules A1, the label 1 found earlier at $i = 0$ is replaced by the sequence 4.5/7, the slash (/) indicating one of the two *dw* attachment sites. The new wall insertion is, by convention, always clockwise with respect to the cell centre. At the same time, label 2 is replaced by label 8, label 3 becomes label 9, and label 5 becomes label 3. A segment with label 4 is replaced by two segments with labels 0/2 which are separated by the second *dw* attachment site. A line is introduced between these two attachment sites to represent the *dw*; it is then given the label 1 in daughter cell *a* with the wall sequence 1. 2. 3. 4. 5, and label 6 in the other daughter cell *b* with the wall sequence 6. 7. 8. 9. 0. The result is a map composed of two cells, a cell *a* with three neighbours and a cell *b* with four neighbours. Other types of cells may appear at subsequent derivation steps, as shown in the next section.

All cells divide at each step into a pair (a, b) of daughter cells; their derivation can be written as cell division rules $a \rightarrow a\ b$, and $b \rightarrow a\ b$. With a finite number of rules, which are applied similarly to all cells, an infinite map derivation can be generated. Six successive wall maps are shown in Figure 3a, one map being derived from a previous map at each step i. Note that in the development of these maps, all the walls are joined cross-wise (*i.e.*, the walls meet at four-way junctions). This contrasts with actual wall networks where there is wall-offsetting. The problem of wall-offsetting will be discussed and resolved later in this chapter.

Theoretically, there are only two simulation systems out of many hundreds of possible systems (Lück *et al.*, 1988) which are able to generate three-sided segmentation: that of A1 (discussed above) and another, A2, which is its symmetric counterpart. The label substitution rules of this last-mentioned system are:

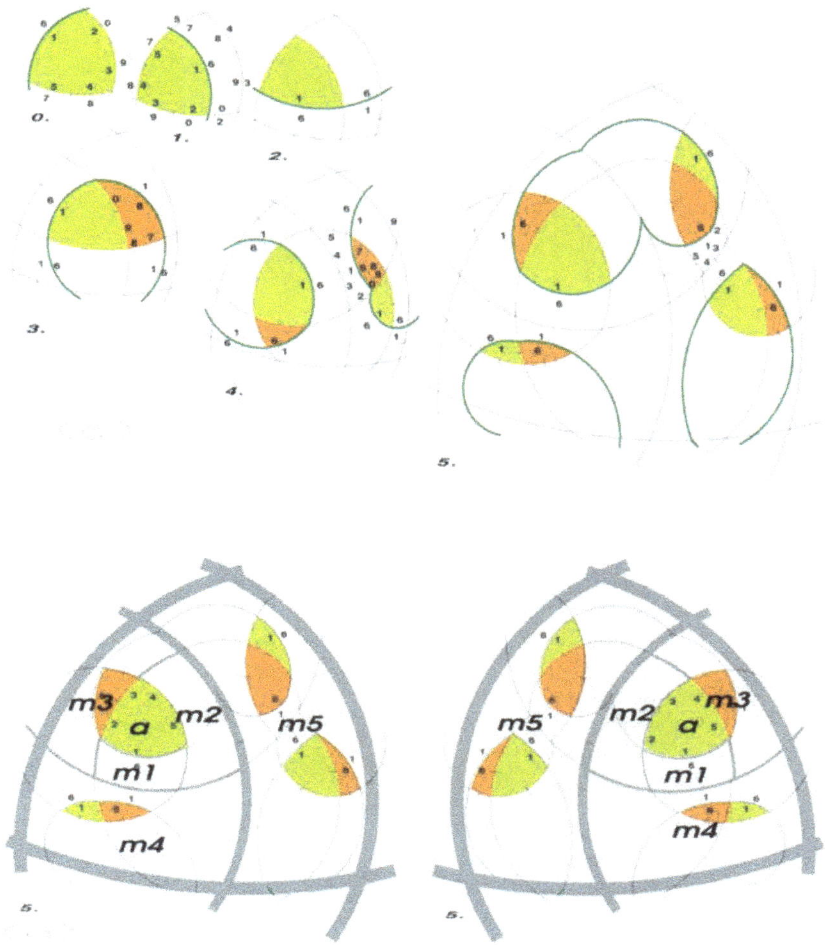

Figure 3. Map derivations for map system S_{5-5}. (a) The initial map is shown at $i = 0$. Subsequent developmental states derived from label substitution rules A1 are represented by maps at steps $i = 1$ to $i = 5$. Wall segments in each cell are labelled in clockwise sequence. At steps $i = 0$ and $i = 1$, a triangular cell a (light-green) shows the wall labelling sequence 1. 2.3.4.5; in the four-sided sister cell, b, the sequence is 6.7.8.90. At subsequent steps $i = 2$ to $i = 5$, all other cells with a division wall labelled 1 are a cells, and all cells with a division wall labelled 6 are b cells. At step $i = 3$, a triangular b cell (orange) with wall sequence 6. 7.8.9.0. emerges in the first merophyte initiated at step $i = 1$ (it corresponds to the second-order a' cell of Fig. 1). Its derivatives (from the next step on) are four-sided a cells with wall sequences 1.2.3.4.5. Two steps later, at $i = 6$(not shown), this first derivative again generates a new triangular first-order a cell. For the complete list of cell types occuring in the S_{5-5} system, see text and Figs 5 and 6. The sequence of merophytes develops in a clockwise sense for each triangular a cell, but in a counter-clockwise sequence for each triangular a' cell, of whatever order it may be. Cell and line colourings are as in Fig. 1. (b) Left: The clockwise sequence of merophytes ($m1 - m5$) in the map at $i = 5$, derived utilizing label substitution rules A1, as in a. Right: The counter-clockwise sequence of merophytes $m1 - m5$ in the map at $i = 5$ are derived utilizing label substitution rules A2. The initial wall sequence is the same as in a, and the wall labelling in the cells is also clockwise, as is conventionally used for all map-systems.

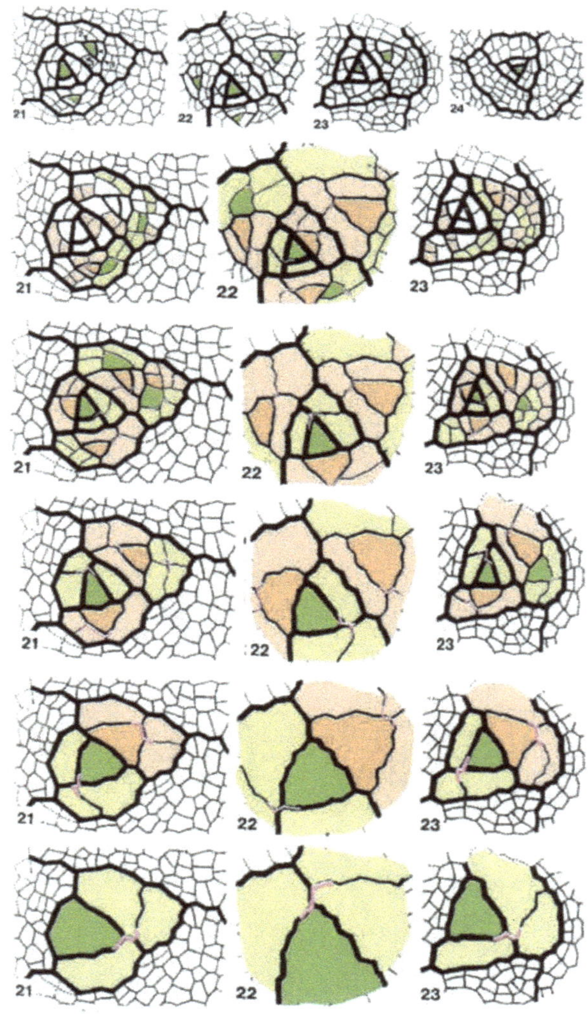

Figure 4. (a) Drawings of four apices of subterranean axes of *Psilotum nudum* (from Takiguchi *et al.*, 1997, with permission of American Botanical Society). Successively formed merophytes, bounded by thick black line, derived from the central, first-order triangular apical cell (*a*) are formed by a 3-sided helicoidal segmentation. The first-order triangular cells, as well as second-order triangular cells, are coloured green. The direction of triangular cell division is clockwise for apices 21, 23 and 24, and counter-clockwise for apex 22. Generally, the division sequence in the merophytes derived from the four-sided *b* cell conform to the scheme shown in Fig. 1, left (see apex 21, where some of the successive division walls within a merophyte are numbered and indicated by arrowheads), but occasionally deviations to the scheme, as shown in Fig. 1, centre, are encountered (*e.g.*, small arrows in apex 22). The cell pattern shown in apex 24 conforms to the pathway shown in Fig. 1, right. (b)–(e) Tetrad relationships within merophytes of apices 21, 22 and 23, which existed one (b), two (c), three (d), or four (e) generations prior to the situation shown in a. Cousin cells within a tetrad are indicated by similar colours (green or pink). The walls involved in the offsetting of division cross-walls which give rise to the *S* or *Z* tetrad types are coloured lilac.

$$
\begin{array}{ll}
1 \rightarrow 0/_2.3 & 6 \rightarrow 9.0/_12 \\
2 \rightarrow 4 & 7 \rightarrow 3 \\
3 \rightarrow 5/_67 & 8 \rightarrow 4 \\
4 \rightarrow 8 & 9 \rightarrow 5/_67 \\
5 \rightarrow 9 & 0 \rightarrow 8
\end{array} \qquad \text{(A2)}
$$

Maps derived from the two systems, A1 and A2, are shown in Figure 3b. Merophytes relating to system A1 wind into the apical cells in a clockwise spiral (Figure 3b, left), whereas merophytes derived from system A2 wind in a counter-clockwise spiral (Figure 3b, right). A first glance at four actual apical patterns of the subterranean axes of *Psilotum nudum* (Figure 4a-originally from Takiguchi *et al.*, 1997) shows that one apex has a counter-clockwise pattern of merophytes (Figure 4a, apex 22) and the other three have clockwise patterns (Figure 4a, apices 21, 23 and 24).

5.2 Identification of cell types

In the simulation system S_{5-5}, 10 different cell types are formed. They are produced over $i = 5$ timesteps. At later steps, only representatives of already-formed cell types are produced. Each cell type is notated as a_y^x or as b_y^x, superscript x indicating the number of cell neighbours or cell walls, and subscript y the order number of the cell, thus differentiating cells with the same number of neighbours but with different wall sequences. Only the surface view of the cells is taken into account.

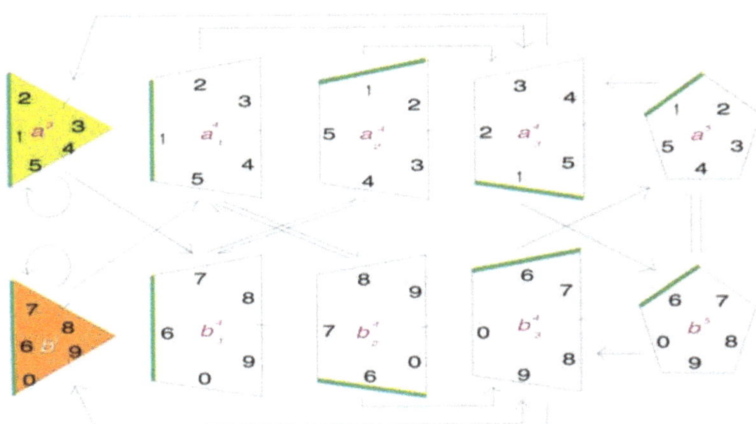

Figure 5. Diagram of the ten cell types. They are derived according to map system S_{5-5} and identified by their labels (purple). Also shown is their approximate shape (the number of walls is also indicated by the superscript number of their label). The labelling of wall segments is according to the development of the map system (see A1). The cell derivation graph is indicated by arrows and shows the cell types taken by daughter cells. Cells a^3 (green) and b^3(orange) are self-reproducing cells and therefore have the potential to become apical cells. The cells shown comprise the basic set of cells. Interpolated α wall segments which generate division-wall off-setting and, hence, cells with additional walls, are not shown.

So, for example, a^3 is a cell of type a, with three neighbours and with wall sequence $(1.\underline{2.3}.\underline{4.5})$, the several segments of one wall being indicated here by underlining (*cf.* Figure 3a.0, state $i = 0$). b_1^4 is a four-sided cell with wall sequence $(6.7.\underline{8.9}.0)$; b_2^4 is another four-sided cell with another wall sequence $(6.7.8.\underline{9.0})$. The forms, wall-labelling and derivations (following substitution rules A1) of each of these 10 types is shown in Figure 5.

With these cell types, **cell** production rules, in contrast to the previously used **wall** production rules, can now be defined. For example: $a^3 \to a^3 b_1^4$ which results from the derivation of the first map shows that a three-sided mother cell divides into an identical three-sided daughter cell (a) and a four-sided second daughter (b).

The set of cell production rules are:

$a^3 \to a^3 b_1^4$	,	a^3	with wall sequence	$(1.\underline{2.3.}\ \underline{4.5})$	
$b^3 \to b^3 a_1^4$	,	b^3		$(6.\underline{7.8}.\underline{9.0})$	
$a_1^4 \to a_3^4 b_2^4$	,	a_1^4		$(1.2.\underline{3.4}.5)$	
$b_1^4 \to a_2^4 b_3^4$	,	b_1^4		$(6.7.\underline{8.9}.0)$	
$a_2^4 \to a_3^4 b_1^4$	,	a_2^4		$(1.\underline{2.3}.4.5)$	
$b_2^4 \to a_1^4 b_3^4$	,	b_2^4		$(6.7.8.\underline{9.0})$	(A3)
$a_3^4 \to a^3 b^5$	,	a_3^4		$(1.2.3.\underline{4.5})$	
$b_3^4 \to a^5 b^3$	,	a_3^4		$(6.\underline{7.8}.9.0)$	
$a^5 \to a_3^4 b^5$	,	a^5		$(1.2.3.4.5)$	
$b^5 \to a^5 b_3^4$	,	b^5		$(6.7.8.9)$	

These rules (A3), therefore, represent a higher-order, more integrated, representation of the same map system, A1. For successive alphabet integration, see Lück and Lück (1987).

5.3 *Merophyte formation by apical divisions*

Except for the problem of wall-offsetting, the comparison of theory and reality revolves around the definition of the boundaries between the merophytes. These boundaries reveal the history of the successive divisions of the triangular apical cell. At some later time in development, a portion of each merophyte is occupied by a second-order triangular apical cell, such as a' in Figure 1, which itself becomes a source of new merophytes. The theory, which consists only of a set of simple wall segment production rules, shows how the original first-order triangular cell, a, is maintained and how a spiral pattern of merophytes descends from the succession of b cells. It also shows how the spiral of merophytes changes direction when it originates from the next-order triangular cell. These nested sets of merophytes of primary and secondary order are highlighted in Figure 3a.4 and Figure 3a.5.

As mentioned, it can be deduced that the self-reproduction of apical cells is described by a cell production step such as $a^3 \to a^3b_1^4$. But there is another cell production, $b^3 \to b^3a_1^4$, which states that b^3 is also a triangular self-reproducing cell, appearing in the next derivation step, $i = 3$, in the first merophyte of the apical cell a^3. Therefore, b^3 cells can be considered as second-order triangular apical cells (denoted as a' in Figure 1). Nevertheless, second-order triangular cells derived from merophytes of b^3 cells are again of type a^3, and so on (see Figure 5). As in other systems besides S_{5-5}, second-order triangular cells may not change their cell type, but presumably go on to behave like the earlier first-order cells.

Although second-order triangular cells have here been denoted as a', this has nothing to do with a or b cells which are labels belonging to the daughter cells of a systems terminology. As can be seen in A3, and also in Figure 5, the first-order s cells (a^3) and the second-order a' cells (b^3) differ with respect to the labelling of their wall segments.

5.4 Wall-offsetting leads to refinement of the wall production system

The most striking difference between the patterns generated theoretically and those realized by the *Psilotum* apices resides in the positioning of adjacent cross-walls. As already mentioned, four-way wall junctions are shown in the theoretical map system but they rarely exist in actual apices. Wherever new division walls attach to an older, common wall in a cell tetrad, they are displaced with respect to each other, creating three-way junctions with either S or Z configurations (Lück *et al.*, 1988; Barlow, 1991). A small segment of older wall intercalated between the attachment sites of two new division walls can produce the desired offsetting. This segment need not be included among the wall segments specified in a given map system (Lück and Lück, 1995). This, of course, is with the proviso that these small segments do not become anchorage sites for any subsequent division wall, which usually seems to be the case. The intercalation of these supernumerary segments often means that hexagonal cells will be realized; previously, when there were four-way junctions, with X-type tetrads, there were only four-sided cells. The resulting S or Z tetrads are not randomly located, however, and their productions can be defined by deterministic rules (Lück *et al.*, 1988; Lück and Lück, 1995).

With this additional information about offsetting, analysis of the drawings of Figure 4a now reveals that the general pattern is for Z tetrads to occur in clockwise segmentation and S tetrads to participate in counter-clockwise segmentation. It also turns out that the division walls attached to a wall which joins cells belonging to two different tetrads are S tetrads if the two contiguous tetrads are sister tetrads, but show a Z configuration if they involve two contiguous cousin tetrads (Figure 4b-e).

These few additional rules permit refinement of the simulation system. It is named S_{5-5off} and its realization, with clockwise merophyte production, is illustrated in Figure 6. The set of cell production rules may also be amended by the introduction of small wall segments, α, that effect the offsetting and which count as cell contacts. Cell type a^3 with wall sequence $(1.\underline{2.3}.\underline{4.5})$ remains unchanged, however. It is an autoreproductive cell. But cell type b^3 with sequence $(6.\underline{7.8}.\underline{9.0})$ now becomes $(6.\alpha.\underline{7.8}.\alpha.\underline{9.0}.\alpha)$ and is transformed from a three-sided cell into a six-sided cell with three new cell contacts provided by the three extra α walls. Therefore, the cell b^3 is now re-named $b^{6-(3)}$ with$^{(}3)$ to remind us of the basic degree. This cell remains autoreproductive, nevertheless.

The new cell type alphabetand the corresponding new set of cell productions can be summarized as follows:

Basic cell type	Basic wall sequence	New cell type	New wall sequence	New cell productions	
a^3	$(1.\underline{2.3}.\underline{4.5})$	a^3	$(1.\underline{2.3}.\underline{4.5})$	$a^3 \to a^3 b_1^{6-(4)}$	
b^3	$(6.\underline{7.8}.\underline{9.0})$	$b^{6-(3)}$	$(6.\alpha.\underline{7.8}.\alpha.\underline{9.0}.\alpha)$	$b^{6(3)} \to a_1^{6-(4)} b^{6(3)}$	
a_1^4	$(1.2.\underline{3.4}.5)$	$a_1^{6-(4)}$	$(1.2.\alpha.\underline{3.4}.\alpha.5)$	$a_1^{6-(4)} \to a_3^{5-(4)} b_2^{7-(4)}$	
b_1^4	$(6.7.\underline{8.9}.0)$	$b_1^{6-(4)}$	$(6.\alpha.7.\underline{8.9}.0.\alpha)$	$b_1^{6-(4)} \to a_2^{5-(4)} b_3^{7-(4)}$	(A4)
a_2^4	$(1.\underline{2.3}.\ 4.5)$	$a_2^{5-(4)}$	$(1.\underline{2.3}.4.\alpha.5)$	$a_2^{5-(4)} \to a_3^{5-(4)} b_1^{6-(4)}$	
b_2^4	$(6.7.8.\underline{9.0})$	$b_2^{7-(4)}$	$(6.\alpha.7.8.\alpha.\underline{9.0}.\alpha)$	$b_2^{7-(4)} \to a_1^{6-(4)} b_3^{7-(4)}$	
a_3^4	$(1.2.3.\underline{4.5})$	$a_3^{5-(4)}$	$(1.2.\alpha.3.\underline{4.5})$	$a_3^{5-(4)} \to a^3 b^{7-(5)}$	
b_3^4	$(6.\underline{7.8}.9.0)$	$b_3^{7-(4)}$	$(6.\alpha.\underline{7.8}.\alpha.9.0.\alpha)$	$b_3^{7-(4)} \to a^{7-(5)} b^{6-(3)}$	
a^5	$(1.2.3.4.5)$	$a^{7-(5)}$	$(1.2.\alpha.3.4.\alpha.5)$	$a^{7-(5)} \to a_3^{5-(4)} b^{7-(5)}$	
b^5	$(6.7.8.9.0)$	$b^{7-(5)}$	$(6.\alpha.7.8.\alpha.9.0.\alpha)$	$b^{7-(5)} \to a^{7-(5)} b_3^{7-(4)}$	

The cell-type alphabet is now composed of one 3-sided cell, two 5-sided cells, three 6-sided cells and four 7-sided cells. This, of course, assumes synchrony of divisions at each derivation step i and the relative numerical proportions of these cells would be slightly modified if different cell generation times were taken into account. The scheme might also have to be modified after systematic analysis of the number of walls possessed by the triangular apical cell, for example.

The basic structure followed by system S_{5-5off} is shown in Figure 6, at $i =$ 5 (which should be compared with Figure 3a, $i = 5$), and is traced out by the broken upper circle of green division walls. This pattern repeats each time that a cell of a tetrad itself becomes a tetrad. Although wall offsetting (marked by the blue interpolated wall segments in Figure 6) has the consequence of creating mostly six-sided cells, it is only a secondary feature of the cellular pattern. The generation of the basic pattern, specified, as we have seen, by system S_{5-5}, remains unmodified. Of course, another S_{5-5off} system is possible if cells a^3 were to be replaced by a cell $a^{4-(3)}$, a situation sometimes found in *Psilotum* apices (Figure 4a, apices 23 and 24; see also Figure 1, right). The evolution of the maps (up to

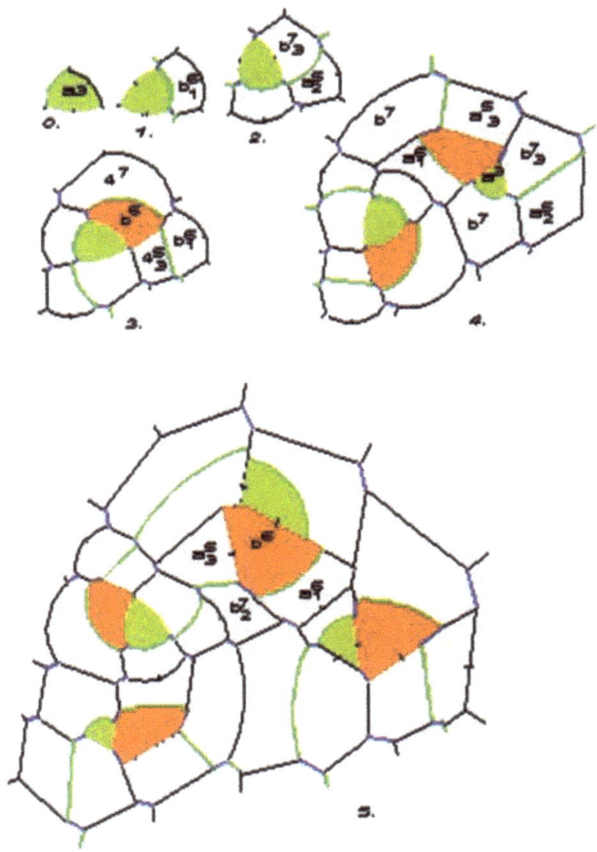

Figure 6. Derivation of ten types of cells according to map system S_{5-5off}. Labelling of walls and cells is according to the production rules A1 and A3. The starting point is an autoreproducing triangular cell, $a3$ (light-green). New division walls are shown as dark-green lines. Interpolated off-setting wall segments are blue lines. Second-order triangular cells (a') are coloured orange.

$i = 5$), now with wall offsetting and identification of the new cell types, is shown in Figure 6.

5.5 *The distribution of apical triangular cells and the problem of dichotomous axis branching*

Such a theoretical derivation of a developmental map-L-system can be of help in the analysis of the actual cellular patterns at the apices of *Psilotum*. First, the theory shows that the emergence of second-order triangular cells (a') is an inherent feature of the system. Second, it makes understandable why there are apices with clockwise segmentation and others with counter-clockwise segmentation, and how both patterns and their associated triangular cells can coexist within the same cellular complex. Third, the spatial distribution of triangular cells upon the apical surface can be predicted by means of the S_{5-5} system. Any departure from the expected structure is easily detectable and would raise questions as to what it might signify in terms of the underlying biology of the apex.

One important question to resolve is whether all triangular cells are potential apical cells. That is, whether cells a^3 and b^3 are biologically equivalent; we have already noted that they differ with respect to their wall labelling. To investigate this, and also to probe the cellular basis of axis branching (*i.e.*, whether it is the result of dichotomy, pseudo-dichotomy, or some other process), we tag triangular a and a' cells with different colors (Figure 3). It is then possible to show which sectors of the apical patterns conform to the theoretical scheme and which do not.

The map representations in Figure 3a, from state $i = 3$ on, represent two apical cells which divide symmetrically. If no difference existed between the a and b cells (a and b being used in the sense of sister cells, as depicted in Figure 1), this could be considered to presage a true dichotomy because the two groups of cells surrounding each of these cells would have exactly symmetrical cell patterns (*i.e.*, cell b_1^4 would already be an a' cell). However, the map derivation shows that the a' cell is a second-order triangular cell which emerges 2 steps after a has divided to give sister cells a and b. Hence, cell a is a first-order triangular cell. Any other derivation is not possible. This situation indicates that, in actual apices, it would be difficult to conclude with confidence that there has been a true dichotomy. In three-sided helicoidal segmentation, where two triangular cells are present, it can be shown that they never share a common mother cell, but, rather, within a given merophyte, one of the two triangular cells is descended from intermediary cells. Such cell patterns are therefore evidence of pseudo-dichotomy of organ axes, not of true dichotomy.

Detailed comparisons can be made between the map derivation in Figure 3a and the *Psilotum* apices analysed in Figure 4a. It turns out that apices 22 and 23, as well as apex 21 (which should be read in the counter-clockwise direction), contain 93 tetrads, 54 of which could be analysed. Of these latter, 43 were concordant with the maps. Apex 24 follows the derivation shown in Figure 1, right. The pattern of cells around its prominent triangular cell conforms more with that around a second-order a' rather than a first-order a cell.

6 Possible regulatory factors underlying cell patterning and its simulation at the *psilotum* apex

6.1 *A System's View of Division Wall Site Selection*

Using system S_{5-5}, it is possible to make the developmentally important link between wall age (or state) and differential wall growth and thereby arrive at a rule for specifying both the orientation of the mitotic spindle and the sites of attachment of a new division wall, dw. As shown earlier (Lück and Lück, 1985), we can represent the double walls and the wall segment productions by means of a derivation graph. Using the rule A1, the graph is as follows:

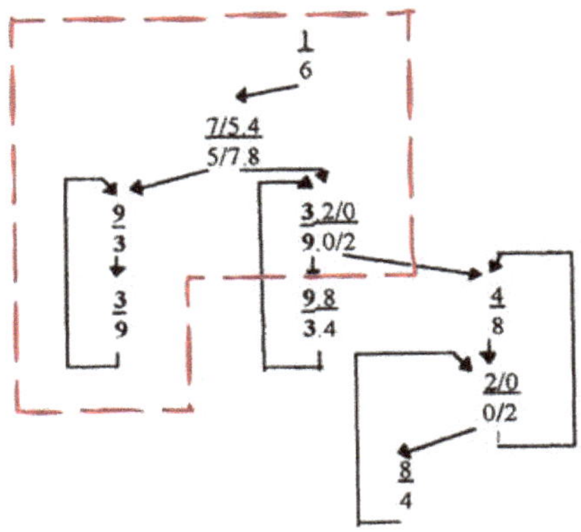

It is possible to select walls which will show different growth rates. This involves a sub-graph structure, delimited within the broken line, which is typical for a three-sided helicoidal segmentation. In the system S_{5-5}, the *dw* of triangular cell *a*, with label 1, is partitioned, one step after its formation, into three wall segments (4.5/7). Two of these segments (5 and 7) will never again increase in length (or increase only slowly), nor will they be further partitioned into further segments; the third segment (4) grows more rapidly and is partitioned into two segments (0/2) at the next cell generation step. As soon as a wall segment labelled 5 or 7 appears, it can only be transformed to accept labels 3 or 9; neither of these segments expands again. Terminal, non-growing segments are represented by black numbers in the derivation graph, while growing segments are shown in green. A magenta label indicates a temporarily non-growing wall.

If we look at cell a^3 (with walls 1.2.3.4.5), we see (Figure 7a.0) that its non-growing segments, 3 and 5, each occupy one half of the two non-division walls. After its division, new *dw*s are attached to the*dw* formed one generation step earlier (Figure 7a.1 and Figure 7a.2). Wall segments 5 and 7 on either side of one of the two *dw* attachment sites (Figure 7a.1) become, according to the wall derivation graph, non-growing wall segments. Wall segment 4, associated with the other attachment site, continues to grow. The *dw* has no other possibility than to attach to segment 4 which, in turn, is now partitioned into two parts, 0 and 2. An alternative possibility is that the *dw* attaches either to segment 2 (now labeled 8), or to that part of segment 1 which becomes the segment labelled 4 at the next step; but these two sites are too close to the first attachment site to provide equitable partitioning of the mother cell. One cell generation step later (Figure 7a.2.), the same operation repeats with respect to the positions of the new *dw*s introduced in map 1. That is, the second attachment site in map 2 will be on the former segment 4 of map 1.

In cell b_1^4 which is the sister of the triangular a^3 cell, the non-growing segment 3 of map 0 becomes segment 9 of map 1. The *dw* in cell b_1^4 is also attached to segment 6 (map 1)-*i.e.*, the *dw* is inserted between non-growing walls 5 and 7 in map 2. Possible alternatives for the second attachment site are the two segments 8 or 0 (map 1). According to the system, it is segment 8 which is selected, whereupon this segment is partitioned into two new segments, 0/2. This could explain how the oblique orientation of the *dw* in cell b_1^4 comes into being since the *dw* has to travel a relatively

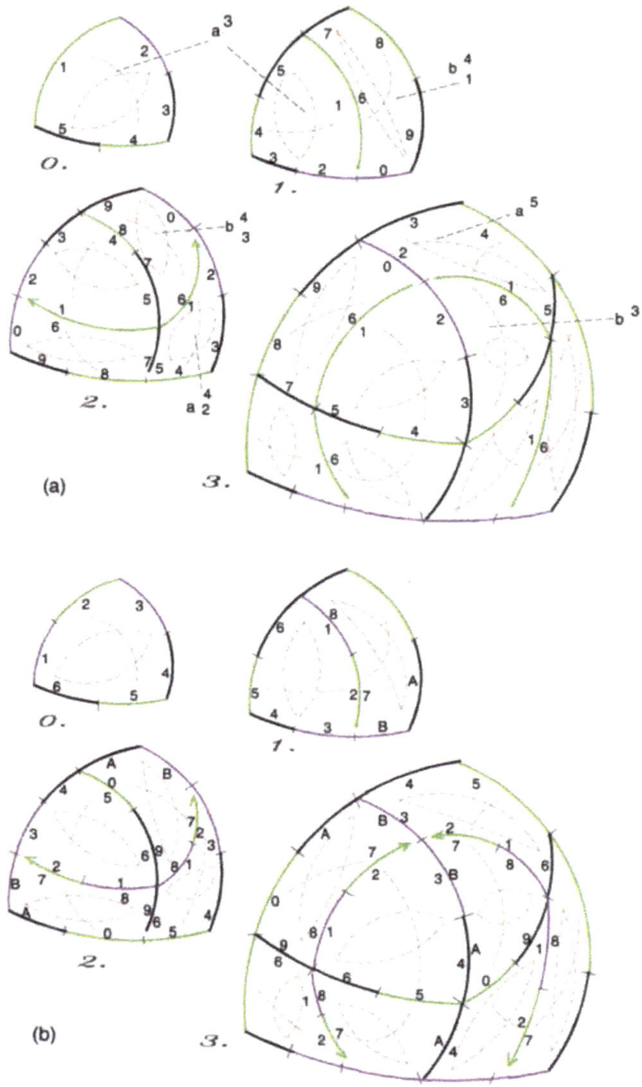

Figure 7. (a) Derivation of four maps (1-3) from an initial map (0) according to map system S_{5-5} and label substitution rule A1, but with the assumption that, at each developmental time-step (cell generation), there are certain walls, or segments of walls, which are either growing (green lines), temporarily non-growing (magenta lines), or terminally non-growing (black lines). In triangular cells, two new non-growing segments are selected at each time-step. Because of symplastic growth of the cellular complex, non-extension of any given wall may be compensated for by a wall sharing a similar non-extending state elsewhere in the map. Within each cell is a figure (broken lines) which represents the spindle-shaped mitotic cell body and the early phragmoplast (or new division wall). The axes of the cell bodies are aligned as though their poles are searching for connection with the centres of non-growing walls. (b) A similar map (at time-step $i = 3$), but where six wall labels have been used in the intial map at $i = 0$ in order to generate another pattern of differential wall growth. This is the outcome of map system S_{6-6}.

long distance to reach the selected attachment site that splits segments 0 and 2. At the next generation step (Figure 7a.3), cell b_3^4 divides and its *dw* attaches to wall 6 of map 2. The second attachment site now has a choice between the segments 0 and 8. According to the system, segment 8 is selected. This leads to the insertion of another oblique *dw* and the emergence of a new triangular cell, b^3. If segment 0 had been selected as the second attachment site, however, the question is whether the resulting division would correspond to the third way of cell wall patterning, given in Figure 1, right.

The proposed derivation graph above is not the only one possible. A second graph, which involves six wall labels for the initial cell (*i.e.*, 1.2.3.4.5.6) instead of five, is as follows:

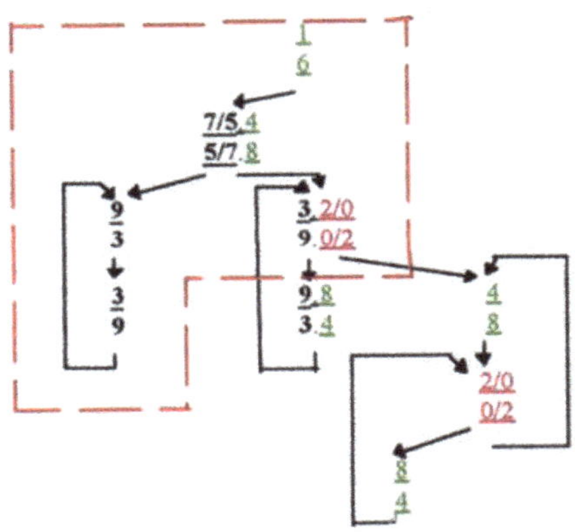

It is the derivation graph of an alternative system, S_{6-6}. The resulting pattern of walls is shown in Figure 7b.

In a two-dimensional representation of map system S_{5-5}, the wall network can be considered as a continuous line which grows and branches from the *dw* (see Figure 7a, and also Lück and Lück, 1981). Not only does the growing line of the *dw* select a precise site on the previous *dw* for its emergence, but the other end of the *dw* is also given the choice of attaching to one of two or three alternative sites on other walls. The circumstances which underlie the eventual choice taken may, as already mentioned, lead to those alternative division patterns which are often found in the *Psilotum* apex (Figure 1). These circumstances may also reflect alterations in the behaviour of the cell body and its interactions with cell walls (see next section), but for this the two-dimensional scheme of Figure 7 has to be translated into the third dimension which applies to the pyramidal cell and its cuboidal derivatives.

With respect to wall offsetting, because the laying-down of two opposite *dws* in two sister cell is not exactly simultaneous (as it is in the synchronous step-wise application of system S_{5-5}), but usually involves cell *b* dividing before cell *a*, the second division wall to attach (in cell *a*) will, apparently out of necessity, avoid the site of the first. Wall-offsetting will be the result. But why this proceeds with either a *Z* or an *S* configuration (*i.e.*, the *dw* in the triangular cell is either to the inside or to the outside of the *dw* in the *b* cell) remains an open question. The avoidance may be because the short segment of wall at the site where a *dw* is already attached is a non-growing, α segment, and *dw*s are constrained to attach only to growing segments.

6.2 Cytological View of Division Wall Site Selection

Cell patterning was interpreted earlier from the point of view of the 'plant cell body', an intracellular apparatus responsible for the growth and morphogenesis of cells. In its guise as a 'mitotic cell body', it is also responsible for the division of the nucleus at mitosis, the inception of the cell plate within a phragmoplast, and the subsequent insertion, at cytokinesis, of this cell plate as a new division wall, *dw*. What features of the plant cell body are relevant to cell patterning of *Psilotum* apices and to its simulation by means of map system S_{5-5}?

The cell patterns of the *Psilotum* apex are especially interesting on account of their apparent direct relationship with organogenesis. In particular, the triangular (pyramidal) cells embedded within the pattern at the apex appear to have organogenetic potential for the development of shoot or subterranean axes. These two organs seem never to develop from a cell which has some other shape. Therefore, the form of the apical cell and the sequence of its divisions, and also those within the merophytes derived from it, are likely to be crucial for organogenesis. Whether an organ has characteristics of a photophobic subterranean axis or of a photophilic shoot probably does not depend upon the cellular pattern *per se*, but is more likely to be a function of internal physiological correlations.

There are two interrelated biological aspects of the cellular patterning system for which there are as yet no satisfactory explanations, although we hope here to supply them. The first concerns the orientation of the mitotic apparatus within the triangular cell in anticipation of delivering one of the two daughter cell bodies to the daughter cell, *b*. The second concerns the rotation, again within the triangular cell, of the sites at which new *dw*s will become attached during successive cytokineses. If the three interior walls of the triangular (or pyramidal) apical cell are referred to as A, B, C, and four successive cytokineses are considered, the pairs of walls (or facets, [ ]) which bear the attachment sites then follow the sequence [A,B], [B,C], [C,A], [A,B], and so on (or the reverse sequence if an opposite sense of division is used). In the schemes shown in Figure 3a and especially in Figure 7a, where relabelling occurs after each division, it is always walls that bear labels (or hold states) 1 and 4 which accept the new *dw*.

Because the triangular cell of *Psilotum* conserves an equilateral aspect, each division plane rotates $\sim 120^{\circ}$ away from that of the previous division plane. Moreover, the sense of the division sequence in a given triangular cell is conserved so that the sequence proceeds either clockwise or counter-clockwise. The 120° displacement may come about because the division plane of the triangular cell is adjusted according to the position of the cell body in relation to certain of its walls. The maintenance of this cell, and the sequential utilization of one of its facets as the outward-facing wall of a new daughter cell, *b*, suggests some physical linkage between the mitotic cell body and either the wall that will be emitted into the *b* cell at cell division or the pair of walls that remain in the *a* cell. Evidence does indeed suggest that there can be wall-nucleus interactions, mediated by the cytoskeleton, which lead to precise intracellular positioning of the nucleus (Grolig, 1998), perhaps due to the necessity of balancing cell body-wall tensions according to tensegrity principles.

Concerning the rotation of the mitotic cell body in anticipation of a predetermined division plane, it is possible that facet of the apical cell which becomes the outer wall of the *b* daughter cell-the facet is represented in two dimensions by the wall segments labelled 2.3 in the map shown in Figure 7a.0-possesses some property which favours the attachment of cytoskeletal structures at its surface. At their other end, these fibres attach to the nucleus (*i.e.*, the fibres could be endoplasmic MTs or actin filaments) and hence the developing mitotic cell body is tethered from one of its poles to wall 2.3. The other pole of the mitotic cell body could be similarly tethered to another wall, for example, to that represented by label 5 (see Figure 7a.0). Cytokinesis follows after this preparatory period of mitotic cell body positioning, and the cell plate is established within the phragmoplast. The phragmoplast invariably aligns perpendicularly to the axis of the mitotic spindle. The cell plate then makes it way by centrifugal growth within the phragmoplast to sites on the wall predetermined by the preprophase band (PPB) of MTs. At the start of the next interdivisional period, the proposed influences of walls 2.3 and 5 pass to the walls which were labelled 4.5 and 1 in the former interdivisional period. Mitotic cell body orientation could, therefore, be the sole factor controlling the division plane. However, the MT connections between cell walls and mitotic spindle poles which have been seen at mitosis may be sufficient only to steady the mitotic apparatus during this phase (De Ruijter *et al.*, 1997); some other force may be required for cell body alignment during the preparatory period for nuclear division. We propose that one such stabilizing system may be akin to the dynactin complex which, in embryonic cells of the nematode, *Caenorhabditis elegans*, determines the alignment of the mitotic apparatus (cell body) in preparation for an alteration to the polarity of growth of a new lineage of cells (Skop and White, 1998).

It is during the preparatory period for cell body division that the PPB forms and then disperses. This band of MTs determines where the cell plate will attach to the parental walls of a mother cell following the completion of cell body division. With reference to map system S_{5-5}, the PPB is presumed to be located at sites on the two walls which are represented by the two line segments 1 and 4 in the map shown in Figure 7a.0. After division of the triangular mother cell, the two putative PPB sites are revealed as being the cell plate attachment sites, /, on each of the two re-labelled walls, 5/7 and 2/0. (Recall that, according to the segment label substitution rules A1, line 1 $\rightarrow$ 4.5/7 and line 4 $\rightarrow$ 0/2). We propose that the PPB is always located on **growing** walls-in this case, on portions of lines 1 and 4-and may even avoid non-growing walls. We further suggest that there are segments of walls in both the triangular apical cell and in its derivatives which are permanently **non-growing**. These are the segments labelled 3, 5, 7 and 9; they are shown in the corresponding derivation graph in Section 6.1 and in Figure 7. Other permanently non-growing segments are those associated with wall junctions. These do not explicitly feature in the simulation maps, but might be considered a type of α wall segment, say $\alpha^{\bullet}$. PPBs would thus avoid such non-growing, $\alpha^{\bullet}$ sites and consequently avert the formation of four-wall junctions, hence favouring the offsetting of *dw*s.

To bring all these observations together, the following hypothesis, within the context of map system S_{5-5}, is proposed in order to account for the division pattern of a triangular a^3 cell. There are portions of walls of this cell which are non-growing. They are represented by labels 5 and 3 in Figure 7a.0. The mitotic cell body of this cells is polarised-it possesses N and S poles-and these poles are attracted to the mid-points of the non-growing wall segments: N, say, is attracted to the mid-point of segment 5, and S to the junction of segment 3 with segment 2 (*i.e.*, the mid-point of temporarily and terminally non-growing wall segments). The attraction of wall and cell body is positive and may be facilitated by a force-generating molecular complex such as dynactin. When the mitotic spindle forms, it does so with its N and S poles tethered to the proposed attraction sites on the non-growing walls. A dynactin complex on the nuclear surface may also be responsible for the segregation of MTs to the N and S nuclear poles prior to the formation of the poles of the mitotic spindle. When the phragmoplast forms within the mitotic cell body, its orientation is initially perpendicular to the spindle. The cell plate develops later within the phragmoplast and may reorient somewhat. The mitotic cell body is also responsible for the development of the PPB band of microtubules. The PPB is positioned as a result of the activity of two sets of endoplasmic microtubules which radiate from each N and S pole of the nucleus and focus at common sites in the cell cortex. There they deposit material for the organization of the MTs of the PPB, and this MT structure continues to direct synthesis of cell wall material. It happens that, according to the system S_{5-5}, and with the present further definition of its wall growth properties according to the derivation graph, the PPB is always located on young growing walls. When the new *dw* forms, it is keyed into these young walls, their juvenile state probably making them more competent to accept the *dw*.

For the sister cell, b_1^4, a similar situation is proposed. Here, the non-growing walls are labelled 7 and 9 in map 1 of Figure 7a, and these, too, have sites attractive to the N and S poles of the mitotic cell body. Because of the shape of this cell and the locations of the non-growing walls relative to each other (see Figure 7a.1), the mitotic cell body lies obliquely within the cell; and because the PPB has to be developed upon a growing wall (the one labelled 8 in map 1), the subsequent *dw* also becomes oriented obliquely. By this means, the sister cells b_3^4 and a_2^4 are created (Figure 7a.2). From the continued application of these principles, a new triangular cell, b^3, inevitably emerges (Figure 7a.3).

Similar considerations, which also accord with this hypothesis, apply in respect to the derivation of the map shown in Figure 7b (in the present, cytological context, it is relatively unimportant that this is based on an alternative system, S_{6-6}) in which non-growing and growing walls are distributed differently in the triangular cell and its descendents.

Because spindle and phragmoplast play such an important role in cell division and *dw* placement, their positions have been explicitly indicated in the maps shown in Figure 7a and b. Their alignments are coordinated with the changing states of the walls which, in turn, are regulated by internal biological analogues of map systems S_{5-5} and S_{6-6}. In fact, in Figure 7b, the maps from system S_{6-6} seem to permit in a more equilibrated disposition of the spindles within the cells. The feature of a non-growing wall which makes it a possible candidate for the attachment of the pole of a mitotic cell body is unknown. However, non-growing $\alpha^\bullet$ wall segments do seem actively to repel PPBs. A three-dimensional view of the cells, such as that shown in Figure 2, would reveal that the lines and segments shown in the two-dimensional maps given in Figure 7 directly correspond to the anticlinal walls of the cell layer (*i.e.*, the superficial cell layer of the merophytes at the *Psilotum* apex). Exactly where the mitotic cell bodies would reside in a three-dimensional representation remains to be established. The simulations presented in Figure 7 remind us that Hébant *et al.* (1978), using fixed and sectioned material, described the polarization of cytoplasm in apical cells of ferns and mosses and related this to the orientation of their mitotic spindles and new *dws*.

7 Was there a precursor of the *psilotum* pattern?

The cell wall patterning system which characterises the *Psilotum* apex and which is simulated by the map L-system S_{5-5}, regularly produces triangular cells which have the potentiality to become new organogenetic centres. However, if all such triangular cells acted in this way there would clearly be much more branching of axes than is actually evident. For a plant as delicate as *Psilotum*, excessive branching may be undesirable since there may be a limit to the number of axes which can be supported within its shoot system. Nor may extra branches be necessary if those which have already formed can liberate adequate numbers of spores for plant reproduction. Thus, a diminution of branching by deviations which ensure that triangular cells do not arise during cellular patterning (see Figure 1, right) could, evolutionarily speaking, have some selective advantage. Moreover, given the general self-similarity of the different levels of biological organization, it is not surprising to find that, just as certain **cell** division patterns fail to form organogenetic triangular cells (*cf.* Figure 1), so some of those **organogenetic** centres which do form, in their turn fail to form new organ axes. Hierarchically-organized 'failures' may be just as important as 'successes' in determining overall biological form, and may here represent a type of self-regulation of axis emergence or self-sterilization of cell reproduction.

Given the already-mentioned morphological similarity between the sporophytes of *Psilotum* and the now extinct *Rhynia gwynne-vaughnii*, it is interesting to enquire whether the cellular patterning process in apices of the more ancient plant are in any way similar to those found in apices of *Psilotum.* If they differ, then we need to know what aspect of cellular behaviour could bring this about. A view into this interesting area of evolutionary developmental biology would be inevitably assisted by comparisons of cell patterns in *Rhynia* and *Psilotum* shoots. Since, in the present case, a developmental genetic investigation of *Rhynia* is, regrettably, impossible, the burden falls on comparisons of simulation procedures to reveal the most probable cell-biological connection between the cellular patterns of these and other evolutionarily-related species.

One published micrograph (Edwards, 1993) of a paradermal section of an apex of *R. gwynne-vaughnii* shows cells which are square in outline and clearly arranged within merophytes (Figure 8a), recognizable on account of their relatively thick boundary walls enclosing other groups of cells separated by walls of different age (evident from their different thicknesses). In one part of the section there seems to be an 'apical cell', though its prominent appearance could be an artefact of sectioning. Despite the uncertain status of this 'apical cell', an arrangement of cells, approximating to that illustrated for the *Rhynia* apex, can be generated by wall re-labelling rules, A5.

$$\begin{array}{llll} 1^S \to 4^W/6^W & 1^W \to 3^E & 1^E \to 8^N/2^N & 1^N \to 3^S \\ 2^S \to 7^W & 2^W \to 4^E/6^E & 2^E \to 3^N & 2^N \to 4^S/6^S \\ 3^S \to 8^W/2^W & 3^W \to 7^E & 3^E \to 4^N/6^N & 3^N \to 7^S \\ 4^S \to 3^W & 4^W \to 8^E/2^E & 4^E \to 7^N & 4^N \to 8^S/2^S \end{array} \quad \text{(A5)}$$

In the resulting scheme (Figure 8b), a four-sided 'apical cell' reproduces itself, to give a new cell *a* and a rectangular daughter cell, *b*, at each of it four walls (WENS) in turn. Actually, all cells behave in this way: that is, all cells, irrespective of whether they are *a* or *b* cells, are four-sided, and all divide with the same realignment of the division plane at every second division. Periclinal divisions could also be interpolated into the scheme so that new cells would regularly be introduced into a sub-epidermal layer.

It is evident that, in the *Rhynia* section shown in Figure 8a, not all the merophytes which derive from the putative apical cell are present in the section.

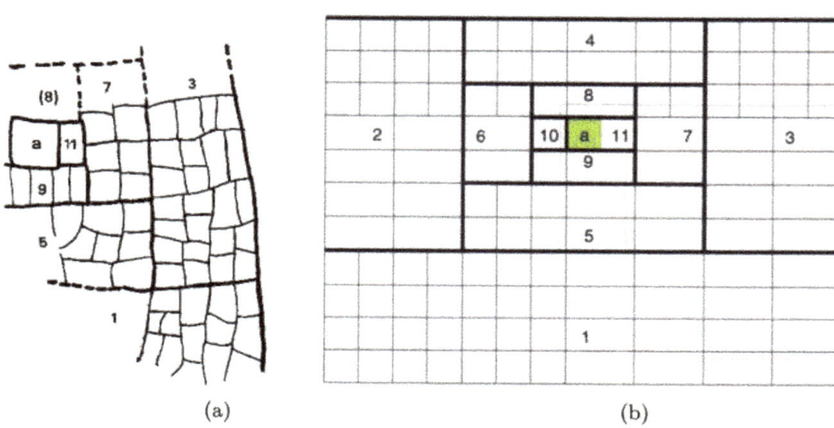

Figure 8. Cellular patterning on a portion of a fossilized shoot apex of *Rhynia gwynne-vaughnii* (a) A cellular pattern drawn from a photograph of a paradermal section of a *Rhynia* apex (see Edwards, 1993). Merophytes, bounded by thick lines, have been identified on the basis of differential wall thickness, indicating origins at different cell generations. Merophytes have also been numbered in a provisional sequence that follows from their production by a putative apical cell, *a*. Missing merophytes and their boundaries are indicated by broken lines and numbers in parentheses. (b) Derivation of a cellular pattern from the map system given in A5 and with the introduction of time-delay for cell divisions (see text). Only the merophyte boundaries are shown. The apical cell, *a*, is coloured light green. The portion enclosed within red lines is that which corresponds to the pattern in a. The correspondence between the two patterns is quite close.

However, the arrangement of cells is sufficiently clear to make plausible the above interpretation, as described by A5 and represented in Figure 8b. Other possible division schemes do not yield such a close resemblance to the observed merophyte pattern.

In order to achieve a satisfactory correspondence between simulation through the above 'WENS' system of A5, it is necessary to introduce time-delays into the overall division process. (Unfortunately, up until now, reasons of space have precluded any discussion of this important topic of division time-delay, which can apply to certain cellular configurations of the *Psilotum* apices.) To introduce a time delay, cells are labelled w, e, n, s, according to the merophyte to which they belong. The successive cell productions are then:

$$a \to aw'_0 \qquad a \to an'_0 \qquad a \to ae_0 \qquad a \to as_0$$

At every third division, all cells with the subscript index 0 do not divide; and all cells with subscript index 1 do not divide every second generation. The prime (′) indicates that it is this marked cell which commences the development of a new W or N merophyte; merophytes E and S commence directly with e_0 and s_0 cells. For example, for the n cells:

$$\begin{array}{ll} n'_0 \to n_0 & n_3^2 \to n_0 n_0 \\ n_0 \to n_0^2 n_1 & n_1 \to n_2 \\ n_0^1 \to n_0^2 n_1 & n_2 \to n_1 n_1 \\ n_0^2 \to n_3^2 & \end{array}$$

All the cells of the *Rhynia* apex-*b*-cell derivatives, as well as the putative 'apical cell' itself-and also all the cells produced by the simulation system of A5, have similar shapes. This is unlike the situation found in the *Psilotum* apex, where there are 10 types of cell (see Figure 5). So, one question is: how, with such conditions of cellular uniformity, could branching take place in the *Rhynia* apex? Implicit in both the concept and the process of branching is the notion of non-uniformity. Therefore, branching is likely to arise because of some type of differential growth. This can be achieved if cells at and around a designated organogenetic centre grow and divide faster than cells located elsewhere. It would have the effect of producing a protuberance upon the apical surface and would herald the initiation of a new organ axis. The thickened and non-growing (*cf.* Figure 7) boundary wall of a maturing merophyte could provide the interface between areas of differential growth and lead to the buckling of the apical dome with the concomitant onset of neo-organogenesis. Moreover, differential wall growth patterns may assist in the branching of *Psilotum* apices also.

Other possible cell-patterning schemes are suggested by recent findings from the apices of *Aglaophyton major*, which has a form similar to *Rhynia* and which, like *Rhynia*, is also a fossil from the Rhynie chert (and may be more primitive than *Rhynia*). Here, triangular, rectangular and other polygonal cells have been observed to co-exist in the surface layer of the sporophyte shoot apex (Prof. H. Kerp, Universität Münster, Germany, *personal communication* to P.W.B.). Assuming that a recognizable organogenetic cellular pattern is absent, the branching of the *Aglaophyton* apex may depend only upon a particular pattern of differential wall growth.

Large areas of apical surface occupied by small rectangular cells is a situation reminiscent of a subterranean apex of *Psilotum nudum* described by Hagemann (1980 -his Figure 2a). Analysis of these particular cellular complexes suggests that the rectangular cells are contained within merophytes which lack triangular cells (*cf.* Figure 1, right) but which, nevertheless, seem to contain organogenetic centres since some of the rectangular cells appear to be organized into spiral patterns. The spiralling rectangular cells may be secondary merophytes which are winding into a single rectangular 'progenitor' cell. A simple map system, A6, which commences with a four-sided cell, generates this pattern of merophytes (here with a clockwise spiral):

$$\begin{aligned} 1 &\rightarrow 4/_{1}2 \\ 2 &\rightarrow 3 \\ 3 &\rightarrow 4/_{1}2 \\ 4 &\rightarrow 3 \end{aligned} \qquad \text{(A6)}$$

Thus, there are a number of systems which could exist in primitive plants. As suggested by *Psilotum nudum* and *Rhynia*, and perhaps by *Aglaophyton*, it is conceivable that a triangular cell is not an absolute precondition for organogenesis. But, conversely, if such a cell is an organogenetic necessity in *Psilotum*, it is not difficult to form a triangular cell from a rectangular cell (or, considering its three-dimensions, to create a four-sided pyramidal cell from a six-sided cuboid). All that is required is for the new division wall to span one corner of the four-sided cell. Nor is it difficult to imagine that, during phylogeny (*i.e.*, *Rhynia*→ → *Psilotum*), a similar cellular transformation could occur by the misalignment of the division apparatus (*e.g.*, displacement of the Rhynialian band of preprophase microtubules). The critical problem is how such an event could then be fixed in time and space so that a pyramidal apical cell would regularly appear in shoot apices composed of cuboidal cells. Genetic mutations can disturb division patterns, and some can have precise spatio-temporal actions (Laux and Jürgens, 1994); so there is no great conceptual difficulty in postulating mutation as a route for the development of a new grade of cells. A further problem, however, is to ensure that the neo-formed pyramidal apical cell(s) would function in the way anticipated for an organogenetic centre. Interestingly, it seems that there is something about the properties of a such a cell which favour its perpetuation. For example, as soon as pyramidal cells become present in callus cultures of the fern, *Pteris cretica*, organogenesis commences (Michaux-Ferrière, 1980). Presumably, the switch from the non-organogenetic to the organogenetic condition is accompanied by a divisional anomaly (an anomaly with respect to the cellular state of the non-organogenetic culture, that is) which creates, from the cuboid cells of the callus, the pyramidal cells capable of organogenetic activity. Such an anomaly may reflect the formation of a 'somatic embryo' since, in the normal embryogeny of *Pteris*, a pyramidal apical cell is formed early on, and then continues to support either shoot or root formation.

8 Prospects

What is the biological significance of cellular patterns and of cell wall patterns? The two patterns need to be considered separately at present because each may have different consequences for organ development: cellular patterns would actually reflect cell-body patterns-and these cell bodies may show particular polarities coordinated with particular cell wall states-whereas cell wall patterns probably have more straightforward interpretations, such being flow-paths for the intra-organ distribution and dissipation of stresses and strains. In the present instance of the *Psilotum* apex, it has been assumed that cellular patterning is more than simply a marker of organogenesis but that, perhaps, because it actually reflects the collective orientation of dynamic cell bodies, it is somehow fundamental to the organogenetic process itself, and hence contributes directly to the development of the plant's form. The link between cell-body and cell-wall patterning processes seems slowly to be becoming clearer; yet, the resulting organogenesis involves numerous sub-sets of developmental processes, including tissue and cellular differentiation: whether these depend upon particular cell shapes (*i.e.*, distribution of walls) is unknown. Differentiation processes are also assisted, in part, by informational inputs from the cell walls and their plasmodesmata, the latter structures being sites in the wall specialized for molecular cell-to-cell communication and which possibly are also constructed by active cell bodies. Moreover, the process of morphogenesis-the development of shape and form-runs in parallel with organogenesis and, while having some association with the biochemical processes of differentiation, may be more immediately regulated by the biophysical properties of the developing whole. Generally, the form of an organ is the result of differential wall growth, especially of the external boundary walls. Some of the properties attending these growth differentials may then impinge upon regulators of gene action, as do the molecules communicated from cell to cell *via* plasmodesmata. Thus, there are two major problems for resolution:

(1) the generation of organ form

(2) the generation of cell body and cell wall patterns within that form.

Of course, these are not new problems (Sinnott, 1940), but are ones which have beset generations of plant developmentalists since the time of Julius von Sachs and his contemporaries (Sachs, 1887, and references therein).

A plausible view, then, is that the form of an organ is the result of cell-body and cell-wall patterning. In the case of multinucleate coenocytic organisms, e.g., *Caulerpa* spp. (Jacobs, 1994) and their algal relatives, it is quite conceivable that their specific and complex forms result from deterministically regulated patterns of cell bodies. Similarly, the development of multicellular organisms-and here *Chara zeylanica*, painstakingly analysed by Sundaralingam (1954), may be taken as a highly illustrative example-also indicates that, at least in apical zones where cell division is taking place, the continuously changing orientation of cell growth and cell division are events arising from alterations to the patterns of differential growth at the organ surface. But, as we have seen (*e.g.*, Figure 7), *differential growth can be built into the division system itself.* Moreover, in the *Chara* apex, once the sub-division of some of the conspicuously large cells of the apex has resulted in the creation of smaller cells (as in the nodal branchlets, for example), then the shape of these smaller cells may also determine their subsequent division pattern. Their patterns, which 'fill-in' the nodal tissue with walls and protoplasts, may well be effected by dynamic and tactile plant cell bodies which are able to sense the topography of their surroundings and, as a consequence, exert a control over the orientation of mitotic spindles and phragmoplasts. Thus, in relation to the zone of form-generation at the shoot apex, de Bary's teasing aphorism is very apposite: 'The plant makes cells, not cells make the plant'. We can now even go so far as to say that 'the plant' is a representation of 'a system'; and so, to paraphrase de Bary, it is 'The system which makes cells...'. We incline to this view since it presents a tightly integrated conception of plant morphogenesis. In this sense, therefore, the form of any plant is the representation of a morphogenetic system. The *Psilotum* apex which we have chosen for analysis *is the representation of one of these archetypal systems, which has been designated* S_{5-5}.

At some stage it will be necessary to fit the patterning of tissue differentiation within the morphogenesis-cell-patterning nexus. To accomplish this in the living organism, it is conceivable that a set of 'second-order' rules is called into play which governs the location of the set of biochemical processes within the cellular pattern that developed according to a set of 'first-order' rules, such as those specified by system S_{5-5}. However, evidence suggests that correct tissue patterns can emerge in plant organs even in circumstances where the cellular patterns have been rendered abnormal by mutation affecting the behaviour of the cell body (Traas *et al.*, 1995). Cellular patterning, therefore, may be relatively unimportant in this context of histogenesis; rather, it is the critical cellular boundaries, and the internal geometric reference points which these boundaries specify, that create informational maps (positional information) which, in turn, regulate tissue differentiation.

Cell walls are evidently not barriers to cell-cell communication-quite the reverse, given that plasmodesmatal microdomains of the wall are specialized for this purpose. Wall patterning and the plasmodesmatal frequencies associated with walls which traverse the different axes of organ symmetry may affect the respective symplasmic fluxes of materials across tissues and hence affect patterns of gene action and tissue differentiation. Presumably, even when mutation affects wall orientations, morphogen fluxes in the radial and vertical directions can remain unaffected because the respective plasmodesmatal frequencies have remained unaltered overall. In this respect, positional information is conserved, the cell wall patterns having little impact on molecular communication-based differentiation programmes. However, should altered wall patterns be associated with altered patterns of biophysical forces which traverse tissues and organs, the consequences for organ development may be quite different on account of the qualitatively different types of information imparted by these forces.

Studies of cellular patterning seek to isolate but one out of a host of sub-processes which make up the more global process of organogenesis. The present analysis of cellular patterning in *Psilotum* strongly argues for its deterministic aspect, and one might speculate that other processes of development have similar underlying determinisms. However, the objective of our investigation is to understand what is going on in the patterning process in terms of cytological and genetic events. While there is some aesthetic appeal in representing the process by the most elegant or simplest model or simulation system, this endeavour might run counter to W.R. Ashby's "Law of Requisite Variety" which, in the cybernetic context for which it was devised, states that the controlling system of a process is as complex as the process itself (Ashby, 1956). We have tried to combine simulation with knowledge of the cytological adjuncts of cell wall patterning. Yet, by so doing, we have done no more than translate the descriptive language of cytology into the more incisive language of systems. Both the cytological and the systems avenues of enquiry aim to illuminate the properties of living organisms and the fundamental cellular units that uphold their existence and which otherwise would remain obscure and mysterious.

Acknowledgments

IACR is grant-aided by the Biotechnology and Biological Sciences Research Council of the UK. We are grateful to Dr František Baluška for some helpful suggestions.

References

1. S. Aldington and S. C. Fry 1993. Oligosaccharins. *Adv. Bot. Res.* **19**:1–101.
2. W. R. Ashby 1956. *An Introduction to Cybernetics.* Methuen, London.
3. F. Baluška, D. Volkmann and P. W. Barlow 1997. Nuclear components with microtubule-organizing properties in multicellular eukaryotes: Functional and evolutionary considerations. *Int. Rev. Cytol.* **175**:91–135.
4. F. Baluška, P. W. Barlow, I. K. Lichtscheidl and D. Volkmann 1998a. The plant cell body: A cytoskeletal tool for cellular development and morphogenesis. *Protoplasma* **202**:1–10.
5. F. Baluška, D. Volkmann and P. W. Barlow 1998b. Tissue- and development-specific distributions of cytoskeletal elements in growing cells of the maize root apex. *Plant Biosystems* **138**:251–265.
6. P. W. Barlow 1991. From cell wall networks to algorithms. The simulation and cytology of cell division patterns in plants.*Protoplasma* **162**:69–85.
7. P. W. Barlow1998. How robust are the development and organization of plant structures in the absence of gravity? *Adv. Space Res.* (submitted).
8. R. C. Brown, B. E. Lemmon and O. A. Olsen 1996. Polarization predicts the pattern of cellularization in cereal endosperm. *Protoplasma* **192**:168–172.
9. M. J. M. De Boer, A. Lindenmayer 1987. Map 0L-systems with edge label control: Comparison of marker and cyclic systems. In: *Graph Grammars and Their Application to Computer Science*, eds H. Ehrig, M. Nagl, G. Rozenberg and A. Rosenfeld, pp. 378–392. Springer-Verlag, Heidelberg. (Lecture Notes in Computer Science, volume 291).
10. N. C. A. De Ruijter, J. Pietrusiewicz, M. B. Montijn, J. H. N. Schel and A. M. M. Van Lammeren, A.M.M. 1997. Spatial limitations induce spindle tilting and result in oblique phragmoplasts in *Vicia faba* L. root tip cells, but do not result in oblique cell walls. *Acta Bot. Neerl.* **46**:279–290.

11. R. M. Duffy 1951. Comparative cellular configurations in the meristematic and mature cortical cells of the primary root of tomato. *Am. J. Bot.* **38**:393–408.
12. D. Edwards 1993. Cells and tissues in the vegetative sporophytes of early land plants. *New Phytol.* **125**:225–247.
13. A. J. Fleming, T. Mandel, I. Roth and C. Kuhlemeier 1993. The patterns of gene expression in the tomato shoot meristem.*Plant Cell* **5**:297–309.
14. F. D. Fracchia 1996. Integrating lineage and interaction for the visualization of cellular structures. In: *Graph Grammars and Their Application to Computer Science*, eds J. Cuny, H. Ehrig, G. Engels and G. Rozenberg, pp. 521–535. Springer-Verlag, Heidelberg. (Lecture Notes in Computer Science, volume 1073).
15. C. Fuchs, M. Broniatowski and Y. Delarue1985. Etude de la distribution spatiale des mitoses dans la feuille du *Tropaeolum peregrinum L. Rev. Cytol. Biol. Végét. — Bot.* **8**:255–264.
16. L. J. Goff and A. W. Coleman 1987. The solution of the cytological paradox of isomorphy. *J. Cell Biol.* **104**:739–748.
17. F. Grolig 1998. Nuclear centering in *Spirogyra*: Force integration by microfilaments along microtubules. *Planta* **204**:54–63.
18. B. E. S. Gunning, J. E. Hughes and A. R. Hardham 1978. Formative and proliferative cell divisions, cell differentiation, and developmental changes in the meristem of *Azolla* roots. *Planta* **143**:121–144.
19. W. Hagemann 1980. Über den Verzweigungsvorgang bei *Psilotum* und *Selaginella* mit Anmerkungen zum Begriff der Dichotomie.*Plant Syst. Evol.* **133**:181–197.
20. C. Hébant, R. Hébant-Mauri and J. Barthonnet 1978. Evidence for division and polarity in apical cells of bryophytes and pteridophytes. *Planta* **138**:49–52.
21. T. E. Holy and S. Leibler 1994. Dynamic instability of microtubules as an efficient way to search in space. *Proc. Natl Acad. Sci. USA* **91**:5682–5685.
22. H. Honda 1983. Geometrical models for cells in tissues. *Int. Rev. Cytol.* **81**:191–248.
23. R. Hooke 1665. *Micrographia or Some Physiological Descriptions of Minute Bodies Made by Magnifying Glasses with Observations and Inquiries Thereupon.* (Facsimile edition) Dover, New York.
24. D. E. Ingber 1997. Tensegrity: The architectural basis of cellular mechanotransduction. *Annu. Rev. Physiol.* **59**:575–599.
25. D. E. Ingber and J. Folkman 1989. Tension and compression as basic determinants of cell form and function: Utilization of a cellular tensegrity mechanism. In: *Cell Shape. Determinants, Regulation, and Regulatory Role*, eds W. D. Stein and F. Bronner, pp. 3–31. Academic Press, San Diego.
26. C. T. Ingold 1973. Cell arrangement in coenobia of *Pediastrum*. *Ann. Bot.* **37**:389–394.
27. W. P. Jacobs 1994. *Caulerpa*. *Sci. Amer.* *271* **6**:66–71.
28. P. Kenrick and P. R. Crane 1997. The origin and early evolution of plants on land. *Nature* **389**:33–39.
29. R. W. Korn 1980. The changing shape of plant cells: Transformations during cell proliferation. *Ann. Bot.* **46**:649–66.

30. R. W. Korn 1993. Evidence in dicots for stomatal patterning by inhibition. *Int J. Plant Sci.* **154**:367–377.
31. R. W. Korn 1994. Pattern formation in plant epidermis through inhibition of immediately adjacent cells by pattern elements. *Protoplasma* **180**:145–152.
32. U. Kutschera 1995. Tissue pressure and cell turgor in axial plant organs: Implications for the organismal theory of multicellularity. *J. Plant Physiol.* **146**:126–132.
33. J. W. La Claire, II 1987. Microtubule cytoskeleton in intact and wounded coenocytic green algae. *Planta* **171**:30–42.
34. T. Laux and G. Jürgens 1994. Establishing the body plan of the *Arabidopsis* embryo. *Acta Bot. Neerl.* **43**:247–260.
35. A. Lindenmayer and G. Rozenberg 1979. Parallel generation of maps: Developmental systems for cell layers. In: *Graph Grammars and Their Application to Computer Science*, eds V. Claus, H. Ehrig and G. Rozenberg, pp. 301–316. Springer-Verlag, Heidelberg. (Lecture Notes in Computer Science, Vol. 73).
36. H. Lodish, D. Baltimore, A. Berk, L. Zipursky, P. Matsudaira and J. Darnell 1995. *Molecular Cell Biology*, 3rd edition. W. H. Freeman, New York.
37. J. Lück and H. B. Lück, H. B. 1981. Proposition d'une typologie de l'organisation cellulaire des tissus végétaux. In: *Actes du 1er Séminaire de l'Ecole de Biologie Théorique du CNRS*, 1-4 Juin 1981, eds H. Le Guyader and T. Moulin, pp. 335–371. ENSAT, Paris.
38. J. Lück and H. B. Lück 1985. Comparative plant morphogenesis founded on map and stereomap generating systems. In: *Dynamical Systems and Cellular Automata*, eds J. Demongeot, E. Golès and M Tchuente, pp. 111–121. Academic Press, London.
39. J. Lück and H. B. Lück 1987. Vers une nouvelle théorie des méristèmes apicaux. In: *Biologie Théorique*, Solignac 1985, ed. H. B. Lück, pp. 197–213. CNRS, Paris.
40. J. Lück, and H. B. Lück 1992. Cellworks: An application to plant morphogenesis. In: *Lindenmayer Systems, Impacts on Theoretical Computer Sciences*, Computer Graphics, and Developmental Biology, eds G. Rozenburg and A Salomaa, pp. 385–404. Springer-Verlag, Berlin.
41. J. Lück and H. B. Lück 1995. Plant assemblage in layers. *Acta Biotheoret.* **43**:95–111.
42. J. Lück, P. W. Barlow and H. B. Lück1997. An automata-theoretical model of meristem development as applied to the primary root of *Zea mays L. Ann. Bot.* **79**:375–389.
43. J. Lück, A. Lindenmayer and H. B. Lück 1988. Models of cell tetrads and clones in meristematic cell layers. *Bot. Gaz.* **149**:127–141.
44. H. J. Marchant and J. D. Pickett-Heaps1972. Ultrastructure and differentiation of *Hydrodictyon reticulatum* VI. Formation of the germ net. *Austral. J. Biol. Sci.* **25**:1199–1213.
45. J. W. Marvin 1939, Cell shape studies in the pith of *Eupatorium purpureum. Am. J. Bot.* **26**:487–504.

46. E. B. Matzke 1949. Three-dimensional shape changes during cell division in the epidermis of the apical meristem of *Anacharis densa* (Elodea). *Am. J. Bot.* **36**:584–595.
47. E. B. Matzke 1950. In the twinkling of an eye. *Bull. Torrey Bot.* Club **77**:222–232.
48. D. Mazia 1993. The cell cycle at the cellular level. *Eur. J. Cell Biol.* **61**:14 (Abstract).
49. N. Michaux-Ferrière 1980. Formation et évolution de cals chez le *Pteris cretica L. Z. Pflanzenphysiol.* **96**:387–405.
50. J. G. Miller 1978. *Living Systems.* McGraw Hill, New York.
51. J. C. M. Mombach, M. A. Z. Vasconcellos and R. M. C. De Alemeida 1990. Arrangement of cells in vegetable tissues. J. Phys D: *Appl. Phys.* **23**:600–606.
52. K. J. Niklas 1992. Plant Biomechanics. *An Engineering Approach to Plant Form and Function.* University of Chicago Press, Chicago.
53. N. Rivier, B. Dubertret and G. Schlieker 1997. The stationary state of epithelial cells. In: *Dynamics of Cell and Tissue Motion*, eds W. Alt, A. Deutsch and G. Dunn, pp. 275–282. Birkhäuser, Basel.
54. J. von Sachs1887. Relations between growth and cell division in embryonic tissues. In: *Lectures in Plant Physiology* (English translation by H. Marshall Ward), pp. 431–459. Clarendon Press, Oxford.
55. A. Siegert 1965. Morphologische, entwicklungsgeschichtliche und systematische Studien an *Psilotum triquetrum* Sw. II. Die Verzweigung (mit einer allgemeinen Erörterung des Begriffes, Dichotomie"). *Beitr. Biol. Pflanzen* **41**:209–230.
56. E. W. Sinnott 1940. The cell-organ relationship in plant organization. *Growth (Suppl.)*:77–86.
57. A. H. Skop and J. G. White 1998. The dynactin complex is required for cleavage plane specification in early *Caenorhabditis elegans* embryos. *Curr. Biol.* **8**:1110–1116.
58. V. S. Sundaralingam 1954. The developmental morphology of *Chara zeylanica Willd. J. Indian Bot. Soc.* **33**:272–297.
59. Y. Takiguchi, R. Imaichi and M. Kato 1997. Cell division patterns in the apices of subterranean axis and aerial shoot of *Psilotum nudum* (Psilotaceae): Morphological and phylogenetic implications for the subterranean axis. *Am. J. Bot.* **84**:588–596.
60. J. Traas, C. Bellini, P. Nacry, J. Kronenberger, D. Bouchez and M. Caboche 1995. Normal differentiation patterns in plants lacking microtubular preprophase bands. *Nature* **375**:676–677.

A CLASSIFICATION OF PLANT MERISTEMS BASED ON CELLWORKS (3D L-SYSTEMS). THE MAINTAINANCE AND COMPLEXITY OF THEIR CELLULAR PATTERNS

JACQUELINE LÜCK[1] AND HERMANN B. LÜCK[2]

[1] *Laboratoire de Botanique analytique et Structuralisme végétal, Faculté des Sciences et Techniques de St-Jérôme, CNRS, ERS 6100, 13397 Marseille cedex 20, France*

[2] *Val d'Arenc, 83330 Le Beausset, France*

Cellworks are L-systems generating 3D networks representative of the network of cell walls in plant tissues. They provide a scaffold for the development of plant forms. The model is a topological, deterministic and interactionless approach to plant morphogenesis with a secondary geometrical interpretation.

By means of cellworks, and using double wall labelling, it is possible to search algorithmically for patterns of cell division activity that lead to particular patterns of cellular arrangements. Within meristems, which are responsible for growth and form of plants, it is shown how the autoreproduction of the shape of apical cells maintains the specific form of plant structure. Furtheron, the way patterns are formed permits the construction of a general classification of plant meristems, internal, apical or in layers, based on a progressive complexification of the shape of apical cells and of the set of cell production rules.

Special attention is given to the geometrical interpretation of the topological cellular arrangements, important for the plant morphologist.

1 Introduction

Classifications of plant forms are based on form families, in relationships based on external features (Philipson, 1990). Often, there is a simple juxtaposition of the description of very complex patterns. Dealing with plant diversity, different recent approaches of classifications (Newman, 1956; Hallé *et al.*,1978; Sattler, 1990; Barlow, 1991 and 1994) have been proposed trying to introduce a principle of formal coherence.

Nevertheless, there exists no classification which provides simultaneously for the emergence, relationships and maintainance of basic patterns in the world of plant variability. If formalized approaches have been nearly impossible for plant morphogenesis, this is because no formal system allowing their construction is available. Some years ago, topological generative systems such as L-systems (Rozenberg & Salomaa, 1980), and particularly multidimensional L-systems (Fracchia & Prusinkiewicz, 1991), have furnished a tool for such an aim. A further requisite is the way these constructions are conceived in order to permit the deduction of relationships and ordering between such systems (Lück & Lück, 1991, 1992, 1993 and 1996).

The classification of plant forms presently developed is considered on a very basic and general level. At first, invariances were searched within the diversity. For this aim, the variable phenomenologies were supposed to be associated to sub-jacent invariant mechanisms. Such invariances have to be few, to avoid again a complex system. The invariances have to permit not only an ordering of phenomenologies, but they may also make sense of their emergence and maintainance. Further, it

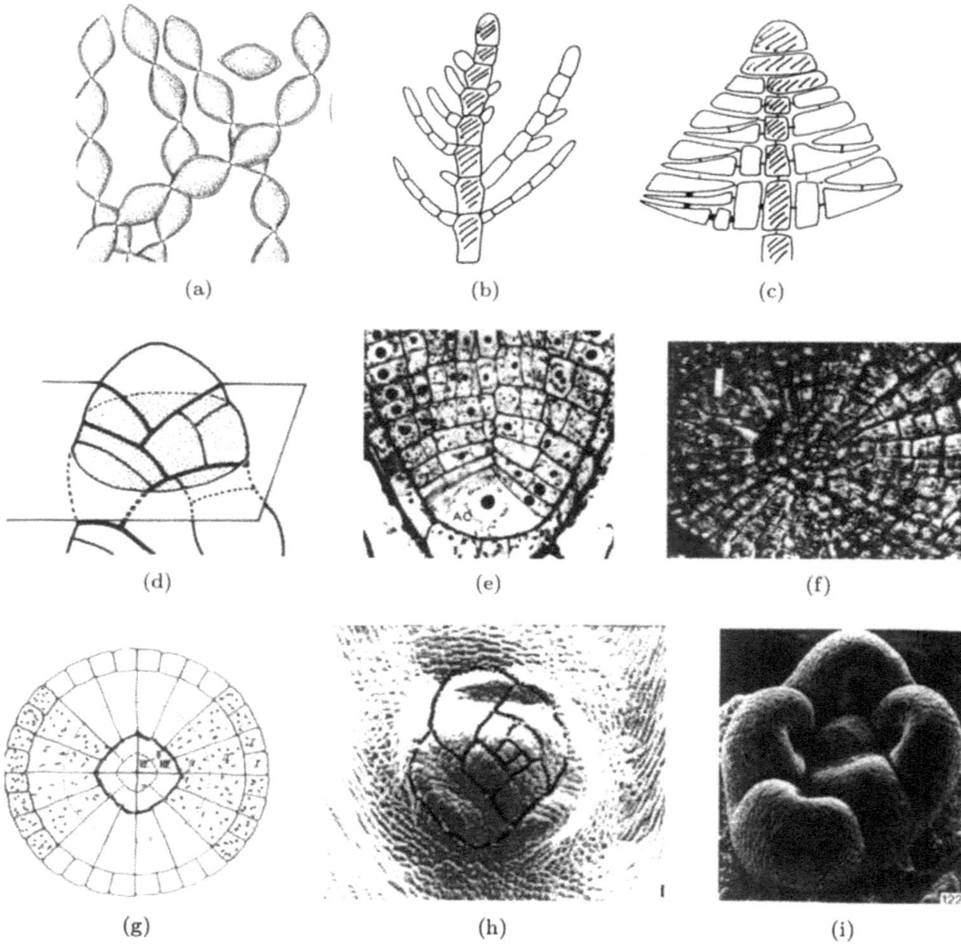

Figure 1. Diversity of cellular organisation in plants. (a): Attached spores in *Monilia fructigena* (Zambettakis) (b): Branched cellular filaments of the algae *Antithamnion plumula* (Lambert, 1991) (c): Connected branched filaments of the alga *Hypoglossum* (Feldman, 1978). (d): Apical tetrahedral cell and merophytes with secondary cell divisions by the fern *Ceratopteris* (Hébant-Mauri, 1977); (e): Apical tetrahedral cell and derivatives in a longitudinal section through the primary root of *Azolla pinnata* (Gunning, 1981). (f): Transversal section of the embryonic radicles of *Epilobium* at the level of initial central cells (g): Schematic drawing of the first cell divisions in the apex of *Epilobium hirsutum* (Bartels, 1956) (h): Apex with 4-sided helicoidal segmentation of the angle shoot by *Selaginella martensii* (Jernstedt *et al.*, 1994). (i): apex of *Zanthoxylum simulans* (van Heel, 1983).

could be added that a theoretical biological framework like that tempted hereby ... "must have a potential to grow and develop into an aesthetically satisfying mathematical theory" (Gromov, 1990).

Our aim was to look for very evident mechanisms, so evident that they may not be recognized as being important. These are three: (1) the division of cells, implying a partition into two parts and not three or more; (2) the divisions are qualitatively unequal as two daughter-cells are at least symmetric with respect to the common division wall; (3) the maintainance of patterns implies some autoreproduction of the form. This is given in a meristem, cambium, etc., i.e. active growing centers in which patterns are maintained during a certain time. The simplest mechanism for reproduction can be found in the filiation of cells, i.e. when the form of the mother cell is reproduced in one of its daughter-cells. Finally, within a given framework of classification arises the problem of discovering all the possible patterns, and then deciding which of them are realized in nature?

Illustrating roughly plant diversity, Figure 1 shows that from spores simply attached by one point, there are filaments of cells, which may be branched, and from which 2D organisations are progressively introduced by secondary contacts. Well known organisations as is the helicoidal 3-sided segmentation of a tetrahedral apical cell by ferns shows some regularities. Such tetrahedral cells are also found in roots. Secondary divisions which mostly occur in the apical derivatives, then called merophytes, may also have repetitive patterns. For higher organisations, regularities are less visible. Nevertheless, there is the possibility of realizing patterns based on a quadrilateral organisation such as is seen in Epilobium, or sequential helicoidal 4-sided segmentation as shown by the apex of Selaginella. For complex phyllotaxis, the limits between merophytes or cell families are no longer observable and patterns are based on leaf distributions. But this does not mean that they do not exist on the cellular level.

2 Cellworks

To generate plant-like patterns, which explain how merophytes or cell families may be organized, we have proposed "cellworks". Other proposals are found in Fracchia & Prusinkiewicz, 1991. These are 3-dimensional L-systems which generate the 3D network of cell walls within a tissue. The net represents the scaffold of the rigid cell wall network in plant tissues. The model is a topological, deterministic and interactionless approach. A secondary geometrical interpretation is necessary for biological meaningfulness.

A cellwork G is a quadruple, $G = \{\Sigma, P, T, c_0\}$, with a *finite alphabet* Σ of n labelled convex polyhedra, representing cells. This alphabet is a union of two sub-alphabets, such as $\Sigma = C \cup F, C = \{c_1, \ldots, c_n\}$ and $F = \{f_1, \ldots, f_n\}$, where the *sub-alphabet* C of labels designates cell types, and the *sub-alphabet* F specifies the corresponding polyhedral representations. An index of c and f indicates the polyhedral degree of the cell. The presently proposed cellworks are characterized by a double labelling of walls, i.e. each wall which separates two polyhedrons is labelled on both of its sides. The faces of the polyhedra of Σ are labelled inside. A set P of *cell productions* such as

$$c \to \omega, c \in C, w \in C^* \text{ and } C \subset \Sigma$$

specifies how cell types are derived one from another, ω being composed of one or two cells. The *starting cellwork* c_0 consists of one or several cells. Further, a *table of cell contacts* T indicates how polyhedra are associated in a 3D-space by help of a colored wall matching. This topological cellular assemblage allows a geometrical interpretation and representation of the cellworks.

Example 1, cellwork IV.C: A first example is a system which describes a 3-sided helicoidal segmentation of an apical tetrahedral cell (Figure 2) such as that indicated for ferns (Figure 1). The alphabet consists of 4 polyhedra, a_4, b_5, b_6 and

Cell alphabet:

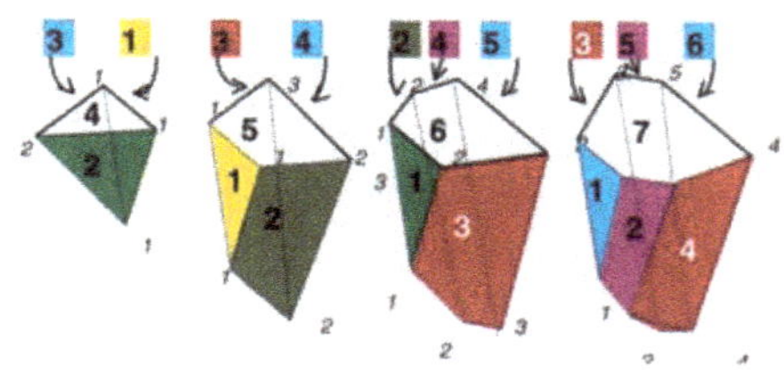

Table of Cell Contacts:

	a_4	b_5	b_6	b_7
b_5	1-111 1-111			
b_6	2-112 1-112	2-1122 2-1122		
b_7	3-112 1-112	3-1123 2-1123	3-[illegible] 3-[illegible]	
b_7		4-223 1-112	4-2234 2-1123	4-[illegible] 3-[illegible]
b_7			5-334 1-112	5-[illegible] 2-[illegible]
b_7				6-445 1-112
e	4 e	5 e	6 e	7 e

Gluing of polyhedra:

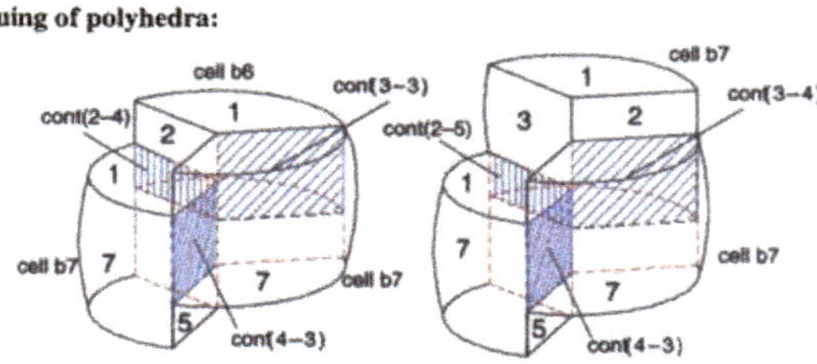

Figure 2. Cellwork with an apical tetrahedral autoreproductive cell simulating the organization of a fern shoot (an amendment of Example 1 in Lück & Lück, 1996). — Alphabet Σ with 4 cells. Sub-alphabet $C = \{a_4, b_5, b_6, b_7\}$, a_4 the apical cell and its derivatives b_5, b_6 and b_7. Sub-alphabet F of polyhedra representing the cell of C (7 colours and labels for walls, e for the environment, and node labels with small characters). Cell productions: $P = \{a_4 \to a_4 b_5, b_5 \to b_6, b_6 \to b_7, b_7 \to b_7\}$. Table of 16 wall contacts: line and column entries: elements of the alphabet; e: environment; bold integers: age of a wall; sequence of small integers: node labels of a wall; each element of the table indicates a *contact* between two walls. Gluing of polyhedra: walls belonging to two cells and having the same colour are associated; contact written, for example, as *cont*(1-6) for wall ages, or *cont(col 4)*, which is equivalent. Fusion of two walls to be associated by bringing into correspondance the cycles of node labels (ex.: for 445 and 112, 4 corresponds to 1 and 5 to 2).

b_7, where the index stands for the number of faces or walls shared with neighbour cells or the environment. Faces are differentiated by colors, which may be numbers or letters. Nodes are also numbered. A set P of production rules indicates that only the tetrahedral cell is dividing. Further, a table of cell contacts indicates which faces of two cells correspond, and node labels indicate the exact positioning of the faces to be glued together.

A cellwork derivation consists of the replacement of each cell by one or two cells in conformity with the production rule. For example, a cell b_6 is replaced by b_7 and wall matchings are assured. All cells present in a cellwork are formally substituted in parallel, e.g. during one discrete timestep (even if the technical execution is rather sequential). After several derivation steps, the system has generated geometrically three compact files of cells, arranged in a stepwise helicoidal way. They are finally all b_7 cells with, at the top, the apical cell a_4. Two intermediary cell states, b_5 and b_6, make the transition between the apical tetrahedral cell and the final b_7 cells. If further cell divisions occured as descents from such b_7 cells, the corresponding merophytes conserve topologically the same mutual disposition.

The example of Figure 2 shows also the gluing principle of polyhedra during one derivation step. A cellwork of three cells is given at $i = 0$, one being a b_6 cell and the two others b_7 cells. After one derivation step, at $i = 1$, the two b_7 cells remain unchanged as $b_7 \rightarrow b_7$, but the b_6 cell is replaced by a b_7 cell conformable to the wall contacts of the table. The resulting cellwork of three b_7 cells shows the helicoidal arrangement of final cell states in the three contiguous cell files.

Example 2, cellwork V.B (cf. p.541): A second example (cellwork V.B of Table 1) shows a 4-sided helicoidal segmentation produced by successive divisions of an apical 5-sided autoreproductive cell. An alphabet of 5 cells, a set P of 5 cell production rules, and a table of wall contacts with 11 colours, lead to the indicated pattern. Again, a stepwise organisation of cells ends at the top with the "bed" of the apical cell a_5 prepared by three intermediary cell states, $b_6 \rightarrow b_7 \rightarrow b_8$. The final construction items are cells of nine faces, b_9. The polyhedra of the alphabet and the geometrical interpretation of the generated cellwork are found in Figure 3.V and in Lück & Lück, 1996.

3 A classification of plant forms

The most important application of the cellwork systems is the possibility they give to classify coherently the basic plant forms, as a direct consequence of the way the systems were constructed. The table of classification (Table 1) starts with column entries concerning the conformation of the initial cell with increasing complexity. As shown for the previously given examples for cases IV and V, n indicates their number of walls, e their number of edges and v their number of vertices. A further indication is the polygonal degree of the dw, i.e. of the last added wall in the polyhedron representative of the initial cell (highlighted by the color yellow). This degree represents the number of anchor faces for a dw on the wall boundary of the apical cell. For the two given examples, this dw degree is equal to 3 and 4, respectively.

The row entry of Table 1 shows 3 groups of patterns, A, B and C, according to the number of faces of the initial cell in contact with the environment. With one environmental face, the previous two examples figure in the group B of Table 1. Their sets of cell type productions show some internal coherence. Accordingly,

Table 1. Classification of basic plant structures: family with *dw* degree $= n - 1$. I-VII classes; A,C:groups. Autoreproductive cells with n number of faces, e number of edges; v number of vertices. Sets of cell type production rules in black: analyzed systems, (red) non-analyzed, (gray background): impossible systems.

Conformation of autoreproductive cell

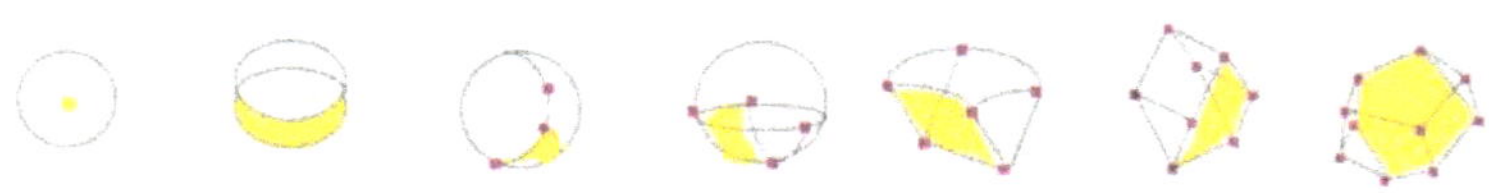

	I.	II.	III.	IV.	V.	VI.	VII.
	Conformation of the initial cell						
	n = 1 e = 0 v = 0	n = 2 e = 1 v = 0	n = 3 e = 3 v = 2	n = 4 e = 6 v = 4	n = 5 e = 9 v = 6	n = 6 e = 12 v = 8	n = 7 e = 15 v = 10
	Polygonal degree of the dw						
	0	1	2 (1+dw)	3 (2+dw)	4 (3+dw)	5 (4+dw)	6 (5+dw)
A	0 environmental faces - **Group of Internal Meristems** - Productions of cell types						
	$a_1 \to a_1 b_2$ $b_2 \to b_2$	$a_2 \to a_2 b_3$ $b_3 \to b_4$ $b_4 \to b_4$	$a_3 \to a_3 b_4$ $b_4 \to b_5$ $b_5 \to b_6$ $b_6 \to b_6$	$a_4 \to a'_4 b_5$ $b_5 \to b_6$ $b_6 \to b_7$ $b_7 \to b_8$ $b_8 \to b_8$ $a'_4 \to a_4 b_5$	$a_5 \to a_5 b_6$ $b_6 \to b_7$ $b_7 \to b_8$ $b_8 \to b_9$ $b_9 \to b_{10}$ $b_{10} \to b_{10}$	$a_6 \to a_6 b_7$ $b_7 \to b_8$ $b_8 \to b_9$ $b_9 \to b_{10}$ $b_{10} \to b_{11}$ $b_{11} \to b_{12}$ $b_{12} \to b_{12}$	$a_7 \to a_7 b_8$ $b_8 \to b_9$ $b_9 \to b_{10}$ $b_{10} \to b_{11}$ $b_{11} \to b_{12}$ $b_{12} \to b_{13}$ $b_{13} \to b_{14}$ $b_{14} \to b_{14}$
	Structure of the generated cellular body						
	Vesicles (invagi-nation)		3-lateral segment-ation	4-lateral segment-ation	non-analyzed	non-analyzed	non-analyzed
B	1 environmental face - **Group of Apical Meristems** - Productions of cell types						
	$a_1 \to a_1 a_1$	$a_2 \to a_2 b_3$ $b_3 \to b_3$	$a_3 \to a_3 b_4$ $b_4 \to b_5$ $b_5 \to b_5$	$a_4 \to a_4 b_5$ $b_5 \to b_6$ $b_6 \to b_7$ $b_7 \to b_7$	$a_5 \to a_5 b_6$ $b_6 \to b_7$ $b_7 \to b_8$ $b_8 \to b_9$ $b_9 \to b_9$	$a_6 \to a_6 b_7$ $b_7 \to b_8$ $b_8 \to b_9$ $b_9 \to b_{10}$ $b_{10} \to b_{11}$ $b_{11} \to b_{11}$	$a_7 \to a_7 b_8$ $b_8 \to b_9$ $b_9 \to b_{10}$ $b_{10} \to b_{11}$ $b_{11} \to b_{12}$ $b_{12} \to b_{13}$ $b_{13} \to b_{13}$
	Structure of generated cellular body						
	Bud (ex-trusion)	Filament	2-lateral segment-ation	3-lateral helicoidal segment-ation	4-lateral helicoidal segment-ation	5-lateral helicoidal segment-ation	2/5-helicoidal segment-ation
C	2 environmental faces - **Group of Plate Meristems** - Productions of cell types						
		$a_2 \to a_2 b_2$	$a_3 \to a_3 b_4$ $b_4 \to b_4$	$a_4 \to a_4 b_5$ $b_5 \to b_6$ $b_6 \to b_6$	$a_5 \to a_5 b_6$ $b_6 \to b_7$ $b_7 \to b_8$ $b_8 \to b_8$	$a_6 \to a_6 b_7$ $b_7 \to b_8$ $b_8 \to b_9$ $b_9 \to b_{10}$ $b_{10} \to b_{10}$	$a_7 \to a_7 b_8$ $b_8 \to b_9$ $b_9 \to b_{10}$ $b_{10} \to b_{11}$ $b_{11} \to b_{12}$ $b_{12} \to b_{12}$
	Structure of generated cellular body						
					Tunica 3-lateral	Tunica 4-lat.+ 3-lat.	non-analyzed

extrapolation to simpler patterns for columns I to III can immediatly been attempted and written down. Looking to which conformation of the initial cell and to which generated plant body such a set of productions corresponds, it results for case III an apical cell with bilateral alternating segmentation giving rise to two contiguous cell files (Figure 3.III). The case II leads to a simple cellular filament

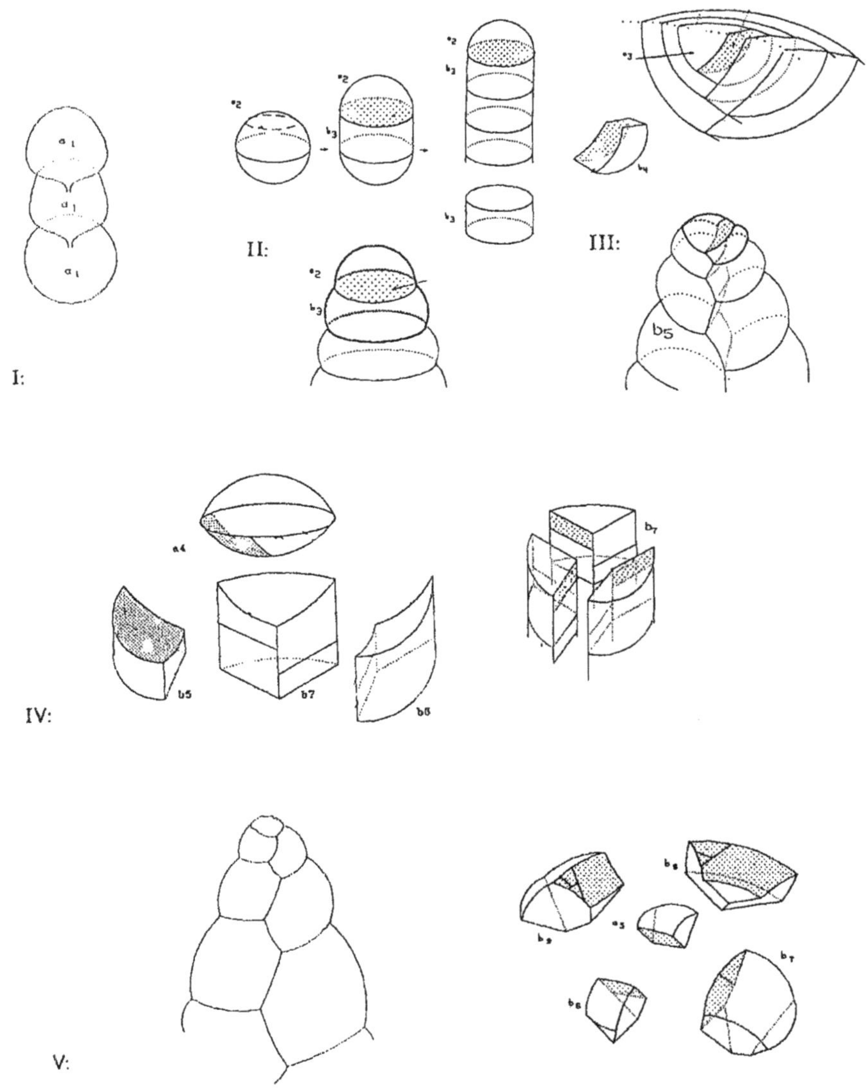

Figure 3. Apical meristems of Group B, Table 1. Geometrical interpretation of cellworks. I: Budding composed of one cell type; II: Cellular filament composed of two cell types; III: 2-sided alternating segmentation; IV: 3-sided helicoidal segmentation (four cell types); V: 4-sided helicoidal segmentation (five cell types).

with an active apical cell (Figure 3.II). The simplest case I with 0 *dw*, i.e. where the *dw* is reduced to a point, can be considered as a budding process (Figure 3.I).

All systems generating cells by the activity of an initial cell entirely embedded into other cells (no face touches the environment, as $e = 0$), belong to the first group, A of patterns (Table 1). These patterns, though topologically realisable for cases II and III (Figure 4.II and Figure 4.III), are nevertheless in biology difficult to find.

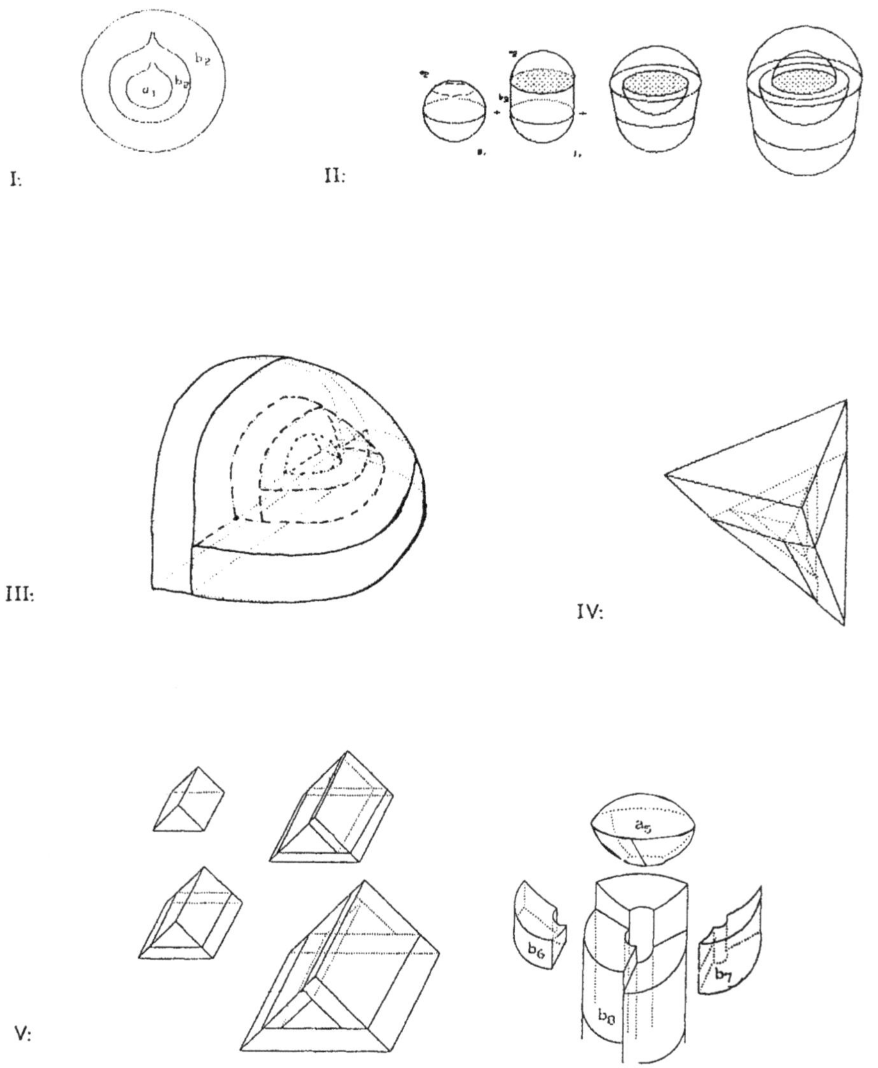

Figure 4. Internal (I-IV) and plate (V) meristems of Groups A and C, respectively (Table 1). I: Vesicle invagination; II: 2-sided alternating internal segmentation; III: 3-sided internal segmentation; IV: 4-sided internal segmentation; V: 3-sided helicoidal segmentation in a layer, represented as a tunica.

The extreme case I with absence of a defined *dw* could represents the invagination or vesicle formation (Figure 4.I). Nevertheless, in case IV.A, the pattern is not exactly constructible as the label distribution in the wall map of the apical cell changes from one cell division to the next one, such as a flip flop between two symmetrical map distributions with regard to wall labels (indicated in Table 1 by the cell types a_4 and a_4').

By investigating all map partitions for the initial cell of conformation V, a third group of patterns, (C), appeared in the corresponding column. It is a 3-sided helicoidal segmentation with three cell files derived from divisions of a central cell in a cell layer, i.e., all cells in the layer have two faces which touch the environment ($e = 2$). When bent like a glove-finger, the generated cell layer remembers tunica-like cell orglanization individualized within a cylindrical plant shoot or root organ. The pattern V.C shows again three cell files arranged in a helicoidal way such as in pattern IV.B, but with a hole inside (Figure 4.V). Interestingly, with an increasing number of environmental faces, the three patterns III.A, IV.B and V.C, which have as a common characteristic the ability to generate 3-sided helicoidal segmentations of an autoreproductive cell, need an initial cell of increasing complexity. But the number of necessary construction items, i.e. the size of the cellular alphabet of cell types, remains the same.

To extrapolate the sets of cell type productions upwards in Table 1, the shape of the initial cell has first to be defined, as the number of faces, edges and vertices may not be sufficient for its specification. For a given system, the next more complex autoreproductive initial cell is obtained by partitioning its wall map so that the partition cycle goes once through *all* regions, thus increasing by one the polygonal degree of the *dw*. In the case of column VI.B ($e = 1$), a 5-sided lateral helicoidal segmentation results in five compact cell-files (Figure not shown). Contrary to what could be expected, the following pattern (column VII) is not a 6-sided segmentation, but, surprisingly, a 2/5 helicoidal segmentation. Just as in a phyllotaxic system, at each time step, it is not the neighbour file on the generative helix, but each second file which receives a new cell at its top as a derivative from the apical cell (Figure 5). This implies, biologically, that long before any bulging out, indicating the formation of a leaf primordium, the structure of the shoot has the predisposition for a given phyllotaxy.

The next case, VI.C, is not trivial either, as on one side the layer cells are arranged like in a 3-sided way of segmentation, and on the other side in a 4-lateral way of segmentation. This recalls situations in plants where, for example, cells or merophytes are arranged in a certain order, and leaves in another way. Related to such a tunica structure, one of its side could be responsible for the cellular organization of the tissue, the other side for the leave positioning. Accordingly, starting with patterns in column IV.B and V.B of Table 1, simpler systems could be defined by going back in the table rows and, on the other side, the table could be filled up to the VIIth column. The complete Table 1 shows, in black, the systems analysed for which a meaningful geometrical representation and its plausible counterpart in plant morphogenesis could be established. All other systems not entirely analyzed until now, have their cell type production sets indicated in grey. Sets in red underline impossible systems.

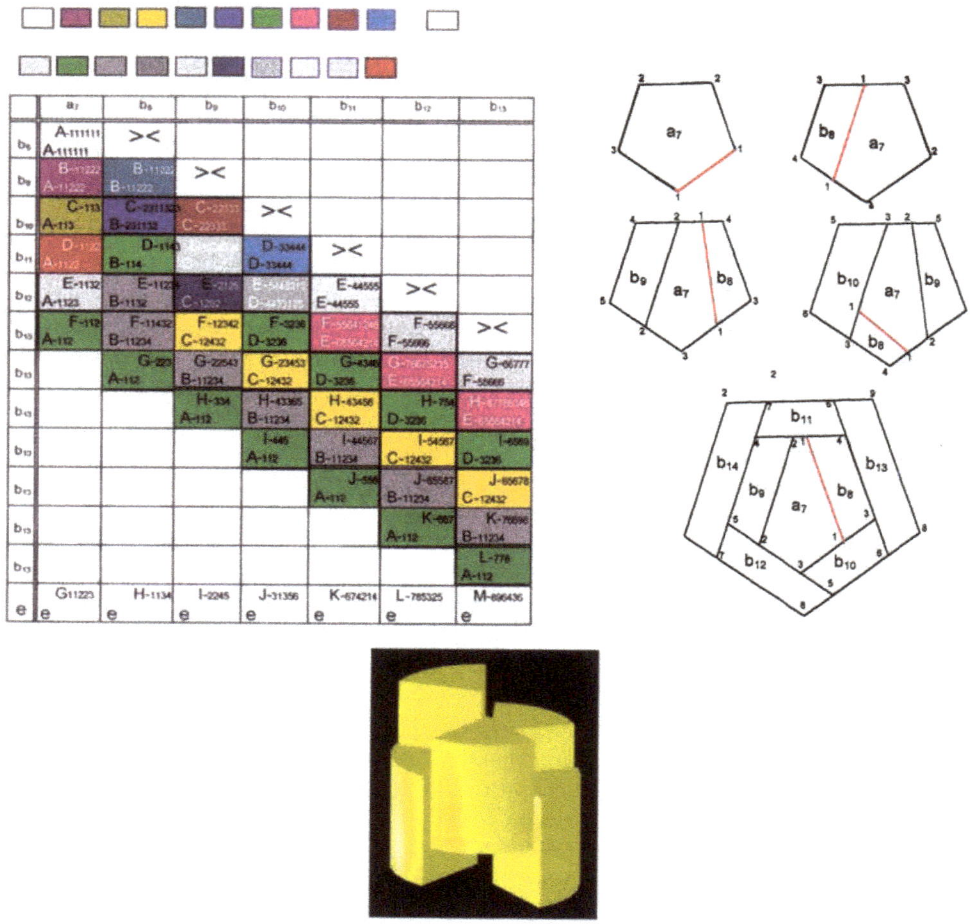

Figure 5. Cellwork VII.B (Table 1) generating the shoot organization for a 5-sided segmentation in a 2/5 phyllotaxic arrangement. Alphabet of 7 cell types as indicated in the entries of the Table of 49 cell contacts between walls, distinguished by 21 colours (or capitals). Node labels as small numbers. — Derivation of the wall G of the cell a_7 facing the environment e. The apical cell a_7 divides at each timestep, 0 divisions of the b cells. Surface view of the cellwork derivation for $i = 0$ to 3 *and* $i = 7$. Cell type inscribed in each surface cellwall with node labels conform to the Table of cell contacts. — Geometrical interpretation of staggered cells in a tier of 5 contiguous cells.

4 Specification of a cellwork system with double labelled walls

In order to construct a cellwork system, the starting point is the choice of an initial cell configuration for which a system allowing autoreproduction of this cell has to be found. A first step is the specification of the face colors of this initial cell. For this purpose we use the generation age of faces in the following way. As each cell boundary has a particular face which is the last added wall during the last cell division, it takes label 1. At the next division, a new division wall (dw) with label

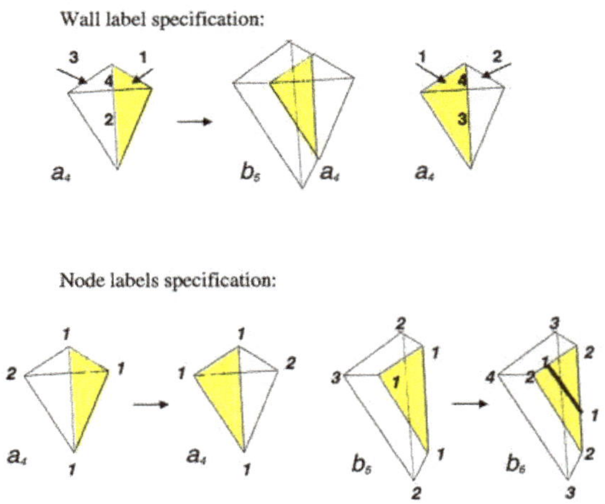

Figure 6. Wall and node label specification during the division $a_4 \rightarrow a_4 b_5$ of a tetrahedron. Division wall (yellow) with age 1 takes in the next step, after division, age 2. Each wall age is incremented by one at each cell division. Wall 4 in contact with the environment. — Node labels of the division wall take the number 1. After each cell division, they are incremented by one until the contact of a b cell with the apical cell is lost.

1 appears. The label of the former dw is incremented by 1, and so on (for example consider in Figure 6 a tetrahedral cell as used in cellwork IV.B). The same strategy is used for node labels. The last dw has node labels 1 which are incremented at each further cell division (cf. Figure 6).

Some faces may be incorporated into lateral derivative cells, the "b" cells (such as, for example, face 3 of the tetrahedron which becomes at the next timestep face 4 of the b cell). Other faces are subdivided by an insertion edge of the dw (for example, face 2 of cell a_4 (Figure 6) furnishes in the next timestep face 3 for the new daughter cell a_4 and face 3 for the other daughter cell b_5). Finally, there are also wall partitions performed by a dw wall edge inserted into a neighbour cell, called induced partitions. An example is wall 1 with the node sequence 111 belonging to cell b_5; at the next timestep this wall becomes partitioned by the edge 11 into two faces, wall 1 with nodes 112 and wall 2 with nodes 1122. The mechanism goes on until new divisions of the initial self-reproducing "a" cell break the contact with the older "b" derivative.

5 How to find a Cellwork system with an autoreproductive initial cell

For biological meaningfulness, we are interested only in systems for which, during successive cell divisions, the configuration of the initial cell is maintained in a lineage of cells. In other words, self-reproduction of the initial cell occurs. For this purpose, an algorithm had been proposed (Lück & Lück, 1996) to find daughter cells which have the same conformation as the mother cell had before division. The wall boundary of the initial cell is opened by removing the last dw. The resulting

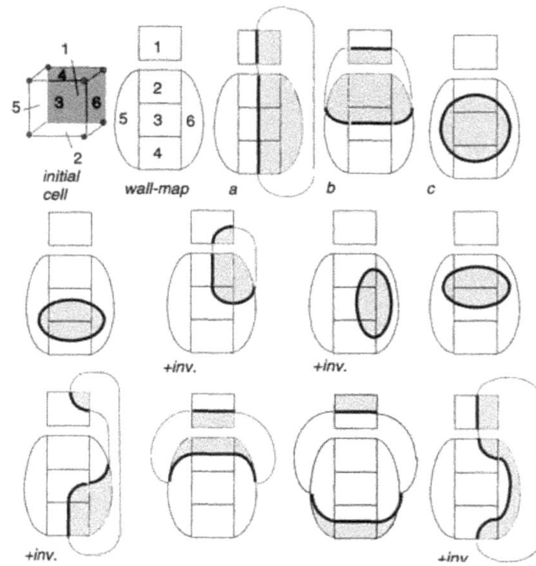

Figure 7. Map partitions of the wall-map of the cubic cell, 15 possibilities with partition cycle of length 4. Grey coloured part of the wall-map belongs the daughter cell a; non coloured part belongs to the derivative cell b. Partitions (a), (b) and (c) allow self-reproduction of the cubic cell with four 4-sided polygonal faces + the dw. (c): reducible system to one wall + the dw (cf. text). Four of the other 8 possible partitions have inverse partitions (+*inv.*). These 12 partitions forbit self-reproduction of the cubic cell.

2D wall map represents the exact neighbourhood of all wall faces as seen from the inside of the cell. In such a map, a new dw traces a cycle of edges by which the dw is attached and cutting some map regions (wall faces) into two parts. Accordingly, we are looking for all possible cycles with length equal to the polygonal degree of the dw. For each cycle, the result is a map partitioned into two submaps. The part of the map which will belong to the new "a" cell has to show the same map structure as the mother cell wallmap, in order to fullfill the condition of autoreproduction. Production rules for walls can be immediately established between labels of regions in the cellwall map before partitioning (the left-hand terms of production rules) and the fate of each region in the partitioned map (the right-hand terms of the production rules). Further, the dw derivation indicates all wall partitions induced by neighbour dw's, and are specified in this way for the table of wall contacts. A detailed representation of wallmap partition can be found for the tetrahedron, the initial a_4 cell of Table 1, in Lück & Lück, 1992, for the a_6 cell of Table 1 in Lück & Lück, 1996, and for the cubic cell in Figure 7.

6 Extended classification of plant forms

A table of classification similar to Table 1 could be constructed beginning with a cube instead of a tetrahedron. The cubic cell, with $n = 6$ faces, $e = 12$ edges and

$v = 8$ vertices would correspond to column VI (Table 1). It initiates a table with a dw degree equal to $n-2$, i.e. the dw is anchored on four of the six faces. The possible wallmap partitions for the cubic cellwall boundary indicate 15 different cycles of length 4, from which 4 are symmetrical partitions (Figure 7). But only three of the 11 partitions create the condition for self-reproduction of the initial cell. One of them generates a 6-sided lateral segmentation. This means that the sister cell of the self-reproducing daughter cell is cut off by dw's which, successively, are parallel to the walls R B D L F U R etc. (for Right, Back, Down, Left, Front and Up). It corresponds accordingly to an inner meristem of group A amy (Figure 8). Again,

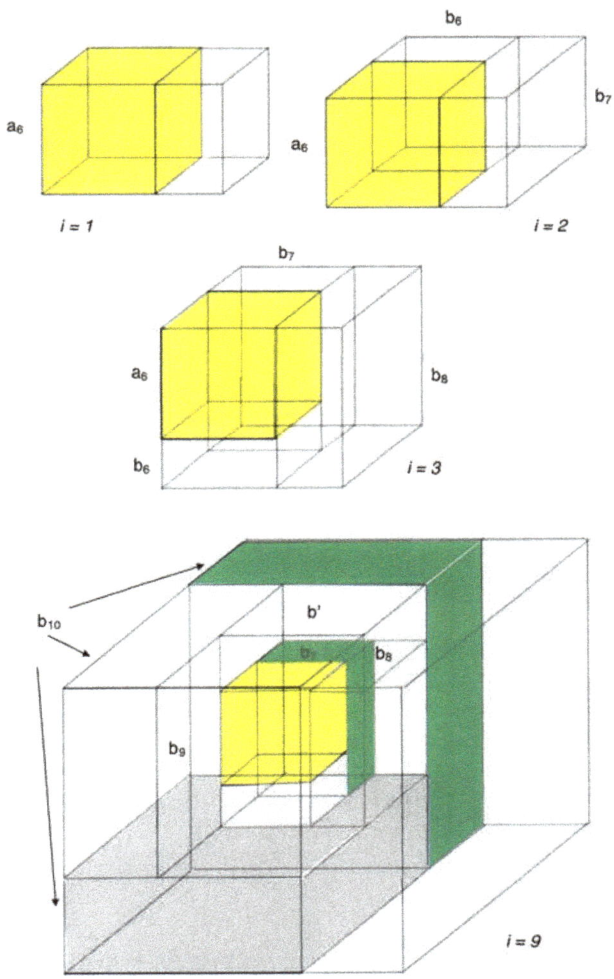

Figure 8. Cellwork VI.A (future Table 2): schematic representation of derivations for steps $i = 1-3$ and $i = 9$. Divisions of the cubic initial cell parallel successively to all six faces. Downwards cutted off merophytes (grey), merophytes in the back (green).

as for the tetrahedron, the initial cell during successive divisions takes alternately symmetrical wall labellings. Is that a propriety of regular polyhedra? A second system allowing a lineage of cubic cells is a 4-lateral segmentation and two faces in contact with the environment. It belongs therefore to meristems in layers or tunicas of group C. The third system generates a filament because the new *dw* is not anchored on the previous *dw*. So it is a system reducible to a simpler one due to a multiplicity of wall faces which cooperate as a unique wall. The cell type productions $a_6 \rightarrow a_6b_6$ and $b_6 \rightarrow b_6$, can be reduced to $a_2 \rightarrow a_2b_3$ and $b_3 \rightarrow b_3$, which is nothing else than the system II.B of Table 1, the fundamental system for filaments with apical transversal divisions.

7 Passage from one generating system to another

It has been shown that for each conformation of an initial cell, the next more complex one can be established within the formal framework of a classification table. But this specification has to go stepwise in increasing complexification. On the other hand, as indicated for a reducible system, at any moment during the division of an initial cell it can switch to another less complex way of division behaviour. Within a group of patterns, the jump to a less complex configuration is possible to any of the patterns in the classification table. This points to the fact that, within the generative patterns, the transformation of a pattern into a more complex one goes stepwise, for example, a 3-sided helicoidal apical segmentation can switch to a 4-sided one. But, within a given pattern, for example, a 3-sided helicoidal segmentation pattern can switch to the formation of a simple filament. Another consequence is the fact that, for each derivative in the Table 1, the "*b*" cell is more complex than the "*a*" cell. This implies that the derivative could give rise to more complicated patterns in lateral organs than are possible for the initial primary organ. There are also relations on the level of apical cells belonging to different tables of classification. From the cube we can switch to the tetrahedron, but not the reverse.

8 Limits of the classification tables

A question arises of how far the tables can be developed? After column VII, there are two other phyllotaxies in the group B ($e = 1$). But the conformation of the "*a*" cell becomes progressively twisted in a geometrical representation. The point is: where is the limit from which on the topological figures can no longer be represented by a geometrical 3D object. Is there a natural limit?

9 From topology to geometry

The way from topology to geometry is very important for imparting biological meaningfulness to the tables. The figure soon becomes very hard to interprete. A way proposed to realize this consist in 2D plans with specified orientations (Figure 9a). Which may cut a cylinder representative of the plant shoot apex (Figure 9b). For example, after each cut, a Boolean substraction gives the division wall inside the

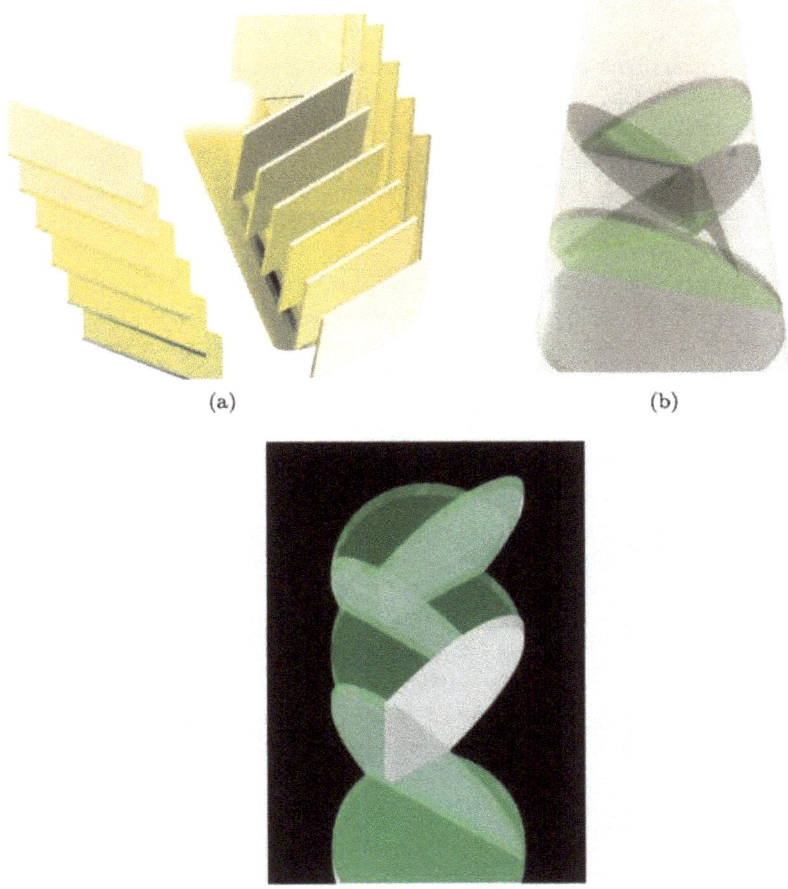

(a) (b)

(c)

Figure 9. Cellwork IV.B (Table 1): geometrical representation of the cellwork. (a): Cutting of a cylinder by parallel inclined sufaces. (b) by Boolean substraction, the cell walls become visible within the cylinder. (c) Isolation of the scaffold of cell walls. A transparence of walls is afforded to visualize wall partition by induction of a division wall edge from a neighbour cell (the non-transparent white wall shows this necessity).

cylinder. Placing together these walls procures directly the scaffold of the inner wallnet (Figure 9c). Then, by substraction of the scaffold from the cylinder, we obtain directly the alphabet of polyhedra, this time with a geometrically possible conformation. So the system could it be defined by the specification of the position of successive wall orientations? The question arises whether cellworks could be specified directly at a geometrical level in stead at a topological level. Such systems may by closer to biological items than are the topological ones.

10 Conclusion

The cellwork system allows an understanding of the construction of plant organs. If a mathematical background can be given for the development of such cellular entities, they can be disussed on a logical basis and propositions for their relationships can be established.

As for the construction of the systems simple biological characteristics have been used, it was hopeful that the classification at their base would have some biological meaningfulness. Further, as the systems are generative, the classification of plant forms goes beyond a morphological description, giving for the first time a framework for morphogenetical aims.

Mathematical results can give a new insight to plant morphogenesis. One of the most striking results is the 2/5 helicoidal segmentation of apical autoreproductive cells. The fact that shoot cellular organization encloses the structure necessary for a typical phyllotaxis, prior to the formation of the leaf primordia, could lead to make a term to the controversal theories such as those based on inhibition zones, minimal areas, and so on. Characteristics specific for this or that theory, can now be seen as being combined and revealed as emergent properties of the same system.

The presence of a unique initial cell from which all other cells would derive can be a problem for biologists. As indicated, such cells are clearly visible in Peridophytes, but not often in higher plants. As map systems (Lück & Lück, 1987), based on synchronic cell divisions, showed that a meristematic cell plate of several cells can be operative on the apex for a 2/5 phyllotaxy, even if the same map is in fact constructed on the base of a central initial cell, confronts us with the role eventually played by an unique initial cell, even without any particular feature pointing to the existence of such an initial cell. The self(reproducing cellular complex is made in evidence by tranformation of the system into another system with integrated items of the alphabet. In other cases, like that of Epilobium or Actinidia, there is a group of 4 central apical cells without any particular sign, from which each procure cells for an apical sector.

Another interesting result is the emergence during the development of new meristematic centers in which the global development takes a new start. Such centers have been shown to be a necessary emergence resulting from a simple repetitive and determinated positioning of division walls in every cell of a tissue (Barlow *et al.*, this volume). Under which conditions such centers actually develop as new organs is another open biological question.

Acknowledgments

Dr P. W. Barlow is gratefully acknowledged for the critical reading of the manuscript.

References

1. P. W. Barlow (1991). From cell wall networks to algorithms. The simulation and cytology of cell division patterns in plants. *Protoplasma* **162**:69–85.

2. P. W. Barlow (1994). Evolution of structural initial cells in apical meristems of plants. *Journ. Theor. Biol.* **169**:163–177.
3. P. W. Barlow, H. B. Lück and J. Lück (1999). Organogenetic cellular patterning in plants. This volume.
4. F. Bartels (1956). Zur Entwicklung der Keimpflanze von Epilobium hirsutum. I. Die im Proembryo ablaufenden Zellteilungen bis zum beginnenden "herzfrmigen Embryo", *Flora* **144**:105–120.
5. J. Feldman (1978). *Les Algues. Prcis de Botanique, 1 — Vgtaux infrieurs*, Masson, Paris: 95–320.
6. F. D. Fracchia and P. Prusinkiewicz (1991). Physically-based graphical interpretation of marker cellwork L-systems. In: H. Ehrig, H. J. Kreowski and G. Rozenberg (eds): *Graph grammars and their application to computer science.* Lecture Notes in Computer Science **532**. Berlin: Springer, pp. 363–377.
7. M. Gromov (1990). Cell division and hyperbolic geometry. IHES/M/90/54: 1–9.
8. B. E. S. Gunning (1981). Microtubules and cytomorphogenesis in a developing organ: the root primordium of *Azolla pinnata.* In *Cytomorphogenesis in Plants*, ed. O. Kiermayer; Springer-Verlag, Wien, Cell biology monographs **8**:301–325.
9. F. Hall, R. A. A. Oldeman and P. B. Tomlinson (1978). *Tropical trees and forest: an architectural analysis.* Springer Verlag, Berlin, Heidelberg, New York.
10. R. Hébant-Mauri (1977). Segmentation apicale et initiation foliaire chez Ceratopteris thalictroides (Fougre leptosporangie). *Can. J. Bot.* **55**:1820–1828.
11. W. A. van Heel (1983). Early development of free carpels. *Blumea* **28**:266.
12. J. A. Jernstedt, E. G. Cutter and P. Lu (1994). Angle shoot development: *Selaginella martensii. Ann. of Bot.* **74**:343–355.
13. C. Lambert (1991). Analyse quantitative de la morphogenèse de l'algue *Antithamnion plumula* (Rhodophycées). Thèse de doctorat. Institut National Polytechnique de Toulouse, pp. 363.
14. J. Lück and H. B. Lück (1991). Vers une nouvelle théorie des méristèmes apicaux des végétaux. *Biologie Théorique*, Solignac 1985. Ed. du CNRS, Paris, 197–213.
15. J. Lück and H. B. Lück (1991). Double-wall cellwork systems for plant meristems. In: H. Ehrig, H. J. Kreowski and G. Rozenberg (eds): *Graph grammars and their application to computer science.* Lecture Notes in Computer Science **532**. Berlin: Springer, pp. 564–581.
16. J. Lück and H. B. Lück (1992). Cellworks: An application to plant morphogenesis. In *Lindenmayer systems, Impacts on Theoretical Computer Science, Computer Graphics, and Developemental Biology*, Rozenberg G., Salomaa A., eds. Springer-Verlag, Berlin, pp. 385–404.
17. J. Lück and H. B. Lück (1993). Parallel rewriting *d*w-Cellwork L-systems for plant development. In *Mathematics applied to Biology and Medecine*, J. Demongeot and V. Capasso, eds. Wuerz Publishing LTD, pp. 461–466.

18. J. Lück and H. B. Lück (1996). Cellworks with cell rewriting and cell packing for plant morphogenesis. In *Graph grammars and their application to computer science*, J. Cuny, H. Ehrig, G. Engels and G. Rozenberg, eds. Lecture Notes in Computer Science, Springer **1073**:536–549.
19. I. V. Newman (1956). Pattern in the meristems of vascular plants. I. Cell partition in living apices and in the cambial zone in relation to the concepts of initial cells and apical cells. *Phytomorphology* **6**:1–19.
20. W. R. Philipson (1990). The significance of apical meristems in the phylogeny of land plants. *Plant Syst. Evol.* **173**:17–38.
21. R. Sattler (1990). Towards a more dynamic plant morphology. *Acta Bioth.* **38**:303–315.
22. G. Rozenberg and A. Salomaa (1980). *The mathematical theory of L systems.* New York, Academic Press.

PLANT MERISTEMS AND THEIR PATTERNS

BEATA ZAGÓRSKA-MAREK
Institute of Botany, Wrocàw University
Kanonia 6/8, 50-328 Wrocàw, Poland

1 Introduction

Specific plant cells positioned in special regions of the multicellular plant body and called meristems assure that the distinctive plant feature, which is indefinite growth, is implemented through their constant divisions. These cells, at least theoretically, are immortal. This is the first reason why plant meristems are so interesting. The oldest plant individuals on Earth are thousands of years old. In hundreds of thousands of their cell divisions many errors might be supposed to accumulate in the original zygotic information. Yet, the immortal meristematic cells, of a tree even a few thousand years old, produce healthy growth increments every year. Fascinating mechanisms protecting the identity of the genetic information, conveyed by these special cells, have been discovered (Klekowski 1988, Kuligowski and Tourte 1979). The value of being conservative is victorious over the value of being innovative. Do these values indeed stand in opposition? On one hand, plant with its indefinite growth has to meet the requirement of harmonious and coordinated development in ontogeny. Shielding genetic blueprint definitely serves this purpose. On the other hand, as an immobile organism being subject to heavy selection in evolution, the plant has to be flexible. This requirement is met by its tolerance to any increase in the information that might help to overcome the environmental stress. This is why polyploidy is not lethal in plants, special secondary pathways of metabolism are present, chimeras are common and speciation is accomplished frequently through hybridization. One, often overlooked, aspect of plant flexibility is the plasticity of morphogenesis (Sinnott 1960, Briggs and Walters 1997). It is manifested attractively in dynamically changing patterns present in, or produced by, plant meristems. The diversity of these patterns and their ontogenetic transformations may represent a compromise between the contradicting necessities for an individual: to have a stable and rich genotype yet to be able to produce a highly adaptable phenotype.

For a plant the origin of meristematic cells is in the embryo. At the beginning of each plant's individual history, the embryo as a whole is composed of cells, which are able to divide. Later in ontogeny some parts of it remain meristematic, whereas other parts differentiate. The derivatives of actively dividing meristematic cells also differentiate, adding constantly new modules to the developing plant organism, which efficiently penetrates and makes use of surrounding space. The differentiation, taking place in the vicinity of the meristem, as well as intense organogenesis associated with its function, is the second reason why meristems, their cellular patterns and organography are intensely studied. It is hoped that the dynamics of developmental changes in the meristem will help to reveal some principal mechanisms governing the process of morphogenesis.

There are many meristems in a plant body: apical and lateral, intercalary meristems, marginal and adaxial meristems of the leaf, meristemoids of adventitious buds, stomatal meristemoids, residual, detached meristems, cork cambium, *et cet.* The most important of these are *apical meristems*, responsible for a constant, polar increase in the length of cylindrical plant organs such as stems or roots, and *lateral meristems* which increase the width of these organs. I take the liberty of arbitrarily selecting two of them: the shoot apical meristem and vascular cambium. It is because most interesting patterns have been found there and substantial progress has been made in recent years in understanding the principles of the formation of these patterns. Last but not least our research group from Wrocàw University has intensively investigated them.

2 Patterns of the shoot apical meristem

2.1 Developmental variation of cellular patterns

The circular paraboloid shape of the apical meristem surface is clearly seen above the youngest organ primordia that bulk at its base in the so-called organogenic region (Figure 1). The cellular pattern visible in top view of this surface is different in lower plants such as ferns or horsetails than in seed plants. The differences,

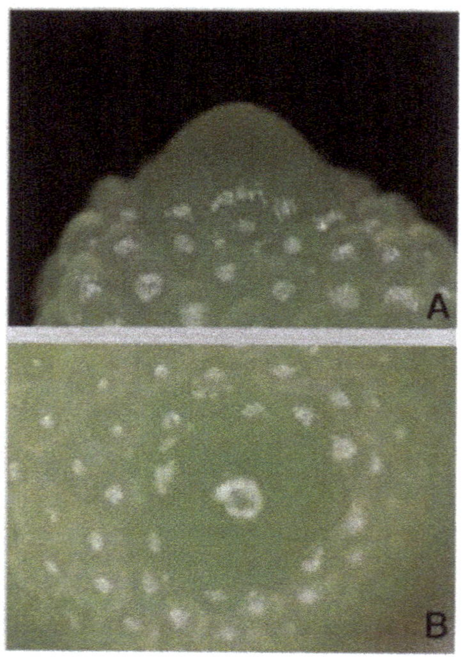

Figure 1. Live apical meristem of spruce (*Picea abies*) as seen from the side (A) and from above (B). Smooth apical dome of the apex rises above the youngest needle primordia arranged spirally in the meristem's organogenic region.

representing evolutionary variation, are based on the nature of the initial cells positioned in the vertex of the meristem axis. These cells are special because all other cells in the meristem are their descendants. In lower plants there is only one initial cell, which is also morphologically different from all other cells of the apex (Bierhorst 1977). In seed plants the number of initials increases. This is advantageous in light of somatic mutations accumulating in the dividing initial, which constantly replicates its genetic information. When there is only one initial, it transmits mutations to all other cells of the shoot. One mutated initial of many transmits its mutation only to the sector of the shoot circumference. The initials in seed plants no longer differ in shape and size from their nearest descendants. Thus their distinction from other cells as well as evaluation of their number and position is difficult. Observations of the mutated sector widths in shoots of the sectorial or mericlinal chimeras suggest the presence of generally few (3 or 4) initials in a surface (Tilney-Bassett 1986, Lyndon 1998). Their identification *in situ* should prove to be even more difficult, if we accept the view that from time to time the stochastic process of selecting initials anew from the whole population of cells occurs. It is postulated to be a mechanism that excludes somatic mutations in higher plants (Klekowski et al. 1985, Klekowski 1988) and has such consequences that there are no permanent initials in their apex. There is no agreement with regard to the permanence or impermanence of initials in seed plants (Soma and Ball 1963, Lyndon 1998). The drift of albino sector in sectorial chimeras of juniper is indirect evidence for initial impermanence (Ruth et al. 1985). The same effect, however, may result from mutation in non-initial cell positioned closely to the vertex of the apical meristem axis. Clonal analysis performed by our research group on meristems of selected representatives of seed plants: *Picea* and *Magnolia* provided the first direct evidence of transient initial cell status in situ (Zagórska-Marek 1998, Zagórska-Marek and Turzańska 1998). Its results suggest that the impermanence of initials is an unavoidable effect of evolutionary increase in their number combined with the condition of isotropic growth of the meristem top surface and minimal energy principle observed by dividing cells. Clonal analysis is a tedious process of identification of all twin cell packets (Puàwska 1986) forming on the surface of the apex due to the divisions and the symplastic growth of cells. Cellular clones that form under such circumstances consist of twin cells and twin subclones of the same lineage. They can be easily identified because a thicker wall that belonged to the ancestral cell surrounds each clone. The younger the walls within the clone, the thinner they are. On the top surface of the apical meristem growth is more or less isotropic. Thus the clones there are isodiametric with a predominantly fourfold radial symmetry: the cell walls of consecutive divisions are at a right angle one to another, in accordance with the minimum energy principle (Figure 2a). On the side of the apical dome where growth becomes polarized and its meridional component prevails due to elongation, clones are oblong. Consecutive cell walls in them are parallel to each other and positioned at a right angle to the direction of growth (Figure 2b). Their surface is again minimal in such configuration.

The search for initials in our study indicated that both in *Picea* and *Magnolia* cellular patterns, in which the whole surface of the meristem is composed of 3 or 4 clones are most frequent. Their borders meet at the geometric center of the apex.

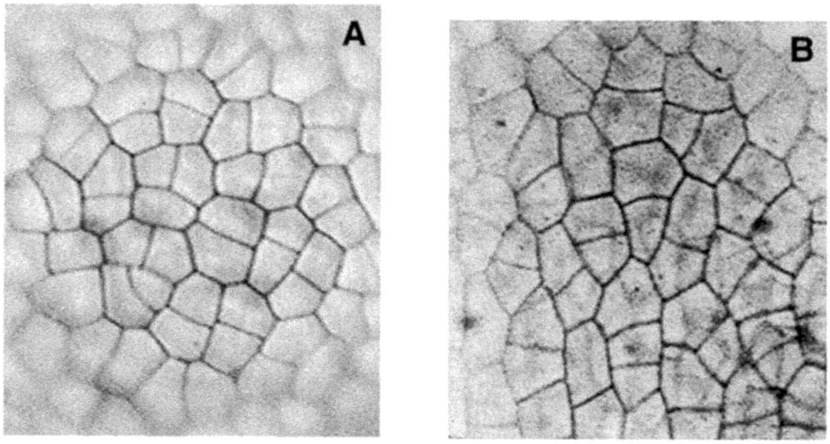

Figure 2. Cellular patterns of the apical dome's surface in magnolia. On the (A), where growth is more or less isotropic, cells of common descent form regular, radially symmetric tetrads. On the side (B) the youngest cell walls are perpendicular to the prevailing (vertical) direction of cell expansion.

The method of determining the center is described in Matkowski *et al.* (1998). Each clone is generated by an initial positioned in its upper corner, at the vertex of the meristem axis. Thus there are 3 or 4 initials in such apices. This is in agreement with suggestions resulting from studies on chimeras. However, apices where, in the geometric center, a young radially symmetric clone consisting of 4 subclonal parts could be found, proved to be most interesting to us (Figure 3). Such clone is evidence that in a previous stage of apex development a single cell was positioned on the top of its axis. Then it started functioning as an initial. Because of growth isotropy in this particular region, two generations of its consecutive divisions were mutually perpendicular. Thus the cell made subsequently a tetrad of clones with 4 initials. Selection of this cell supports the hypothesis of initial impermanence. There is however yet another important aspect of its existence. Walls of the second generation of divisions within a tetrad, in principle, do not meet — the levels of their insertion pass one another along the wall of the first generation. The initially small distance between them must increase, as the apex grows, ultimately separating opposite clones of the tetrad. When they drift apart there are no longer 4 neighboring initials on the top of the apex. The decision "has to be made" which group of 3 neighboring cells will become the new initial group. Selecting any of two inner corners of a tetrad means that ever since that moment there will be 3 initial cells instead of 4. The triad, contrary to the previous pattern, should be very stable and able to function at the center of the apex for a long time. There is of course a possibility that, at that critical moment of change, a single cell will be selected as previously. In such case the whole sequence of events is repeated.

The proposed scenario of developmental changes in cellular patterns of the apex explains why 4 or 3 initials are most frequent. Most importantly it shows that the dynamics of changes in the number of initials and their position is something that

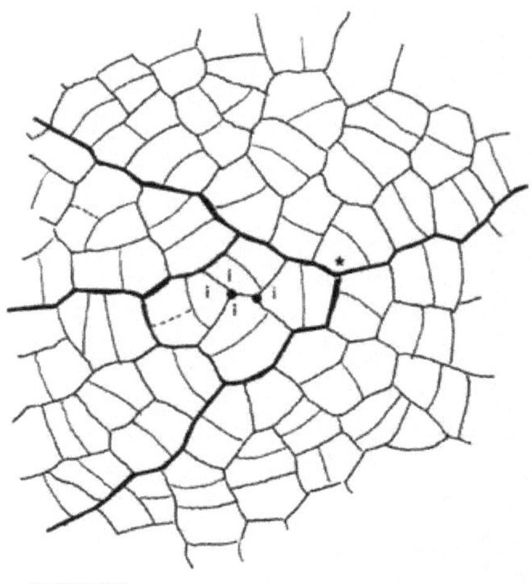

Figure 3. Map of the cell arrangement on the top of the apical meristem of spruce, in which one cell, positioned some time ago in the present geometric center, formed a young, radially symmetric tetrad. The previous cellular center, where clonal borders of three sectors produced by three former initials meet, is already shifted towards the side of the meristem. The position of the present geometric center of the apex is indicated by the concentric arrangement of the youngest cell walls (red) on the periphery of the meristem; *i* — the initial cells presently functioning at the apex, bar length is 30 μm.

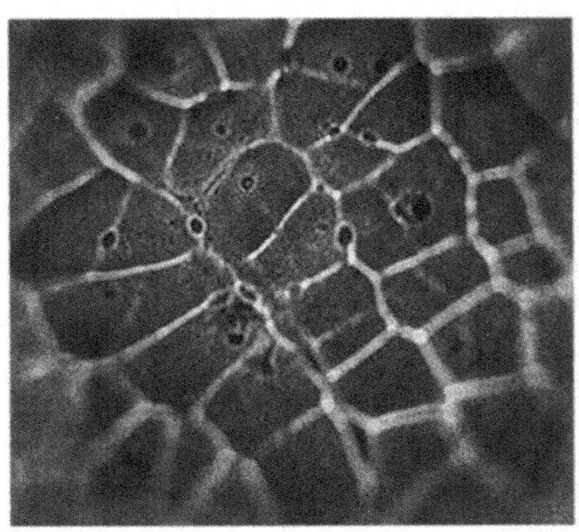

Figure 4. Microphotograph of the top-surface of peculiar spruce meristem, where recently introduced polarity of growth has been manifested in cell walls of dividing cells paralleling each other.

should be expected: it is a necessary consequence of the principles the dividing plant cells must obey. Although the sequence presented above seems to be the most common, it does not explain all patterns present in apices. Some of them indicate that growth of the apical surface may be at least temporarily polarized (Figure 4).

2.2 The geometry of a growing apex

Despite constant cell divisions, which make the paraboloid apex grow, its shape is generally maintained in development (Lyndon 1998, Romberger, Hejnowicz and Hill 1993). An interesting question arises: is geometry maintenance a rule? Are any disturbances in parameters of growth possible? It is known that in *Picea* the apical meristem annually undergoes seasonal changes in the profile of the apical dome and in its size. Similar changes occur in vegetative apices of angiosperms induced to flowering or when their growth conditions alter. These changes in general are radially symmetric: the circumference of the apex increases or decreases evenly, relatively to its height. We will call this type of change a *continuous* one.

A shift in the position of the vertex of the meristem axis, associated with the processes described in the previous section, may have different consequences for the geometry of the apex. After the change all cells become positioned differently with respect to the principal directions of growth (Hejnowicz 1984). Radial symmetry of circumferential growth may be affected, local disturbances of the surface may appear because in growing clones on one side of the meristem the cells are now closer, whereas on the opposite side they are farther from the newly established geometric center. At present we know very little about possible effects of such a situation.

Writing a specific computer program is the best way of acquiring knowledge about any set of principles (Dawkins 1996). A program written by Radosàw Karwowski from our research group allows us to make a plant apical meristem of any shape and size from a small group of initial cells (Figure 5). Their contribution to the circumference of developing apex may thus be investigated. It appears that it is very different depending upon the initials' quantity, size, shape, number of contacts *et cet.* Furthermore, the program shows how this contribution changes with deviation of the meristem axis, as in the development of a stochastic meristem where initials are impermanent. What is the behavior of a clone, which contributed to the circumference according to the former pattern of principal directions of growth when its situation changes? Do frequencies of cell division in it alter? The general question here is whether the cells of the clone are a totally passive subject of genetically controlled supracellular mechanisms maintaining a "given" geometric form of the apex or if they have some autonomy in making this form. The cell cycle itself might be the key to some limited autonomy of dividing cells: being in different stages of their cycle they may react differently to the situation in which there is a rapid change in the position of the apical dome axis. As it was already mentioned, when they become differently positioned than before in respect to the principal directions of growth, some of them, closer to the new geometric center are advanced in their activity, others being farther, at the same time are delayed. However, only few of them might, at the moment, be in G1 phase of the cell cycle, the length of which phase can be regulated. Other cells have to divide, finishing their cycle, although geometrically there is no need for it. It is possible that before new equilibrium is reached some irregularities may appear along the apex circumference. These changes in apex geometry, if indeed they occur, should be called *discontinuous*.

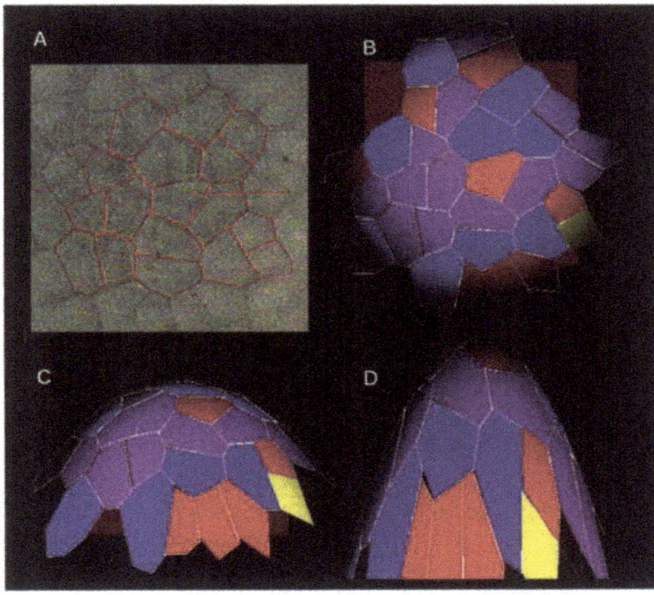

Figure 5. Computer simulation of the apical dome's different shapes (C, D) created from the same initial group of cells (A) mapped from the real meristem. Clones developing from these cells contribute differently to the apex circumference depending upon their original size, shape, number of contacts *et cet.*. Cells are colored according to the number of their neighbors: 4 — yellow, 5 — red, 6 — purple, 7 — blue.

Our computer program colors initials differently depending upon the number of their contacts. This makes the identification of clones originating from the initials easier. Cells constituting a layer are usually 6-sided in a top view. The nearly ideal pattern of hexagonal wax cells in a flat sheet of honeycomb provides an attractive analogy. The surface of the meristem, however, is not flat. The cells here predominantly have 6 contacts with other cells of the superficial layer but sporadically cells with 7 or 5 contacts occur. The presence and distribution of pentagonal cells is especially interesting. They may play the same role as pentagonal facets in C_{60} fullerens or geodesic domes constructed by Richard Buckminster Fuller. How many of them are on the surface of a paraboloid meristem? In Fuller's geodesic dome 12 pentagons are required for the closure of the surface into a sphere. Intuition says that in the shoot apical meristem their number should depend on how flat the surface of the apex is. Some plant meristems are slim with the curvature of the surface well pronounced like in *Chamaecyparis*, *Hordeum* or *Hippurus*; some are almost flat like in *Pterocarya* or *Actinidia*. A certain number of pentagonal cells, most probably, neutralize the presence of cells having 7 contacts. A surplus of them, however, should be held responsible for the curvature of the superficial cellular layer of the meristem. The distribution of pentagons on the surface of the real meristem is shown in Figure 5. At present we are in the course of exploring all these possibilities empirically — using real apices with their true cellular patterns, and theoretically — in models. It is worth noticing that fulleren-like structures can be found elsewhere in the plant body, also on levels of its composition, which are entirely different hierarchically. The clathrin cage constructed of triscelions strengthens the coated vesicles within a cell. The frame of this spherical skeletal cage is composed of hexagonal and pentagonal facets. This observation brings about the reflection on universality of some principles governing pattern formation in nature.

2.3 Phyllotaxis

Beautifully regular phyllotactic patterns generated by the growing apex in the course of organogenic processes are diverse. Their quality depends upon the relationship between the size of initiated organ primordia and the size of the lateral surface of the meristem in the organogenic region where primordia are packed. In general, patterns are qualified in two major categories: *chiral* patterns, otherwise called spiral or helical, distinguished by having mirror-like symmetry, and *achiral* to which all whorled patterns belong. This classification appears artificial in the light of recent studies on the phenomenon of ontogenetic phyllotactic transformations. During the transition one type of pattern may easily transform into another and there seems to be no limitation with regard to the type of the patterns involved. Chiral patterns change into achiral and vice versa (Meicenheimer 1981, 1982), often sequentially, as it has been demonstrated in *Anagallis* (Kwiatkowska) and recently in *Juniperus* (Szostak and Zagórska-Marek 1998). Moreover a chiral pattern of one type may transform into another — also chiral. Transitions of this type have been discovered in *Abies balsamea* (Zagórska-Marek 1985). In conifers they are infrequent and thus difficult to study. A beautiful object, in which not only the diversity of phyllotaxis is exceptionally rich but also transitions are very common, is the *Magnolia* flower (Zagórska-Marek 1994), shown here in Figure 6. The transitions between chiral patterns usually result in dramatic changes in such classic descriptive criteria of phyllotactic pattern as the chirality of the ontogenetic

Figure 6. Inside of live cucumber magnolia flower (*Magnolia acuminata*). In this species an extremely rich qualitative diversity of phyllotaxis is present in the spiral pattern of carpels arranged on the sides of elevated generative shoot.

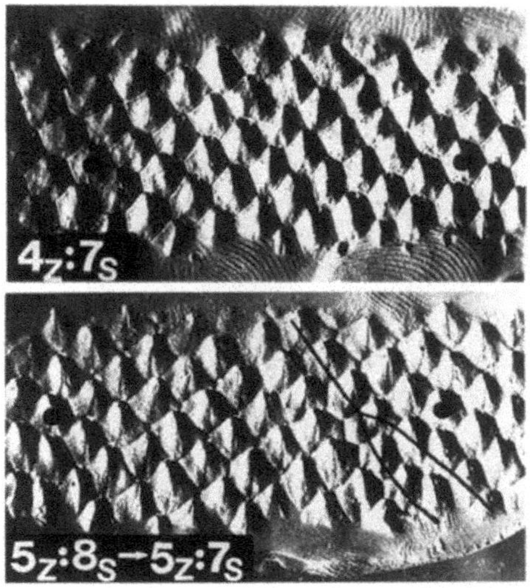

Figure 7. Modeling clay replicas of the lateral surface of magnolia generative shoots, made by rolling each shoot over the surface of clay. The pattern of the unrolled surface resembles crystal lattice: arrangement of carpels is very regular. Between surface markers (black dots labeling the same carpel replicated again when the shoot was rolled over 360 degree) the number of lines connecting all pattern elements can be counted. In the upper photograph there are 4 lines inclined to the right (Z) and 7 lines inclined to the left (S). This means that in the replicated shoot the Lucas pattern of phyllotaxis was present in its pure form. In a lower photograph a dislocation is present — one line missing from the original Fibonacci pattern means that ontogenetic transformation of phyllotaxis took place during the process of carpel primordia formation. An interesting visual effect can be appreciated on such pictures — carpel impressions can be perceived either as elevations or as impressions depending upon the position of imaginary source of light (modified from Zagórska-Marek 1994).

helix, the divergence angle, the jugacity of the pattern. What changes least is the contact parastichy pattern. Because of regularity in pattern element distribution (leaves, florets, carpels) in the pattern it very much resembles the lattice of a crystal (Figure 7). Transformation was noticed to be marked by the defect of the lattice called, in crystallography, a dislocation (Zagórska-Marek 1987). Phyllotactic dislocations appeared where one or more parastichy were added or lost in the pattern (Figure 7B). Interestingly this route leads to the formation of chiral patterns, which are impossible, according to some theoretical models (Jean 1994). They exist well and healthy in *Magnolia* floral apices, where they have been documented on SEM microphotographs (Figure 8).

It is a rather trivial constatation, intuitively acceptable, that stable patterns of phyllotaxis are produced by apical meristems with stable parameters of growth. Then, what is it, that changing ontogenetically in the apex geometry causes dislocation and qualitative transformation of phyllotaxis? The first thought is such: because the dislocation is strictly localized along the circumference of the apex, it most probably results from *discontinuous* changes: the local disturbances of its circumferential growth. These should produce either a deficit or a surplus of the

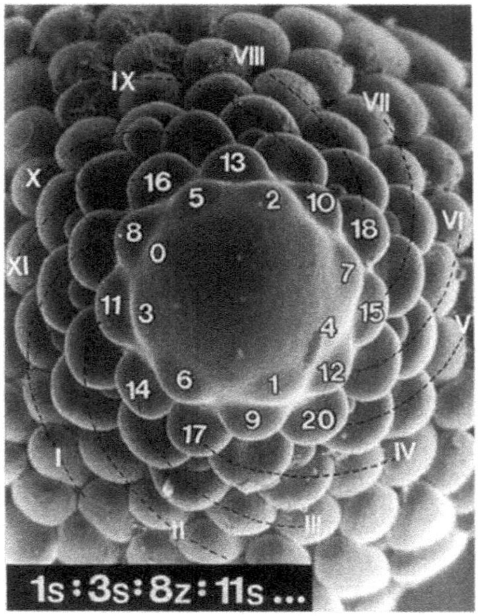

Figure 8. Apical meristem of *Magnolia acuminata* with very special, unique pattern of phyllotaxis, seen in scanning electron microscope. The spiral lines (parastichies), connecting carpel primordia, belong to the mathematical series shown at the bottom of the photograph. Unlike in Fibonacci series, 3 and 8 numbers refer to parastichies of opposite orientation. Before, this pattern existence in nature has been questioned from the theoretical point of view (with permission of Acta Societatis Botanicorum Poloniae).

surface, on which organ primordia are initiated, causing in turn either parastichy loss or addition. Observations of primordia, which frequently fuse atop of dislocations seemed to support this assumption. However, in their elegant theoretical study Douady and Couder (1996) have demonstrated that qualitative changes of phyllotaxis may also result from the *continuously* changing ratio of the size of initiated primordia to the size of apex circumference in the organogenic region. A simulation of this effect was made in our lab with the use of a computer program written by Johannes Battjes (Figure 9a). It was performed under following assumptions: continuous change in primordia size, a stable circumference of the apex, primordia initiated in the first available space. The ratio of course would be affected as well by abrupt sectorial circumferential changes postulated earlier. This was demonstrated in a second simulation made with application of another variant of Battjes' program (Figure 9b). The complicated transitions engaging simultaneously many dislocations, either in one or in both sets of contact parastichies, represent yet another challenge in modeling (Figure 10). They look much more drastic than those obtainable in simulations, where first changes seem to occur in this set in which the parastichies are steeper.

Empirical studies are now necessary to show, which variant is employed by real apices in nature. It seems that the first variant is most likely to occur in apices,

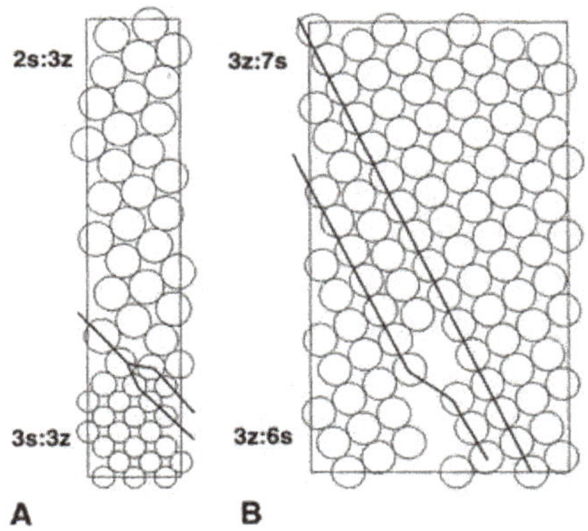

Figure 9. Two examples of computer-made phyllotactic transformations, in which dislocation effect appears either due to continuous (A) or discontinuous (B) change in parameters of growth. In the first case the continuous increase in a size of pattern elements packing on all the same size surface, according to the principle of filling the first available space, causes the loss of one parastichy and transition from achiral tricussate pattern (3:3) to spiral Fibonacci pattern (2:3). In the second case, the sectorial increase of the surface on which the packing occurs, shown in the simulation by preceding pattern elements drifting apart, causes an addition of one parastichy to the set of 6. This effect marks qualitative change from trijugate Fibonacci pattern (3:6) to unusual 3:7 pattern belonging, according to Jean (1994), to the family of "impossible" solutions, quite common, however, in magnolia flowers.

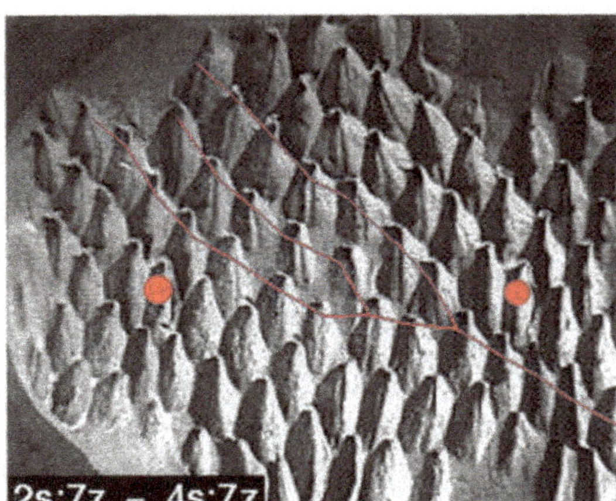

Figure 10. Unusual phyllotactic transformation in magnolia generative shoot marked by dramatic increase from 2 to 4 parastichies oriented to the left (two almost simultaneous dislocations). This occurs *without* any change in the set of opposite 7 parastichies. In computer simulations first changes usually occur in the set of more numerous (steeper) parastichies.

where primordia are relatively big, thus the expression of phyllotactic pattern (i.e. the numbers of contact parastichy pair) is low. A particular developmental sequence of achiral and chiral patterns in juniper, showing that phyllotaxis in this genus can be spiral (!), suggests constant and most likely continuous gradual decrease of the ratio mentioned above. The sequence proceeds from the initial and most common in juniper 2:2 decussate phyllotaxis (achiral) through the low order spiral 2:3 Fibonacci pattern (chiral) to the 3:3 tricussate pattern (again achiral). The tricussate pattern further changes into the 3:4 Lucas phyllotaxis (chiral) and this ultimately transforms into the final 4:4 tetracussate pattern (achiral).

In apices with a higher expression of phyllotaxis (floral apices of *Helianthus* or *Magnolia*, vegetative apices of *Picea* or *Abies*), where primordia are relatively small, fluctuations of the meristem surface along its circumference, which make possible the insertion of new parastichy or subtraction of the existing one, are more likely. *It cannot be excluded that the changes in meristem axis inclination, associated with the phenomenon of initial impermanence, and described in the previous chapters, may create developmental instability of apex surface leading eventually to dislocation.* Nevertheless, we explore the possibilities of the variant proposed by Douady and Couder. Battjes' program has been modified by Radosàw Karwowski in order to allow changes in the parameters of a simulation in the course of pattern growth. We want to apply it to recreate the developmental situation in the *Magnolia* floral apex, which makes initially very large primordia of petals in whorls of three. Then it produces an undetermined number of very small, spirally arranged primordia of anthers, ending its organogenic activity by setting carpel primordia bigger than former ones but also positioned in a spiral. It can be speculated, that the accumulation of subsequent, genetically programmed changes in sizes of primordia of different identity, ultimately produces an outstanding diversity of phyllotaxis present in a carpel zone of the floral apices of *Magnolia*. To make our task even more difficult the capricious *Magnolia* shows that this diversity is not merely an unpredictable outcome of the changes. The spectrum of phyllotactic diversity is typical for an individual tree and repeated in consecutive years of blooming. Also different species of *Magnolia* differ in the degree of diversity. The genetic factor determining quite precisely the direction and extent of changes is most probably involved. This conclusion is supported by our investigations of phyllotaxis in other plants: different species of club mosses and cacti also differ in the type of the phyllotactic spectrum (Gola 1996, 1997). A special clone of *Anagallis* has been found in which transformations of phyllotaxis are more frequent than in any other clone (Kwiatkowska 1997). Perhaps, in the near future, no longer will a phyllotactic pattern itself be a diagnostic feature for a taxonomist, but rather a whole spectrum of phyllotactic patterns in a particular plant species.

3 Cellular patterns of vascular cambium

Cambium is the meristematic tissue responsible for secondary growth in plants (Romberger, Hejnowicz, Hill 1993). It develops in woody plant cylindrical organs such as the stem or root at the end of the primary growth phase as a continuous cylindrical envelope composed of embryonic cells — cambial initials. The most important task of cambium is the production of secondary conducting tissues: xylem and phloem. The derivatives of dividing cambial cells when deposited externally (centrifugally) differentiate into phloem cellular elements, when internally (centripetally) — into xylem elements. All the derivatives of a single cambial cell are lined up along the radius of a cylindrical organ — somewhere between phloem cells on one side and xylem cells on the other side, sits an initial — a "mother of all".

There are two different conducting systems in a cylindrical organ of a woody plant — vertical and transversal. The first conducts carbohydrates from the crown to the root system on the phloem side, and water and minerals from the roots to the crown on the xylem side. Its elements are produced by elongated cambial cells, which are called the fusiform initials. The second system is initiated by isodiametric cambial initials grouped together on a surface of the cylindrical cambial envelope in the form of cambial rays. It is a system of parenchyma rays, which connects the phloem and xylem cells positioned on both sides of the cambium. These connections are horizontal and run along the radius of the root or the stem.

Fusiform initials are special: not only do they decide about cambial morphology (storied versus nonstoried cambium) but above all they exhibit the phenomenon of handedness, which isodiametric cells lack. The basis of this unique ability to recognize right and left directions in cambium is unknown. It is nevertheless manifested in many developmental processes, especially in oriented divisions and oriented growth associated with a periodic cell inclination change due to the presence of cambial morphogenic waves (Hejnowicz 1974).

3.1 Storied pattern

Cambial fusiform initials producing vertical elements of secondary conducting tissues are elongated. Their longitudinal dimension largely exceeds the radial and tangential dimensions. These cells have an axis of polarity and their arrangement on the surface of cambium in different plant species is not the same: the cells are either lined up horizontally or not. Strangely enough this arrangement resembles that of elongated, polarized molecules in *liquid crystals*, the specific properties of which depend upon this feature. In species advanced evolutionarily, representing few families of flowering plants, storied cambium is present — fusiform cells form along its circumference a pattern of horizontal tiers (Figure 11a). Exactly the same line up of molecules is observed in smectic liquid crystals. In gymnosperms and flowering plants less advanced in phylogeny, the arrangement of fusiform initials in cambium is nonstoried thus analogous to the arrangement of molecules in nematic crystals (Figure 11b). In plants the storied structure represents a progressive character in evolution. It is of polyphyletic origin: developed several times independently in different lineages of angiosperms. We do not know, what is so advantageous for these cells in being storied. This structure was nevertheless selected more than once, despite the fact that the initial pattern is almost completely lost in secondary phloem and xylem due to differentiation. It would be hard to say of course, which structure: nematic or smectic is better for a liquid crystal.

As xylem derivatives of cambium accumulate in the center of a growing organ and push outwards, the process of cambial circumferential expansion counteracts by an increase in the number of initials. Multiplicative divisions that shorten cell length (pseudotransverse divisions) and are followed by the intrusive elongation of

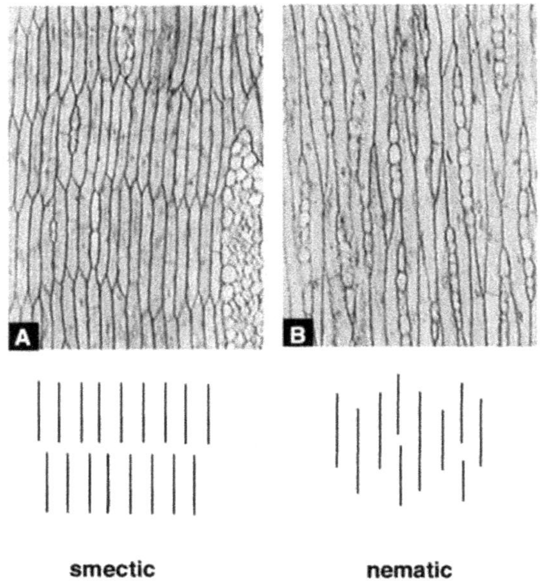

Figure 11. Two patterns of cell arrangement on the surface of cambium: storied (A) and nonstoried (B). In the first type elongated cambial initials are of similar length and the ends of neighboring cells are positioned at the same level. In liquid crystals this type of line-up of elongated molecules is called smectic. In the second type cells are generally longer, they are of different length and the cell ends are positioned on different levels of cambial surface, representing attractive analogy to the structure of nematic liquid crystals.

the descendant cells, promote the nonstoried structure of cambium. For a long time it was believed that the storied pattern is a rigid structure that develops in absence of cell elongation due to the longitudinal divisions reducing cell width (Bailey 1923, Beijer 1927). Let us stress that such divisions require much more energy for making a new partition than the shortening divisions. It is thus even a greater puzzle, why this particular feature was promoted in selection. As the descendants of a dividing cell inherit its length, there is no necessity for their subsequent elongation. Cells behaving in that manner form the clone — a storey, in which they all, originating from one ancestral cell, are lined up side by side. According to this traditional concept the more divisions in the ontogeny of cambium the more distinct its storied pattern is. Young cambia may not exhibit tier-like structures. Limited growth of cell ends in storied cambium is nevertheless required in a process of cell inclination change (Hejnowicz and Zagórska-Marek 1974). Interestingly the "expensive" longitudinal divisions are replaced by the more ancient, energy saving shortening divisions when storied cambium is wounded — in such cases its storied structure disappears.

In the early eighties it was discovered, in *Tilia*, that shortening divisions and excessive elongation — cellular events typically occurring in nonstoried cambia — may not only be present in storied cambium, but may also contribute significantly to its pattern maintenance (Zagórska-Marek 1984). The disruptive effect of excessive elongation, resulting in the production of two storey high giant cells, was subsequently corrected by the shortening division of these cells. Both descendants of such a cell, after division, contributed to two *different* storeys. One of the cells positioned in a storey other than the mother cell storey (either upper or lower) disrupted the continuity of its lineage. Similar effects of repair followed the occurrence of shortening divisions in cells of normal length. The subsequent elongation of descendant cells either placed them precisely side by side within the same storey (an effect undistinguishable from the outcome of longitudinal division) or brought one of them to a neighboring storey. Yet another interesting observation was made: whenever cellular ends growing intensively exceeded the level of the storey border, sooner or later they were shortened to that level. These facts are of special interest because they indicate that:

— a storey is not a clone but a mosaic structure composed of cells of different lineages.

— a storied pattern is dynamically maintained, any errors are quickly fixed.

— the extent of intrusive elongation is controlled on supracellular level.

— a storied pattern represents a type of a morphogenetic map in which the storey borders are recognizable by growing cell ends, that adjust to their level.

It is especially worth noting that atavistic cellular events may act assertively for, or be subjected to, the requirements of the progressive condition of being storied.

3.2 The pattern of growth activity

Bannan (1956) noticed that in nonstoried cambium some neighboring fusiform initials elongated in one direction. At the same time, in their vicinity, elongation of other cells was opposite. This means that there is a spatial diversification of the cambial surface with regard to the intrusive growth. When cells are orderly arranged, like in storied cambium, such differences appear to take on the form of a very regular *growth activity pattern* (Zagórska-Marek 1975). The pattern is particularly attractive and available for study in such cambium, where intense cell inclination changes occur (Figure 13). These are associated with the presence of a vertically moving cambial morphogenic wave. The domain pattern (Hejnowicz and Romberger 1973) is the first manifestation of this wave. Within one domain cell inclination change is unidirectional — to the right (Z domain) or to the left (S domain). Domains are distributed evenly one under the other. Borders between successive domains run transversely to the vertical axis of the cambial cylinder. In the storeys that are within the domain at a given moment, the growth activity pattern is composed of neighboring groups of cells, the ends of which grow or do not grow (Figure 12). These cells are therefore active or inactive in the process of cell inclination change. The type of domain defines the direction of the change. There is evidence of lateral shifts of such growth activity (Zagórska-Marek 1975, Wòch and Zagórska-Marek 1982). As a result, the cells, which were so far inactive, are activated, and the active cells are switched off. A most interesting situation is observed

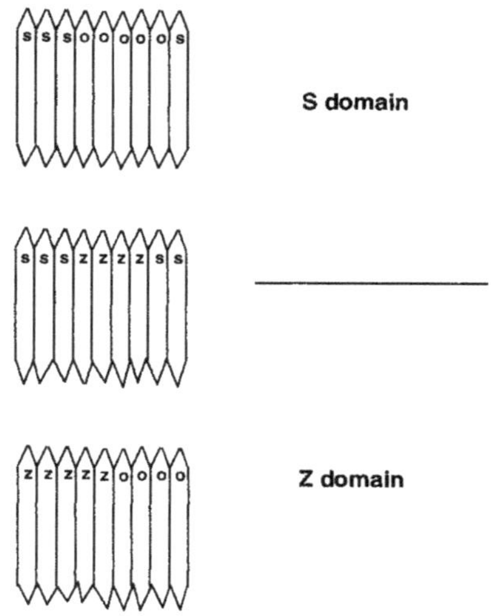

Figure 12. Growth activity pattern in storied cambium — in a storey positioned within S domain, where cells change orientation of their longitudinal axes to the left, cells at the present moment either grow to the left (s) or do not grow at all (o). On the border between domains pattern comprises of groups of cells growing in opposite directions, in Z domain growth activity pattern is the same as within S domain, with that difference, that actively growing cells reorient to the right (z). Pattern horizontal shifting reassures reorientation of all cells of the storey in sufficiently long period of time.

on the border between domains. There the growth activity pattern transforms into a pattern in which some groups of cells change inclination to the right and others change inclination in opposite direction. Along each storey one group follows another. This minute, probably also mobile, transverse domain pattern is very difficult to study — it disappears as quickly as the domain border passes a particular level in cambium and it changes into the common growth activity pattern.

Modeling this particular phenomenon reveals that recognition of pattern periodicity can be difficult, even impossible, when the time period, in which the extent of cell inclination change is investigated, is not properly balanced with other parameters of the model, such as the rate of cell inclination change and the velocity of the field providing the signal for the cell ends to grow (Figure 13).

3.3 Structural wavy patterns

Cambial structure is sometimes wavy. In such cases there are areas on the cambial surface in which cells are inclined to the right and others in which they are inclined to the left. Wavy cambium produces figured wood highly prized for its beauty. Its attractive patterns are the result of different light reflection from the surface to which xylem elements are differently inclined. According to Hejnowicz (1973), the first to discover and study its nature, the wavy structure of cambium is a result of the presence of a special morphogenetic field called the cambial morphogenetic wave. The wavy character of this field was proved by such findings as the pulsations

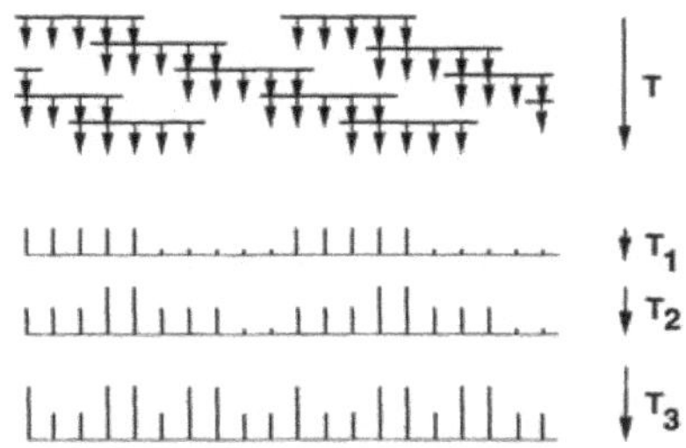

Figure 13. This model shows, that periodicity of growth activity pattern can be sometimes undetectable. The effects of the presence of hypothetical field, traveling horizontally within a storey and providing the signal which makes reorienting cells grow, are dependent upon the length of time period in which the extent of cell inclination change had been measured for each cell (T_1, T_2, T_3). Each arrow indicates the time, in which the cell, induced by the signal, reorients by one unit of cell inclination change. Histograms show data gathered for each cell of the storey. In T_1 periodicity of the growth activity pattern is most obvious, in T_3 it practically disappears! In a case of lower velocity of the field, the periodicity would persist longer.

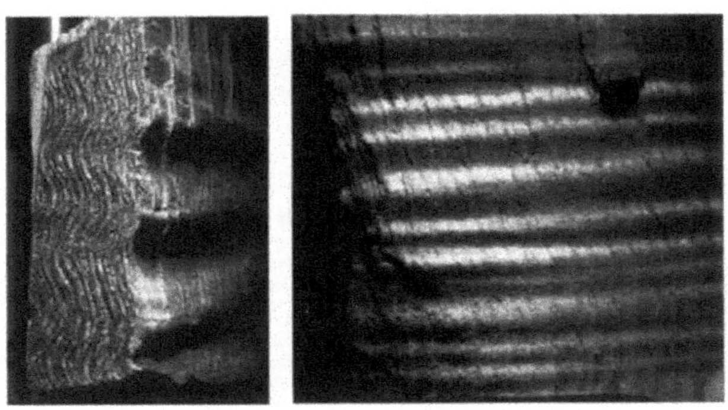

Figure 14. Wavy wood of oak (left) and ash (right) produced by cambium with structural waviness present in the arrangement of the initial cells. Wavelength is smaller in ash than in oak. Waviness in cambium is not stationary, it moves vertically and this is reflected in the oblique course of ripples formed on the radial face of wood.

of domain height in *Fraxinus* — an effect explainable only under the assumption of cambial wave superposition (Hejnowicz 1974). The wave induces oscillations in the orientation of cambial cells. The period (T) of these oscillations equals about 20 years and shows that we are dealing with definitely the longest biological rythm known on Earth. The summary of our knowledge on the phenomenon was recently given in the Encyclopedia of Plant Anatomy (Zagórska-Marek 1995).

Structural waviness develops when the amplitude of the inducing wave increases to a value sufficient to make the cell deviations visible. We do not know what triggers this process. As a matter of fact, a strong asset of this particular morphogenetic field is that the effects of its action are visible: are structural not physiological. This made it exceptionally attractive to study and allowed us to establish many of its properties. This field is probably a manifestation of a more general phenomenon.

However, periodicity associated with minute morphogen concentrations changing in time and in space, as postulated by Turing or Meinhardt, if not inaccessible at all for investigation *in vivo*, is certainly much more difficult to study.

As it is shown in Figure 15, the way the amplitude of the inducing wave increases in time and the way the oscillations are initiated in space are decisive for the quality

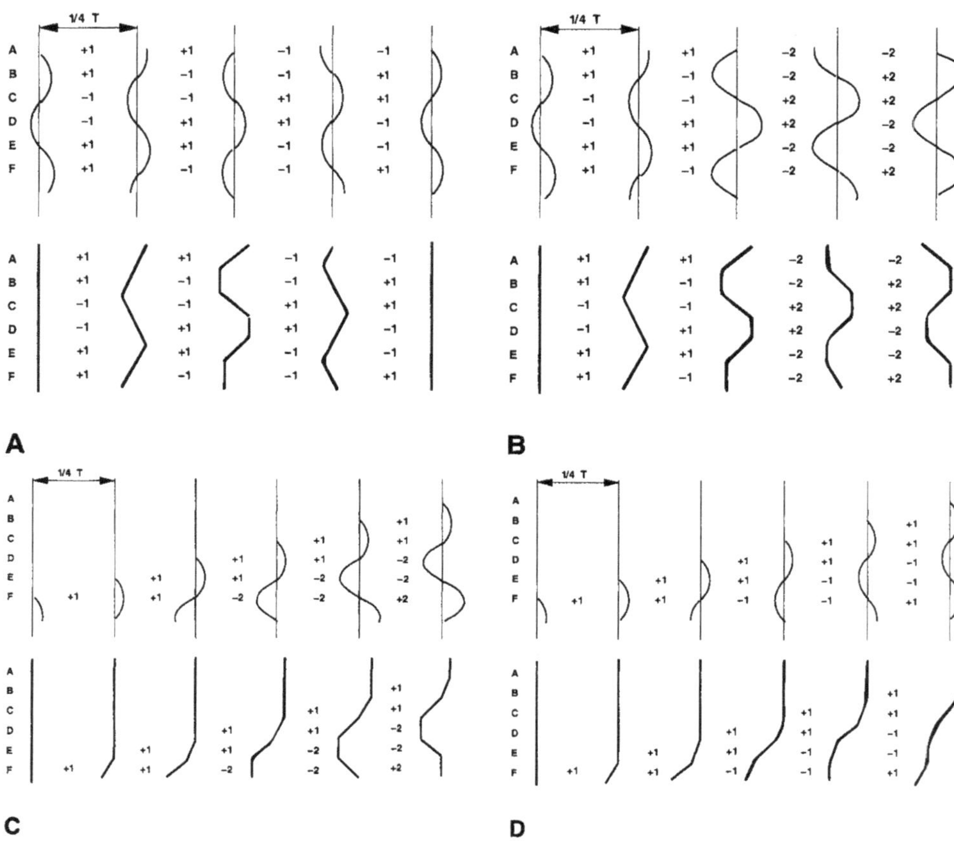

Figure 15. Spatial and temporal changes in cambial wavy structure (lower graphs) induced by cambial morphogenetic wave (upper graphs) shown in $1/4T$ intervals, where T is a period of morphogenetic wave. *Modulated waviness* (A), which periodically disappears, travels vertically over the cambial surface (levels A, B, C...) twice as slow as inducing it cambial wave. The wave has constant amplitude from the moment of its appearance and triggers cell oscillations on all levels simultaneously. Slow increase in this wave amplitude during the first period (B) stabilizes oscillations, changing structural wave from *modulated* to *permanent*. The latter travels with the same velocity as morphogenetic wave. When cambial wave, initiated locally is propagated vertically towards new levels of the cambial surface the type of developing structure depends upon whether the amplitude of the wave increases or is constant. In the first case (C) it will be *permanent*, vertical, undistinguishable from the type shown in B, in a second — it will take a form of *inclined waviness* (D). Overall inclination of this waviness to the right or to the left (the latter not shown here) depends upon the direction of the first cell inclination change. It is likely that this is the way the spiral grain develops in trees.

of structural waviness appearing in cambium. Three different types have been distinguished:

— *modulated waviness* (Figure 15A), disappearing periodically, found in african mahogany *Entandrophragma*;

— *slanted waviness* (Figure 15D), inclined to the right or to the left, most probably responsible for the effect of spiral grain in *Platanus* and *Aesculus*;

— *permanent vertical waviness* (Figure 15B,C), the most frequent in trees.

The first and the second type develop when the amplitude of the inducing wave increases from 0 to its ultimate value so rapidly that it is practically constant from the moment of its appearance in cambium. Modulated waviness is created when the oscillations start in a certain region of cambium simultaneously at all its levels, while slanted waviness forms when they start on successive levels in consecutive time intervals. Oscillations are released by the approach of the wave front. Permanent waviness is produced when the amplitude increases slowly during the first period of oscillations. For a full description of these patterns and their properties see Zagórska-Marek (1995). What is most important in this discovery, is that the same basic phenomenon — the cambial wave — may be responsible for different morphological patterns of cambium. Phenotypic diversity mentioned at the beginning of this article is created here on the simple basis of the ability of fusiform initials to oscillate.

4 Conclusions

Many patterns, to which the reader has been introduced in this article, are excellent examples of developmental variation in plants because they are dynamic and changeable. As Briggs and Walter point out: "Plasticity remains a comparatively neglected area in the study of plant variation." Patterns in plant meristems, presented here, were relatively easy to investigate because they are visual, represent the morphological category of patterns. Physiological aspects of similar developmental variation are yet to be disclosed. We know nothing about the subcellular and molecular basis of the 20 year long cambial cell cycle; we have not recognized the nature of morphogens that might interact in the diffusion-reaction mechanism postulated to be involved in phyllotaxis. What factor makes phyllotactic diversity richer or dislocations more frequent in some species than in others, also remains to be seen. Another aspect of pattern formation in nature is the universality of its principles. Dislocations, fulleren-like structures, the crystalline line up order, wave — are phenomena intrinsic to the process of pattern formation. As we see, they are present in non-living matter as well as in living organisms, on various levels of their construction — from the molecular composition of the cell to organ architecture and distribution.

Glossary

ZYGOTIC INFORMATION: genetic information, specific for each living individual, encoded in a sequence of nucleotides of DNA molecules present in the nucleus of a zygote. The latter is a cell resulting from gamet's fusion (syngamy) and represents the first stage of each individual development.

ONTOGENY: life history or development of an individual organism, from the zygote to the individual's death.

POLYPLOIDY: status on the cell nucleus, in which the chromosome number has been multiplied and is greater than two of the basic chromosome number.

ONTOGENETIC TRANSFORMATION: developmental qualitative change of any characteristic of the living organism, taking place in a course of its ontogeny.

VASCULAR CAMBIUM: meristematic (embryonic) tissue producing cells differentiating ontogenetically into the vascular i.e. conducting tissues.

ORGAN PRIMORDIA: the first, earliest rudiments of such organs as leaves or flower parts, appearing in a form of discrete elevations on the lateral surface of the plant apical meristem.

SECTORIAL/MERICLINAL CHIMERAS: plant apical meristems composed of genetically different cells. The differences may result from mutations, they can be intraspecific — the cells belong then to different species, or even intrageneric. The cells of a given type occupy the sector of the meristem circumference; if this condition is present in one layer of cells only (in stratified meristems of angiosperms) the chimera is called mericlinal.

ALBINO SECTOR: in a chimera a sector of white cells, that, due to mutation, have lost the ability to produce chlorophyll, green photosynthetic pigment.

ISOTROPIC GROWTH: the expansion of growing cells that is equal in all directions of space.

PARASTICHY: the spiral (within the apex) or helical (on the stem) secondary line connecting primordia or mature organs according to their ontogenetic sequence of appearing in plant body.

TRICUSSATE PATTERN: whorled pattern of phyllotaxis in which there are three plant organs in a whorl and in subsequent whorls these organs alternate. As the result there are 6 vertical lines around the stem connecting all organs present in a pattern.

SECONDARY GROWTH: circumferential growth of cylindrical plant organs such as stem or root leading to the increase in their width, due to cell divisions of secondary meristems such as cambium or phellogen. This activity starts when differentiation of all cells produced by primary apical meristems (primary growth) is completed.

XYLEM AND PHLOEM: plant tissues conducting water and minerals (xylem) or products of photosynthesis (phloem).

STOREY: horizontal tier of cells of the same length arranged in a tissue side-by-side.

References

1. I. Bailey (1923). The cambium and its derivative tissues. IV. The increase in girth of the cambium. *Amer. J. Bot.* **10**:499–509.
2. M. W. Bannan (1956). Some aspects of the elongation of fusiform cambial cells in *Thuja occidentalis* L. *Canad. J. Bot.* **34**:175–196.
3. J. J. Beijer (1927). Die Vermehrung der Radialen Reihen in Cambium. *Rec. Trav. Bot. Neerl.* **24**:631–786.
4. D. W. Bierhorst (1977). On the stem apex, leaf initiation and early leaf ontogeny in filicalean ferns. *Amer. J. Bot.* **64**:125–152.
5. D. Briggs and S. M. Walters (1997). *Plant variation and evolution.* 3rd edition. Cambridge University Press. Cambridge.
6. R. Dawkins 1996. *Climbing mount improbable.* Polish Edition.
7. S. Douady and Y. Couder (1996a). Phyllotaxis as a self-organizing iterative process. I. The spiral modes resulting from time periodic iterations. *J. Theor. Biol.* **178**:255–274.
8. S. Douady and Y. Couder (1996b). Phyllotaxis as a self-organizing iterative process. II. The spontaneous formation of a periodicity and the coexistence of spiral and whorled patterns. *J. Theor. Biol.* **178**:275–294.
9. S. Douady and Y. Couder (1996c). Phyllotaxis as a self-organizing iterative process. III. The simulation of transient regimes in ontogeny. *J. Theor. Biol.* **178**:295–312.
10. E. Gola (1996). Phyllotaxis diversity in *Lycopodium clavatum* L. and *Lycopodium annotinum* L. *Acta Soc. Bot. Poloniae* **65**:235–247.
11. E. Gola (1997). Phyllotactic spectra in cacti: *Mamillaria* species and some genera from *Rebutia* group. *Acta Soc. Bot. Poloniae* **66**:237–257.
12. Z. Hejnowicz (1973). Morphogenetic waves in cambia of trees. *Plant Sci. Let.* **1**:359–366.
13. Z. Hejnowicz (1974). Pulsation of domain length as support for the hypothesis of morphogenetic waves phenomenon. *Acta Soc. Bot. Poloniae* **43**:261–271.
14. Z. Hejnowicz (1984). Trajectories of principal directions of growth, natural coordinate system in growing plant organ. *Acta Soc. Bot. Poloniae* **53**:29–42.
15. Z. Hejnowicz and J. A. Romberger (1973). Migrating cambial domains and the origin of wavy grain in xylem of broadleaved trees. *Amer. J. Bot.* **60**:209–222.
16. Z. Hejnowicz and B. Zagórska-Marek (1974). Mechanism of changes in grain inclination in wood produced by storeyed cambium. *Acta Soc. Bot. Poloniae* **43**:381–398.

17. R. V. Jean (1994). *Phyllotaxis. A Systematic Study in Plant Morphogenesis.* Cambridge University Press. New York.
18. E. J. (Jr.) Klekowski (1988). *Mutation, developmental selection, and plant evolution.* Columbia University Press. New York.
19. E. J. (Jr.) Klekowski, N. Kazarinova-Fukshansky and H. Mohr (1985). Shoot apical meristems and mutation: Stratified meristems and angiosperm evolution. *Amer. J. Bot.* **72**:1788–1800.
20. J. Kuligowski and Y. Tourte (1979). La détermination des cellules apicales chez une Ptéridophyte; rôle particulier du génome paternel. *Bull. Soc. bot. Fr.* **126**. *Lettres bot.* **5**:491–505.
21. D. Kwiatkowska (1995). Ontogenetic changes of phyllotaxis in *Anagallis arvensis* L. *Acta Soc. Bot. Poloniae* **64**:319–325.
22. D. Kwiatkowska (1997). Intraspecific variation of phyllotaxis stability in *Anagallis arvensis. Acta Soc. Bot. Poloniae* **66**:259–271.
23. R. F. Lyndon (1998). *The shoot apical meristem: its Growth and Development.* Cambridge University Press. Cambridge.
24. R. D. Meicenheimer (1981). Changes in *Epilobium phyllotaxy* induced by N-1-naphthylphthalamic acid and α-4-chlorophenoxyisobutyric acid. *Amer. J. Bot.* **68**:1139–1154.
25. R. D. Meicenheimer (1982). Change in Epilobium phyllotaxy during reproductive transition. *Amer. J. Bot.* **69**:1108–1118.
26. A. Matkowski, R. Karwowski and B. Zagórska-Marek (1998). Two algorithms of determining the middle point of the shoot apex by surrounding organ primordia positions and their usage for computer measurements of divergence angles. *Acta Soc. Bot. Poloniae* **67**:151–159.
27. Z. Puàwska (1986). The sequence of cell divisions in the I tunic layer of *Actinidia arguta* Planch in light of the development of twin cell complexes. *Acta Soc. Bot. Poloniae* **55**:171–179.
28. J. A. Romberger, Z. Hejnowicz and J. F. Hill (1993). *Plant Structure: Function and Development.* Springer-Verlag. Berlin, Heidelberg, New York.
29. J. Ruth, E. J. (Jr.) Klekowski and O. L. Stein (1985). Impermanent initials of the shoot apex and diplontic selection in a juniper chimera. *Amer. J. Bot.* **72**:1127–1135.
30. E. W. Sinnott (1960). *Plant Morphogenesis.* McGraw-Hill. New York.
31. K. Soma and E. Ball (1963). Studies on the surface growth of the shoot apex of *Lupinus albus. Brookhaven Symposia in Biology* **16**:13–45.
32. M. Szostak and B. Zagórska-Marek (1998). *The new patterns of phyllotaxis in juniper. Proceedings of 51 Meeting of Polish Botanical Society*, p. 481. Gdañsk 1998.
33. R. A. E. Tilney-Bassett (1986). *Plant Chimeras.* Arnold. London.
34. W. Wòch and B. Zagórska-Marek (1982). Reconstruction of storeyed cambium in the linden. *Acta Soc.Bot.Poloniae* **51**:215–228.
35. B. Zagórska-Marek (1975). Growth activity of fusiform initials in storeyed cambium. *Acta Soc. Bot. Poloniae* **44**,
36. B. Zagórska-Marek (1984). Pseudotransverse divisions and intrusive elongation of fusiform initials in storeyed cambium of *Tilia. Canad. J. Bot.* **62**:20–27.

37. B. Zagórska-Marek (1985). Phyllotactic patterns and transitions in *Abies balsamea*. *Canad. J. Bot.* **63**:1844–1854.
38. B. Zagórska-Marek (1987). Phyllotaxis triangular unit; phyllotactic transitions as the consequences of the apical wedge disclinations in a crystal-like pattern of the units. *Acta Soc.Bot.Poloniae* **56**:229–255.
39. B. Zagórska-Marek (1994). Phyllotaxic diversity in *Magnolia* flowers. *Acta Soc. Bot. Poloniae* **63**:117–137.
40. B. Zagórska-Marek (1995). Morphogenetic waves in cambium and figured wood formation. In: *The cambial derivatives.* M. Iqbal (Ed.). pp. 69–92. *Encyclopedia of Plant Anatomy.* Gebruder Borntraeger. Berlin — Stuttgart.
41. B. Zagórska-Marek (1998). Clonal analysis provides an evidence of transient initial cells in the stochastic plant meristems: The discussion of consequences. *Folia Morphol. (Warsz.)* **57**, pp. 75.
42. B. Zagórska-Marek and M. Turzańska (1998). Cellular pattern of the shoot apical meristem provides an evidence of impermanent initials in seed plants. *Proceedings of 51-st Meeting of Polish Botanical Society*, pp. 547, Gdańsk.

MECHANICAL STRESS PATTERNS IN PLANT CELL WALLS AND THEIR MORPHOGENETICAL IMPORTANCE

ZYGMUNT HEJNOWICZ
Department of Biophysics and Cell Biology
Silesian University
Katowice, Poland

1 Introduction

Often morphogens and their gradients are considered as sole bearers of positional information. In this article, another point of view is taken which emphasizes mechanical factors rather than chemical ones. Assume, for instance, that the core of a turgid plant tends to expand more than peripheral tissue does. As a consequence, the peripheral tissue is brought into tensile stress, while the core brings itself into compressive stress (superimposed on the turgor-caused tensile stress of the cells). The field of mechanical forces and stresses in such a structure have a great potential to serve as a morphogenetical field. This field depends strongly upon actual geometry of the stressed body and may provide information for its development.

An organ in which one part is under tension while the other is under compression is an example of prestressed structures. Prestressing (tensegrity) is a common feature of organisms on each level of plant and animal organization. On the subcellular level it occurs in the cytoskeleton (Ingber 1997), on cellular level — in cell wall or extracellular matrix. On the organ level prestressing occurs in plants as tissue stresses in wilting stems or growth stresses in woody stems (Hejnowicz 1997), and in animals (vertebrate) as tension in muscles and compression in bones.

This article aims to show factors underlying anisotropy (at a point) and non-uniformity (at different points) of stresses in cell walls in plant stems, and the importance of stress anisotropy and stress distribution for self-organization of the stem.

2 Stress anisotropy

Mechanical stress is a tensor quantity; for every direction in space there is a value at a point in the tensor field. In three dimensions there are three principal directions in which the quantity attains extreme values (maximal, minimal, and of saddle type). The directions are mutually orthogonal. In two dimensions (cell wall surface) which suffice to describe the stress as a point in cell wall, there are two directions in which the normal stress, σ, (defined as the force per unit of transverse area) attains maximal and minimal value, unless the stress is the same in all directions in the plane tangent to the surface (isotropy case, in which every direction is principal). Normal stress may be tensile (conventionally taken as positive) or compressive (negative).

The cell wall of each living cell is under a tensile stress (σ). The stress depends on turgor pressure (P), cell wall thickness (t), cell shape, and the direction in which stress is measured. For example, in a cylindrical cell with radius r, the longitudinal in spherical cell

$$\sigma = \frac{P \pi r^2}{2 \pi r t} = \frac{P r}{2 t}$$

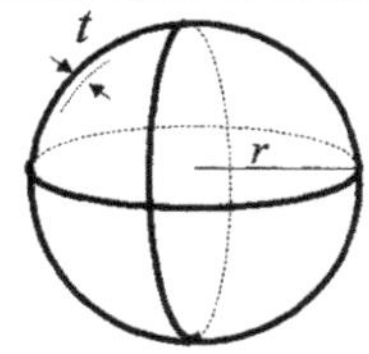

in cylindrical cell

$$\sigma_{long} = \frac{P \pi r^2}{2 \pi r t} = \frac{P r}{2 t}$$

$$\sigma_{trans} = \frac{P\, 2\, r\, l}{2\, l\, t + 4\, r\, t} \approx \frac{P r}{t}$$

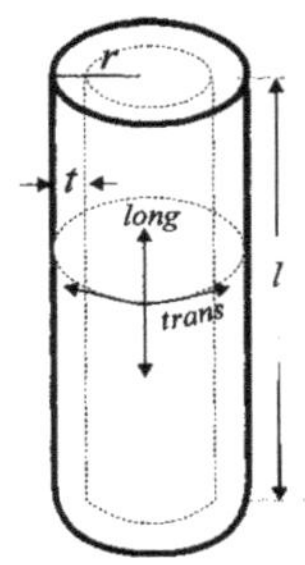

Figure 1. Scheme illustrating turgor-caused tensile stress (σ) in the cell walls in orthogonal directions in a spherical and cylindrical cells. In the cylindrical cell there is anisotropy of stress characterised by the stress ratio (longitudinal stress to transverse stress) 1:2.

stress (σ_{long}) is given by $\sigma_{long} = \mathrm{Pr}/2\mathrm{t}$, but the transverse (circumferential) stress (σ_{trans}) is $\sigma_{trans} = \mathrm{Pr}/\mathrm{t}$ (Figure 1). In general, there is anisotropy of stresses in the walls; only in spherical cells the stresses are the same in all directions. In free cells, this anisotropy may offer the cell important information regarding its shape and especially major and minor axes.

In the case of cells in an organ, the pattern of stress anisotropy is changed due to: (i) packing of cells which results in their polyhedric shape, and (ii) the occurrence of tissue stresses in an organ. The pattern offers additional information to a cell related to the fact that it is not alone in an organ.

In a polyhedric cell the stresses at the edges and vertices are lowered in the direction in which curvature is increased due to packing. In general, however, the differences in anisotropy are not large and the average anisotropy in a cell is similar when it has uniform curvature being free, or when it is polyhedric being packed. Much higher changes in anisotropy appear due to tissue stresses.

Tissue stresses (TSs) represent prestresses in turgid plant organs. In a turgid stem they can be easily demonstrated by stem dissection; some layers or strands become shorter and the other longer than the original length of the dissected segment. Until recently, TSs were considered as a result of differential growth of tissues in a stem (for references see Peters and Tomos, 1996). Differential growth was supposed to result from either the differences in the auxin response of different tissues or from the different auxin contents in different tissues. Because the differential growth was not a predictable phenomenon, TSs were treated as many other unpredictable growth phenomena, and deserved little general attention, though at the end of the last century they used to belong to most vividly discussed phenomena.

It has been shown that TSs are unavoidable physical consequence of the variation in mechanical parameters of cell layers or strands within a turgid stem (which mechanical stability is due to turgor pressure). The most dense (with respect to cell walls) layer is under maximal tensile TS while the least dense is under maximal compressive TS (Hejnowicz and Sievers, 1996a). These structure-based TSs, of course, may be modulated by a differential growth of tissues, as well as by stresses caused by external forces (e.g., in bending, in penetrating soil, etc.). It is thus convenient to distinguish two types of TSs according to their origin: structural and growth differential (the stresses caused by external forces do not belong to prestresses).

Measurements of the mechanical parameters of sunflower hypocotyls allowed us to calculate the structural TSs (Hejnowicz and Sievers 1996a). Direct measurements of the TSs occurring in the hypocotyls showed that the measured stresses fit well the stresses calculated from the model. Also the TSs determined in *Reynoutria* stems fit their structural characteristics (Hejnowicz and Sievers 1996b). The evidence obtained so far allows us to conclude that the stresses originating from structural differences of tissues in a turgid stem readily explain TSs. Differential growth, i.e., differential unreversible expansion of tissues, is not required to explain the TSs in straight stems.

It is interesting and important that the structural TSs may attain relatively high values (Figure 2). This is especially true for the tensile TS. Though the tensile and compressive forces involved in the generation of TSs have the same magnitude, usually the cross-sectional area of the tissue in tensile TS is much smaller than that under compression, which results in a higher tensile TS. Both in sunflower hypocotyl and in the *Reynoutria* stem the tensile TS is approximately five times higher than osmotic pressure (which cannot be lower than turgor pressure). As

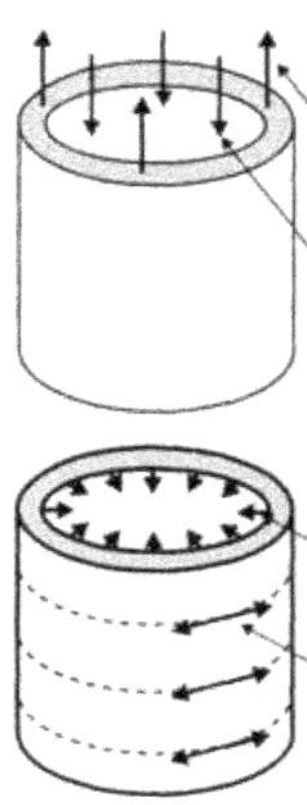

what and where	forces [N]	stresses [MN·m^2]
longitudinal tensile in the epidermis	+0.35	2.50
longitudinal compressive in the ground tissue	- 0.35	0.15
transverse-radial compressive in the ground tissue	0.01 (per 1 mm of length)	0.43
circumferential tensile in the epidermis		0.06

π=0,55MPa

Figure 2. Scheme of tissue stresses and their numerical characterisation in sunflower hypocotyl. For comparison the osmotic pressure of cell sap (π) is also indicated (MPa is formally equal to MN·m^{-2}).

indicated, the tensile stress in the cell wall depends on turgor pressure and on the ratio r/t. Thus if there were no TSs, the longitudinal stress in cells with different diameters and wall thicknesses (in the same stem) would differ, because usually there is little variation in turgor pressure.

It is also known that, in the case of isolated cells or tissues, when wall thickness is higher than 20% of cell radius, the internal pressure plays little or no role in dictating the tensile stress within the wall. Indeed, isolated collenchyma immersed in water is flaccid, in spite that it develops full turgor pressure. However, *in situ* there is a high longitudinal tensile TS in the collenchyma that compensates for the weak longitudinal stress directly caused by turgor pressure.

In the model for structural TSs, inherent is the assumption that longitudinal elastic strain is the same in all longitudinal walls in a stem, which means that the longitudinal stress is the same in all cell walls whatever is the ratio R/t (assuming that Young's modulus of cell wall material is the same). In a stem, even a thick-walled cell with low turgor is under the same longitudinal tensile stress as that in a thin-walled cell with high turgor. This is achieved by superimposition of the TS on the stress caused directly by turgor pressure in the wall of a particular cell; the longitudinal tensile stress caused directly by turgor is increased or lowered depending on whether the cells belong to a tissue which is under tensile or compressive TS (Figure 3). Since the longitudinal TSs result from the interaction of all cells in the stem and their values depend on the histological status of the stem, the TS acting on a particular cell changes the stress ratio in the longitudinal wall of this cell

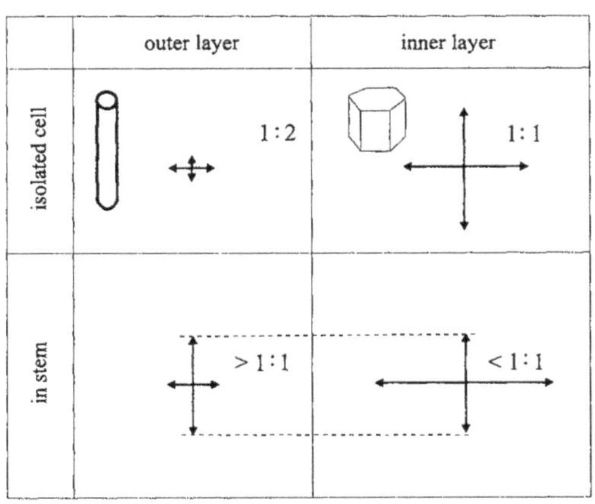

Figure 3. Scheme illustrating the stress ratio ($\sigma_{long}/\sigma_{trans}$) in the walls of turgid cells when isolated and in situ in a stem. Two types of cells are considered: (i) cylindrical, thick-walled, low-diameter cell of the epidermis or the collenchyma type that occur in the outer tissue of the stem, and which in situ is under a high longitudinal tensile tissue stress, and (ii) isodiametric, thin walled, large-diameter cell of the parenchyma type that occur in the inner part of the stem, and which in situ is under moderate compressive longitudinal tissue stress. It is assumed that the cells do not differ in turgor pressure. Observe that in situ the longitudinal stress is the same in both cell types, and the stress ratio differs considerable for the thick-walled cell when isolated and in situ.

depending on the histological properties of the particular cell and the histological properties of all other cells in the stem (at the considered level). Tissue stresses integrate the cells in an organ to such an extent that they may be considered as a primary holistic feature of the organ. In contrast to the longitudinal stress in cell walls, the transverse stress may be variable in the stem. This stress depends not only on the histological status of the stem but also on the radial coordinate in the stem. For instance, when inner thin-walled cells elastically expand in transverse direction due to their turgor, this expansion must be accommodated by tangential expansion of the more outer cells which may have thicker walls, and therefore a tangential tensile TS may appear in them. Consequently, transverse TS may also provide positional information to cells.

Considering the cumulative effect of longitudinal and transverse TSs on stress anisotropy, we can conclude that the anisotropy for a wall of a particular cell depends on histology of the stem and on the position of the cell in the stem.

The effect of position may be accentuated when a stem gently sways, as usually happens. Namely, swaying means that at a particular side of a stem the longitudinal tensile stress in cell wall changes cyclically due to changing curvature of this side. Especially, swaying results in an oscillation of the stress anisotropy in this layer which provides most of the strength, i.e., which is on stem periphery, and being composed of cells with low r/t, is under a high tensile TS.

How can the stress anisotropy in cell wall be detected by cells? Surely, it cannot be detected by means of strain anisotropy in a growing cell because plastic (unreversible) strain, which represents growth, is not related in any unique way to the stress causing it. However, even in human everyday use, the stress is measured by different methods; by means of elastic strain as in a spring-type balance, or by means of oscillation frequency as during violin tuning. Other methods are based on piezoelectric, or optical (birefrigency) effects. It is thus possible that also cells have ways to detect stress, or stress anisotropy, independently of the accompanied strains. Experiments on *Nitella* (Richmond 1983) support this expectation.

Williamson (1990) postulated that in the plasma membrane there are receptors which transmit the information about stress anisotropy in cell wall across the plasma membrane to orient cortical microtubules in the direction of maximal stress. That the microtubules are sensitive to what happens specifically in the wall they underlie is supported by the fact that in epidermal cells, which has an external and an internal cell wall usually of different thickness, experimental factors that reorient the microtubules cause a specific reorientation at a particular wall instead of at all walls of a particular cell.

What may be the importance of the stress anisotropy for cells? The microtubules determine the orientation of the cellulose microfibrils to be deposited that in turn affects the cell expansion. So, a feedback based on the stress, relating future cell expansion to present cell shape and other histological features of the particular cell and of all other cells in the stem (through TSs) can work.

It is noteworthy that in a tissue which typically is under tensile TS, and thus is characterized by an increased stress ratio, like the epidermis and collenchyma, a helicoidal microfibrillar texture occurs in cell wall, while in the tissues which typically are under compressive TS, like the parenchyma, the multinet texture with initially transverse microfibrils occurs. (However, the relationship between tensile longitudinal TS and helicoidal texture cannot be reversed; e.g., such a texture occurs in root hairs where it is self-assembled.)

Recently, we have been performing studies on microtubule orientation in the epidermis and in thin walled parenchyma of sunflower hypocotyl (not yet published). In the outer epidermal wall (that is known to be characterized by helicoidal texture) which *in situ* is under the high tensile force of TS type, the microtubules underlying this wall are oriented steeply oblique with a differential inclination in different cells. It looks so as if the inclination of microtubules changed cyclically with time, and what one observe in different cells at the instant of fixation of the hypocotyl are different phases of this cycle. In contrast, in cortical cells, which are under compressive TS, the microtubules are oriented transversely. When the peripheral layer, consisting of epidermis and two or three layers of parenchyma, is isolated from the hypocotyl by peeling, and incubated into a turgid state, the orientation of the microtubules under the external wall, that previously was under high tensile TS, becomes much less inclined or even transverse. However, when the isolated peel is incubated being stretched by force substituting for the missing tensile TS, the microtubules maintain their varying steep inclination.

Tensile stress in cell walls provides the "driving force" for cell growth, and the growth rate is proportional to the stress. Theoretically, higher growth rates may be expected in the direction of maximal stress than in the direction of the minimal stress in the cell wall unless this direction is reinforced by the cellulose microfibrils. Indeed, this occurs in the tissue which is under the high tensile longitudinal TS in the stem; the helicoidal texture in the cell walls of such a tissue enables growth in the direction of maximal stress. However, one should keep in mind that helicoidal structure is unique in the respect that the longitudinal direction (the direction of maximal tensile TS) is the mean direction of microfibrils, though relatively few microfibrils if any are oriented in this direction.

Very often the higher growth occurs in the direction of minimal stress. This is due to the fact that microtubules, oriented in the direction of the maximal stress, cause similar orientation of microfibrils which reinforce this direction to such an extent that the threshold for irreversible strain surpasses the stress, while the threshold in the perpendicular direction (the direction of minimal stress) is lower than the stress in this direction. Such is the growth of stem parenchyma *in situ* which being under compressive TS is characterized by maximal stress in transverse direction. One should also keep in mind the restriction imposed on the growth of stem parenchyma by the rule of symplastic growth in plant organs; the growth of a particular cell is subordinated to the growth of the whole organ. In the stem, the longitudinal growth is controlled mainly by the peripheral layer where the maximal stress in cell wall is longitudinal.

3 Stress distribution

Often the stress is distributed non randomly on cell surface, i.e. the value of a particular principal stress changes with the position on cell surface. As already mentioned the stress decreases when the curvature of cell wall increases, so the vertices and edges of a polyhedric cell are characterized by lower stress than that on facets.

It is interesting that higher growth of cell surface occurs at the sites of lower stress. This is evident in tip growing cylindrical cells where the stress attains minimum at the tip, and in a polyhedric cells where it is the vertex or an edge that is the site predisposed for growth. An example of growth at vertices is provided by the development of stellate parenchyma; files of flat cells are transformed into the parenchyma by the growth initially at vertices (Korn, 1984). Another example, occurring often, is provided by intrusive growth, a sort of growth which cannot be accommodated by the old concept of sliding growth which postulated shearing of plasmodesmata. The growing edge, or vertex, seemingly releases enzymes that promote swelling and weakening of the middle lamella, along with deletion of plasmodesmata in advance of the intruding edge.

Surely, the growth at edges or vertices, which is involved in intrusive growth, is not realized in the way of a physical push of the neighboring cells by a wedge action, thus putting the neighboring cells under compression, because the edge is the least stiff portion of a cell (local minimum stress). A precondition for growth at the edges seems to be a tension across the middle lamellas adjacent to the intruding edge, which means a tensile stress perpendicularly to the direction of intrusive growth. Such growth occurs especially in the vascular cambium (Figure 4) and may be responsible for changing the mean angle between cambial cells and stem axis (Figure 5) (Hejnowicz and Zagórska-Marek, 1974). An interesting case is the development of vessel members in secondary xylem (Figure 6).

In pressurized solids, distribution of irreversible (plastic) strain usually corresponds to the distribution of stress. However, in an expanding cylindrical balloon this relationship does not hold; in a not fully extended cylindrical balloon (Figure 7A) its tip is the least stressed part but at the same time it accommodates the most rapid expansion when air is pumped into the balloon. Tip growing cells

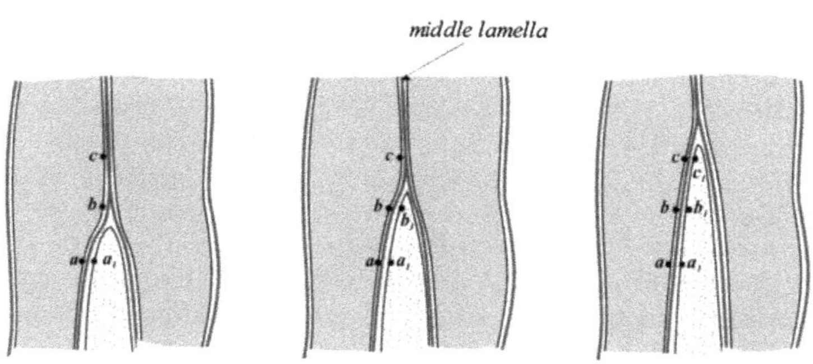

Figure 4. Scheme illustrating intrusive growth: the expansion of the clear cell occurs only at its tip (a vertex or an edge perpendicular to the drawing plane). Existing contacts between the cell and its neighbours beyond the tip (including new contacts established by the growing cell) do not change. A,B,C — consecutive stages.

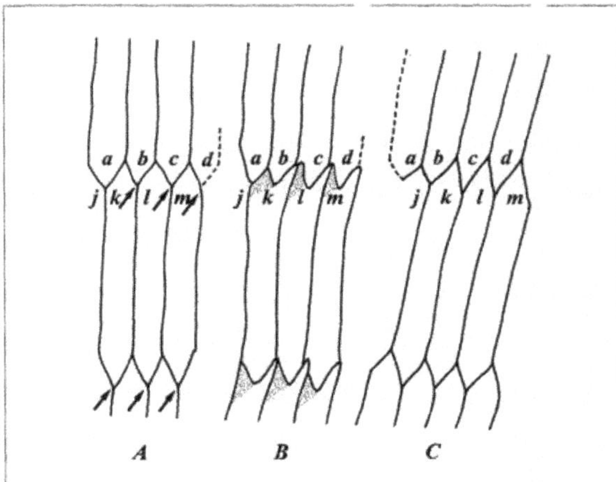

Figure 5. Scheme illustrating how chirally oriented intrusive growth allows to change the orientation of cambial cells. The vertices (edges) growing intrusively are indicated by arrows, the cell parts which are eliminated are indicated by shading). A,B,C - consecutive stages.

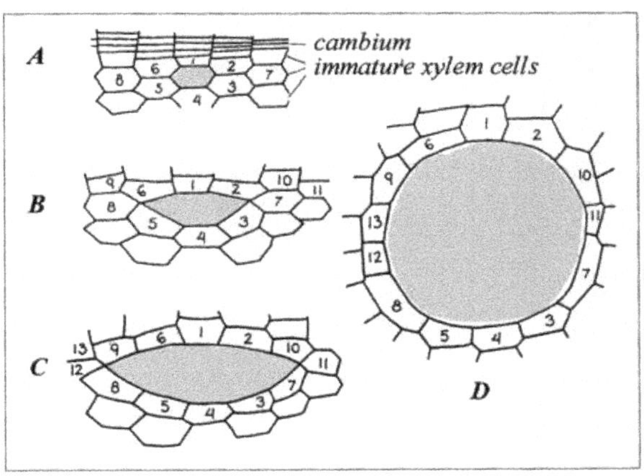

Figure 6. Scheme illustrating how intrusive growth at longitudinal edges perpendicular to the drawing plane of a cell (shaded) in developing xylem leads to a wide vessel in ring-porous wood. In the cambium and the young xylem there is tensile tissue stress in radial direction during during the formation of wide vessels. A-D consecutive stages.

like hyphae, rhizoids (Figure 7B), root hairs, may be considered as those realizing the mechanical model of the expansion of a cylindrical balloon at the weak-stressed tip. Maybe, in the evolution, hypha-type cells "learned" to grow at the tip where curvature was maximal and tensile stress minimal. Thereafter, the cells "realized" such growth also in the situation which did not conform to the mechanical model of expansion.

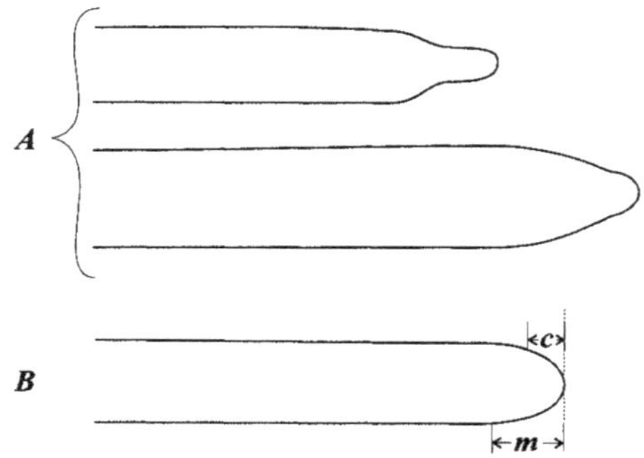

Figure 7. Schemata of: A — a cylindrical balloon in two stages of inflation in which the tip portion is not extended but the stresses there, due to higher curvature, are lower than in the inflated portion, and B — a rhizoid in which the circumferential growth is limited to the portion c, and the latitudinal growth to the portion m of the tip.

4 Conclusions

The anisotropy of tensile stress in cell wall of a turgid cell depends on cell shape, but in multicellular organ is affected by tissue stresses acting in the organ. In a stem, the tissue stress acting on a particular cell depends on mechanical characteristics of all remaining cells. In the longitudinal direction in a stem there is tensile tissue stress in cells with thicker walls and/or smaller diameter, and compressive tissue stress in cells with thinner walls. The tissue stresses are superimposed on the tensile stress acting in cell wall of a particular cell and due directly to turgor pressure in this cell. In a longitudinal cell wall in a stem, the tensile stress component due to tissue stress may be much higher than that due directly to turgor pressure. The maximal stress at a point in a cylindrical cell is then reoriented from transverse (without tissues stress) to longitudinal. Ultimately, the stress anisotropy in cell walls depends (through the tissue stresses) on histology of the stem and position of the wall in the stem. In a polyhedric cell in a tissue, the stresses at the edges and vertices are lowered in the directions in which curvature of cell wall is increased. Consequently, the stress in cell wall is distributed non uniformly on the cell surface even in isodiametric cells; the vertices and edges are characterised by lower stress than that on facets. The sites of lower stress are predisposed for growth, as exemplified by tip growing cylindrical single cells, and by intrusive growth of cells in a tissue.

Glossary

AUXIN: A class of plant hormones.

CELL WALL: A specialized structure accrued just outside the plasma membrane of a plant cell. It consists of long microfibrils composed of cellulose which are embedded in an amorphous matrix of carbohydrate and proteinaceous materials.

COLLENCHYMA: A primary, subepidermal supporting tissue in stems and leaf petioles and vein ribs.

CYTOSKELETON: A complex network of protein filaments (actin filaments, microtubules and intermediate filaments) that extends throughout the cytoplasm.

DIFFERENTIAL GROWTH: The growth of one part of a cell, a tissue, or an organ occurring at a rate different from that of another part, involved in a wide variety of morphogenetic phenomena in plants.

EPIDERMIS: The outermost layer of cells on the primary plant body.

GROUND TISSUE: Any tissue other than the epidermis and the vascular tissue in the young stems, roots and leaves; mostly the parenchyma.

HELICOIDAL WALL STRUCTURE: The structure of a wall consisting of a stack of lamellae with microfibrils parallel within each lamella but with a small angle between microfibrils of successive layers.

HYPOCOTYL: The portion of the embryonic plant axis between the cotyledons and root.

MICROFIBRILIS : Long bundles of cellulose chains in the cell wall.

MICROTUBULES: Long, stiff polymers of tubulin that extend throughout the cytoplasm. In plant cells, the *m.* in the cortical cytoplasm govern the orientation of microfibrils polymerized on the outer face of the plasma membrane.

MIDDLE LAMELLA: A layer cementing the primary walls of neighboring cells.

MORPHOGEN: Any endogenous compound that exerts a morphogenetic effect at a low concentration.

OSMOTIC PRESSURE (OF A SOLUTION): The hydrostatic pressure which would be required to prevent the net flow of water across a membrane which separates the solution from pure water.

PARENCHYMA: A tissue consisting of living cells with relatively thin walls abundant in the primary plant body.

POSITIONAL INFORMATION: Any mechanism which causes cells in an embryo or an organ to respond to their position in the developmental field. The control of development by *p.i.* requires a method of assigning a positional value to each cell in the field, and a means by which each cell can interpret the assigned positional value in terms of an inherent reference library.

RHIZOID: Tip growing cylindrical cell which helps to hold the plant to a substrate.

SLIDING GROWTH: Refers to an old (unrealistic) concept which postulated sliding of plant cells with respect to each other during growth of a tissue.

SYMPLASTIC GROWTH: The coordinated growth of cells within plant organ; contiguous walls do not slide or slip with respect to each other.

TENSEGRITY: A building system based on continuous series of tension elements (e.g. elastic strings) with a discontinuous series of compression-resistant struts; a term coined from tensional integrity.

TURGID PLANTS: Plants (their organs) that wilt when deprived of water. The mechanical stability of *t.p.* depends on cell walls stretched by turgor pressure when water supply is abundant.

TURGOR: Elastic distention (strain) of a plant cell wall by the turgor pressure. Often used as synonym of turgor pressure.

TURGOR PRESSURE: The hydrostatic pressure developed by the fluid content of a turgid cell and causing tensile stress (and elastic strain) of the cell wall.

VASCULAR CAMBIUM: A layer of cells between the phloem and xylem that generates new xylem and phloem cells and therefore is responsible for stem or root thickening.

VESSEL: An axial file of special xylem cells (vessel members) having perforations at the junctions, and thus arranged in the form of a pipeline conduit.

XYLEM: The principal water-conducting tissue composed of tracheary elements (tracheids and/or vessel members), fibers and parenchyma.

References

1. Z. Hejnowicz and A. Sievers (1996a). Tissue stresses in organs of herbaceous plants. III. Elastic properties of the tissues of sunflower hypocotyl ans origin of tissue stresses. *J. Exp. Bot* **47**:519–528.
2. Z. Hejnowicz and A. Sievers (1996b). Tissue stresses and their graviresponsive changes in stems of *Reynoutria japonica* Houtt. *J. Plant Physiol.* **149**:91–101.
3. Z. Hejnowicz (1997). Graviresponse in herbs and trees: a major role for the redistribution of tissue and growth stresses. *Planta* **203**:136–146.
4. Z. Hejnowicz and B. Zagórska-Marek (1974). Mechanism of changes in grain inclination in wood produced by storeyed cambium. *Acta Soc. Bot. Pol.* **43**:381–398.
5. D. E. Ingber (1997). Tensegrity: the architectural basis of cellular mechanotransduction. *Ann. Rev. Physiol.* **59**:575–599.
6. R. W. Korn (1984). Cell shapes and tissue geometries. In *Positional controls in plant development.* P. W. Barlow & J.Carr (Eds.) p.33–42, Cambridge Univ. Press, Cambridge-Sydney.

7. W. S. Peters and A. D. Tomos (1996). The history of tissue tension. *Ann. Bot.* **77**:657–665.
8. P. Richmond (1983). Patterns of cellulose microfibril deposition and rearrangement in *Nitella: in vivo* analysis by birefrigence index. *J. Appl. Polym. Sci. Symp.* **37**:107–122.
9. R. E. Williamson (1990). Aligment of cortical microtubules by anisotropic wall stresses. *Austr. J. Plant Physiol.* **17**:601–613.

TENSORIAL MODEL FOR GROWTH AND CELL DIVISION IN THE SHOOT APEX

JERZY NAKIELSKI
Department of Biophysics and Cell Biology
Silesian University, Katowice, Poland

Plants, unlike animals, grow throughout their lifetimes. The plant body enlarges by the continuous activity of a special localized regions called meristems which are centers of cell division and cell expansion. A plant grows in length by its apical meristems, of the stem and root, and in width by a sheath of lateral meristems. Organs such as leaves, flowers, and fruits, however, rarely have sharply localized meristems but enlarge throughout much of their extent, as an animal body does, until growth ceases. When they reach maturity all their tissues stop growing and there is no meristematic region set apart by which further growth may be accomplished. Meristems are of much interest to workers of plant morphogenesis. They provide a "continuous embryology" (Sinnott 1960) for the plant and offer an important point for studies of problems of plant development. This paper deals with the meristem of the shoot apex. It generates cells, which subsequently elongate and differentiate into the primary tissues of the stem and leaves. The apical part of the shoot apex has the form of a regular dome situated, mostly, above the level of the youngest leaf primordia. The dome is the ultimate source of all the cells of the stem. I have worked out a simulation-based model for growth and cell divisions in the shoot apical dome described, mathematically, by the growth tensor (Hejnowicz and Romberger, 1984). Here, the model is considered in two cellular layers, one related to the axial section and the other related to the surface of the dome. In both, cell patterns coming from the apex of spruce seedlings will be developed.

1 Symplastic Growth and the Growth Tensor

In most plants, the typical mature cell is surrounded by a cell wall. Because meristematic cells generally are cemented together along their walls, plant organs grow symplastically (Priestley 1930; Erickson 1986). The symplastic growth is defined as the coordinated growth of cells within the organ. During such growth the neighbouring cells do not slide or glide with respect to each other and the organ as a whole maintains its physical integrity. Thus, the organ growth is considered continuous.

A meristematic tissue may be visualized as a three-dimensional network in which the filaments of the mesh are the contact edges of the cells (Romberger et al. 1993). During growth the network is extended and, commonly, deformed. Obviously, the network does not submit passively to this growth as sponge rubber does to bending and stretching. Rather, active cellular processes are involved in growth, new molecules are introduced between the old in the walls, and new cells are formed within the spaces previously occupied by parent cells. After many generations of cell divisions, the arrangement of cells are seen. They often manifest some symptoms of the symplasticity. In the axial sections of shoot and root meristems, cell walls are arranged into two types of mutually orthogonal lines, so called "periclines" and "anticlines" (Sachs 1887). The lines, considered smooth regardless of the zig-zag resulting from the shape of individual walls, preserve orthogonality during growth. The walls tangent to them, though displaced with respect to the tip, retain a recognizable periclinal and anticlinal orientation. Also newly formed partition walls are either anticlinal or periclinal. The walls thus intersect each other in an orthogonal manner and, because this orthogonality is preserved during growth, the cellular pattern remains steady as the individual cells move down through it. The situation is similar to that in well-designed water fountain, in which the display pattern is steady because the individual water jets or droplets follow the same routes as did their predecessors.

One of the most obvious facts which a study of growth reveals is that it does not proceed at the constant rate. Furthermore, maintaining anticlinal or periclinal orientations of cell walls during growth demands a special distribution of growth rates within the organ. The rate can be different in different parts of the organ and change depending on the direction; however, they are coordinated with, and subordinated to, the realization of the overall pattern and shape of the organ that is being generated. Among other things for this property, the symplastic growth conveniently lends itself to mathematical description by means of the growth tensor, GT (Hejnowicz and Romberger, 1984).

The growing organ is characterized by the continuous field of the displacement velocity, **V**. From the knowledge of **V**, the GT is calculated. It follows directly from the definition of the relative elemental rate of growth, $RERG_l$, (Richards and Kavanagh, 1943, Silk and Erickson 1979), namely $RERG_l$ in a given direction $\mathbf{e}_s$, can be expressed as (Hejnowicz and Romberger 1984) $RERG_l(s) = T^q_p \, e^p e_q$ where T^q_p is defined as the growth tensor (for p,q=1,2,3 related to the components of $\mathbf{e}_s$, the summation convention is used). Calculation has shown (Hejnowicz and Romberger 1984) that T^q_p corresponds to the covariant derivative of **V**.

The GT generates the field of growth rates in the organ . Inherent characteristics of the field are principal directions of growth, PDGs (Hejnowicz and Romberger

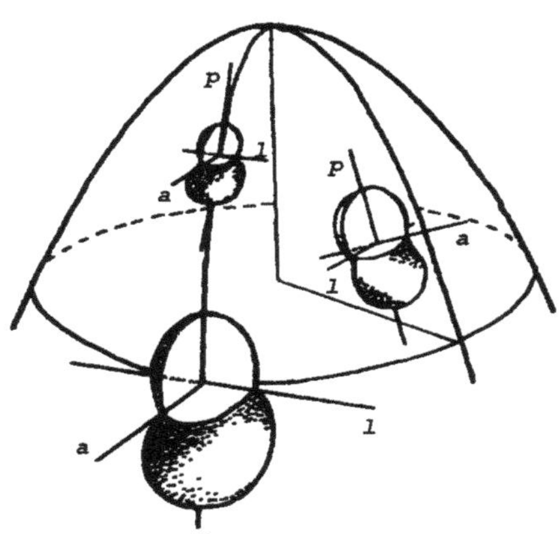

Figure 1. A diagram showing the directional variation of $RERG_l$ in some points of GT field in the shoot apex. The values of $RERG_l$ are arranged into the 3d figures. At each point, three principal directions of growth (PDGs): *p* (*periclinal*), *a* (*anticlinal*), *l* (*latitudinal*) are shown. Pairs of PDGs form the principal planes. In each figure, its upper part limited by the planes has been cut out.

1984, Hejnowicz 1984), Figure 1. At a given point are three mutually orthogonal PDGs unless growth is isotropic when every direction is principal. These directions can be called: *periclinal*, *anticlinal* and *latitudinal*, Figure 1. In two of them $RERG_l(s)$ attain extreme values and the third one is perpendicular to the mentioned two. Between successive points of the field, PDGs change in a continuous way giving a pattern of PDG trajectories. Through every point passes three PDGs and three trajectories of PDGs. Each pair of PDGs defines a principal plane, each pair of PDG trajectories defines a principal surface. If GT field is maintained over time, the PDG trajectories are steady and can be approximated by lines of a natural curvilinear coordinate system (Hejnowicz, 1984). In this system, GT and the **V** field on the base of which GT is calculated, are given in a relatively simple form.

From tensor analysis of the organ growth, we know that if the pattern of PDG trajectories is steady over time , then segments joining recognizable points of the wall mesh-work and chosen so as to be tangent to the PDG trajectories, maintain their orthogonality during growth. Hence, the hypothesis has been proposed (Hejnowicz 1984, 1989) that periclines and anticlines represent PDG trajectories and planes of cell divisions are typically normal to PDGs.

Study of cellular patterns is an important issue in developmental biology. In theoretical approaches, algorithms describing growth and tissue geometry usually emphasize rules acting at the cellular level (reviewed by Prusinkiewicz 1994). Meanwhile, in a light of what was mentioned above, a rigorous algorithm should invoke also rules related to the pattern of PDGs existing at the organ level. Here, I show a simulation model including such the algorithm. It will be used to animate growth and cell division in the shoot apical dome.

2 Model for Growth and Cell Division in the Shoot Apex (2-dimensional)

Consider the growing shoot apical dome. Assume that GT field with PDGs, as well as cell pattern coming, for example, from a section of the dome, are known. Postulates are the following:

1. Cells in the section are assumed to be polygons. The cell pattern is represented by a polygon network in which edges meet at vertices.

2. In order to simulate growth, the GT acts on the network. Growth of individual cells is determined by position of the cells in GT field.

3. Cells displace along PDG trajectories (Figure 2). To obtain position of a given cell at $(t_0 + \Delta t)$, differential equations for the displacement velocity $\mathbf{V}$ are solved, and using coordinates of the cell at t_0, new locations of the cell vertices are found.

4. Cell growth and cell multiplication are separated, superimposed processes. During growth, when the network is extended and deformed, new cells by divisions are formed. Then, a parent cell is replaced by two daughter cells.

5. Cells divide only when a critical value of their area, A_{cr} , is exceeded. At a given time-step of the simulation, a cell divides only once.

6. To determine the location of the partition wall within the space of the dividing cell, the cell center (as the center of area) and the distance, d_{min}, from the center to the nearest edge, are calculated. The wall is represented by a line segment which passes through the point M (Figure 2) found randomly, inside the circular region drawn from the center with the radius $r = d_{min}/k$ where $k = const.$

7. Two variants of the orientation of the division wall are considered (Figure 3):

   — in variant I, the wall is attached to the nearest edge at the right angle, and extended to the edge located on the opposite side of the point M.

   — in variant II, the wall relates to PDGs found for the point M, among two PDGs, the one for which the wall can be shorter, is chosen.

   In both, to avoid a four-way junction which are rather unusual (Barlow 1991, Lloyd 1991), if the end of the wall happens at the existing vertex, the point of the attachment of the wall is displaced by a short segment $\varepsilon = const$, the same for the whole apex.

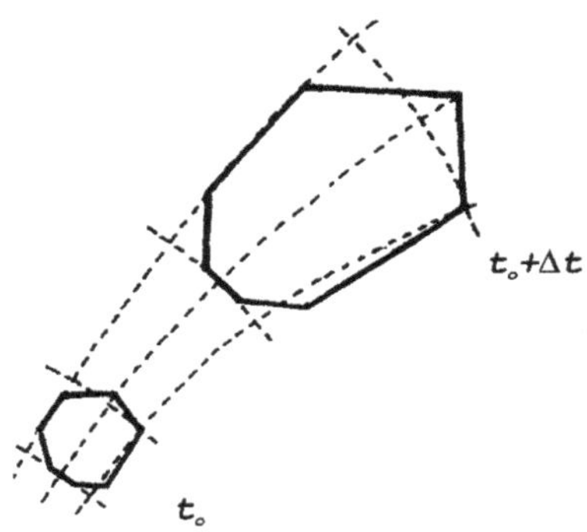

Figure 2. An example of the cell growth — broken lines represent trajectories of PDGs.

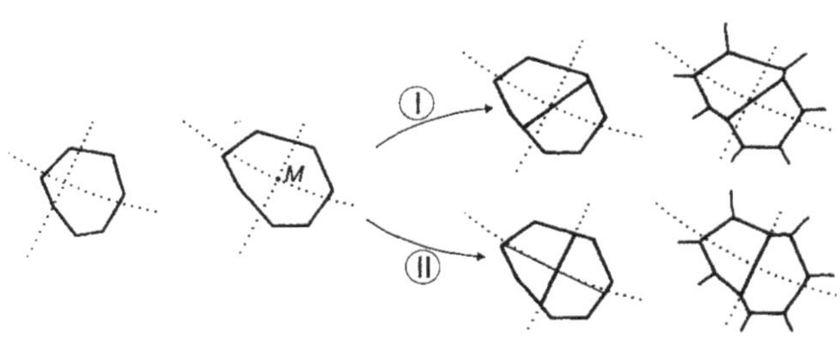

Figure 3. Schemes for cell division in the variants I (top) and II (bottom). The division wall passes throughout the point M found randomly near to the cell center. In the variant I, the wall is perpendicular to the nearest wall, in the variant II, the wall is oriented along one of two PDGs, the direction for which it can be shorter, is chosen. Broken lines represent PDG trajectories.

8. To have more realistic angles at each new vertex, just at the moment of the creation of the new wall, the network is locally modified by diminishing of the partition wall by $1/m$ of its former length, where $m = const$ (the same for the whole apex). This rule relates to the difference in strength between the wall of mother cell and the cell plate. The difference is suggested to yield perpendicular junctions, which, when new plate gains strength, allows the three facets to re-arrange, forming angles which tend to be equal (Korn 1980).

For studies of the shoot apex, two cellular patterns are interesting, one related to the axial section, the other related to the dome surface. If the dome is paraboloidal and grows steadily without a rotation around axis, it is convenient to assume a paraboloidal coordinates as the natural coordinate system (Figure 4). Then, in the axial plane, PDG trajectories are represented by the u and v lines, the periclinal PDG is tangent to $v = const$, the anticlinal PDG is tangent to $u = const$, whereas vs corresponds to the profile of the dome (Figure 4a).

In the case of the dome surface, a projection of the surface on the plane tangent at the dome tip is needed (Figure 4b). Then, in the top view, the periclinal and anticlinal PDGs are represented by the radial direction whereas the latitudinal PDG remains the same.

The GT appropriate for the paraboloidal shoot apex was defined by Hejnowicz *et al.* (1984). It has been found for the displacement velocity $\mathbf{V}$ having only one component, namely $V_u = c\sqrt{\frac{u^2+v^2}{u^2+v_s^2}}$ where v_s represents the profile of the dome and $c = const$. The distribution of RERG$_l$ obtained for this case in the previous paper (Nakielski and Rumpf 1991), is shown in Figure 5. It can be seen that the values

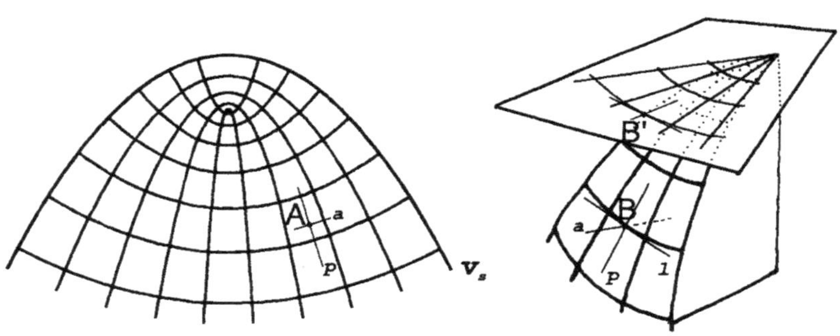

a b

Figure 4. The paraboloidal coordinate system (u, v, φ) describing PDG trajectories in the shoot apex: (a) the axial plane, (b) a part of the surface of the apex. In (a), the u and v lines are confocal parabolas, the outermost line, v_s , corresponds to the surface of the dome; at the point A, two PDGs, p and a, are shown. In (b), a projection of the surface on the plane tangent at the dome tip is shown — the point B is represented by the point B', three PDGs, p, a, l at the point B are represented at B' by two orthogonal lines, one oriented radially, the other oriented tangentially.

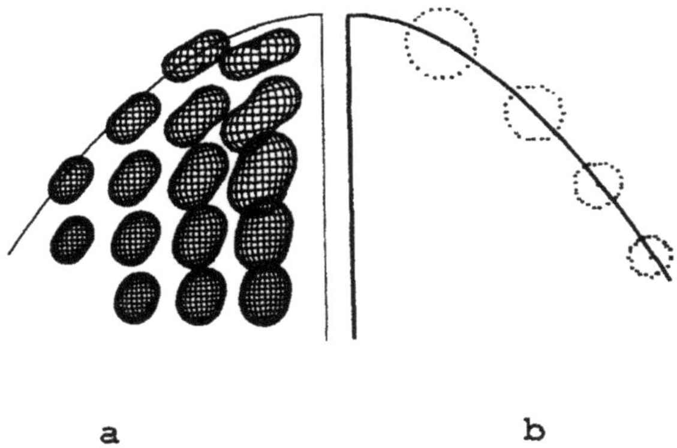

Figure 5. The distribution of $RERG_l$ in the shoot apex for the case when growth at the dome surface is isotropic: (a) growth in the axial plane (b) growth in the planes tangent to the dome surface. In (a), 3d indicatrices of $RERG_l$ obtained by the finite element method for points lying in the nodes of the rectangular net are shown. In (b), 2d plots for points located at the surface with increasing distance from the tip are shown, the orientation of each plot is changed by the rotation by 90° around the periclinal direction.

of $RERG_l$ in every direction tangent to the dome surface are equal, i.e., the surface growth is isotropic (Figure 5b), whereas in the axial plane, a distinct anisotropy of growth rates occurs (Figure 5); at each point, $RERG_l$ in the periclinal direction is greater than $RERG_l$ in the anticlinal direction. In both, the dome surface and the axial plane, $RERG_l$ decreases, at the surface, with increasing distance from the tip, and in the axial plane, with increasing distance from the axis of the dome. With reference to the model, it means that cell pattern specified for the dome surface will be simulated under the assumption of the isotropic growth, whereas the other cell pattern, specified for the axial section, will be simulated assuming the anisotropic growth.

To see how the algorithm works, let us simulate growth of single cells at first. Figure 6 and Figure 7 show two types of sequences, one for cells in the axial section (Figure 6), the other for cells at the top view (Figure 7), both developed in the variants I and II.

In Figure 6, the incoming cells, located in different part of the dome, are identical. In the period from $t = 0$ to $t = 1$ they have grown only a little. However, at $t = 1$, because A_{cr} is exceeded, all cells divide. In the variant I, each new wall, as postulated, is attached to the nearest wall at the right angle. Since the cells, initially identical, have similar shape, the partition walls show similar orientation irrespective of the position of the cell within the dome. Though PDGs are generally ignored, one cell, at the right side of the dome (Figure 6, $t = 1$) divides exactly periclinally because its nearest wall is oriented anticlinally. In the variant II, in contrast, every cell division relates to PDGs. Since PDGs change throughout the apex, every new wall shows different orientation in space, at a given position, it is either periclinal or anticlinal.

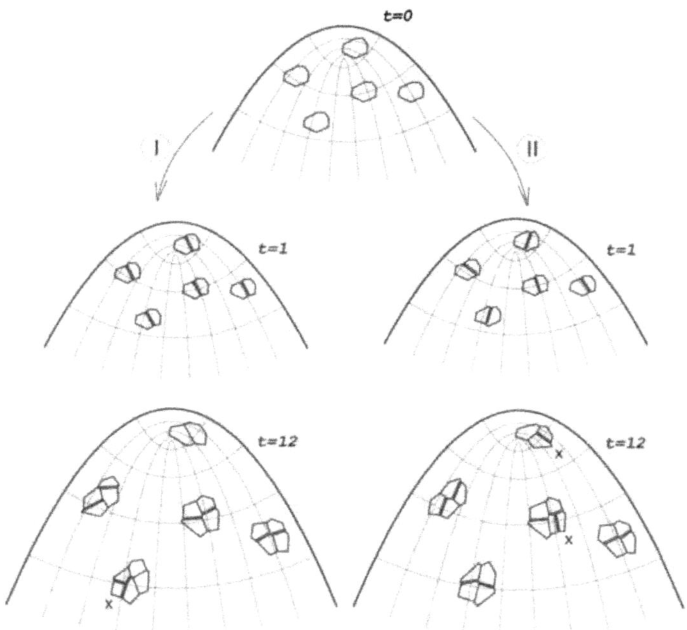

Figure 6. Sequences of growth and cell divisions in the axial plane of the apex for the variants I (left) and II (right). The group of initially identical cells located in different regions in the apex is developed. PDG trajectories (green) are shown, the newly formed walls (red) are marked. At $t = 1$, the orientation of cell divisions is similar, in the variant I, and different, depending on PDGs, in the variant II. At $t = 12$, all new walls show different orientation. However, in the variant II, they consequently relate to PDGs. For "x" see text.

In the period from $t = 1$ to $t = 12$, a significant growth has occurred. As a result, at $t = 12$, all cells are extended and deformed. In the used GT field, the periclinal growth dominates. Thus, new divisions are mostly transversal to the direction of the maximal growth and perpendicular to the wall formed previously. In the variant I, new walls are closer to PDGs then, at $t = 1$, only two walls in the cell tetrad located at the left side of the dome are clearly oblique with respect to PDGs. In the variant II, as before, all new walls are located along PDGs always perpendicular to the wall formed previously. In both variants, in the cell packets marked by "x" the next division is formed. Notice that if the division occurs in the variant I, there is no such division in the corresponding cell in the variant II and vice versa.

The simulation for the dome surface is shown in Figure 7. The incoming pattern consists of three cells located near the tip of the dome. The cell including the tip maintains its position, two other cells displace during growth in the radial direction. The simulation is limited to the distal portion of the apex containing 5-6 cell rows, for this region the deformation of cell shape related to the projection, is of little importance. According to the assumed GT field, growth at the dome surface is isotropic. However, the isotropic growth is fully expressed only on the

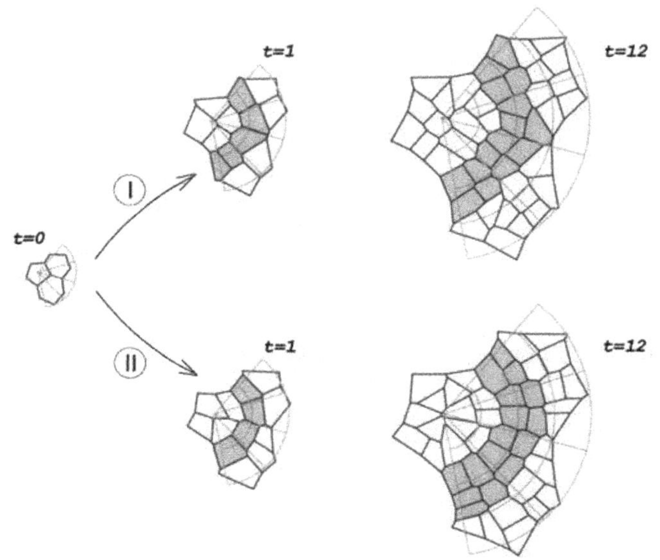

Figure 7. Sequences of growth and cell divisions at the top view of the apex for the variants I (top) and II (bottom). The group of three cells initially located near the tip is developed. PDG trajectories (green) are shown. Notice the marked row of cells (blue) formed by daughter cells obtained as a result of divisions in two cells at the right side. In the variant I, it is irregular, whereas, in the variant II, it maintains symmetry related to PDGs.

outer surface of the protodermal cell layer, whereas on the inner surface of this layer, growth departs from isotropy. It means that at least on the inner surface of the protodern, PDGs can be seen. For these reasons, in modeling work both variants of the orientation of cell divisions are considered.

It is seen that in both variants cell patterns develop initially in a similar way (Figure 7). However because of difference in the orientation of cell divisions they progressively come to differ from each other in details. In the variant I, the cell pattern is irregular and some new walls are oblique with respect to PDGs. As a result, the marked cell row is rather sharp-edged. In contrast, in the variant II, a radial symmetry of the cell pattern is manifested. Every new wall is formed either radially or latitudinally. Thus, cell rows and cell files intersect each other in an orthogonal manner.

3 The Example of Application of the Model

It has been found that the GT field assumed here occurs in the shoot apex of spruce seedlings in the vegetative phase of the development (Nakielski 1987). At least in the apical portion of the dome surface, there is indeed the isotropic growth — cell pattern at the top view is shown in Figure 8a. Notice that cells grow more or less uniformly without domination of any direction. Thus, they are arranged into three more or less equal sectors originated by initial cells near the tip of the dome.

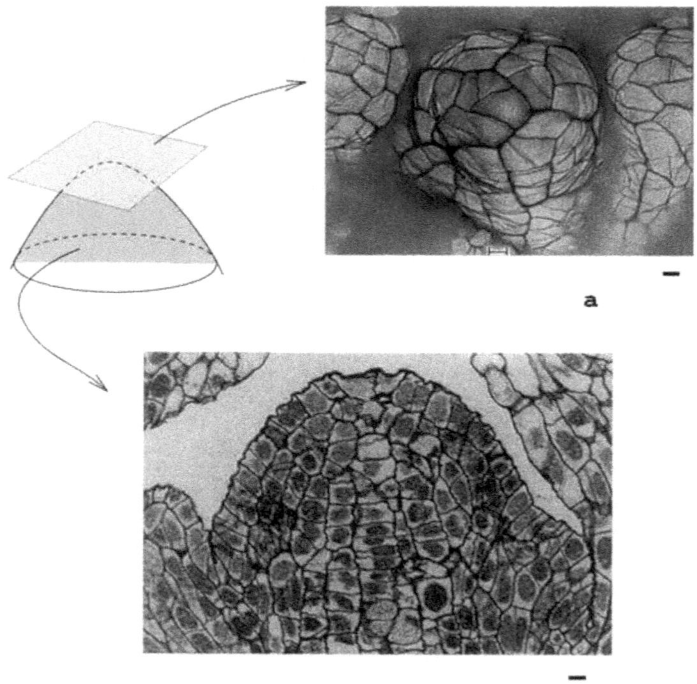

a

b

Figure 8. Cellular patterns in the shoot apical dome of a spruce seedling: (a) the top view, (b) the axial section. The apices are at the age: 12-plastochrons in (a) and about 100-plastochrons in (b). Bar = 10 μm.

Figure 8b shows the axial section of the apex. Here, the cell pattern can be considered typical for the anisotropic growth. In the inner part of the apex corresponding to the corpus, though the orientation of cell divisions is unrestricted, all distinguishable cell packets are extended in the periclinal directions. The outermost layer of cells corresponding to the tunica shows a predominance of anticlinal divisions.

The cell patterns shown in Figure 8a,b were assumed as incoming data for the computer simulation presented on Figure 9 and following figures. The modeled structure was obtained by digitalization of the cell pattern. Individual cells were considered as convex polygons without limitations related to the number of edges. They formed a polygonal mesh in which two neighboring cells had a common edge which joined two vertices. At a vertex only three edges could meet (three-way junction).

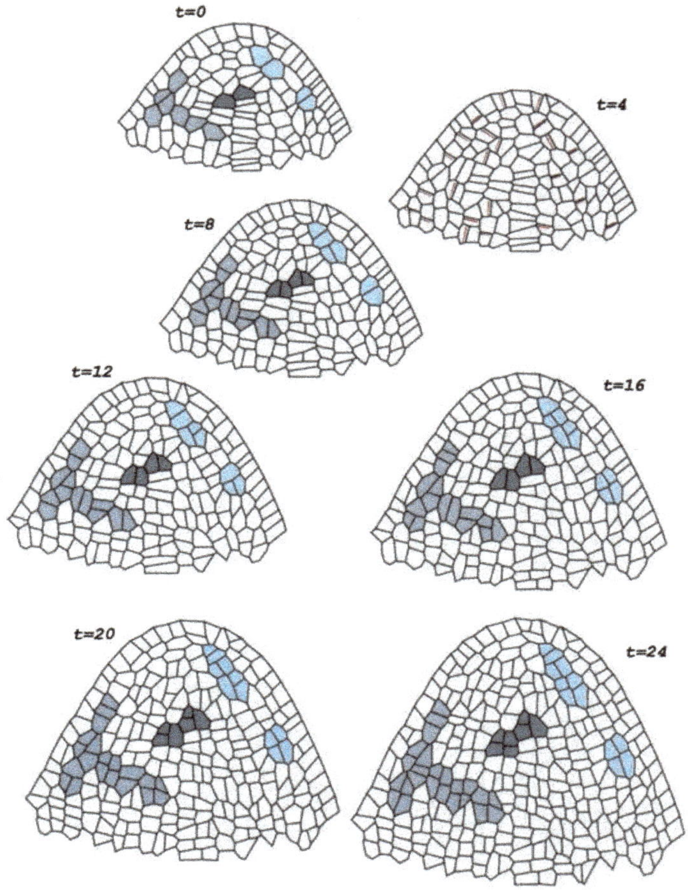

Figure 9. Sequence of growth and cell division in the axial plane obtained for the cell pattern coming from Figure 8b. The variant II was assumed.. At $t = 0$, chosen cells are marked by different shade of blue, the cells are developed in successive times except $t = 4$ in which all division walls formed in the period from $t = 0$ to $t = 4$, are denoted (red).

In order to run a simulation, the specification of parameters is needed. The value of the critical cell area, A_{cr} was estimated in relation to the average cell area, A_{av} of the incoming cell patterns, it was assumed that $A_{cr} = 150\%\ A_{av}$. The other parameters were the following: $k = 4$, $m = 14$, $\varepsilon = 0.05\sqrt{A_{av}}$, $\Delta t = 0.05$ (all determined heuristically). In the paraboloidal system, the profile of the dome was described by $v_s = 3.0$ and for V_u , the constant $c = 1.0$ was taken.

Consider how the cell-wall patterns develop over a number of time-steps. The sequence of growth and cell divisions for the apex in the axial plane is shown in Figure 9. The network increases maintaining the profile of the dome. Individual cells divide both, periclinally (formative divisions) generating new files and anticlinally (proliferative divisions) multiplying the number of cells within a given file.

Cell packets are regular and extended, as in Figure 8b, mainly along the periclinal direction. Similarly to real tissues, characteristic zig-zag lines described by periclines and anticlines are formed. The cell pattern remains more or less steady as the individual cells move down through it. To a first approximation, it can be considered as self-perpetuating.

The sequences of growth and cell divisions at the surface of the apex obtained for the variants I and II are shown in Figure 10 and Figure 11. The cell patterns are initially the same in both cases. However, the two patterns progressively come to differ from each other because of the different rule for orientation of cell divisions. As expected, the pattern generated by the variant I is less regular than the pattern generated by the variant II. In the case of the variant II (Figure 11), the radial symmetry of the cell pattern becomes more and more visible and, at $t = 12$, a characteristic arrangement of cells into sectors is seen. Notice the orientation of cell

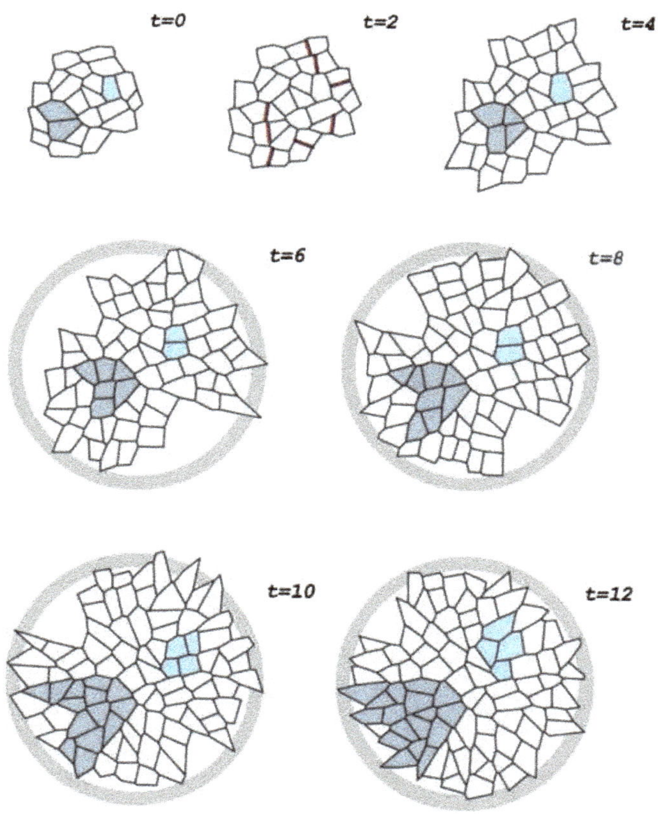

Figure 10. Sequence of growth and cell division obtained at the top view for the cell pattern coming from Figure 8a. The variant I was assumed. At $t = 0$, chosen cells are marked by different shades of blue, the cells are developed in successive times except $t = 2$ in which all division walls formed in the period from $t = 0$ to $t = 2$, are denoted (red). From $t = 6$, the cells which were displaced beyond the indicated circular limit, were omitted from the print-out.

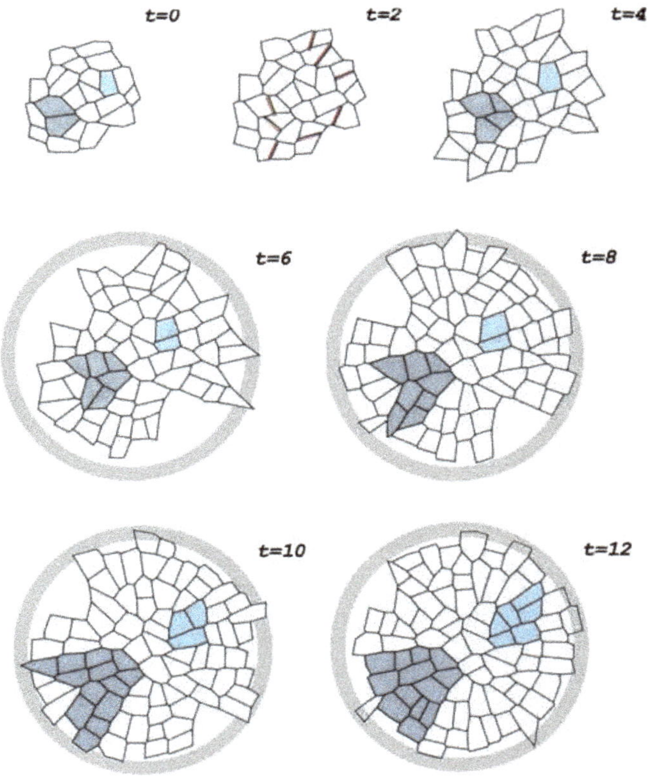

Figure 11. Sequence as in Figure 10, for the variant II. The explanations are the same.

divisions at $t = 2$. In the variant I, new formed walls are relatively well adjusted to the existing cell pattern, whereas in the variant II, some new walls are clearly oblique with respect to the existing walls.

4 Discussion

This paper presents a new approach for the simulation of growth and cell division in multicellular organs. In contrast to discrete models based on L-systems (Lindenmayer 1968, reviewed by Prusinkiewicz 1994), the formalism of continuous functions is used. Growth of the organ is determined by the field of the GT and its PDGs. By application of the field to the cell-wall network, a developmental sequence of the growing organ is obtained. In the algorithm, the orientations of division walls are specified relative to the pattern of PDGs. This is the first case in which the pattern of PDGs existing at the organ level has been taken into account.

In modeling, two variants of the algorithm were used. The orientations of division walls in the variant I depend on geometry of the cell, while those in variant II depend on PDGs. The algorithm used in the variant I can be less reasonable because the division plate of this variant favors the edge located at the minimal distance from the point M, whereas, in fact, both points of the attachment of the division plate are important in the same degree. It should be noticed, however, that the situation in which the newly formed wall is rigorously perpendicular to the nearest edge at its one end, and oblique, quite often, to the other wall at the opposite end, occurs only at the start of the procedure. Later, in the same time-step of the simulation, the wall is diminished according to the rule 8 and then, the angles at both points of the attachment of the wall are corrected. This makes the variant I acceptable.

At the outer surface of the apex where, as assumed, growth is isotropic, PDGs are indistinguishable. What was the purpose in using variant II in relation to the dome surface? Firstly, it was mentioned that the isotropic growth is limited only to the outer surfaces of the proptodermal cell layer, whereas on the inner surface of this layer PDGs can be visible. Secondary, empirical data relate mainly to the distal portion of the dome surface, i.e., the part in which only all the cells near the tip including the initial cells and their most immediate derivatives are seen. In this portion of the spruce apex, there is indeed isotropic growth. It is possible, however, that beyond this part, in sub-distal portion of the surface, growth changes to anisotropic. Then PDGs could be visible also on the outer cell layer. The results of the simulation have shown that cell pattern obtained for the distal portion of the apex, limited to 2-3 cell rows, is more realistic in the variant I then in the variant II. However, with increasing distance from the tip, for next 2-3 cell rows, a domination of the variant II generating a typical sectorial symmetry, is seen. The question needs further anatomical studies.

Many problems related to the application of the model are open for further research. One refers to the parameters of the algorithm. Here, as in the previous paper devoted to the modeling of growth in root apices and limited to the variant I (Nakielski and Barlow 1995), all parameters were used globally. For this reason some differences between tissues of the apex, for example those related to cell size, have been disregarded. It is possible, however , to assume different values of the parameters and consider particular tissues. The other problems relate to the shape of the dome and type of GT field. The paraboloidal dome is well described in terms of the GT, it would be interesting, however, to apply the model, for example, to rather elliptically-shaped dome, at the surface of which, except small region near the tip, the anisotropy of growth occurs, as in the shoot apex of barley (Hejnowicz *et al.* 1986). Also the first attempts have shown that the model is useful in modeling of transformations in cell pattern related to the change of positions of initial cells.

A good correspondence of the model with cell-systems (de Boer and de Does, 1990, de Boer *et al.* 1992) can be seen. For example, both stress periclinal and anticlinal orientation of cell divisions as important for the formation of cellular patterns. Both intend to have the angles distributed between edges at the point of the attachment of the division wall, as evenly as possible — my rule [8] coincides with the center-of-gravity algorithm. Furthermore, the pattern of PDGs used here, can be considered as a kind of a "directional morphogenetic field" postulated in cell-systems to specified orientations of division walls. This suggests a link between these two approaches, even if their bases are so different.

Glossary

ANISOTROPIC: Having different properties along different directions. Opposite of isotropic.

ANTICLINAL: Perpendicular to the surface of the organ.

APICAL MERISTEM: A zone of continual growth located at the tips of shoots and roots.

CELL DIVISION: The separating of a cell into two daughter cells.

CELL PACKET: A group of cells enclosed within a wall that at one time was the wall of a single cell. From this cell all the cells of the packet were originated by successive divisions.

INITIAL CELL: A meristematic cell that gives rise to two cells, one remaining in the meristem, the other added to the plant body.

ISOTROPIC: Referring to material whose properties as measured in any direction, are equivalent.

LATERAL MERISTEM: Area of actively growing cells in the stem and roots of plants that cause these areas to thicken.

MERISTEM: Plant tissue containing cells that can continually divide throughout the plant's life.

MERISTEMATIC: Having the capacity to produce new cells by cell division.

ORGAN: A distinct and visibly differentiated multicellular part of a plant.

PERICLINAL: Tangential to the surface of the organ.

PLASTOCHRON: A period of time between the initiation of two successive leaves.

PRIMARY TISSUES: Tissues derived from the embryo and the apical meristems.

PROTODERM: A tissue of the apical meristem that gives rise to the outer layer of cells (epidermis) covering the plant body.

ROOT APEX: The apical meristematic end of a root.

SHOOT APEX: The apical meristematic end of shoot.

SPRUCE SEEDLING: A seedling of *Picea* species.

TUNICA: External layer (or layers) of cells in the shoot apex that gives rise to the protoderm.

References

1. P. W. Barlow (1991). From cell wall network to algorithms. The simulation and cytology of cell divisions in plants. *Protoplasma*, **162**:69–85.
2. M. J. M. De Boer and M. De Does (1990). Relationship between cell division pattern and global shape of young fern gametophytes. I. A model study, *Bot. Gaz.*, **151(4)**:423–434.
3. M. J. M. De Boer, F. D. Fracchia and P. Prusinkiewicz (1992). A model for cellular development in morphogenetic fields. In: G. Rosengerg and A. Salomaa, eds. Lindenmayer systems: *Impact on theoretical computer science, computer graphics, and developmental biology*, Springer-Verlag, Berlin, 351–370.
4. R. O. Erickson (1986). Symplastic growth and symplasmic transport. *Plant Physiol.* **82**:1153.
5. Z. Hejnowicz (1984). Trajectories of principal growth directions. Natural coordinate system in plant growth. *Acta Soc. Bot. Pol.* **53**:29–42.
6. Z. Hejnowicz (1989). Differential growth resulting in the specification of different types of cellular architecture in root meristems. *Environ. Exp. Bot.* **29**:85–93.
7. Z. Hejnowicz and J. A. Romberger (1984). Growth tensor of plant organs. *J. Theor. Bot.* **110**:93–114.
8. Z. Hejnowicz, J. Nakielski and K. Hejnowicz (1984). Modeling of spatial variation of growth within apical domes by means of the growth tensor. II. Growth specified on the dome surface. *Acta Soc. Bot. Pol.* **53(3)**:301–316.
9. Z. Hejnowicz, J. Nakielski, W. Wloch and M. Beltowski (1988). Growth and development of shoot apex in Barley. III. Study of growth rate variation by means of the growth tensor. *Acta Soc. Bot. Pol.*, **57(1)**:31–50.
10. R. W. Korn (1980). The changing shape of plant cells: transformations during cell proliferation. *Ann. Bot.* **46**:649–666.
11. A. Lindenmayer (1968). Mathematical models for cellular interaction in development, Parts I and II. *J. Theor. Biol.*, **18**:280–315.
12. C. W. Lloyd (1991). How does the cytoskeleton read the lows of geometry in aligning the division plane of plant cell? *Development Supplement* **1**:55–65.
13. J. Nakielski (1987). Variations of growth in shoot apical domes of spruce seedlings: A study using the growth tensor. *Acta Soc. Bot. Pol.*, **56(4)**:625–643.
14. J. Nakielski and P. W. Barlow (1995). Principal direction of growth and the generation of cell patterns in wild-type and gib-1 mutant roots of tomato (*Lycopersicon esculentum* Mill.) grown in vitro. *Planta* **196**:30–39.
15. J. Nakielski and M. Rumpf (1991). Growth in apical meristems of plants. Visualization tools and growth tensor method. SFB 256, no. 11, Bonn University.
16. J. H. Priestley (1930). Studies in the physiology of cambial activity. II. The concept of sliding growth. *New Phytol.* **29**:96–140.
17. P. Prusinkiewicz (1994). Visual models of morphogenesis. *Artificial Life* **1**:61–74.
18. O. W. Richards and A. J. Kavanagh (1943). The analysis of the relative growth gradients and changing form of growing organisms: Illustrated by Tobacco leaf. *Amer. Nat.* **77**:385–399.

19. J. A. Romberger, Z. Hejnowicz and J. F. Hill. *Plant Structure: Function and Development.* Springer-Verlag, Berlin, 1993.
20. J. Sachs (1887). Lecture XXVII. Relations between growth and cell-division in the embryonic tissues. In: *Lectures in plant physiology* (translated by H. Marshall Ward), 431–459, Clarendon Press, Oxford.
21. W. K. Silk and R. O. Erickson (1979). Kinematics of plant growth. *J. Theor. Biol.* **76**:481–501.
22. E. W. Sinnott, *Plant morphogenesis*, McGraw-Hill, New York, 1960.

DNA and Genetic Control

DNA NANOTECHNOLOGY: FROM TOPOLOGICAL CONTROL TO STRUCTURAL CONTROL

NADRIAN C. SEEMAN
Department of Chemistry, New York University
New York, NY 10003, USA
E-mail: ned.seeman@nyu.edu
http://seemanlab4.chem.nyu.edu

DNA nanotechnology entails the exploitation of DNA molecules for non-biological purposes. The properties that make DNA such a successful molecule in living systems are also very powerful when applied in chemical systems; DNA nanotechnology has resulted in the assembly of polyhedra, topological figures, nanomechanical devices and crystals. Unusual motifs of DNA, particularly branched molecules and structures derived from them, can be combined with sticky-ended association to direct the assembly of multiply-connected species. Branched junctions have been combined in this fashion to create molecules whose helix axes have the connectivities of a cube and of a truncated octahedron. A solid support-based methodology has been developed to facilitate the construction of DNA objects. Owing to the plectonemic nature of the DNA double helix, DNA polyhedra are complex catenanes of cyclic single strands. A half-turn of double helical DNA has a structure that permits it to behave as a unit tangle in topological construction. In addition to catenanes, it is straightforward to prepare specific DNA knots; Borromean rings have also been constructed from DNA. The goal of directing the assembly of periodic matter from DNA components requires a rigid species, for which the DNA double crossover molecule has been employed. Predictable patterns result when the double crossover motif is incorporated into triangular molecules which are ligated in one dimension. Two-dimensional crystals of double crossover molecules have been assembled and shown to possess specific programmed topographic patterns. It is possible to alter the patterns enzymatically and chemically, so that features may be added or subtracted readily. The rigidity of the double crossover molecule has been combined with the B-Z transition, thereby creating a DNA nanomechanical device. Future applications of DNA nanotechnology entail the assembly of three-dimensional crystalline arrays, the use of DNA to direct the assembly of other molecules, and the applications of unusual DNA motifs to DNA-based computing.

1 DNA Nanotechnology

DNA is well-known to all of us as the genetic material of living systems. Knowledge of its double helical structure (Watson & Crick, 1953) has had an enormous impact on our understanding of life's processes. The double helix may well come to symbolize our civilization to future civilizations, in much the same way that the pyramids of Egypt or North America, the cathedrals of Medieval Europe, the Great Wall of China, the Parthenon and the Taj Mahal are structural emblems of the societies that built them. However, these cultural icons represent the apotheoses of their respective civilizations, whereas the DNA double helix represents the first step in the development of molecular biology. In the 45 years since the announcement of its structure, the fundamentals of molecular genetics and gene expression have been worked out, the biotechnology industry has developed, diagnostics and criminology have been revolutionized, and the sequences of several bacterial genomes, as well as a portion of the human genome have been determined.

In this chapter, I will describe the exploitation of DNA molecules for non-biological purposes. The very properties that make DNA such a successful molecule in living systems are also very powerful when applied in chemical systems. Although I will introduce this article by a short description of linear double helical DNA, I will move quickly to other motifs of DNA, particularly those containing branches. The biotechnology industry is predicated on the ability to tack together DNA molecules in an arbitrary fashion (Cohen *et al.*, 1973). The products are linear or cyclic genetic elements that direct the syntheses of commercially valuable gene products. When this same strategy is applied to branched molecules, one can make polyhedra, topological figures, nanomechanical devices and crystals. This type of molecular construction is called DNA nanotechnology (Seeman, 1998).

Of what value is it to make such products, from DNA or from other materials? Owing to the author's background as a crystallographer, the key inspiration for this program has been the desire to produce crystalline material in a predictable fashion. Currently, the difficulty and unpredictability of the crystallization step is the greatest impediment to determining the 3-D structures of biological macromolecules and macromolecular complexes. DNA offers a system whereby one can direct the interactions of crystal components to form periodic matter. If these components contain ordered guests, they too will be crystalline, and their structures can be determined (Seeman, 1982). Of course, the ability to order guests in a lattice is not restricted to ordering biological macromolecules. The same techniques could be used in a bottom-up approach to the assembly of nano-circuitry and extremely dense memory devices (Robinson & Seeman, 1987).

2 Double Helical DNA

The DNA-based progress in the second half of the twentieth century is a consequence not only of the knowledge of the DNA structure and its implications, but of the simplicity of the way in which genetic information is stored and expressed. The Schrödinger equation has been known since the 1920's, but accurate wave functions are available only for relatively small molecules, because of the difficulties of calculating them. By contrast, the complementary pairing relationship between bases on the two strands of double helical DNA is understood easily by schoolchildren: Adenine (A) pairs with thymine (T); guanine (G) pairs with cytosine (C). Genetic information is encrypted in the linear sequence of these bases on each strand.

These hydrogen bonded interactions between the planar bases are illustrated in Figure 1. It can be seen that there are two hydrogen bonds between A and T, but that there are three hydrogen bonds between G and C. If one compares the upper two hydrogen bonding positions that the two base pairs have in common, the patterns are reversed: The A-T pair contains a clockwise cycle of hydrogen bonds and the G-C pair contains a counterclockwise cycle. This difference provides one bit of information to specify the pairing of the four bases. However, this is not enough information to establish the specific pairs; it is well known that in solution every base can pair with every other base, including itself (Voet & Rich, 1970). The other bit of information that is needed derives from their relative

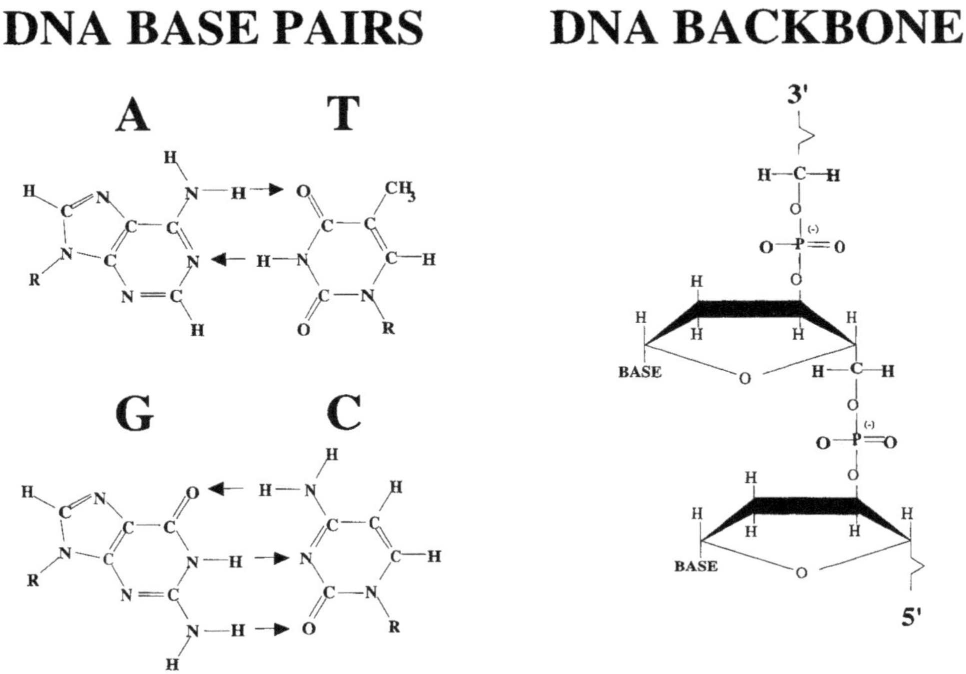

Figure 1. The DNA Base Pairs and Backbone Structure. The four bases are shown on the left in their Watson-Crick pairing arrangement, and labeled by their conventional abbreviations: Adenine (A), Thymine (T), Guanine (G) and Cytosine (C). Covalent bonds are indicated by single or double lines, and hydrogen bonds are indicated by arrows from the hydrogen atom to the acceptor. The strength of hydrogen bonds is 2-5 kcal/mol, and the strength of covalent bonds is 80-150 kcal/mol. Consequently, hydrogen bonded structures can be disrupted by much milder conditions. The usual chemical abbreviations, the first letters of atom names, are used for hydrogen (H), carbon (C), nitrogen (N) and oxygen (O); R represents the point at which the base is attached to the deoxyribose sugar of the DNA backbone. Note that the A-T pair is the same size as the G-C pair, but their hydrogen bonding schemes are different. The DNA backbone structure is shown on the right. The backbone consists of alternating phosphate groups and deoxyribose sugars. The phosphate groups (a phosphorus atom, P, surrounded by four oxygen atoms) are shown to carry a negative charge, because they are ionized at neutral pH. The sugars contain five carbon atoms, as well as hydrogen and oxygen atoms. Four of the carbon atoms and one oxygen atom form a five-membered ring. The two ends of the backbone are indicated as the $5'$ end and the $3'$ end. There is a chemical polarity to the backbone, because chemically inequivalent atoms are encountered in a different order as one traverses the chain from $5'$ to $3'$ or from $3'$ to $5'$; by convention, the $5' \rightarrow 3'$ direction is used in naming sequences.

sizes: A and G are roughly twice the size of T and C. How is size imposed on two pairing bases? Therein lies the importance of the double helix. DNA is a polymer, and the backbones of two molecules form the famous double helix. The bases must fit inside the double helix, and there is just room for one large base (called a purine) and one small base (called a pyrimidine), without distorting the backbone. Thus, the double helical backbone is an inherent part of the way that one strand of DNA recognizes its complementary partner through Watson-Crick base pairing.

I have skipped over a key difference between A-T and G-C pairing, the third hydrogen bond in the G-C pair. Of course, this hydrogen bond provides free energy that makes the G-C pair more stable than the A-T pair. The strands of DNA separate at a temperature which is a direct function of their G-C content (Marmur & Doty, 1962). Of the many forms of base pairing available with the four conventional bases, only the Watson-Crick G-C base pair has three hydrogen bonds. This may be one of the factors contributing to the observation that in conventional conditions heterogeneous sequences of DNA associate preferentially to form Watson-Crick base pairs in double helical structures.

It is worth making a few comments about the DNA backbone. It consists of deoxyribose sugar residues that alternate with phosphate groups. The base-sugar-phosphate unit is called a 'nucleotide'. The phosphate groups are ionized at physiological pH (about neutral), so DNA is a long polyanion that must be neutralized by metallic or organic cations. The sugar molecules are all in the D-configuration, so that DNA is chiral, like many biological molecules. The base to base distance along the helix axis is about 3.4 Å, and there are about 10.5 residues per turn. Thus, the helical repeat is about 36 Å; the width of the helix is about 20 Å. It is important to recognize that the backbone strands have a structural polarity. The chemical details of this polarity are not relevant here, but its implications are. The two ends of the molecule are called the $5'$ and $3'$ ends; as a consequence of the polarity, the sequence $5'-\mathrm{GAC}-3'$ is not the same as $5'-\mathrm{CAG}-3'$. By convention, sequences are written in the $5' \to 3'$ direction. This fact is clearly important for the ability of DNA to store information. The two strands of the double helix are oriented with their polarities opposite. The structural consequence of this arrangement is that perpendicular to the helix axis are dyad axes that relate the atoms in the backbones, but not the bases. Whereas the base pair planes are also perpendicular to the helix axis, a dyad axis goes roughly through the base pair plane; as with any periodic system, a second, independent dyad axis exists half-way between the base planes, and a half-residue rotation (ca. 17°) from the axes lying within the base planes.

3 Genetic Recombination and the Holliday Junction

Is DNA always in a double helical state? Certainly not. It would not be possible to replicate or to transcribe information from DNA, if it were impossible to unpair the bases and open up the double helix; unpairing is necessary to provide access to the information on the Watson-Crick pairing surfaces of the bases. In addition, there are other structures of DNA that exist as ephemeral intermediates in the processes of replication and recombination. For our purposes, the most interesting of these is the Holliday (1964) intermediate of genetic recombination. Genetic recombination is a phenomenon that occurs universally in living organisms, from viruses to humans. A great deal of DNA metabolism is devoted to ensuring the stability necessary for its role as genetic material, but recombination is a process that affords the flexibility necessary for the adaptation of species to a changing environment. The basic molecular feature of recombination is the interaction of two pieces of DNA to yield new genetic material that may incorporate information

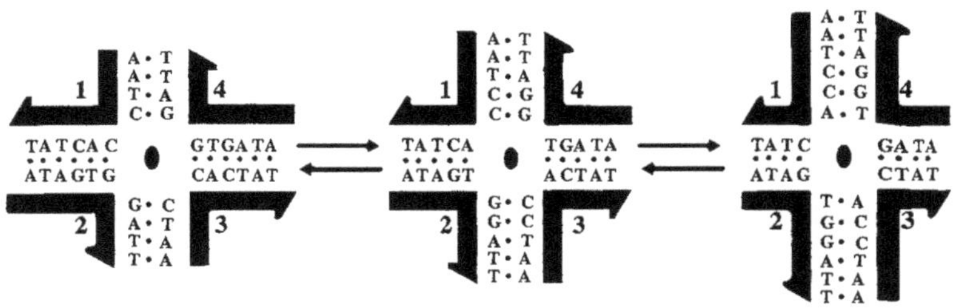

Figure 2. Branch Migration of the Holliday Junction. This drawing illustrates two steps of branch migration by a 4-arm DNA branched junction having twofold sequence symmetry. The half-arrowheads indicate the 3′ ends of the strands, and the small dots indicate Watson-Crick base pairing between the strands. The twofold sequence symmetry is indicated by the elliptical object at the center of each junction. Starting from the left, in the first step, the C's and G's that flank the branch point on the horizontal arms pair across the junction and become members of the vertical arms. In the second step, the A's and T's that flank the branchpoint in the central molecule undergo the same pairing isomerization and also become members of the vertical arms. This process can iterate in either direction, until the junction resolves to linear duplex molecules.

from both interacting molecules. The resulting DNA may show such alterations as insertions, deletions, changes of sequence, rearrangements, or exchanges of flanking markers.

The Holliday junction is formed from two double helices that exchange strands. Thus, there are four strands involved, and they produce four double helical arms that flank a branch point. The exchange of strands takes place between two double helices that have the same sequence, so that the sequences that flank the branch point are related by twofold symmetry, as shown in the central molecule of Figure 2. The presence of the symmetry destabilizes the position of the branch point, because the molecule can isomerize to move the branch point in either direction. This phenomenon is called branch migration (Hsieh & Panyutin, 1995); it is illustrated in Figure 2, where a single step of branch migration is illustrated in both directions. If branch migration continues stochastically, the four strands will eventually resolve to two double helices, because the formation of the branch point costs free energy (Lu *et al.*, 1992).

4 Immobile Branched Junctions

It is often inconvenient to study the properties of molecules that are capable of branch migration, because the population will be heterogeneous. In synthetic systems, it is possible to produce immobile junctions, with fixed branch points, if one eliminates the twofold symmetry (Seeman, 1982). An example of such a molecule is illustrated in Figure 3; this is the well-characterized molecule, J1 (Seeman & Kallenbach, 1983), containing 8 base pairs in each arm. In addition to eliminating the twofold symmetry, it is useful to take further steps to ensure the spontaneous association of the component strands to produce the target complex. After all, this is an excited state, and a solution containing 10^{13} copies of each strand may happen upon other arrangements than the one desired. The antithesis of control over

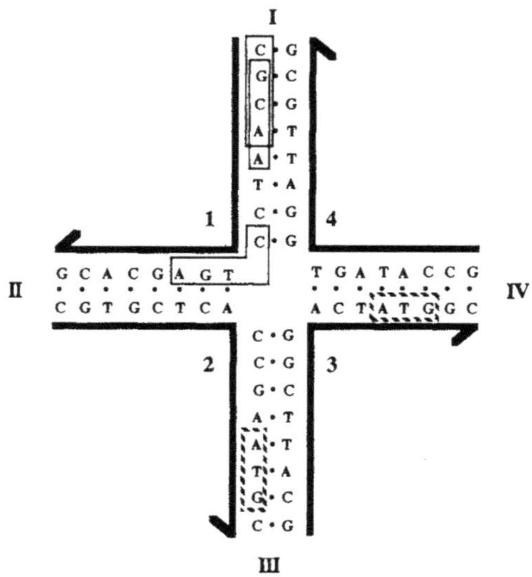

Figure 3. A Stable DNA Branched Junction. The junction shown is composed of four strands of DNA, labeled with Arabic numerals; arms are labeled with Roman numerals. The 3′ end of each strand is indicated by the half-arrows. Each strand is paired with two other strands to form a double helical arm. There is no homologous twofold sequence symmetry flanking the central branch point, thereby stabilizing its position. Tetrameric elements are boxed; trimeric elements are in dashed boxes.

chemical systems is symmetry, in its broadest sense, i.e., multiple outcomes that are equivalent, or nearly so, from the standpoint of the free energy of the system. Thus, one must differentiate the target structure from any other possible alternative structures. The alternatives must be made sufficiently unfavorable, relative to the target, that only the target will form in appreciable quantities.

Ideally, one would like each of the 32 base pairs of the J1 molecule to be unique, so that no other options would be available to the complex. This is not possible, but fortunately the pairing of DNA is cooperative, so one can consider short segments as distinct units (Seeman, 1982, 1990). For example, each 16-mer strand of the J1 junction shown in Figure 3 can be treated as a set of 13 overlapping tetramers. The first two of these on strand 1 are boxed, CGCA and GCAA. One can insist that each of these be unique. In addition, one can eliminate the complements of all tetramers flanking a bend, so they do not have the option of being double helical. For example, TCAG is not present in the molecule, because it is the complement of the boxed sequence CTGA at the bend of strand 1.

By applying these rules to the selection of sequences consistent with the target architecture, one eliminates competition from sequences of length four or greater. Thus, only trimers (e.g., the ATG sequences in dashed boxes) are the largest units that can compete with the octamer arms of J1. If one used trimers instead of tetramers as the basic design elements, competition would only be from duplicated dimers; however, there would be 56 trimers in the J1 molecule (see Figure 3), and their total diversity is only 64. It is much easier to design this junction from tetramers (Seeman & Kallenbach, 1983), and it is known to form well (Kallenbach *et al.*, 1983). When designing larger complexes, pentamers or hexamers are used, rather than tetramers. Often, one must include sequences that violate the principles described above; for example, analysis of products may require the insertion of a symmetric sequence that is the recognition site for a restriction enzyme. When this is the case, it is usually easy to design sequences around those sites.

It is important to recognize that the work described here depends strongly on the ability to produce DNA molecules of appropriate lengths that contain the sequences that optimize the design properties discussed above. Fortunately, convenient automated DNA synthetic chemistry has been available for more than a decade (Caruthers, 1985). It is simple to produce DNA molecules containing over 100 nucleotides in less than a day. Very long molecules often contain synthesis flaws that have little impact on nanotechnological assembly. However, these flaws can impede the use of synthetic strands in biological systems such as *in vitro* transcription; this problem can be remedied by using the Polymerase Chain Reaction (PCR) (Saiki *et al.*, 1986) to generate molecules without faults (Wang *et al.*, 1998).

5 Characterization of Unusual DNA Motifs and Structures

The DNA branched junction is the first of several unusual DNA motifs that I will mention in this chapter. It is important to discuss the evidence that they form as we say they do. J1 and other immobile junctions have been characterized extensively (Lilley & Clegg, 1993; Seeman & Kallenbach, 1994), because they are close analogs of the Holliday recombination intermediate. Conventional physical techniques that are commonly used by chemists, e.g., NMR, have only recently yielded interesting data about these molecules (Miick *et al.*, 1997); none of the species here has been crystallized by itself yet, although a low-resolution crystal structure of a Holliday junction complexed with Cre protein has been reported recently (Hargreaves *et al.*, 1998). In general, we characterize a new motif first by means of non-denaturing gel electrophoresis, in order to establish whether the complex forms cleanly, and with an appropriate molecular weight. On non-denaturing gels, hydrogen bonded species migrate as a function of the molecular weight of the complex, as well as factors associated with the secondary structure, such as the surface area. Bands appearing below the apparent monomer indicate instability (Wang *et al.*, 1991), and bands appearing above it indicate molecular strain that can be relieved by making a multimer (Fu & Seeman, 1993). Discrete bands indicate a closed complex (e.g., for a 4-arm branched junction, molecule 1 hydrogen bonds to molecule 2, 2 to 3, 3 to 4 and 4 to the same copy of molecule 1), but a smear of bands can indicate an open complex (1:2:3:4:1:2:3...), with no discrete closure.

For overall physical analysis, a Ferguson plot is run, to measure the friction constant of the complex, and to establish the dependence of mobility on gel concentration. Thermal denaturation profiles provide an estimate of the stability of the species and of its component parts. The primary structural tool for analyzing these complexes is Fe(II)EDTA^{2-} autofootprinting (Churchill *et al.*, 1988): The pattern of chemical cleavage caused by hydroxyl radicals generated by this reagent is monitored. The extent of cleavage is a function of hydroxyl radical accessibility to the residue (Balasubramanian *et al.*, 1998). The key to the analysis is that a radioactively labeled strand is complexed with its Watson-Crick linear duplex complement, and, in a separate experiment, with the other strands in the complex. Strands whose cleavage pattern is identical in both experiments are assumed to be in a double helical conformation; we have found empirically that nucleotides that flank crossovers between helices typically are protected, relative to the duplex pattern.

DNA nanotechnology produces species far more complex than simple branched junctions. Complex systems discussed below, including a DNA cube and a DNA truncated octahedron, knots and Borromean rings, are analyzed on denaturing gels. This is a system in which molecules of the same topology separate from each other as a function of their molecular weight; secondary structures are destroyed in denaturing conditions. Knotted molecules are compared with marker cyclic species in order to characterize their linking: The more tightly knotted a given strand is, the more rapidly it will migrate through a gel. All of these species contain restriction sites, susceptible to restriction nucleases. In knots, lack of digestion by restriction endonucleases is one of the demonstrations that a domain is in the Z-conformation, and therefore contains positive nodes (see below). Catenated (topologically linked) molecules are broken down at specific sites by restriction enzymes to produce target catenanes or circles, for which markers can be prepared independently. Denaturing gel electrophoresis is very sensitive to topological differences: Two links differing only in their linking number will separate readily (Fu *et al.*, 1994); likewise, a linear catenane and a cyclic catenane of the same molecular weight migrate differently (Chen & Seeman, 1991b).

6 DNA Ligation and Cohesion

Complementary interactions are the key to the role of DNA in biological systems. They are also central to biotechnology and to DNA nanotechnology. We have already established that two complementary strands will cohere to form a double helix. We have also shown that multiple strands can be induced to flank a unique point of branching. In addition to the blunt-ended species that we have already discussed, it is also possible to produce double helical DNA molecules that have single-stranded overhangs called 'sticky ends', shown at the top of Figure 4. In the same way that two complementary strands will form a double helical molecule, two DNA molecules with complementary sticky ends will cohere to form an intermolecular hydrogen bonded complex; this association is shown in the middle of Figure 4. If desired, it is also possible to convert this association to a covalent linkage, by the addition of DNA ligase and appropriate cofactors; as shown at the bottom of Figure 4, the original two double helices plus sticky ends are converted to a single double helix. The ability to splice together DNA molecules with complementary sticky ends is arguably the fundamental reaction of biotechnology.

From a chemical perspective, the cohesion of sticky ends is a remarkable phenomenon. It is rare to be able to predict the specificity and geometry of intermolecular interactions in chemical systems. Not only is it possible to program sticky ends to cohere, but the diversity available is huge. For a sticky end 6 nucleotides long,

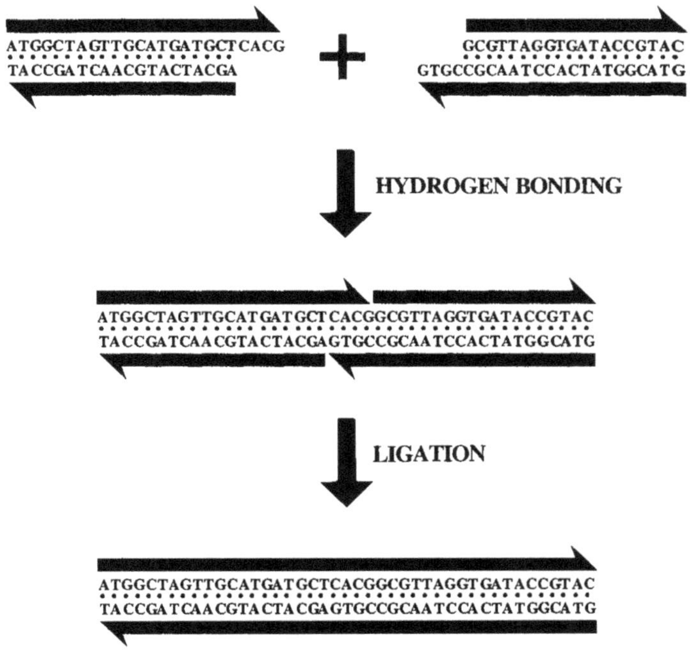

Figure 4. Sticky-Ended Cohesion and Ligation. Two linear double helical molecules of DNA are shown at the top of the drawing. The antiparallel backbones are indicated by the black lines terminating in half-arrows. The half-arrows indicate the $5' \to 3'$ directions of the backbones. The right end of the left molecule and the left end of the right molecule have single-stranded extensions ('sticky ends') that are complementary to each other. The middle portion shows that, under the proper conditions, these bind to each other specifically by hydrogen bonding. The bottom of the drawing shows that they can be ligated to covalency by the proper enzymes and cofactors.

there are 4032 (4^6 possible sticky ends less 4^3 self-complementary sequences) sequences and 2016 unique pairs that can be designed to associate. In addition to predicting preferential molecular associations, the local geometry of the interaction is known: Sticky ends associate to form B-DNA (Qiu *et al.*, 1997), the familiar DNA structure in which the base pairs are roughly perpendicular to the helix axis. The existence of predictable DNA intermolecular recognition has facilitated the development of DNA-based computing (e.g., Adleman, 1994; Lipton, 1995; Winfree, 1996), which relies on the programmable associations of molecules.

Is the fidelity perfect? The pairings and/or mispairings of DNA molecules are chemical reactions, just like all other chemical reactions. Reactions are characterized by free energy differences between products and reactants; in turn, these are related to the equilibrium distribution of target products and unwanted byproducts. Furthermore, the outcome in a given competitive reaction is a function of conditions, including DNA concentrations, ionic strength, temperature and pH. The ultimate success of DNA-based computing may well turn on maximizing the relative free energy differences between DNA pairs and mispairs under appropriate conditions in solutions containing great molecular diversity. The thermodynamic data needed to evaluate the relative free energies of mispairs are not yet in hand. DNA nanotechnology is more robust than DNA-based computing in this regard, because relatively few asymmetric sticky ends are needed to generate target products.

7 Branched DNA and Sticky Ends

The outcome of the molecular recognition depicted in Figure 4 is a linear DNA molecule, in the sense that it is unbranched. The axes of the two helices shown at the top can be regarded as line segments; joining line segments produces longer lines, and, given flexibility, can lead to circles or perhaps knots. From a simple geometrical perspective, the missing elements in this system are vertices. Fortunately, the branched DNA molecules discussed above can be used to provide this system with vertices. If one adds sticky ends to a branched molecule, a macromolecular valence cluster is created that contains specifically addressable ends (Seeman, 1982). This idea is illustrated in Figure 5, where a 4-arm branched junction with two complementary pairs of sticky ends self-assembles to produce a quadrilateral. The vertices of this quadrilateral are the branch points of its four component junctions, and the edges consist of double helical DNA. The open valences on the outside of the quadrilateral suggest that it might be possible to tile the plane with this arrangement.

The product shown in Figure 5 appears to be as simple to generate as an object on the macroscopic scale. However, an implicit assumption in that assessment is that the individual subunits are rigid components. How rigidly does branched DNA behave in solution? When the type of assembly illustrated in Figure 5 is attempted, one discovers that individual junctions are not completely rigid. We can expect that the double helical edges of each object are torsionally and flexurally stiff, because the persistence length of DNA is usually taken to be approximately 500 Å(Hagerman, 1988): This value leads to a 66.5 ÅRMS end-to-end distance for two turns of DNA, nominally 68 Ålong. Thus, on a small scale, say using 2-turn edges, the rigidity of linear DNA appears to be adequate for nanoconstruction. However, the 'valence angles' between the double helices that flank junctions appear to be flexible (Ma *et*

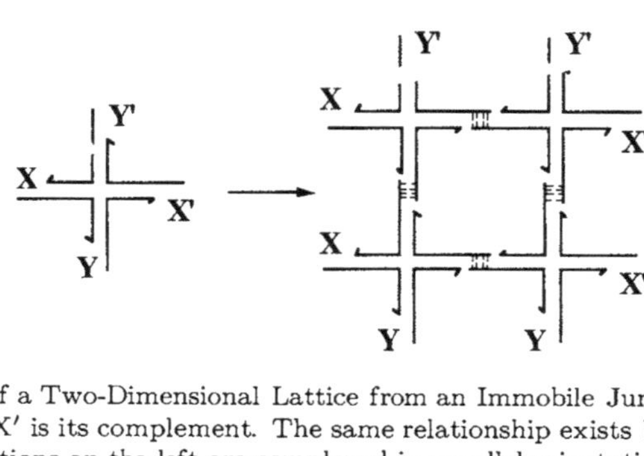

Figure 5. Formation of a Two-Dimensional Lattice from an Immobile Junction with Sticky Ends. X is a sticky end and X′ is its complement. The same relationship exists between Y and Y′. Four of the monomeric junctions on the left are complexed in parallel orientation to yield the structure on the right. X and Y are different from each other, as indicated by the pairing in the complex. As shown in Figure 4, ligation by DNA ligase can close the gaps left in the complex. The complex has maintained open valences, so that it could be extended by the addition of more monomers.

al., 1986; Eis & Millar, 1993; Liu *et al.*, 1994). If a 3-arm junction is oligomerized to yield oligolaterals with two turns between branch points, it readily forms a series of cyclic products: trimers, tetramers, pentamers and so on; similar results have been seen with 4-arm junctions (Petrillo *et al.*, 1988). Thus, the flexibility of DNA objects corresponds to multiple minima in the potential energy surface; these multiple minima impede the use of oligomerization to generate a given cyclic polygon, or to generate the lattice suggested by Figure 5.

8 Connectivity

In addition to the 4-arm branched junction, it is also possible to make 3-arm (Ma *et al.*, 1986), 5-arm and 6-arm (Wang *et al.*, 1991) immobile branched junctions. There may be no limit to the number of double helical arms that can flank a branched junction (Seeman & Kallenbach, 1988). This is an important point, because the connectivity of an object or array (Wells, 1977; Williams, 1979) produced by DNA nanotechnology is limited by the number of arms that flank the junctions that comprise it. For example, it is possible to design a cube from 3-arm branched junctions, because it is 3-connected; this means that all of its vertices are connected through edges to three other vertices. However, an octahedron must be made from 4-arm junctions, and an icosahedron requires 5-arm junctions. A simple cubic lattice is 6-connected, and would require 6-arm junctions to produce it.

9 Topological Features of DNA Constructs

It is often convenient to represent the double helix as a ladder-like molecule, as we have done in Figures 2–5. However, this is really a simplification that both obscures key features of DNA constructs and masks useful properties. The strands of the double helix wrap around each other, and it is important to realize the

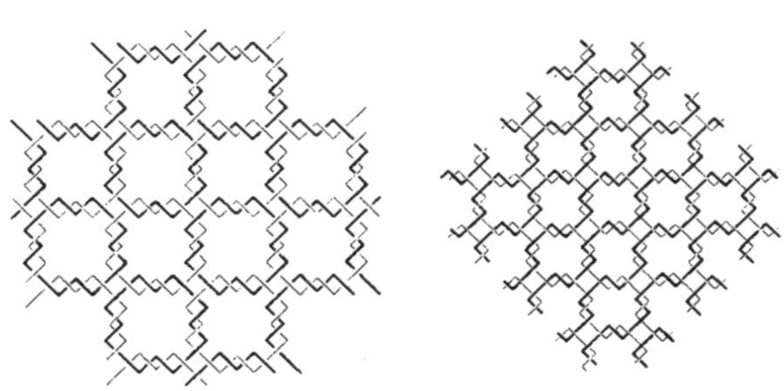

Figure 6. Topological Consequences of Ligating DNA Molecules Containing Even and Odd Numbers of DNA Half-Turns in Each Edge. These diagrams represent the same ligation shown in Figure 5. However, they indicate the plectonemic winding of the DNA, and its consequences. The DNA is drawn as a series of right-angled turns. In the left panel, each edge of the square contains two turns of double helix. Paired strands are drawn with alternate thick and thin lines. These are seen to form a structure like molecular chain mail. In the right panel, each edge of the square contains 1.5 turns of DNA. Therefore, the strands do not form cycles, but extend infinitely in a warp and weft meshwork.

implications of that fact for DNA nanoconstruction. Figure 6 illustrates in two variations extended versions of the products assembled in Figure 5. On the left, the branch points have been separated by two turns of DNA. The product is a series of linked cyclic molecules, much like molecular chain mail. On the right, the same motif has been drawn, but the vertices are now separated by 1.5 turns of DNA. This product consists of linear strands of DNA that are interwoven like the threads of a textile produced on a loom.

The difference between the integral and half-integral separations of branch points arises repeatedly in the design of DNA objects. In a sense, the double helical half-turn is the quantum of single-stranded DNA topology. It corresponds to a crossing or node in a DNA catenane or knot. Sometimes this feature is called a 'unit tangle' (Sumners, 1990). We will see below that the equivalence between a double helical half-turn and a unit tangle can be very useful in the construction of topological targets (Seeman, 1992).

10 Construction of DNA Polyhedra

One can circumvent the flexibility problem if one restricts one's targets to molecules with relatively few sticky ends (plus their complements) that are all unique. The first interesting molecule we made of this sort was a molecule whose helix axes are connected like those of a cube or a rhombohedron (Chen & Seeman, 1991a). Each

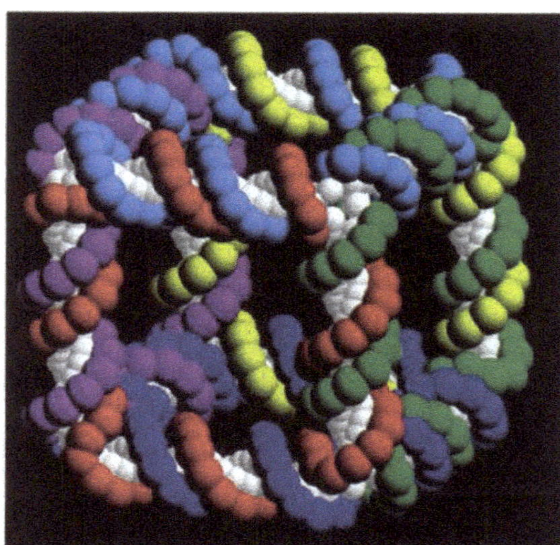

Figure 7. A DNA Cube. Each nucleotide is represented by two balls, a colored ball for the backbone, and a white ball for the base. Each edge of this molecule contains two turns of double helical DNA, and consequently each face corresponds to a cyclic single strand. For example, the front face corresponds to the red strand, which is linked twice to the four strands (green, cyan, magenta and blue) flanking it; it is linked only indirectly to the yellow strand on the back face. The double helical edges are about 2 nm wide, and about 7 nm long; their thickness leads to a linear dimension of roughly 10-12 nm for the object if it adopts a cubic shape.

vertex of the cube is the branch point of a 3-arm junction, and each of its 12 edges consists of two turns of double helical DNA. A schematic drawing of the cube is shown in Figure 7. Each face of the molecule consists of a cyclic strand of DNA, as a consequence of the exact number of turns per edge. Each cyclic strand is linked twice (because of the two turns) to its four nearest neighbors. For example, the

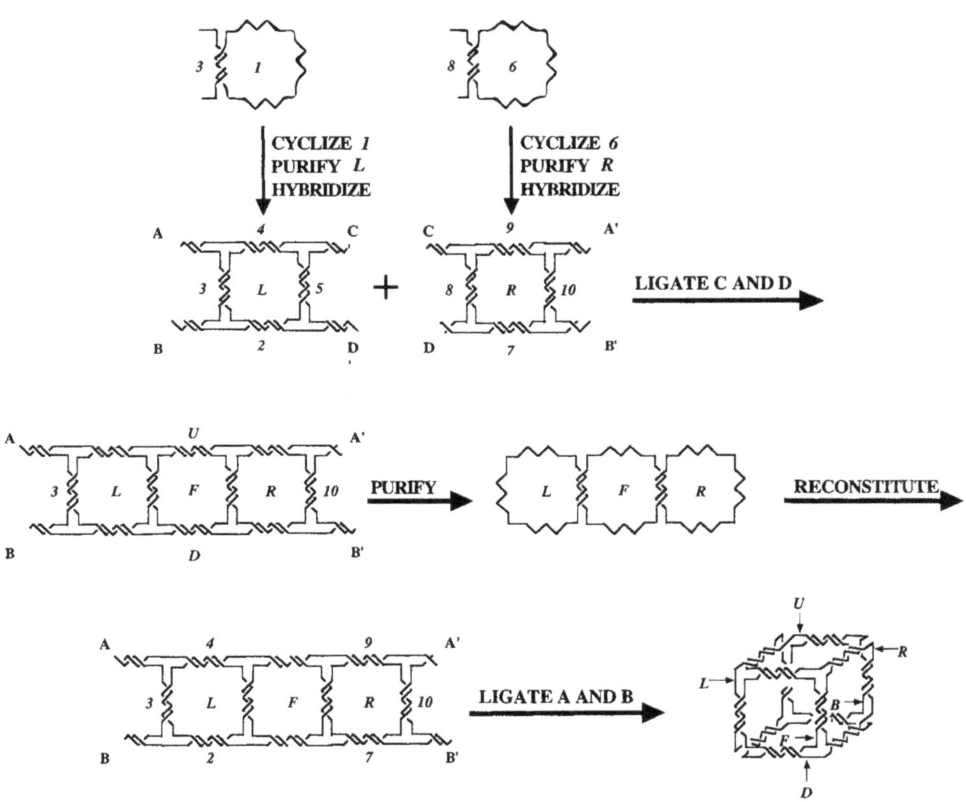

Figure 8. The Synthetic Scheme Used to Synthesize the Cube-Like Molecule We have used the convention that as a new strand is formed in this figure by ligation, its identification changes from one or more numbers to a letter corresponding to its position in the final object. For example, strand 1 is synthesized as a linear molecule, but is referred to as L, once it is cyclized. The six final strands in the object are referred to as L (left), R (right), U (up), D (down), F (front) and B (back). This scheme depicts 5 steps (separating 6 stages) in the synthesis of the 3-D, 3-connected DNA object, after individual strands have been purified. The first step has two parts, the cyclization of the full-length strands 1 and 6 to form L and R, respectively. These two cycles are then hybridized with strands 2-5 and 7-10, respectively, to form the constituent squares shown in the second step. These squares are ligated together at the complementary sticky ends C' and C (5 base 5′ overlap) and D and D' (4 base 3′ overlap). This reaction forms strands U, F and D from their constituents. U and D are discarded in a purification on a denaturing gel that isolates the L-F-R triple complex. The L-F-R-2-3-4-7-9-10 complex is then reconstituted and the final ligations are performed to close sites A and A' (4 base 5′ overlap) and B and B' (5 base 3′ overlap), as well as sealing the nicks in U (4-9) and D (2-7). This series of reactions forms the entire 3-D, 3-connected object.

red strand on the front face is linked to the green strand on the right, the cyan strand on the top, the magenta strand on the left, and the dark blue strand on the bottom. It is linked only indirectly to the yellow strand on the back face.

Although the molecule is drawn cube-shaped, we do not know the geometry of the molecule. Our synthetic control is on the topological level, and so must be our proof of synthesis. The synthesis of the molecule is illustrated in Figure 8. Two squares are ligated in solution to form a ladder-like intermediate in small yield. In order to purify the intermediate, it must be denatured to yield the linear triple catenane corresponding to the ultimate left, front and right faces of the cube. The ladder-like intermediate is then reconstituted, and the ligation completed. Each edge of the cube contains a recognition site for a different restriction endonuclease that will cleave the molecule only at that position. This feature presents many different opportunities to challenge the correctness of the structure. The most robust of our proofs is the cleavage of the two front vertical edges, the only ones present in the linear triple catenane. When cleaved, the product is the linear triple catenane corresponding to the top-back-bottom of the molecule.

The synthesis shown in Figure 8 is somewhat cumbersome, so we developed a solid-support methodology to facilitate the synthesis of individual objects (Zhang & Seeman, 1992). Solid-support synthesis always enables the convenient removal of reagents and catalysts from the growing product. The features special to this procedure include: (1) Each ligation cycle creates a robust intermediate object that is

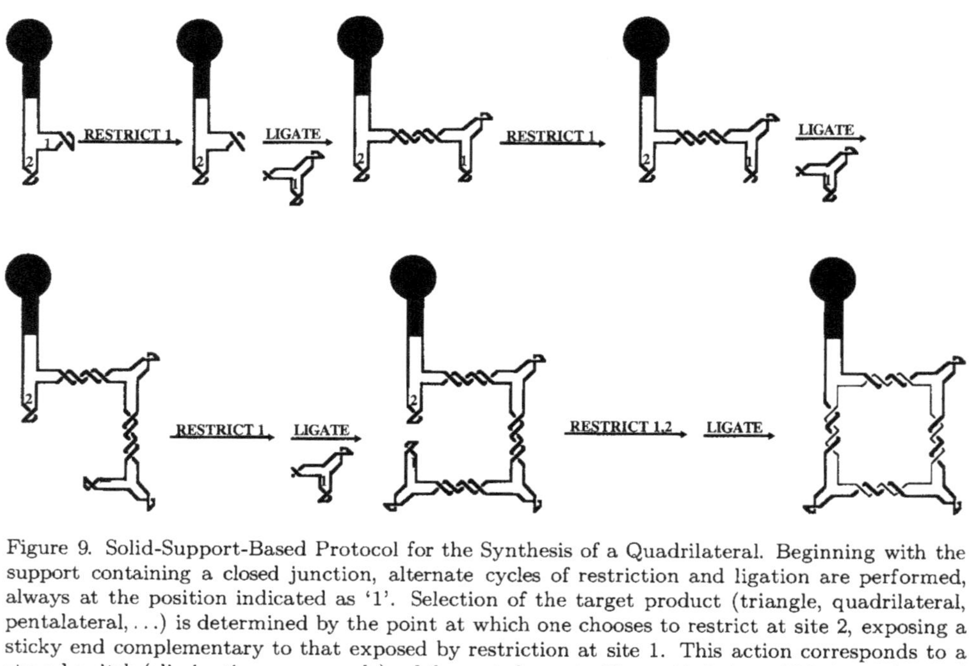

Figure 9. Solid-Support-Based Protocol for the Synthesis of a Quadrilateral. Beginning with the support containing a closed junction, alternate cycles of restriction and ligation are performed, always at the position indicated as '1'. Selection of the target product (triangle, quadrilateral, pentalateral, ...) is determined by the point at which one chooses to restrict at site 2, exposing a sticky end complementary to that exposed by restriction at site 1. This action corresponds to a strand switch (eliminating a zero node), of the sort shown in Figure 11, below. This is emphasized by the lines of different thickness with which the 'square' catenane is drawn.

covalently closed and topologically bonded together. (2) Consequently, successfully ligated intermediates are immune to exonucleases, but failures can be destroyed while still on the support. (3) Control derives from the restriction of hairpin loops forming each side of a new edge. The method permits one to build a single edge of an object at a time. Thus, it is possible to perform intermolecular ligations under conditions different from those used for intramolecular ligations. Intermolecular reactions are done best with asymmetric sticky ends, to generate specificity. Sequences are chosen in such a way that restriction sites are destroyed when the edge forms. The method is outlined in Figure 9, which shows the construction of a quadrilateral. An initial 3-arm branched junction is attached to the support, then one arm is restricted, and a second junction is added. The process is repeated twice more, and then closure is achieved by restricting two opposing hairpins and ligating the exposed sticky ends to each other.

One of the major advantages of using the solid support is that the growing objects are isolated from each other. This permits the use of symmetric sticky ends in intramolecular ligations, without intermolecular ligation occurring. More generally, the solid support methodology permits one to plan a construction as though there were only a single object to consider. Many of the differences between a single molecule and a solution containing a picomole of molecules disappear when the molecules are isolated from each other on a solid support.

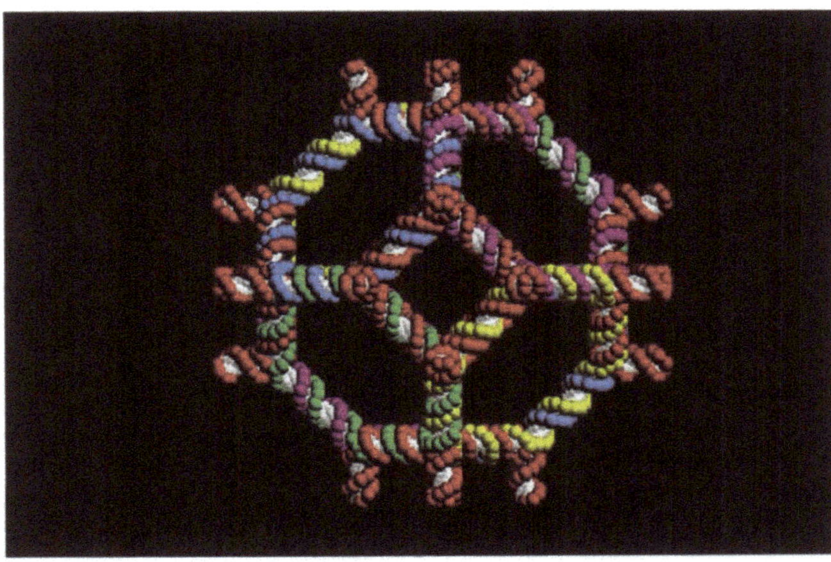

Figure 10. Front View of a DNA Truncated Octahedron. The same conventions apply as in Figure 7. This molecule has two turns per strand, just as the cube does. Hence, it is a 14-catenane. The truncated octahedron is a 3-connected object, but each vertex is flanked by a 4-arm branched junction, so there are 24 extra arms pointing outwardly. The strands corresponding to the square faces are drawn in red, and all of the extra arms are derived from these strands. The strands corresponding to the hexagons are drawn in magenta, cyan, green and yellow. In this conformation, the square-to-square distance across the object is about 23 nm.

We have used the solid support methodology to build a DNA truncated octahedron (Zhang & Seeman, 1994). This Archimedean solid has 14 faces, 36 edges, and 24 vertices. Each edge contains two turns of DNA, so the molecule is a complex 14-catenane. The truncated octahedron is a 3-connected polyhedron, but each of its vertices is flanked by a 4-arm branched junction. Although we were not able to make enough of the molecule to use as a starting material, the extra arms at each vertex were included with the aim of using them to connect truncated octahedra together in a zeolite-like lattice (Williams, 1979). All of the extra arms are parts of the strands corresponding to the square faces, as shown in Figure 10. Rather than individual junctions, whole squares and a tetrasquare complex were added in the intermolecular part of this construction (Zhang & Seeman, 1994). It is straightforward to demonstrate that all of the square strands are in place, by monitoring increases in denaturing gel mobility as each of them is restricted on their extra arms. The hexagons have been demonstrated by showing that the squareless octacatenane could be broken down to a series of tetracatenanes of hexagons that flank the squares; markers for these tetracatenanes could be constructed independently, so the comparison is a valid test of synthesis.

11 DNA Catenanes and Knots

Both of the polyhedra discussed above have been characterized because they are catenanes of single strands of DNA. Kool (1998) has reported a deliberately formed single strand catenated with a double stranded circle. Double stranded catenanes of DNA are known to be the components of kinetoplast DNA (Chen *et al.*, 1995a, 1995b). It is well-known that an intimate relationship exists between catenanes and knots (White *et al.*, 1987). This relationship is shown in Figure 11(a). The object on the far left is a 5_1 knot, which has been drawn with an arbitrary strand polarity, indicated by the arrowheads. Every topological object is characterized by the nature and placement of its unit tangles (Sumners, 1990), or nodes, which appear as crossings in projection. One can imagine that each strand involved in a node consists of two parts, one entering the node and one leaving it. The transition from the knot on the left of Figure 11(a) to the catenane at its center involves destroying a node by switching the entering and leaving parts of the two strands, while preserving the polarity. Sometimes this operation is called creating a zero node (White *et al.*, 1987). The second transition in Figure 11(a) involves destroying a second node by replacing it with another zero node, thereby converting the catenane to a trefoil (3_1) knot.

Figure 11(b) puts this operation in a DNA context. On the left is a (negative) node formed by two strands of right-handed B-DNA. The helix axis is indicated by the horizontal double-headed dashed arrow, and the dyad normal to it is drawn as a vertical dashed line terminating in two ellipses. On the right is the product of the reaction. Note that we have replaced the absolute operation of Figure 11(a) with a chemical equilibrium symbol, because the reaction can go both ways. Figure 11(c) shows the same reaction in a DNA molecule that has been rotated 90° about the horizontal helix axis, so that we are looking down the dyad axis. It is clear that the product of the forward reaction is a pair of opposed hairpins. However, the reverse

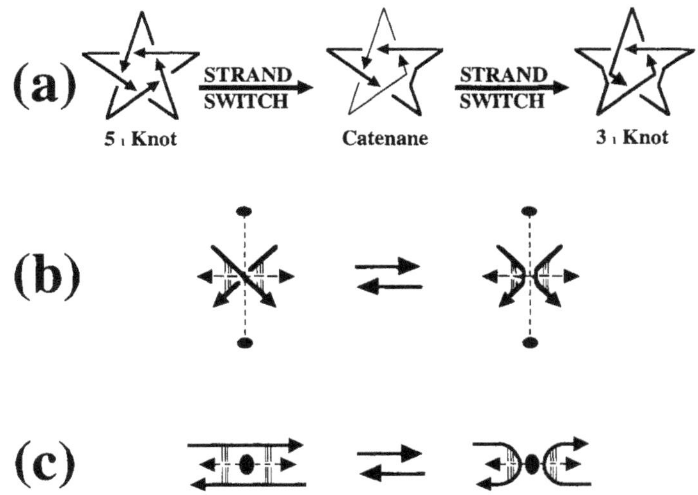

Figure 11. Zero Node Operations. (a) Strand Switch to Remove Nodes in Knots and Catenanes. On the left is a 5-noded knot (5_1), with its polarity indicated by arrowheads. Passing to the middle, one strand switch has been performed; this operation converts the knot to a catenane, which is drawn with lines of two different thicknesses. On the right, another strand switch has been performed, making a new 3-noded knot. (b) A Strand Switch in a DNA Context. Backbones are indicated by thick arrows, held together by three base pairs on each side. The helix axis is horizontal, and the dyad axis is vertical. The strand switch reconnects the strands, but maintains polarity. The reaction symbol replaces the right directional in (a). (c) View Down the Dyad Axis. The view in (b) has been rotated 90° about the horizontal axis. The hairpin nature of the product is clear here. It should be evident that the leftward reaction shown here is identical to the ligation shown in the last step of Figure 9.

reaction is a synthetic step that joins the two hairpins. We have seen this reverse reaction in another context in Figure 9, when the quadrilateral was closed: A knot (a circle, or trivial knot, in this case) is converted to a catenane; it is possible to draw any knot with a single line, but two lines (drawn with two different thicknesses in Figure 9) are needed to draw the product catenane.

Thus, we have been dealing with these types of transitions already. However, there is another reason for knots to be an inherent part of DNA nanotechnology. Ultimately, one would like to reproduce DNA molecules by the same efficient processes catalyzed by DNA polymerase in the cell, either *in vivo*, or *in vitro* by means of PCR (Saiki *et al.*, 1986). Unfortunately, it is not possible to reproduce a branch by DNA polymerase-catalyzed replication, because the enzyme tracks along the strand. Figure 12(a) shows that the replication of a branched molecule leads to two different duplex molecules, rather than two branched molecules. Figure 12(b) illustrates a potential solution to this problem, which involves converting a DNA catenane to a DNA knot, using the same approach as in the conversion of the knot to a catenane shown in Figure 9. A pentagonal dodecahedron is represented by its Schlegel diagram, and each edge has been decorated with two turns of DNA, which would lead to a 12-catenane. However, an exocyclic helix has been added to each cyclic strand, and then these exocyclic helices have been joined together (by the

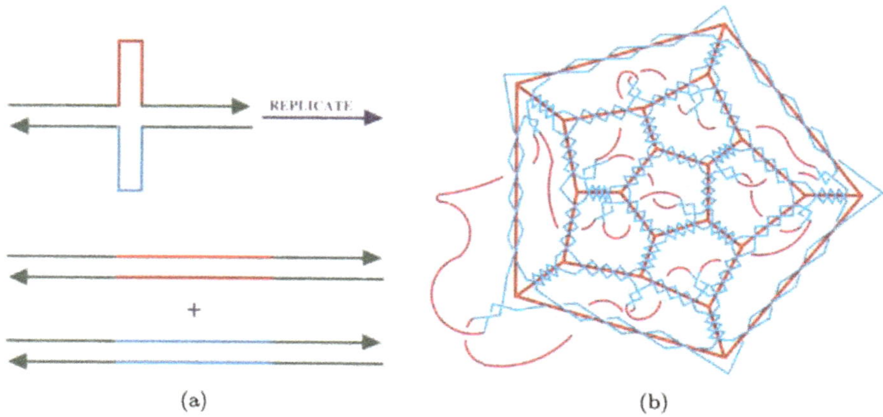

Figure 12. The Biological Generation of Branched DNA. (a) The Problem with Replicating Branches. A stable branch is shown, indicated by a dotted backbone on one strand and a solid one on the other. Replication by DNA polymerase leads to two different duplexes, rather than reproduction of the branched molecule. (b) A Single-Strand Strategy to Produce Branched Structures by Biological Replication. A Schlegel diagram of a pentagonal dodecahedron is shown, with two turns of DNA per edge. The 5′ end of the strand is indicated by a filled circle, and the 3′ end is indicated by the arrowhead. Each cyclic face contains an exocyclic helix, and these have been connected together to produce a single long strand that could be replicated. If it folded properly, it could be trimmed by restriction enzymes to remove all of the connecting (curved) DNA. The product would be the target polyhedron, ready to ligate to other polyhedra.

curved lines). Thus, the molecule is a single long knotted strand of DNA whose sequence could be reproduced by DNA polymerase. The exocyclic helices would be designed to contain restriction sites, so that the excess connecting DNA can be trimmed away after folding (Seeman, 1991).

12 Construction of DNA Knots

The folding of the complex molecule shown in Figure 12(b) is loaded with pitfalls, so we have practiced control of DNA folding on simpler knotted molecules. DNA is arguably the most favorable molecule for the construction of topological targets, such as knots. The reason for this is illustrated in Figure 13(a). This is a drawing of a trefoil knot in a DNA context. The strand has been given an arbitrary polarity (corresponding to the $5' \rightarrow 3'$ polarity of DNA), and a box has been drawn about each of its three unit tangles (Sumners, 1990) or nodes. The strands within the box act as its diagonals, dividing the box into four regions, two between parallel strands and two between antiparallel strands. The strands of the double helix are antiparallel, 6 base pairs (about a half turn of DNA) have been drawn to join the two strands in each box. The helix axis and the primary dyad axis of the region have also been drawn. Thus, a half-turn of DNA is the equivalent of a unit tangle in a topological figure (Seeman, 1992). The center of the knot is just a 3-arm DNA branched junction, and the outside strands can be connected easily by unstructured DNA; oligo-dT (e.g., T_{15}) works well in this role.

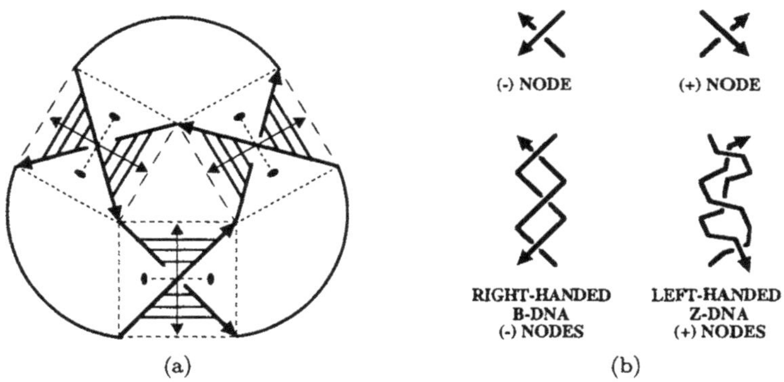

Figure 13. DNA Topological Methodology. (a) The Relationship Between a Half-Turn and a Node. A trefoil knot has been drawn with thick lines; its polarity is shown by the arrowheads along the knot. A dashed box has been drawn about each node, so that the strands of the knot divide the boxes into four regions, two between parallel strands and two between antiparallel strands. A half-turn of base pairs is drawn between antiparallel strands; the helix axes and dyad axes are indicated. (b) The Signs of Nodes. A negative node is on the left and a B-DNA helix below it; it is clear that right-handed B-DNA produces negative nodes. A positive node is on the right, and a Z-DNA helix with positive nodes below it. The zig-zag nature of Z-DNA is indicated.

We have been successful in assembling a trefoil knot from this motif (Du & Seeman, 1994).

Nodes in topological species can be of two signs. The top part of Figure 13(b) illustrates negative nodes and positive nodes, showing that they are mirror images of each other. The bottom part of Figure 13(b) illustrates the fact that right-handed B-DNA leads to negative nodes. What about positive nodes? There is another structure of DNA, called Z-DNA (Rich *et al.*, 1984). It has a zig-zag backbone structure whose helix is left-handed, even though it contains D-sugars. Z-DNA requires both special sequences and special conditions, usually associated with high ionic strength. A schematic diagram of Z-DNA in Figure 13(b) shows that it leads to positive nodes. Of the 'proto-Z' sequences (those that can be converted to Z-DNA), some convert under milder conditions than others. Thus, it is possible to produce different knotted topologies, by changing solution conditions. An example is shown in Figure 14, where the same strand can be used to generate a circle, a trefoil knot with negative nodes, a figure-8 knot and a trefoil knot with positive nodes, respectively, by increasing the ionic strength of the solution (Du *et al.*, 1995a). It is possible to interconvert these different species by DNA topoisomerases (Du *et al.*, 1995b), and the existence of RNA topoisomerase activity has been demonstrated by interconverting a circle and a knot made from RNA (Wang *et al.*, 1996, 1998).

13 DNA Borromean Rings

Borromean rings are a rich family of topological structures (Liang & Mislow, 1994) whose simplest member [Figure 15(a)] appears on the coat of arms of the Borromeo family, prominent in the Italian Renaissance. Although linked together, removal of

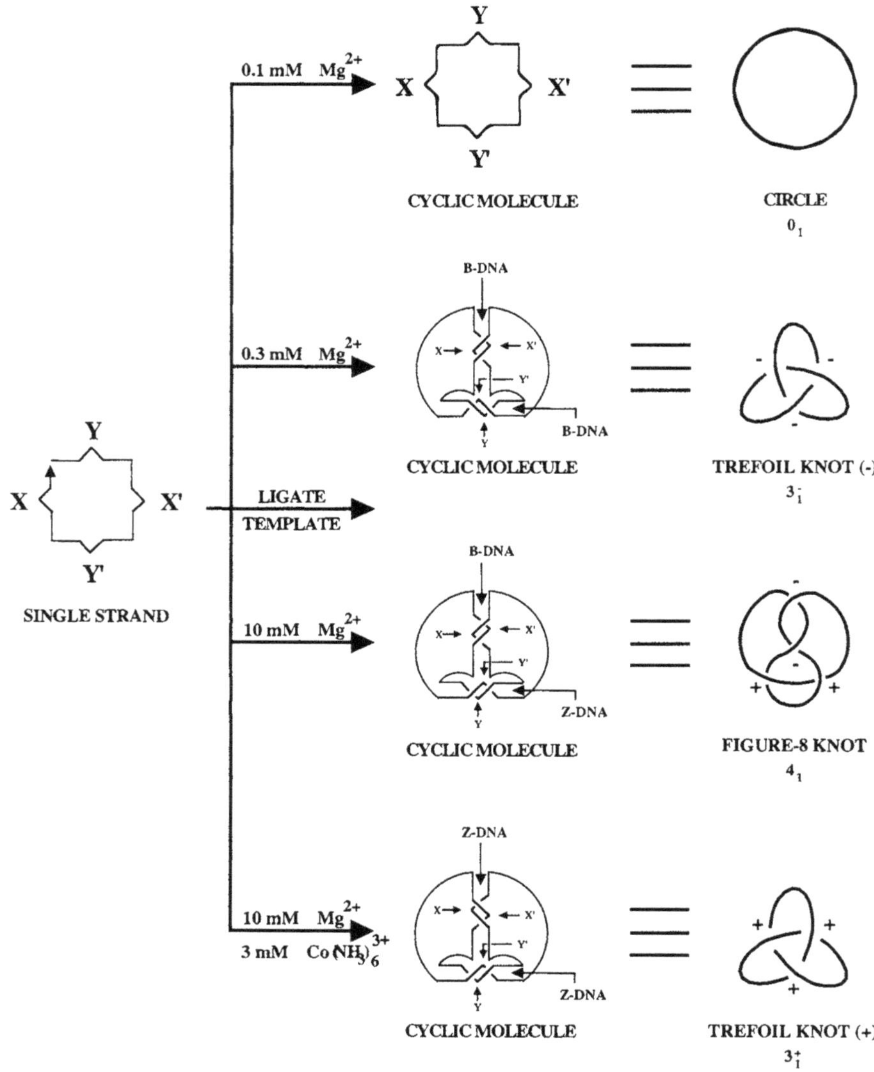

Figure 14. The Synthetic Schemes Used to Produce Four Topologies. The left side of this synthetic scheme indicates the molecule from which the target products are produced. The four pairing regions, X and its complement X′, Y and its complement Y′ are indicated by the bulges from the square. The 3′ end of the molecule is denoted by the arrowhead. The four independent solution conditions used to generate the target products are shown to the right of the basic structure. The pairing and helical handedness expected in each case is shown to the right of these conditions, and the molecular topology of the products is shown on the far right of the figure.

any individual circle unlinks the remaining rings. Molecules with the Borromean property present a formidable synthetic task, because their assembly entails placing nodes with equal numbers of positive and negative signs specifically about a link.

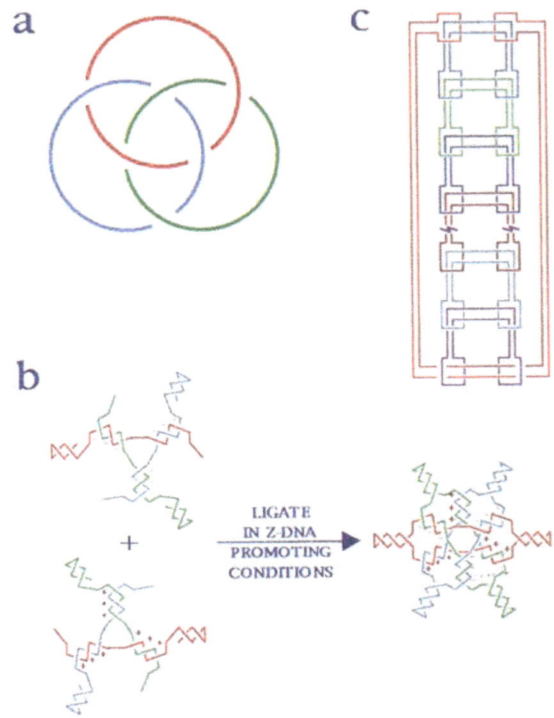

Figure 15. Borromean Rings. (a) Traditional Borromean Rings. The rings are drawn in three different colors. Note that no two rings are linked. (b) The Synthesis of Borromean Rings from DNA. A B-DNA branched junction (top) and a Z-DNA junction (bottom) are shown, each with 1.5 turns per arm; the signs of are nodes noted. The three strands are drawn with lines of different colors. At the ends of each strand are partial hairpins (one almost complete, one that supplies the missing portion to another incomplete hairpin). When ligated in Z-promoting conditions, the Borromean arrangement on the right is obtained. (c) There Is No Limit To the Number of Rings in a Borromean Link. An arbitrarily large number of rings is shown linked in a Borromean arrangement, so that cleavage of any one of them results in the unlinking of the entire complex. Each ring is colored differently, but one of them is broken by thunderbolts. This symbol indicates that any number of rings could be included at that point.

The signs of the three inner nodes in Figure 15(a) are opposite from those of the three outer nodes. The relationship between nodes and half-turns of DNA is the key to constructing this link. The three inner nodes constitute a 3-arm branched junction, just as they did in the knot of Figure 13(a). What about the three outer nodes? The target here is topological, so we can think of the drawing in Figure 15(a) as being distorted in the same way as a polar map of the Earth; the center corresponds to the North Pole, and the circumference corresponds to the South Pole. Thus, both sets of nodes flank poles, and both can be constructed from 3-arm branched junctions, one of B-ZNA and the other of Z-DNA.

It is often inconvenient to work with half-turns of DNA, so we have replaced the single nodes with three half-turns of DNA; this change does not affect the

Borromean property of the link. It is also convenient to add some hairpins to the "equator" of the molecule for convenience in construction and analysis. The ultimate construction is illustrated in Figure 15(b), in which two 3-arm branched junctions, one of B-DNA and one of Z-DNA are ligated in Z-DNA conditions to produce a link with the Borromean property. Proof of synthesis is performed by restricting the three rings individually, and showing that the products are only single circles (Mao *et al.*, 1997).

There is no limit to the number of links that can be combined to exhibit the Borromean property. This point is illustrated in Figure 15(c), in which an arbitrarily large number of rings are linked in this fashion. It may be possible to use Borromean rings as a convenient chemical .AND. test: The truth of a series of N propositions could be represented by the cyclic closure of molecules representing them in a Borromean context. In that case, the completely true set of propositions would have a much larger molecular weight under denaturing conditions than any set containing a false statement. The completely true set would be linked, but a single false statement would correspond to the same molecules broken into N-1 individual circles and a linear fragment. Thus, any separation method based on molecular mass under denaturing conditions would separate the molecules representing only true statements from the others, with minimal ambiguity. Denaturing gel electrophoresis would certainly be able to effect such separations. The physical separation of molecules with molecular masses proportional to N or N-1 becomes harder as N increases. The system described here would not suffer from this problem, because there would be only molecules of masses proportional to N or to unity.

14 Rigid Components

The work discussed up to this point consists of constructions that are topological in nature, without a strong geometrical basis. The cube and the truncated octahedron are really complex catenanes, and the shapes of the knots and the Borromean rings are only incidental to their key feature, the linking of their strands. Nevertheless, it is clear that the real targets of DNA nanotechnology must demonstrate strong geometrical control, not merely topological control. Consequently, several years ago we sought to determine whether there were any motifs based on DNA branched junctions that behave as though they are rigid in the oligomerization assay; only rigid components can provide geometrical control.

DNA double crossover (DX) molecules have been suggested as intermediates in recombination involving double strand breaks (Thaler & Stahl, 1988; Sun *et al.*, 1991), and they are known to be intermediates in meiosis (cell division that gives rise to germ cells) (Schwacha & Kleckner, 1995). They consist of two 4-arm branched junctions that are joined at two adjacent double helical arms. When modeled in small systems (Fu & Seeman, 1993), there are five different motifs, three of which have parallel helix axes, and two have antiparallel helix axes (Figure 16). Both types are further subdivided by whether their crossover points are separated by an even (DPE or DAE) or odd (DPO or DAO) number of double helical half-turns; there are two types of DPO molecules, depending on whether the extra half turn corresponds to a wide (DPOW) or narrow (DPON) groove. It is also possible to

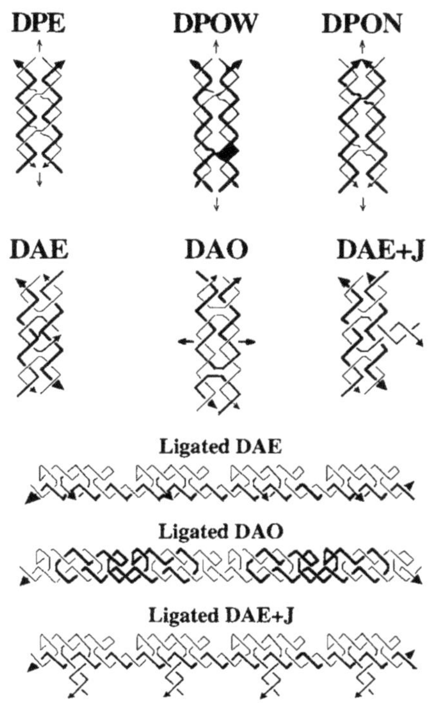

Figure 16. Double Crossover Molecules and Their Ligation Products. The top row contains the three parallel isomers of double crossover (DX) molecules, DPE, DPOW and DPON; 'P' in their name indicates their parallel structure. Arrowheads indicate 3′ ends of strands. Strands drawn with the same thickness are related by the vertical dyad axis indicated in the plane of the paper. DPE contains crossovers separated by an even number (2) of half-turns of DNA, DPOx by an odd number; in DPOW, the extra half turn is a major groove spacing, in DPON, it is a minor groove spacing. The middle row illustrates two other DX isomers, DAE, and DAO. The symmetry axis of DAE is normal to the page (the symmetry is broken by the nick in the central strand); the symmetry axis of DAO is horizontal within the page; in the case of DAO, strands of opposite thickness are related by symmetry. DAE+J, in the second row is a DAE in which an extra junction replaces the nick shown in DAE. Ligation products of the DAE, DAO and DAE+J molecules are shown at the bottom of the drawing, where one domain has been capped by hairpins. Ligation of DAE and DAE+J leads to a reporter strand, but ligation of DAO leads to a polycatenated structure.

make a derivative of DX molecules by including one or more junctions in the motif; Figure 16 shows a molecule in an extra junction has been added to a DAE molecule, to produce a DAE+J molecule (Li *et al.*, 1996).

Parallel DX molecules are not well behaved in model systems, even though they are the DX molecules relevant to biological systems, where they are usually much larger. Hence, antiparallel molecules have received most of the attention in this work. The oligomerization assay usually relies on the presence of a single strand, called a reporter strand (Li *et al.*, 1996), whose fate reflects that of the complex in the oligomerization assay: If it remains linear, so does the complex; if it cyclizes, so does the complex. The bottom of Figure 16 shows that ligating both the DAE molecule and the DAE+J molecule results in the presence of a reporter strand, drawn more darkly than the other strands. However, the ligation of DAO produces only a catenane, which is harder to analyze. Thus, the assays reported so far have been restricted to DAE-like molecules. Fortunately, these molecules pass the oligomerization assay successfully (Li *et al.*, 1996), suggesting that it is possible to construct DNA nanotechnological systems with predictable geometrical properties.

15 1-D and 2-D Periodic Arrays

I mentioned above that key goals for DNA nanotechnology include the construction of periodic matter. The inherent translational symmetry of periodic matter requires that the components used for its assembly be rigid, because flexible components can lead to cyclization and undesired assemblies (Liu *et al.*, 1994). Ultimately, we would like to achieve periodicity in three dimensions, so as to construct designed crystals (Seeman, 1982, 1985) and dense memory devices (Robinson & Seeman, 1987). However, we have begun with one dimensional and two dimensional systems involving DX molecules. The one-dimensional system entails the inclusion of a DX molecule along one edge of a triangle, as shown in Figure 17(a). The triangle has been selected (Yang *et al.*, 1998b) because it is an inherently rigid component (e.g., Kappraff, 1990). The length of the DX domain through which ligation occurs is 4.5 helical turns long. The odd number of half-turns means that the ligation products will have triangles pointing in alternate directions from a central axis. When the ligated molecule is visualized by atomic force microscopy (AFM), one sees a long line with small bumps on either side. The spacing of triangular points is roughly the expected 32 nm. It is always wise to challenge any image derived from a microscopic technique. Consequently, we have performed the experiments shown in Figures 17(b) and 17(c). Figure 17(b) illustrates a ligation in which a simple DX molecule is interspersed with the triangle. The repeat should increase to about 60.5 nm, and it does when visualized in the AFM. Finally, Figure 17(c) shows that adding a half turn (five pairs in this case) to the interspersed DX molecule is expected to change the image dramatically: The spacing should revert to 32 nm, but should be restricted to one side of the molecule; again, this is the result obtained. Scaled pictures of the products drawn in Figure 17 can all be superimposed on the AFM images obtained. No circular species are seen in these experiments, and no bias in curvature is observed. The molecules obtained can be 85 or more triangles in length. Thus, it appears that the DX motif can be added successfully to a DNA polygon, without decreasing its structural integrity.

What about using the DX molecules by themselves? Winfree (1996) has identified four topologies that can be formed by homogeneous antiparallel DX molecules. These topologies differ by separating their components by an even number or an odd number of half turns. Thus there is a lattice, DAE-O formed by DAE molecules separated by an odd number of half turns, and another, DAE-E, in which the DAE molecules are separated by an even number of half turns. Likewise, there are analogous lattices, DAO-E and DAO-O. DAE-E is interesting, because its ultimate topology is a type of molecular chain mail; DAE-O produces reporter strands both vertically and horizontally. The DAO lattices produce only vertical reporter

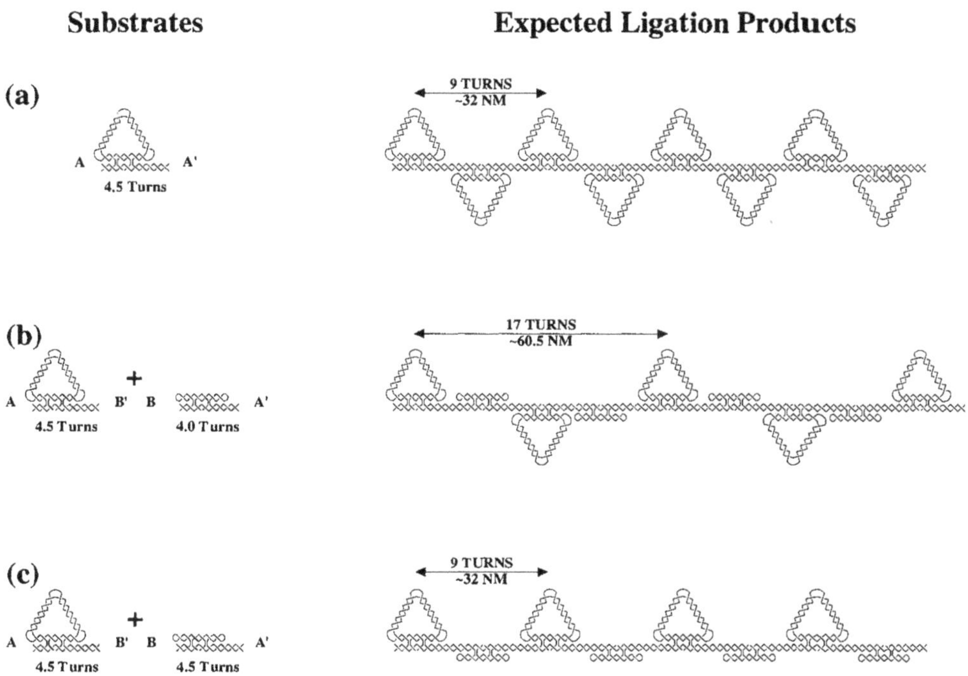

Figure 17. Atomic Force Microscopy Ligation Experiments. Three ligation experiments are shown in this diagram. The substrates are shown on the left, and the expected ligation products are shown on the right. (a) The triangle is prepared with complementary sticky ends, and ligated to produce a linear array in which the vertex opposite the DX points alternately in opposite directions, because the length of the ligated helix is 4.5 turns of DNA; successive vertices on the same side of the central helix are separated by about 32 nm. The experiment in panel (b) illustrates the ligation of the same triangle, but this time alternating with a DX molecule whose ligation domain is 4.0 turns. One domain on the DX molecule is sealed with hairpin loops. The sticky ends on the triangle (represented as A and B′) and on the DX (represented as B and A′) enforce this strict alternation. This structure spreads out the separation of successive vertices on the same side of the central helix to about 60.5 nm. The experiment in panel (c) is similar to the middle experiment, except that the ligation domain of the DX molecule is 4.5 turns. This ligation produces a product in which all of the triangles are on the same side of the central helix.

strands. These topologies are illustrated at the top of Figure 18. In addition, it is possible to mix DAO and DAE molecules, as shown at the bottom of Figure 18.

It is possible to arrange DX molecules into a 2-D lattice that can be visualized in by atomic force microscopy (AFM). This has been done both with DAO molecules and with DAE molecules (Winfree *et al.*, 1998). We have used a DX+J type motif slightly different from the one shown in Figure 16 to provide a topographic label visible in the AFM. We use two junctions (DX+2J) and orient the extra arms (whose ends are closed in a hairpin loop) to be perpendicular to the plane of the DX helix axes, so that there is one arm pointing out of the plane, and one pointing into it. The top of Figure 19 shows an experiment we performed with the DAE motif. A DAE molecule, A, and a DAE+2J molecule, B*, self-assemble by

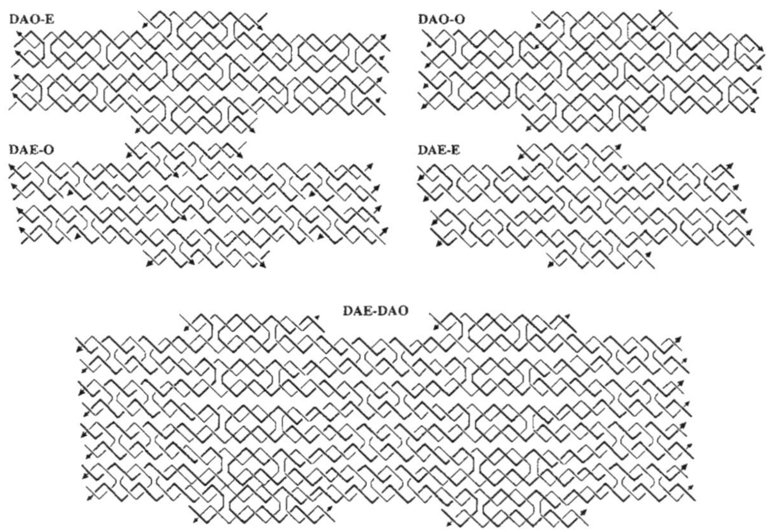

Figure 18. Topologies of Two-Dimensional Lattices Composed of Antiparallel DX Molecules. The four arrays at the top are built from homogeneous DAO or DAE molecules; crossings are separated by an even (E) or odd (O) number of double helical half-turns. Strand polarities are shown by the arrowheads on their 3′ ends. Strands are drawn with different thicknesses for clarity. Both DAO lattices contain vertical zig-zag strands, only. DAE-E consists of molecular chain mail, and DAE-O produces reporter strands in both the horizontal direction (straight) and the vertical direction (zig-zag). The large array at the bottom, DAE-DAO, contains both DAE and DAO molecules.

hydrogen bonding to produce a 2D array. Each molecule is about 16 nm long and 4 nm wide. The extra arms in B* are seen in the AFM as stripes, because the AFM cannot resolve the separation between hairpins in the 4 nanometer direction. The rows alternate, because the sticky ends direct their association to do so. The stripes appear to be about 32 nm apart in the image we obtain. To ensure that the spacings are not artifacts, we have conducted another experiment, shown at the bottom of Figure 19. This time we used the same types of components, but now there were four molecules, three DAE molecules, A, B and C, and one DAE+2J molecule, D*. This time the stripes are expected to be about 64 nm apart, and indeed they are (Winfree *et al.*, 1998).

One could generate a large number of 1-D patterns from this approach, just by increasing the number of individual components. In principle, the same could be done in two dimensions. However, each component consists of four or five strands of DNA, and it might be expensive to produce as many DX molecules as needed for a diversity of patterns. We have developed another method of diversifying the patterns (F. Liu, R. Sha & N.C. Seeman, submitted for publication). This is done by including restriction sites in the extra arms, or by ligating new hairpins to sticky ends in the array. The top of Figure 20 illustrates a 32 nm striped array containing four components, A, B^, C and D*, in which B^ differs from D*, because it contains a restriction site lacking in D*. When the array is digested by this restriction enzyme, it is converted from one with 32 nm stripes to one with 64 nm stripes.

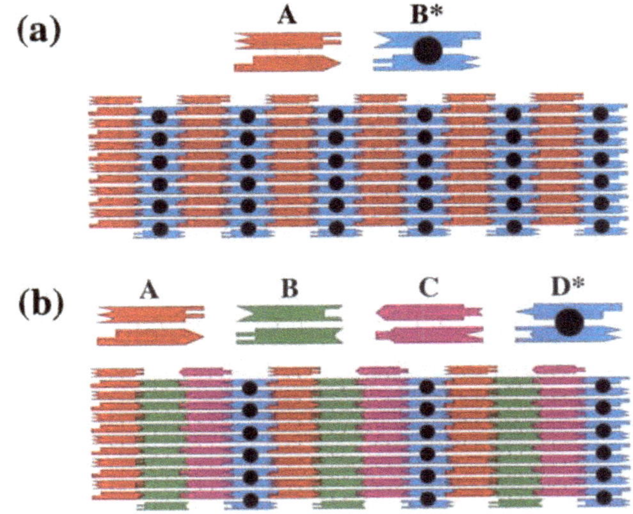

Figure 19. Construction of a Two-Dimensional Array from DAE and DAE+2J molecules. The component molecules of each array are shown above it schematically at double size. Individual components are shaded differently for clarity. Sticky ends and their complementarities are represented geometrically at the ends of the rectangles. (a) Two molecules are used to make the array. Both molecules, A and B*, are represented by two rectangles held together by short lines. A is a DAE molecule, but B* is a DAE+2J molecule that contains protrusions perpendicular to the plane of the helix axes. The protrusions are represented by the dark circle at the center of the molecule. When a lattice is formed from these molecule by satisfying the complementarity of the sticky ends, the array contains rows of protrusions that appear as stripes in the AFM. These stripes are separated by double the molecular length, roughly 32 nm. (b) Four molecules are used to make the array. The array has four components, A, B, C and D*. Only D* is a DAE+2J molecule, so the stripes are separated by four times the molecular length, roughly 64 nm.

The converse experiment is shown at the bottom of Figure 20. Here, a 64 nm striped array (A, B°, C and D*) contains a small extra arm (only weakly visible in the AFM) in its B° unit that ends in a sticky end. It is possible to add a hairpin to that sticky end, thereby converting the 64 nm array to a 32 nm striped array. Thus, with a relatively limited number of components, it is possible to generate a variety of patterns. The number of 1-D patterns is dependent on phasing the pattern by some sort of border. In general, the number of patterns in an array phased by a border will be 2^p for p-hairpin patterns that are not otherwise distinct to the AFM, but unphased patterns are markedly fewer; the number of unphased patterns is the same as the number of different necklaces that be made with p beads of 2 colors, a much lower number. For example, if p = 4, there are 16 phased patterns, but only 6 unphased patterns.

16 Nanomechanical Devices

In addition to static devices, it would be valuable to produce mechanical motion by combining the structural work described above with DNA transitions. We have

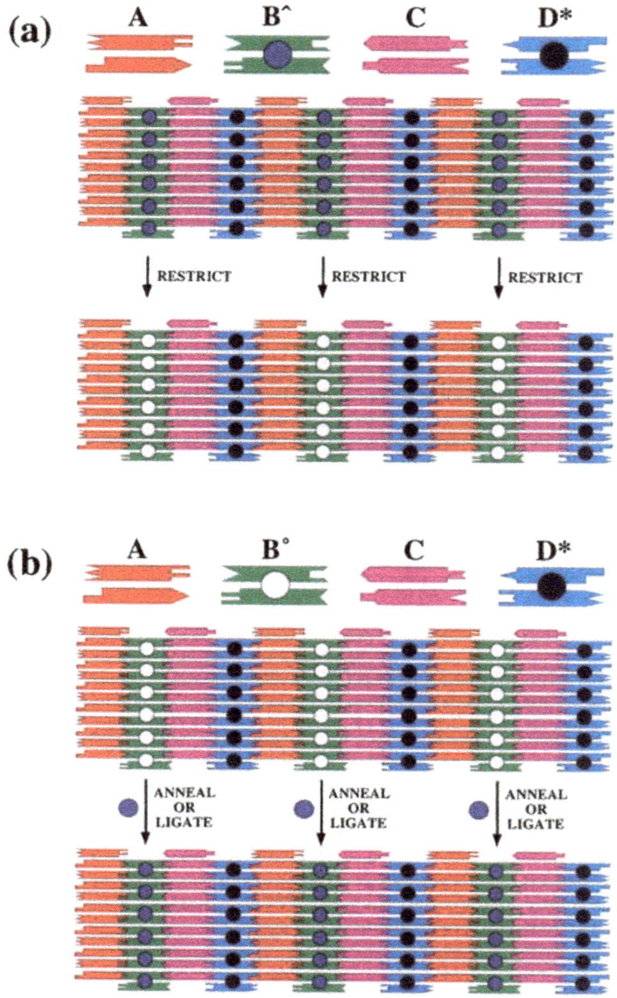

Figure 20. Modification of DAE Lattices. The same conventions are used as in Figure 19. (a) Restriction of a Crystalline Array. A and C are DX molecules, and B^ and D* are DX+2J molecules. The different circle fillings indicate that the hairpins differ between the two DX+2J molecules. The bottom part of the diagram illustrates the effect of removing the hairpin of B^ by restriction: The prominent stripe is replaced with a much less intense feature. (b) Ligation to a Crystalline Array. The same conventions apply to this panel as to (a). The difference here is that B° replaces B^. Its out-of-plane components contain short arms ending in sticky ends that do not produce an intense feature in the AFM. The drawing illustrates that the ligation of hairpins to this array produces a pattern similar to the starting pattern of (a).

managed to control branch migration (Yang *et al.*, 1998a), but the system is cumbersome. The most successful of our attempts is the use of the B-Z transition to produce a nanomechanical device. This molecule is illustrated in Figure 21. It consists of two DAO double crossover molecules that are coupled by a piece of

DNA, shown in yellow, whose sequence permits it to undergo the B-Z transition. I pointed out above that the conversion of B-DNA to Z-DNA has two requirements: There must be a sequence that can be transformed to Z-DNA, and there must be conducive solution conditions to promote the transition. The first requirement can be used to limit the transition in space: Only the twenty nucleotides pairs shown in yellow are capable of undergoing the transition. The second requirement can be used to limit the transition in time. One can cycle solution conditions from those that promote B-DNA to those that promote Z-DNA, and then revert back. We have demonstrated the effectiveness of this approach to nanomechanical motion by adding fluorescent dyes to the two DAO domains; the distance between the dyes changes from one state to the other, and this can be measured by fluorescent resonance energy transfer (Mao *et al.*, 1998). Although Figure 21 indicates that the molecule rotates a half-turn, the real change is 3.5 turns. The largest excursion of atoms from one state to the other is about 60 Å. We have measured conversion between two equilibrium states, and we do not yet know whether this device is capable of transmitting a force.

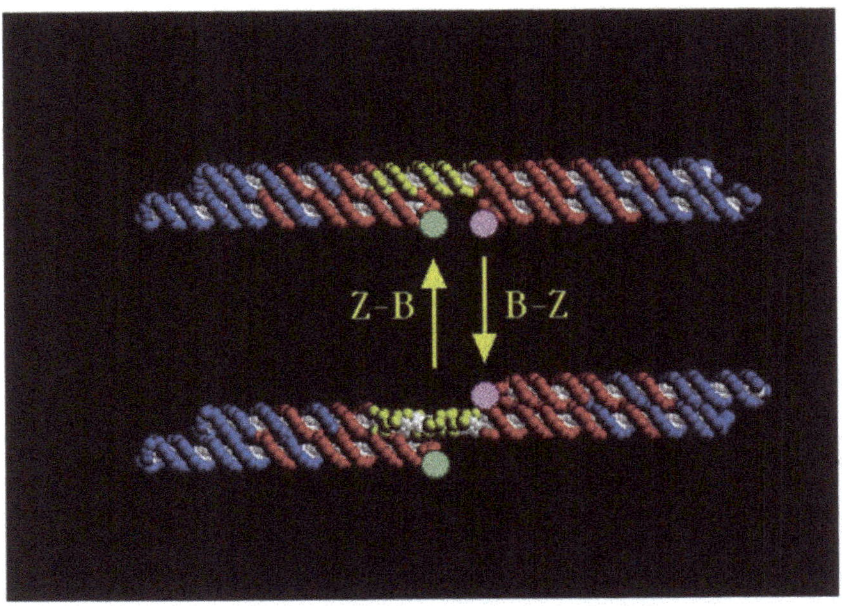

Figure 21. A DNA Nanomechanical Device. The same conventions apply as in Figures 7 and 10. The device consists of two DAO double crossover molecules (blue and red strands) connected by a piece of DNA that can be converted to Z-DNA; although it is part of the red strand, this section is drawn in yellow. The transition to Z-DNA can be effected by the addition of $Co(NH_3)_6Cl_3$, and reversion to B-DNA can be produced by its removal. The motion is detected by the use of fluorescence resonance energy transfer, which is sensitive to the separation of two fluorescent dyes. The dyes are represented by green and magenta stippled circles. The largest excursions of atoms in this system are about 60 Å.

17 Prospects

Where is DNA nanotechnology going? The key goal remains the extension of the successes in two dimensions to three dimensional systems. Figure 22(a) shows the way in which a 6-connected simple cubic lattice can organize macromolecular guests. If the ordering is high enough, the resulting complexes can be used for the X-ray crystallographic determination of 3-D molecular structure. In addition to being the first goal enunciated for this program, there is another important reason to pursue this end: This is a goal compatible with the properties of DNA. In contrast to some of the materials applications that come to mind, it does not entail drying or high temperature or very large arrays. Ordered arrays containing 10^{11} or fewer unit cells (Bilderback *et al.*, 1994) would be adequate for the purpose.

It is unlikely that the simple cubic lattice of Figure 22(a) would be formed successfully into a 3-D lattice, because 6-arm junctions are not well-structured (Wang *et al.*, 1991). The success with DX molecules suggests that they can be combined usefully with an inherently rigid geometrical motif. In two dimensions, we have worked with triangles, as indicated in Figure 17. The three-dimensional equivalent is the deltahedron, which is also inherently rigid (e.g., Kapraff, 1990). Figure 22(b) indicates that DX molecules could be incorporated into three edges of an octahedron. These three edges span 3-space, and the resulting lattice would

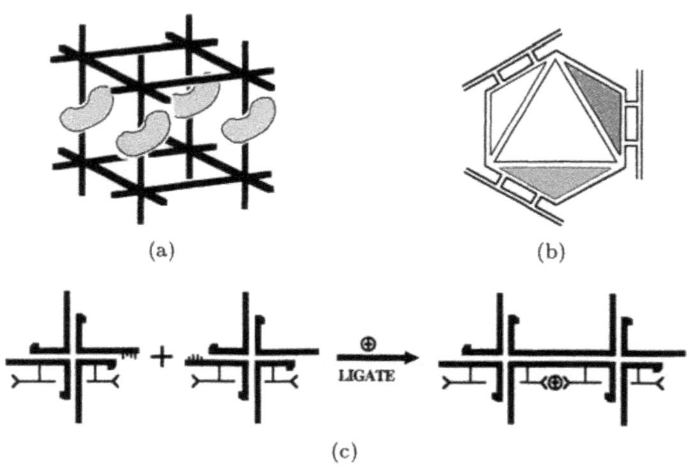

Figure 22. Future Applications of DNA Nanotechnology. (a) A Guest in a Simple Cubic Lattice. The DNA lattice is drawn as a portion of a simple cubic lattice made from 6-arm junctions. The guests are represented by the kidney bean-shaped features in every unit cell. If the guests can be aligned parallel to each other, their structures can be determined by X-ray crystallography. (b) An Octahedron in which Three Edges are Double Crossover Molecules. The front view of the octahedron is shown. The three edges are not coplanar, so they span 3-space. Connecting the inherently rigid octahedra together would produce a trigonal array. (c) DNA as Scaffolding. Two branched junctions are shown, and a molecular wire is attached to them. When the two junctions cohere with each other, so does the molecular wire, which forms a synapse. This drawing represents one way that DNA could be used to organize molecules whose properties are best suited to a particular molecular task.

be related to a cubic-close-packed array, but would be only trigonally symmetric, rather than cubic.

The ability to pattern materials to a resolution of about 10 nanometers would clearly have impact on the fabrication of useful devices, if the applications were compatible with the physical properties of DNA. It may be possible to overcome the difficulties presented by the environmental requirements of DNA if it is used only as scaffolding to direct the assembly of other polymers whose properties are more appropriate for the applications in mind. An example of this type of scaffolding is illustrated in Figure 22(c), where a molecular synapse is produced by two DNA molecules to which molecular wires have been attached. Once the other molecules were assembled, the DNA could be eliminated.

Winfree (1996) has suggested that DNA double crossover molecules could be used as one-dimensional cellular automata in DNA-based computing. In addition to the ability to carry out computations, this approach also leads to the possibility of arbitrary programmed patterns within the materials. As noted above, DNA-based computing and related enterprises require control of intermolecular interactions over a much greater diversity of sequences than does DNA nanotechnology.

An intermediate goal would be to make periodic complex patterns that are programmed by sticky ends. Figure 23 depicts schematically a 2-D array of 100 DNA DX molecules arranged as tiles. Each DX molecule consists of five boxes,

Figure 23. An Inexpensive Route to Diverse Patterns. This array contains 100 tiles, each of which represents a DNA DX molecule. The sticky ends of each tile are color coded, and the two opposite ends of each diagonal are complementary. There are five colors repeated twice in each direction, so the drawing contains four unit cells of the repeat. If each color represents a strand, then 40 strands could make this array from DAO components. If diversity to produce DX or DX+2J molecules is taken into account, another 20 strands would be needed, for a total of 60. However, to program this array with unique sticky ends, 150 strands would be needed. For an array of N x M, with S strands needed per tile for diversity, SNM strands are needed for complete specification of sticky ends, but S(N+M) strands are needed for this scheme. Nevertheless, this method may be much harder to implement.

one clear one in the center and four colored ones flanking it on either end of each diagonal. The colors represent sticky ends; those on opposite ends of each diagonal in each tile represent complements. There are 10 molecules in each direction. There are five colors in each direction, so the drawing represents four unit cells. If the DX molecules arrange themselves along any of the colored diagonals, there is nothing to prevent them from permuting themselves to generate a large number of random patterns. However, if both sets of sticky ends can be made to have equal impact, then only the pattern shown will result. Of course, it is possible to program this pattern with 25 different DX molecules, but that is much more expensive than this approach.

In addition to 2-D patterns, one can imagine extending this scheme to 3-D, to produce complex crystals with designed structural features at specific locations. For example, one could imagine coupling the nanomechanical properties of the device described above with the static properties of designed arrays. Activation of the device could act as a robot arm to bring reactants into proximity, and then move them along for further processing. Thus, ultimately, one could imagine the preprogrammed assembly and functioning of nanoscale factories.

These dreams are still a long way from realization. To accelerate progress, it is important that greater numbers of investigators enter in this field. For many years, ours was the only laboratory working in DNA nanotechnology, but others are now getting involved. Mirkin *et al.* (1996) have attached DNA molecules to colloidal gold, with the goal of assembling nanoparticles into macroscopic materials and for used in diagnostics (Elghanian *et al.*, 1977). Likewise, Alivisatos *et al.* (1996) have used DNA to organize nanocrystals of gold. Niemeyer *et al.* (1994) had earlier used DNA specificity to generate protein arrays. More recently, Shi and Bergstrom (1997) have attached DNA single strands to rigid organic linkers; they have shown that they can form cycles of various sizes with these molecules. Braun *et al.* (1998) have reported using DNA to template an electric circuit element. Likewise, Nilsen *et al.*'s (1997) predictions about the nature of DNA dendrimers have recently been reduced to practice (Wang *et al.*, 1998b).

The ideas behind DNA nanotechnology have been around for a long time (Seeman, 1982). However, their realization has been slowed by the realities of the experimental situation: The establishment of the conditions, constraints and protocols that convert ideas into experimentally demonstrated facts. It appears that many of these are in place now, and that the increasing activity in the field will help to achieve its central goals in the near future. I hope that reaching those milestones will be just the beginning of an extensive and burgeoning area of investigation.

Glossary

BACKBONE: Biological molecules such as proteins and nucleic acids are polymers of different monomeric elements in a specific order. The monomers of proteins are amino acids and the monomers of DNA are nucleotides. They have very specific properties based on the order of the elements. Each of these monomers has a constant repeating portion, and a side group that differentiates it from other monomers. For example, the four bases, A, T, G and C are the side groups of nucleotides. The monomers are linked together through the constant portion to form the polymer of interest. The constant portion is referred to as the backbone. The alternating sugar-phosphate structure shown in Figure 1 is the backbone of DNA.

MOLECULAR CONFIGURATIONS (D AND L): Four different substituents can be bonded to tetrahedral atoms, such as carbon atoms, in two different configurations. These two configurations are mirror images of each other. Carbon atoms of this sort are called asymmetric centers. For example, in Figure 1, one of the carbon atoms of the sugar is bonded to the base, a hydrogen atom, a carbon atom and an oxygen atom. Molecules containing asymmetric centers can rotate the plane of plane polarized light in one direction and their mirror image molecules rotate it the same amount in the opposite direction. Molecules that rotate it to the right, and those related to them, are called dextrorotatory, and are said to be in the D-configuration; those that rotate it to the left, and those related to them, are called levorotatory and are said to be in the L-configuration. The sugars of DNA are in the D-configuration, but the amino acids of proteins are in the L-configuration.

POLYMERASE CHAIN REACTION (PCR): The polymerase chain reaction is a laboratory technique for amplifying particular sequences of DNA. It uses an enzyme, called DNA polymerase, that is capable of making a complementary copy of a DNA molecule, and does so during cellular replication. A heat-resistant version of DNA polymerase is used. DNA polymerase requires a primer to replicate DNA; this primer is a short synthetic complement to the first part of the DNA to be replicated. The DNA molecule and the polymerase are added to a solution that contains primers for both strands in vast excess over the long sequence. The DNA is heated, the two strands separate, the solution is cooled, and the primers bind to the two strands. When the primers are bound, DNA polymerase makes a copy of each strand. Twice as much DNA is present as before the polymerase made the copies. The solution is heated again, the strands separate again, new primers bind, and new copies are made. This process continues iteratively until the primers are used up.

GEL ELECTROPHORESIS: The DNA molecules discussed in this article are most frequently separated by impelling them through a gelatin-like material, polyacrylamide. The polyacrylamide is cast into thin slabs, and the DNA is loaded on one side of the slab. The cathode (negative electrode) is attached to this side of the slab, and the anode (positive electrode) is attached to the opposite side. DNA is negatively charged, so it migrates from the cathodic side to the anodic side; this process is called electrophoresis. The DNA interacts with the acrylamide during migration, and this affects it migratory rate. The gel may contain denaturants that disrupt hydrogen bonding, so that the only properties of the DNA that affect its migration are the molecular weight and the topology. This is called denaturing gel electrophoresis. Molecules of the same topology migrate as a function of the logarithm of their molecular weight in denaturing gel electrophoresis. The gel may contain factors that promote hydrogen bonding, and this type of electrophoresis is called non-denaturing gel electrophoresis. Factors such as the shape and surface area affect the migratory rates of the hydrogen bonded complexes examined by non-denaturing gel electrophoresis.

RESTRICTION ENZYMES: Bacteria contain enzymes that cleave at particular sequences of DNA, usually containing between four and eight nucleotides. The cuts produced may be blunt or staggered. The sequences are often self-complementary; for example EcoR I produces a staggered cut by cleaving GAATTC between the G and the A. Some restriction enzymes recognize a sequence, but cut in its vicinity; for example, Bbs I produces a staggered cut 2 and 6 nucleotides 3′ to GAAGAC. Some recognize interrupted cuts; for example, Dra III cuts between the third N (any base) and the G of CACNNNGTG. Their diversity and fidelity make restriction enzymes central tools in both genetic engineering and DNA nanotechnology.

DNA HELICAL SENSES: Normally, DNA is in what is known as the B-conformation. This is the familiar helical structure, and is the structure believed to be assumed by DNA within the cell. This is a right-handed structure. We are all familiar with helical staircases that contain a central pole and a banister. If one walks up one of these staircases and has one's right hand on the banister and the left hand on the pole, it is a right handed helix; it is also right handed if one walks down the staircase with one's right hand on the pole and one's left hand on the banister. The opposite cases, up with one's left hand on the banister or down with one's right hand on the banister are left-handed helices. Special sequences of DNA, particularly repetitive CG sequences, are capable of forming an unusual DNA structure, called Z-DNA, in conditions of high ionic strength. Z-DNA is left-handed, and has a zig-zag backbone structure.

OLIGOLATERALS: We use the word 'oligolateral' to indicate a figure containing a small number of edges, but not constrained to be planar. Polygons are usually taken to be planar, but we do not know the shapes of DNA figures with more than three sides. Consequently, we have adopted this term to specify them.

EXONUCLEASES: Cells contain enzymes that can degrade DNA; these are called DNA nucleases. Some of these enzymes can only degrade DNA from one end or the other. These are called exonucleases, to contrast them with endonucleases, which can cleave DNA anywhere along the chain. Exonucleases are useful in differentiating cyclic molecules from non-cyclic (linear) molecules: Cyclic molecules cannot be digested by exonucleases, because they lack an end.

HAIRPIN LOOPS: Normally, we think of the DNA double helix to be composed of two different molecules. However, the two molecules can be joined at their ends to create a hydrogen bonded single molecule that has a short loop on the end, called a hairpin loop. For example, the sequence GTCGATGACTTTTGTCATCGAC would for a hairpin loop structure, because it contains two complementary sequences linked by four T's that could form the loop.

KINETOPLAST: The kinetoplast is the mitochondrion of the Trypanosome, a protozoan organism. Its genome is a polycatenane of DNA, rather than a linear molecule.

TOPOISOMERASE: Topoisomerases are enzymes that catalyze strand passage operations in nucleic acids. There are two types of topoisomerases, Type I and Type II. Type I topoisomerases catalyze single strand passage reactions, so that the linking number of the substrate will change in unit steps in each reaction. Type II topoisomerases catalyze double strand passage events, so that the linking number of the substrate will change in steps of two in each reaction. Some topoisomerases are associated with an energy source, so they can catalyze strand passage operations that require energy, such as supercoiling closed circular double helical DNA molecules; other just catalyze the approach to equilibrium. Topoisomerase III from E. coli can catalyze strand passage operations in RNA.

IONIC STRENGTH: The ionic strength of a solution is related to the concentration of ions in the solution, but it also takes into account the fact that some ions have greater charges than others. The formula for ionic strength is I.S. $= 0.5 * \sum_i [c_i][z_i]^2$, where $[c_i]$ are the concentrations of every cation or anion, and $[z_i]$ is the charge on the ion. Thus, a Mg^{2+} cation contributes four times the ionic strength of a Na^+ cation.

Acknowledgments

This research has been supported by grants GM-29554 from the National Institute of General Medical Sciences, N00014-98-1-0093 from the Office of Naval Research, NSF-CCR-97-25021 from the National Science Foundation/DARPA and F30602-98-C-0148 from the Information Directorate of the Air Force Research Laboratory located at Rome NY. The experimental work reported here summarizes research performed in our laboratory at New York University by Junghuei Chen, Yuwen Zhang, John Mueller, Shouming Du, Hui Wang, Tsu-Ju Fu, Yinli Wang, Xiaojun Li, Xiaoping Yang, Jing Qi, Bing Liu, Hangxia Qiu, Furong Liu, Chengde Mao, Weiqiong Sun, Ruojie Sha, Lisa Wenzler, Zhiyong Shen, Hao Yan, Man Hoi Wong, Phiset Sa-Ardyen and Siwei Zhang. I have also discussed research performed by Erik Winfree at the California Institute of Technology. I am grateful to all of these people for their collaboration in this work.

References

1. Adleman, L. (1994). Molecular computation of solutions to combinatorial problems. *Science* **266**, 1021–1024.
2. Alivisatos, A. P., Johnsson, K. P., Peng, X., Wilson, T. E., Loweth, C. J., Bruchez, M. P. Jr and Schultz, P. G. (1996). Organization of 'nanocrystal molecules' using DNA. *Nature* **382**, 609–611.

3. Balasubramanian, B., Pogozelski, W. K. and Tullius, T. D. (1998). DNA strand breaking by the hydroxyl radical is governed by the accessible surface areas of the hydrogen atoms of the DNA backbone. *Proc. Nat. Acad. Sci.* (USA) **95**, 9738–9743.
4. Bilderback, D. H., Hoffman, S. A. and Thiel, D. J. (1994). Nanometer spatial resolution achieved in hard x-ray imaging and Laue diffraction experiments. *Science* **263**, 201–203.
5. Braun, E., Eichen, Y., Sivan, U. and Ben-Yoseph, G. (1998). DNA-templated assembly and electrode attachment of a conducting silver wire. *Nature* **391**, 775–778.
6. Caruthers, M. H. (1985). Gene synthesis machines: DNA chemistry and its uses. *Science* **230**, 281–285.
7. Chen, J. and Seeman, N. C. (1991a), The synthesis from DNA of a molecule with the connectivity of a cube. *Nature* **350**, 631–633.
8. Chen, J. and Seeman, N. C. (1991b). The electrophoretic properties of a DNA cube and its sub-structure catenanes. *Electrophoresis* **12**, 607–611 (1991).
9. Chen, J., Rauch, C. A., White, J. H., Englund, P. T. and Cozzarelli, N. R. (1995a). The topology of the kinetoplast DNA network. *Cell* **80**, 61–69.
10. Chen, J., Englund, P. T. and Cozzarelli, N. R. (1995b). Changes in network topology during the replication of kinetoplast DNA, *EMBO J.* **14**, 6339–6347.
11. Churchill, M. E. A., Tullius, T. D., Kallenbach, N. R. and Seeman, N. C. (1988). A Holliday recombination intermediate is twofold symmetric. *Proc. Nat. Acad. Sci.* (USA) **85**, 4653–4656.
12. Cohen, S. N., Chang, A. C. Y., Boyer, H. W. and Helling, R. B. (1973). Construction of biologically functional bacterial plasmids in vitro, *Proc. Nat. Acad. Sci.* (USA) **70**, 3240–3244.
13. Du, S. M. and Seeman, N. C. (1994). The construction of a trefoil knot from a DNA branched junction motif. *Biopolymers* **34**, 31–37.
14. Du, S. M., Stollar, B. D. and Seeman, N. C. (1995a). A synthetic DNA molecule in three knotted topologies. *J. Am. Chem. Soc.* **117**, 1194–1200.
15. Du, S. M., Wang, H., Tse-Dinh, Y.-C. and Seeman, N. C. (1995b). Topological transformations of synthetic DNA knots. *Biochem* **34**, 673–682.
16. Eis, P. S. and Millar, D. P. (1993). Conformational distributions of a four-way DNA junction revealed by time-resolved fluorescence resonance energy transfer. *Biochemistry* **32**, 13852–13860.
17. Elghanian, R., Storhoff, J. J., Mucic, R. C., Letsinger, R. L. and Mirkin, C. A. (1997). Selective colorimetric detection of polynucleotides based on the distance-dependent optical properties of gold nanoparticles. *Science* **277**, 1078–1081.
18. Fu, T. J. and Seeman, N. C. (1993). DNA double crossover molecules. *Biochem* **32**, 3211–3220.
19. Fu, T. J., Tse-Dinh, Y. C. and Seeman, N. C. (1994). Holliday junction crossover topology. *J. Mol. Biol.* **236**, 91–105.
20. Hagerman, P. J. (1988). Flexibility of DNA. *Ann Rev Biophys & Biophys Chem* **17**, 265–286.

21. Hargreaves, D., Rice, D. W., Sedelnikova, S. E., Artymiuk, P. J., Lloyd, R. G. and Rafferty, J. B. (1998). Crystal structure of E.coli RuvA with bound DNA Holliday junction at 6 A resolution. *Nature Str. Biol.* **5**, 441–446.
22. Holliday, R. (1964). A mechanism for gene conversion in fungi. *Genet. Res.* **5**, 282–304.
23. Hsieh, P. and Panyutin, I. G. (1995). DNA branch migration. *Nucleic Acids & Mol. Biol.* ed F Eckstein, DMJ Lilley, 9, 42-65, Berlin: Springer-Verlag.
24. Kallenbach, N. R., Ma, R. I. and Seeman, N. C. (1983). An immobile nucleic acid junction constructed from oligonucleotides. *Nature* **305**, 829–831.
25. Kappraff, J. (1990). *Connections.* New York: McGraw-Hill. p. 273.
26. Kool, E. T. (1998). Recognition of DNA, RNA and Proteins by Circular Oligonucleotides. *Accts Chem Res* **31**, 502–510.
27. Li, X., Yang, X., Qi, J. and Seeman, N. C. (1996). Antiparallel DNA double crossover molecules as components for nanoconstruction. *J. Am. Chem. Soc.* **118**, 6131–6140.
28. Liang, C. and Mislow, K. (1994). On Borromean links. *J. Math. Chem.* **16**, 27–35.
29. Lilley, D. M. J. and Clegg, R. M. (1993), The structure of the four-way junction in DNA. *Ann. Rev. Biophys. Biomol. Str.* **22**, 299–328.
30. Lipton, R. J. (1995), DNA solution of hard computational problems. *Science* **268**, 542–545.
31. Liu, B., Leontis, N. B. and Seeman, N. C. (1994). Bulged 3-arm DNA branched junctions as components for nanoconstruction. *Nanobiol* **3**, 177–188.
32. Lu, M., Guo, Q., Marky, L. A., Seeman, N. C. and Kallenbach, N. R. (1992). Thermodynamics of DNA branching. *J. Mol. Biol.* **223**, 781–789.
33. Ma, R. I., Kallenbach, N. R., Sheardy, R. D., Petrillo, M. L. and Seeman, N. C. (1986). Three arm nucleic acid junctions are flexible. *Nucl. Acid. Res.* **14**, 9745–9753.
34. Mao, C., Sun, W. and Seeman, N. C. (1997). Assembly of Borromean rings from DNA, *Nature* **386**, 137–138.
35. Mao, C., Sun, W., Shen, Z. and Seeman, N. C. (1998). A nanomechanical device based on the B-Z transition, *Nature*, in press.
36. Marmur, J. and Doty, P. (1962). Determination of the base composition of deoxyribonucleic acid from its thermal denaturation temperature. *J. Mol. Biol.* **5**, 109–118.
37. Miick, S. M., Fee, R. S., Millar, D. P. and Chazin, W. J. (1997) Crossover isomer bias is the primary sequence dependent property of immobilized Holliday junctions. *Proc. Nat. Acad. Sci.* (USA) **94**, 9080–9084.
38. Mirkin, C. A., Letsinger, R. L., Mucic, R. C. and Storhoff, J. J. (1996). A DNA-based method for rationally assembling nanoparticles into macroscopic materials. *Nature* **382**, 607–609.
39. Niemeyer, C. M., Sano, T., Smith, C. L. and Cantor, C. R. (1994). Oligonucleotide-directed self-assembly of proteins. *Nucl. Acids. Res.* **22**, 5530–5539.
40. Nilsen, T. W., Grayzel, J. and Prensky (1997). Dendritic nucleic acid structures. *J. Theor. Biol.* **187**, 273–284.

41. Petrillo, M. L., Newton, C. J., Cunningham, R. P., Ma, R. I., Kallenbach, N. R. and Seeman, N. C. (1988). Ligation and flexibility of four-arm DNA junctions. *Biopolymers* **27**, 1337–1352.
42. Qiu, H., Dewan, J. C. and Seeman, N. C. (1997). A DNA decamer with a sticky end. *J. Mol. Biol.* **267**, 881–898.
43. Rich, A., Nordheim, A. and Wang, A. H.-J. (1984). The chemistry and biology of left-handed Z-DNA. *Ann. Rev. Biochem.* **53**, 791–846.
44. Robinson, B. H. and Seeman, N. C. (1987). The design of a biochip. *Prot. Eng.* **1**, 295–300.
45. Saiki, R., Bugawan, T. L., Horn, G. T., Mullis, K. B. and Ehrlich, H. A. (1986). The analysis of enzymatically labeled b-globin DQa DNA with allele-specific oligonucleotide probes. *Nature* **324**, 163–166.
46. Schwacha, A. and Kleckner, N. (1995). Identification of double Holliday junctions as intermediates in meiotic recombination. *Cell* **83**, 783–791.
47. Seeman, N. C. (1982). Nucleic acid junctions and lattices. *J. Theor. Biol.* **99**, 237–247.
48. Seeman, N. C. (1985) Macromolecular design, nucleic acid junctions and crystal formation. *J. Biomol. Str. & Dyns.* **3**, 11–34.
49. Seeman, N. C. (1990). De Novo design of sequences for nucleic acid structural engineering. *J. Biomol. Str. & Dyns.* **8**, 573–581.
50. Seeman, N. C. (1991). The construction of 3-D stick figures from branched DNA. *DNA & Cell Biol.* **10**, 475–486.
51. Seeman, N. C. (1992). The design of single-stranded nucleic acid knots. *Molecular Engineering* **2**, 297–307.
52. Seeman, N. C. (1998). DNA nanotechnology: Novel DNA constructions. *Ann. Rev. Biophys. Biomol. Str.* **27**, 225–248.
53. Seeman, N. C. and Kallenbach, N. R. (1983). Design of immobile nucleic acid junctions. *Biophys. J.* **44**, 201–209.
54. Seeman, N. C. and Kallenbach, N. R. (1988). Nucleic acid junctions: A successful experiment in macromolecular design. In: *Molecular Structure: Chemical Reactivity and Biological Activity,* ed by JJ Stezowski, JL Huang, MC Shao, Oxford University Press, Oxford (1988) pp. 189–194.
55. Seeman, N. C. and Kallenbach, N. R. (1994). DNA branched junctions, *Ann. Rev. Biophys. Biomol. Struc.* **23**, 53–86.
56. Shi, J. and Bergstrom, D. E. (1997). Assembly of novel DNA cycles with rigid tetrahedral linkers. *Angew. Chem. Int. Ed. Eng.* **36**, 111–113.
57. Sumners, D. W. (1990). Untangling DNA. *Math. Intelligencer* **12**, 71–80.
58. Sun, H., Treco, D. and Szostak, J. W. (1991). Extensive 3′-overhanging, single-stranded DNA associated with the meiosis-specific double strand breaks at the ARG4 recombination intitation site. *Cell* **64**, 1155–1161.
59. Thaler, D. S. and Stahl, F. W. (1988). DNA double-chain breaks in recombination of phage l and of yeast. *Ann. Rev. Genet.* **22**, 167–197.
60. Voet, D. and Rich, A. (1970). The crystal structures of purines, pyrimidines and their intermolecular complexes. *Prog. Nucl. Acid. Res. & Mol. Biol.* **10**, 183–265.

61. Wang, H., Di Gate, R.J. and Seeman, N. C. (1996). An RNA topoisomerase. *Proc. Nat. Acad. Sci.* (USA) **93**, 9477–9482.
62. Wang, H., Di Gate, R. J. and Seeman, N. C. (1998a), The construction of an RNA knot and its role in demonstrating that E. coli DNA topoisomerase III is an RNA topoisomerase. In *Structure, Motion, Interaction and Expression of Biological Macromolecules*, vol. 2, ed by R. H. Sarma & M. H. Sarma, Adenine Press, NY, 103–116.
63. Wang, J., Jiang, M., Nilsen, T. W. and Getts, R. C. (1998b). Dendritic nucleic acid probes for DNA biosensors. *J. Am. Chem. Soc.* **120**, 8281–8282.
64. Wang, Y., Mueller, J. E., Kemper, B. and Seeman, N. C. (1991). The assembly and characterization of 5-Arm and 6-Arm DNA junctions. *Biochem.* **30**, 5667–5674.
65. Watson, J. D. and Crick, F. H. C. (1953). A structure for deoxyribose nucleic acid. *Nature* **171**, 737–738.
66. Wells, A. F. (1977). *Three-Dimensional Nets and Polyhedra.* John Wiley & Sons, New York.
67. White, J. H., Millett, K. C. and Cozzarelli, N. R. (1987). Description of the topological entanglement of DNA catenanes and knots by a powerful method involving strand passage and recombination. *J. Mol. Biol.* **197**, 585–603.
68. Williams, R. (1979). *The Geometrical Foundation of Natural Structure.* Dover, New York.
69. Winfree, E. (1996). On the computational power of DNA annealing and ligation. In *DNA Based Computing*, ed. EJ Lipton, EB Baum. Providence: Am Math Soc, 199–219.
70. Winfree, E., Liu, F., Wenzler, L. A. and Seeman, N. C. (1998). Design and self-assembly of two-dimensional DNA crystals. *Nature* **394**, 539–544.
71. Yang, X., Liu, B., Vologodskii, A., Kemper, B. and Seeman, N. C. (1998a). Torsional control of double stranded DNA branch migration. *Biopolymers* **45**, 69–83.
72. Yang, X., Wenzler, L. A., Qi, J., Li, X. and Seeman, N. C. (1998b). Ligation of DNA triangles containing double crossover molecules. *J. Am. Chem. Soc.*, in press.
73. Zhang, Y. and Seeman, N. C. (1992). A solid-support methodology for the construction of geometrical objects from DNA. *J. Am. Chem. Soc.* **114**, 2656–2663.
74. Zhang, Y. and Seeman, N. C. (1994). The construction of a DNA truncated octahedron. *J. Am. Chem. Soc.* **116**, 1661–1669.

3D DNA PATTERNS AND COMPUTATION

NATAŠA JONOSKA
Department of Mathematics, University of South Florida, Tampa, Florida 33620
E-mail: jonoska@math.usf.edu

We propose use of three dimensional graph structures for solving computational problems with DNA molecules. Building blocks of k-armed branched junction molecules are used to form graphs. Algorithms to solve the Hamiltonian Cycle problem and the 3-vertex colorability problems are presented. Both of these algorithms require constant number of laboratory steps regardless of the size of the graph. Two mathematical problems that raise from this approach are initiated.

1 Introduction

The first thoughts that come to mind when the subject of mathematical biology is mentioned concern such mathematical models as the stochastic processes, differential equations and statistical methods that are used to solve problems in genetics, population growth or epidemiology. But since recently, with the seminal article of L. Adleman in 1994 [1] a new direction within the intersection of mathematics and biology has been introduced. L. Adleman [1] solved a small instance of the well known Hamiltonian Path Problem solely by using DNA molecules and techniques from molecular biology. Although the instance solved was tiny, the potential for using DNA molecules for computing opened a field for new ideas in both mathematics and biology. Many authors have built upon these ideas since [17,18,24,25,26].

The observation that the complexity of each living being is a result of several simple biological processes (transcription, replication, recombination...) that are guided by the information that is encoded in DNA reminded many mathematicians to a well known fact in recursion theory. Every computable function can be obtained by operations applied to a basic set using a small number of simple functions [8]. This observation accounts for the appearance of many theoretical models of DNA computers, capable of performing the complex computations of a universal Turing machine (for example [2,10,14,22,23,27,38]).

Like Turing machines, which operate on strings and accept languages, most of the DNA computer models use this linearity as a main force. The information in a DNA molecule is stored in a sequence of nucleotides, also called bases, A,G,C,T (adenine, guanine, cytosine and thymine) joined together by phosphodiester bonds. A single strand of DNA has also a "beginning" (usually denoted by $5'$) and an "end" (denoted by $3'$). Hence the molecule is oriented. The nucleotides also tend to seek their complements (A is complementary to T and C is complementary to G). This is the well known Watson-Crick complementarity. Two single stranded DNA molecules with complementary sequences of nucleotides and opposite orientation join together through hydrogen bonds and form a double stranded molecule which in space appears as a double helix. When double stranded molecules are heated to 95^oC, they disentangle (denature) into single stranded molecules. If single stranded molecules are cooled, they seek their complement and re-anneal into double stranded form. These properties are used by many models of DNA computers. A four letter alphabet can be used $\{A, G, C, T\}$ (in comparison with $\{0, 1\}$) and simple operations on the DNA strings can be performed using a few commercially available enzymes that act on the molecules in different ways.

With this article we would like to present a different approach to computing. Instead of strings and languages, we can use a capability of DNA that is unavailable to our electronic computers: the formation of three dimensional structures and the preservation of information within three dimensional structures in space. There are several motivations for this approach.

In a cell, where the DNA molecule appears naturally, the molecule is stored by intertwining millions of times, subjected to four or five successive coilings around proteins. The molecular biologists have studied supercoiling and catenanes of DNA; the actions of enzymes such as recombinases and topoisomerases that change the three dimensional structure, the topology of DNA ([7,37]). Most of this work was done in collaboration with mathematicians. Today it is well accepted that the successful transfer of information within a cell and the well being of the organism also depends on the so called topology of the DNA ([7,19,37]). Mathematicians have also been intrigued by these facts and have studied the action of enzymes on DNA and its topology [36] and have modeled RNA folding [15,16].

On the other hand, we are witnessing an explosion in new laboratory techniques and products of biotechnology. Chemistry laboratories are using synthesized DNA for designing and assembling unusual three dimensional DNA structures [30]. Junction molecules, used in the following section, are fairly well understood. These molecules were used to construct DNA polyhedra, a quadrilateral, a truncated octahedron and a cube [4,6,34]. Knots have been shown to be rather easy constructions (compared to other structures). In theory, virtually any knot can be constructed using right-handed B-DNA for negative crossings and left-handed Z-DNA for positive crossings [33,30,34,9]. Many catenanes and linkages of DNA molecules are known, but just recently, using B and Z DNA, the first Borromean DNA rings were assembled [21]. Another very important step towards constructing three dimensional DNA crystals was made by the design and the assembly of a two dimensional lattice made of DNA tiles [35,39].

Last, but certainly not least, the inspiration for using three dimensional graph structures as computational tool comes from the knot theory. Since the beginning of knot theory mathematicians have struggled to distinguish the equivalence classes of knots (up to moves and twists in space without cutting or gluing). Many invariants have been introduced such as the polynomials of Alexander, Jones, Vassiliev. We still do not know of a computable invariant that is a complete invariant, i.e. an invariant that uniquely determines the knot up to its equivalence class. The capability of DNA to form knots and three dimensional structures provides another way of manipulating information in a potentially much more powerful way.

The following section of this article describes algorithms for two NP-complete problems, the Hamiltonian Cycle problem and the 3-vertex colorability problem. These and similar problems make large demands of computational time on current computer systems such that graphs of even a moderate size become practically unmanageable. Here, it is proposed that they be solved by forming three dimensional graph structures with DNA molecules. The algorithms require only constant

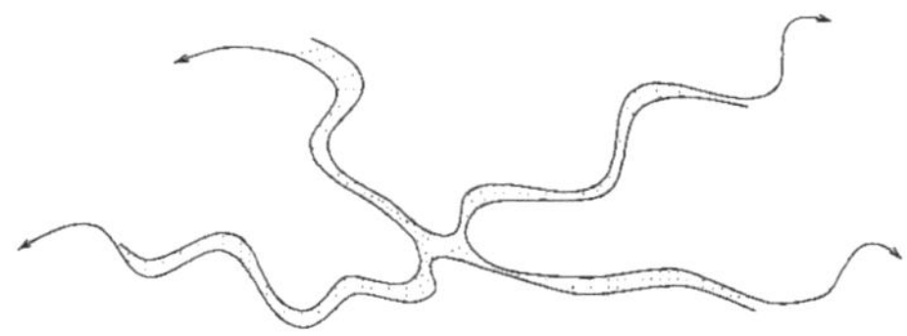

Figure 1. A four armed branched junction molecule.

number of laboratory steps regardless of the size of the graph. The purpose of this article is to point out another approach to DNA computing, therefore, scaling up of these algorithms to larger size of graphs is not discussed. The third section proposes two mathematical problems that have risen from these algorithms and the manipulation of DNA molecules.

2 Constructing Graphs with k-Armed Branched Molecules

In this section we will show how theoretically we can construct three dimensional (3D) structures, graphs in particular, and use this structures for computational purposes. Since the whole DNA computing fever started with the Adleman's solution of a small instance of the Hamiltonian Path Problem, we choose the Hamiltonian Cycle Problem (HCP) as the first problem for our description of three dimensional computations. The second problem that we describe is the three-vertex-colorability (3VC) problem which is also a well known NP-complete problem.

The k-armed branched molecules (k is a natural number ≥ 2) seem to be suitable for graph construction. An example of 4-armed branched molecule is presented in Fig. 1. In this figure, the double helix of the molecule is not presented. Hydrogen bonds between the anti-parallel, complementary Watson-Crick (WC) bonds are depicted as dotted segments between the strands. Polarity of the DNA strands is indicated with arrowheads being placed at the $3'$end. The angles between the "arms" and known to be flexible. If we allow each "arm" to be 200 to 300 base pairs long, then the "arms" of this molecule become rather flexible too and we deliberately show them curved. The $3'$ends are extended such that each arm ends with a single strand. This single stranded part will join its WC complement once placed in a test tube.

Construction and properties of these molecules are fairly well understood [31,32]. By making the body of the connection between the arms larger, it seems that we do not have to put any restrictions on how many arms a molecule can have [29]. In the following we show how these molecules can be used as a computational tool.

2.1 The Hamiltonian cycle problem

Let G be a finite graph with $V(G)$ the set of vertices and $E(G)$ the set of edges. A *Hamiltonian cycle* c of G is a cycle that goes through every vertex exactly once. The *Hamiltonian cycle problem (HCP)* asks whether a given graph G has a Hamiltonian cycle.

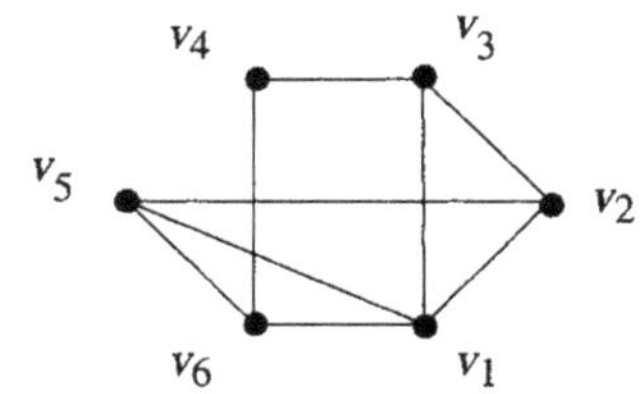

Figure 2. An example of a graph.

We will consider the graph presented in Fig. 2 as an example. The cycle $v_1v_6v_4v_3v_2v_5v_1$ is a Hamiltonian cycle. Of course, there are many other cycles that are not Hamiltonian, for example, $v_1v_6v_5v_1$ or the loop $v_2v_5v_1v_2v_5v_1v_2$.

We construct the graph by using k-armed branched molecules for each k-degree vertex. For the example presented in Fig. 2, we will need one 2-armed branched molecule (which is just a regular double-stranded DNA molecule), four 3-armed branched molecules and one 4-armed branched molecule. The $3'$end extension of each arm (representing one end of an edge) is complementary to the $3'$end of the arm (representing the other end) of the same edge. Since a cycle passing through a vertex uses only two edges incident to the vertex, there are $\binom{k}{2}$ possible ways for a cycle to pass through a vertex. Every path may appear in two opposing DNA orientations within the molecule, hence, for each k-degree vertex we form $2\binom{k}{2}$ k-armed branched molecules that represent that vertex. Each one of these molecules will represent one way a cycle could pass through the vertex. To ensure that only that path will be used, we phosphorylate the $5'$end of that connection. The ligase enzyme bonds, "ligates" the beginning of one DNA strand to the end of another DNA strand, but would not ligate (connect) ends if the $5'$end is not phosphorylated at the appropriate place. We illustrate this in Fig. 3 where we show six molecules representing six different paths through a vertex of degree 4. The stars indicate the $5'$end that will be phosphorylated. There will be six other molecules that establish that same connections between the edges but with the other orientations.

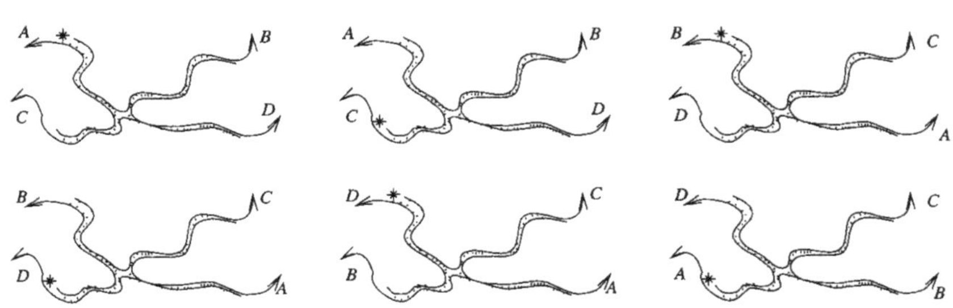

Figure 3. Six molecules representing a vertex with degree 4. The four codes A, B, C, D within the single stranded $3'$ends encode the four edges incident to the vertex.

The molecules are designed such that each single strand of the k-armed branched molecule has a constant length (say n). All vertex building blocks are combined and their compatible ends are allowed to form double-stranded DNA. The resulting molecules that have no open ends represent graphs; copies of the original graph, or covering graphs of the original graphs.

Proposition. Let $C = \{G_1, \dots, G_m\}$ be the set of graph structures G_i that are obtained by adding all building block molecules in a test tube and allowing them to join. Assume that C contains all possible graph structures. Let c be a cycle in the graph G. Then there is $G_j \in C$ that contains a circular single-stranded DNA molecule that encodes c.

The proof of this proposition follows from the design of the vertex building blocks.

The algorithm for the HCP has the following four steps:

- Combine all vertex building blocks in a single test tube and allow the compatible ends to hybridize and to form double-stranded DNA. The "nicks" that have 5′end phosphorylated are ligeted by a ligase. At this point, graph structures that represent covering graphs of the original graph need to be removed. Covering graphs are much more complex and heavier by molecular weight, hence their removal can be done by gel electrophoresis. (The problem with the covering graphs is addressed in the next section.) Gel electrophoresis is a technique that discriminates molecules of different sizes and different geometry and topology of DNA ([3,7,37,36]).

- Heat the molecules to 95^0C degrees. Molecules are denatured and the only circular molecules that are left are cycles of the graph. We remove non-circular DNA strands from the test tube. There are several ways that this can be performed for example by an exonuclease enzyme that recognizes the open ends or by gel electrophoresis. The exonuclease virtually destroys to single nucleotides molecules that are not circular.

- We choose a vertex v and use as PCR primers oligos that are complementary to the strands forming the building blocks of that vertex. The PCR amplifies the molecules and generates double stranded molecules that encode cycles that pass through v. Polymerase chain reaction (PCR) is a standard technique in molecular biology that cycles about 30 times through two or three temperature changes, uses DNA polymerase enzyme and produces an exponential increase of a given target molecule. We refer the reader to Ausubel et al. [3] for details.

- By gel electrophoresis we select molecules that are mn in length, where m is the number of vertices in the graph and n is the length of the single strands used for the vertex building blocks. If any molecules are recovered, then the answer to the HCP is "yes" otherwise is "no". We note that in parallel with the main experiment another experiment used as a positive control is carried out. This experiment is performed for a graph for which we know a Hamiltonian

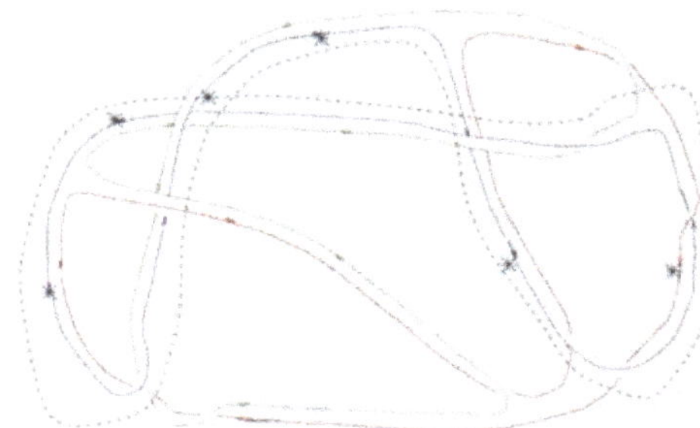

Figure 4. A graph structure that contains a Hamiltonian cycle.

cycle exists. Its success will confirm that our main experiment, producing an answer "no", is successful. This technique for carrying out positive and (or) negative control experiment is a standard technique in molecular biology and is an experimental necessity since many laboratory protocols are not 100% efficient.

One DNA graph structure for the graph in Fig. 2 that would be obtained after the first step of the algorithm is presented in Fig. 4. The colored lines indicate the three DNA single stranded molecules that make up the graph structure. The blue strand indicated with a dotted line is a circular single-stranded molecule that encodes the Hamiltonian Cycle. The stars indicate nicks that are ligated.

3 The Three Vertex Colorability Problem (3VC)

Let G be a graph with vertices V and edges E. The graph G is *3-vertex-colorable* if there is a surjective (onto) function $f : V \to \{a, b, c\}$ such that if two vertices $v, w \in V$ are adjacent (connected by an edge) then $f(v) \neq f(w)$. This means that out of three colors we assign one color to each vertex such that no two adjacent vertices are colored with the same color. The n-colorability is defined similarly.

In Fig. 5, we use the same example of a graph with a possible 3-vertex-coloring. As in the previous application, if a vertex has degree k then a k-armed branched molecule is used for its building block. As before, for the example presented in Fig. 5, we will need one 2-armed branched molecule, four 3-armed branched molecules and one 4-armed branched molecule (see Fig. 6). The 3′ends of the

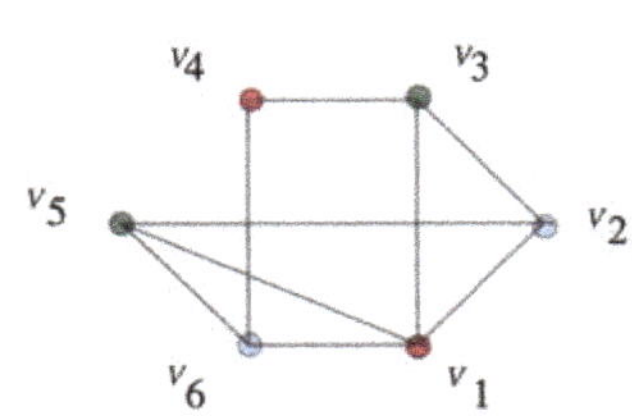

Figure 5. Coloring of a graph.

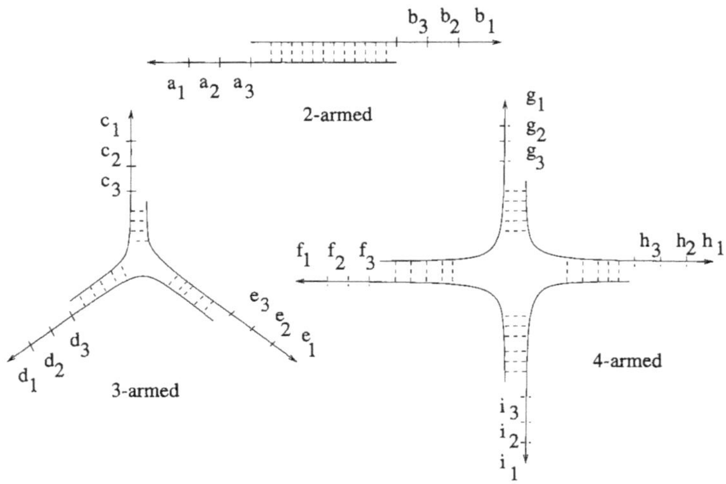

Figure 6. Building blocks for vertices.

Figure 7. Edge building block.

k-armed branched molecules end with single stranded extensions. These extensions are 30 to 45 nucleotides long and consist of three parts, each 10 to 15 nucleotides long. The first and the third parts (for example x_1 and x_3 where $x \in \{a, \ldots, i\}$ in Fig. 6) are specific encodings for the edge that is represented by the given arm of the molecule. The middle part of the encoding is the same for all arms of the vertex molecule and represents the color of the vertex. For each vertex three molecules are needed (each corresponding to one of the three possible colors of the vertex).

Each edge is a regular double-stranded molecule. The $3'$ends of the molecule end with single stranded segments that are complementary to the corresponding "arm" of the vertex that is incident to the edge. Hence, the first and the third part of the encoding at the $3'$end are $\bar{x}_1$ and $\bar{x}_3$ (see Fig. 7) which are WC-complementary to x_1 and x_3 ($x \in \{a, \ldots, i\}$) of the corresponding "arm" of the incident vertex molecule. The middle part of the encoding $\bar{x}_2$ is complementary to the color of the incident vertex, but here, the color sequence $\bar{x}_2$ at one $3'$end is different from the color sequence $\bar{y}_2$ at the other $3'$end.

For each edge we have exactly six double stranded molecules, each representing a pair of distinct colors at the endpoints of the edge molecule.

As above mentioned, to form the graph, all edge molecules and all vertex building blocks are combined and their compatible ends are allowed to form double-stranded DNA. Once formed, the edges are locked together by sealing all open "nicks" in the DNA strands with DNA ligase.

Proposition. For a given graph G, a graph structure can be formed by vertex building blocks and edge molecules if and only if G is 3-vertex-colorable or 2-vertex-colorable.

The proof of this proposition follows from the construction of the vertex and the edge building blocks.

The algorithm for the three vertex colorability now has the following four steps:

1. Combine all vertex building blocks with all edge molecules in a single test tube and allow the complementary ends to hybridize and be ligated.
2. Remove partially formed 3D DNA structures with open ends that have not been matched. This could be done by using an exonuclease enzyme.
3. From the graphs formed in the above steps, remove by gel electrophoresis the graphs that are larger than the original graph. This step will be explained in the next section.

   ll

4. If there are graph structures remaining in the test tube, then we conclude that the graph is 3-colorable. Note that all 2-colorable graphs also are 3-colorable. The conclusion is confirmed with a positive control experiment.

The number of laboratory steps in this procedure again does not depend on the number of vertices (or edges) in the graph. Once the building blocks are formed, the procedure needs only four steps to perform.

In Fig. 8 we show a possible formation of a graph structure. The graph structure depicted forms a knot as a single stranded DNA. Such knot structures have been used to study recombination enzymes [37], and similar techniques for detecting the knot structures might be useful here. There are graphs, however, that do not form a single stranded knot, but rather form links (multicomponent circles), and partially formed graphs may contain smaller knots and links as well. We address this issue in the next section.

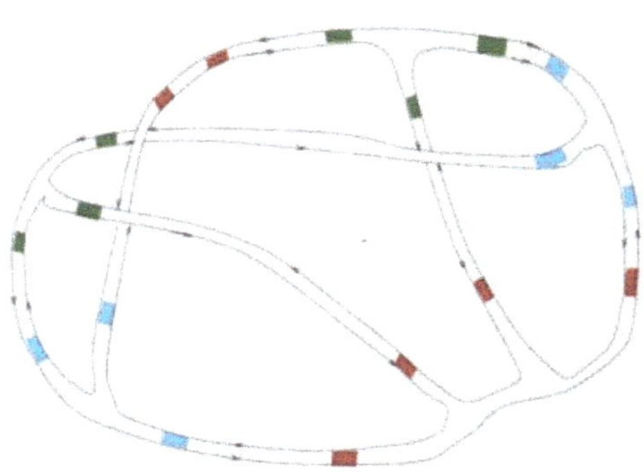

Figure 8. DNA graph structure of Fig. 5. Colored shades represent DNA sequences that correspond to the three colors.

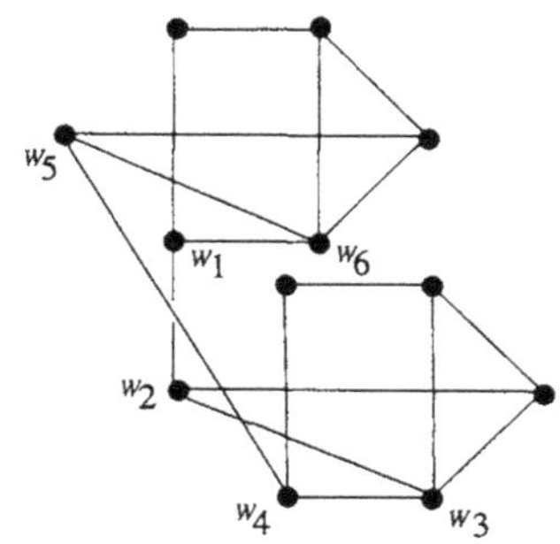

Figure 9. Double cover of the graph in Fig. 5.

4 Recognition of Graph Structures

At the end of the procedure for the 3VC problem we are faced with the problem of recognizing the formed graph structure. In both procedures there is also a possibility that instead of the graphs that we want, the covering of the graphs will be formed. These graphs can be formed because the graphs are built with vertex and edge building blocks containing the information about the graph locally. Covering graphs locally are the same as the original graph. An example of a double cover of the graph in Fig. 2 is depicted in Fig. 9.

The molecules representing covering graphs are larger and heavier then the graph structures that we are seeking. There have been many gel-electrophoresis techniques that have been developed that together with electronic microscopy can be used in recognizing different three dimensional DNA structures [5,4,7,37]. Many of these techniques have already been established and molecular biologists and chemists are confident that different graph structures made of DNA are distinguishable [4,7,37,39]. These techniques have been the main investigation of the so called biochemical topology of DNA where topological methods are used to deduce enzymatic mechanism to substrate DNA structure from a change in DNA topology associated with a reaction. The whole field has been enriched by mathematical knot theory and as Cozzarelli [7] says, most of this work required a collaboration with mathematicians and he believes that "mathematics will make an increasingly important contribution to biology".

4.1 Covering graphs and complexity

We mention here that another classical NP-complete problem, the satisfiability (SAT) problem could be solved by using three dimensional graph structures [12]. SAT asks whether for a given logical formula α there is an assignment ({true, false}) to the variables in α that would give α the value "true". The algorithm for this problem is based on an equivalent description through graphs and contact networks [20]. The number of steps in this algorithm increases linearly with the number of variables in α. An interesting observation is worth mentioning.

Using 3D DNA structures we can describe algorithms for solving three very well known NP-complete problems, the Hamiltonian Cycle problem (HCP), the 3-vertex colorability problem (3VC) and the Satisfiability problem (SAT) ([12]). For all of these problems, it is possible that graphs that are multiple covers of the original graph are formed by the DNA. But, only for SAT, the multiple covers of the original graph do not produce a false answer. The molecules that encode solutions in the covering graphs correspond to solutions in the original graph ([12]). This certainly is not the case with HCP and 3VC. For example, in Fig. 9 the cycle $w_1w_2 \cdots w_6w_1$ passing through exactly six vertices is a cycle in the double cover of the graph that corresponds to the non-Hamiltonian loop $v_6v_5v_1v_6v_5v_1v_6$ in the original graph in Fig. 2. A similar simple example can be found, where a double cover of a graph can produce a wrong answer for the 3VC problem. On the other hand, the algorithms for the HCP and the 3VC require only a constant number of steps in the laboratory. We have been unable to find such a simple algorithm for SAT. We mention here that it has been shown that theoretically "one pot" biochemical reactions have a large potential for computation [41]. In particular, SAT could be solved by this technique. But the algorithm still requires a linear increase in number of sequential interactions of the molecules.

The above observations raise a question about the complexity of these problems within a 3D setting. In the complexity theory, all three problems are in the class of NP-complete. Due to an essential difference between the classical notion of computation and the computation with 3D structures, we cannot apply the classical notion of complexity to 3D algorithms. This calls for a a definition of complexity of 3D algorithms in conjunction with a notion of complexity of DNA computations.

Another question is also raised: are the potential problems that may appear by formation of the covering graphs some sort of "trade off" for reducing the computational steps from linear increase (in SAT) to constant (in 3VC and HCP)?

4.2 Graphs made of one single stranded DNA molecule

Besides many developed techniques in working with DNA the researchers are still more confident that a complex graph structure made of DNA would be much simpler to detect if it is made of one single stranded DNA molecule [4,5,28]. If a circular double stranded molecule is heated and denatured, the result is two linked loops made of single stranded DNA. This may cause difficulties in separating the desired molecules. In the previous section we described two algorithms for solving computational problems by DNA molecules where in both cases during the process of computation a graph structure was constructed. In the algorithm for HCP, the final solution did not depend on whether the graph structure could be distinguished or not, but in the case of 3VC, the final result depends on how well we can determine the formation of a DNA graph structure that corresponds to our original graph. As a mathematical problem we are faced with a question: can we characterize the graphs that can be constructed by one single stranded DNA molecule?

We can state the problem mathematically. Let G be an undirected graph with set of vertices V and set of edges E. Let $\bar{G}$ be a directed graph obtained from G by replacing each edge $e \in E$ with a pair of directed edges e^+, e^- with opposite orientation. If e is incident to vertices v_e, w_e then we write $e^+ = (v_e, w_e)$ and $e^- = (w_e, v_e)$.

A path $p = e_1 \cdots e_n$ in $\bar{G}$ is a *strand path* if p is a loop (starts and ends at the same vertex) such that $e_i \neq e_j$ if $i \neq j$ $(i, j = 1, \ldots, n)$. In particular, e^+e^- is a strand path for each edge $e \in E$.

The graph G is *s-DNA covered* if there are s strand paths $p_1, \ldots, p_s$ in $\bar{G}$ such that for every edge e of $\bar{G}$ there is a unique i such that p_i passes through e.

Using the construction from the previous section we can easily show that every graph can be s-DNA covered for some s. In the DNA graph structure every DNA strand corresponds to a strand path p. We denote with $c(G)$ the minimum of all s such that G is s-DNA covered.

Problem 1: given a graph G, determine $c(G)$.

Problem 2: characterize graphs G such that $c(G) = 1$.

We refer to the graph in Fig. 2. Two different DNA structures of the same graph are presented in Fig. 4 and in Fig. 8. In the first case, G is 3-DNA covered and in the second case it is 1-DNA covered. So $c(G) = 1$.

Every vertex in a graph (with degree greater than 2) can be described by 3-junction molecules (by extending the body of the central part of the vertex molecule) [28]. In Fig. 10 two 3-junction molecules presenting one vertex are depicted. The difference between these two molecules is in the connection between the edges (arms) that is established. The molecule (b) is essentially same as the molecule (a). If we twist the sides B and C in (b) so that C comes at the top and B comes on the right, we have the same structure molecule as in (a). [We note that "twist"is used only intuitively for the purpose of the theoretical argument. Twist might not be a good way to show differences in configuration for DNA [28]. If we set a reference point at A, a half-turn difference in the length of the double helix at side A reverses the positions of B and C in space.]

When the 3-arm branched molecule becomes part of a 3D DNA structure, the three single strands can be part of three, two or one circular single stranded molecule. If they are part of three circular single stranded molecules then, as shown in Fig. 11 (i.a), by changing the connection from (a) to (b) in Fig. 10, the three circular single stranded DNA molecules become one. After the change in the connection, one circular single stranded DNA molecule either becomes three (as in Fig. 11 (i.a)) or remains to be one. (Fig. 11 (i.b)). If we have two circular single

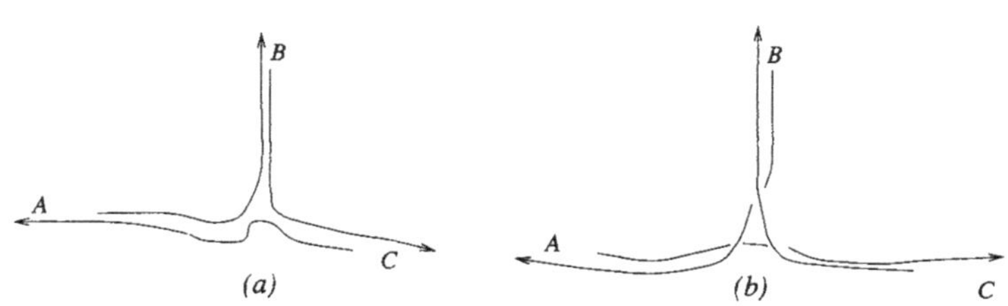

Figure 10. Different connections in a vertex of degree 3.

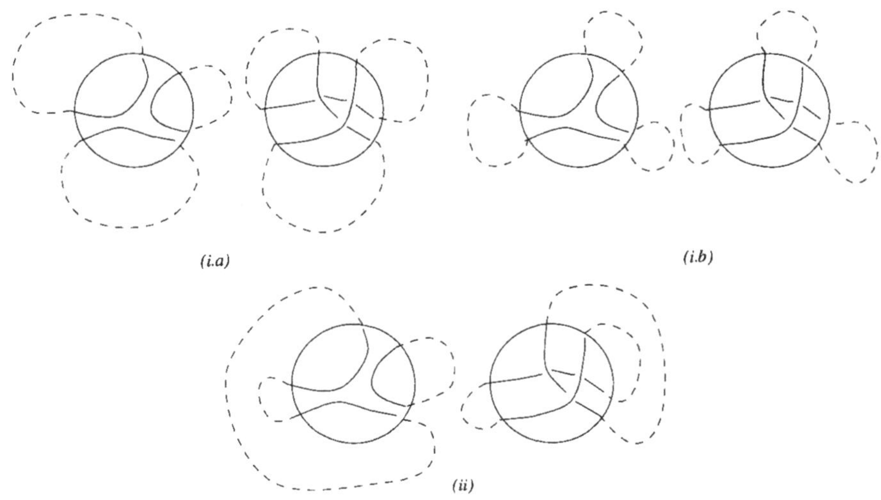

Figure 11. Cases of moving between connection (a) and (b) in Figure 10.

stranded DNA molecules, by changing in between connections ((a), (b) in Fig. 10) the number of strands remains at 2 (Fig. 11 (ii)).

We say that a graph is *weakly connected* if there is an edge e such that if we remove e from G then G becomes disconnected. Using the change of connections within 3-armed branched molecules as shown in Fig. 11 we can show the following:

Proposition. If G is a planar graph that is not weakly connected, then $c(G) \leq 2$.

There are examples of planar weakly connected graphs with $c(G) > 2$. We are not sure whether the above proposition holds for graphs with arbitrary genus.

5 Final Remarks

The purpose of this article is, through two examples, to propose investigations of the use of 3D structures in DNA computing, both in theory and experiments. The ideas were motivated by the work of N. Seeman and his laboratory group who demonstrated not only that 3D DNA structures appear naturally, but it is possible to design, assemble and construct 3D DNA structures in the laboratory. The most recent success is a construction of a molecular mechanical device attributed to the properties of B-Z transition of DNA [35]. This opens a horizon for design of computational 3D DNA objects that can participate in biochemical processes, and at the same time perform controlled mechanical movements.

We believe that the theoretical algorithms and laboratory procedures described here are empirically possible. At the moment, the 3D DNA structure manipulations described here have not been tested in the laboratory. The success of such protocols will require extensive experimentation. The first step of the procedures described in

Section 2 assumes that all possible combinations of graph structures are generated. This step is similar to the first step in many DNA based algorithms as well as in the Adleman's algorithm [1]. Studies have shown, however, that even in a fairly simple mixture, appropriate reaction conditions for annealing are important and difficult to determine [11]. To date, we are unaware of laboratory experiments of this type being performed on 3D DNA structures. It is our belief, however, that as polymer technology develops, constructing the 3D DNA structures needed for this approach should become easier.

Formation of covering space structures remains possible. They themselves carry a substantial amount of information and may be useful in many computational problems.

Acknowledgment

The ideas presented here are a product of valuable conversations with Tom Head, Nadrian Seeman, Erik Winfree, Junghuei Chen, Masahico Saito, Stephen Karl. Since my graduate student days, Professor Tom Head has continually provided me with advice, support, guidance and encouragement. The idea for the HCP algorithm presented here is a result of my conversation with N. Seeman and E. Winfree in search for simplification of the structures that were introduced earlier [13]. Junghuei Chen has been a valuable support with his confidence that the algorithm for 3VC can be a doable experiment. I have had many mathematical conversations with my colleague Masahico Saito. The problem of characterizing the graphs that can be constructed by one single stranded DNA molecule is one of our common interests and the proposition in Section 3 is a product of our conversations. The molecular biologist, Stephen Karl, has been the most instrumental person in my efforts to learn molecular biology.

References

1. L. Adleman, *Molecular computation of solutions of combinatorial problems*, *Science* **266** (1994), 1021–1024.
2. L. Adleman, *On constructing a molecular computer*, in R. Lipton, E. Baum, editors *DNA based computers*, DIMACS: Series in Discrete Mathematics and Theoretical Computer Science, **27** Providence RI, American Mathematical Society (1996), 1–21.
3. F. M. Ausubel, R. Brent, R. E. Kingston, D. D. Moore, J. G. Seidman, J. A. Smith, K. Struhl, P. Wang-Iverson and S. G. Bonitz. *Current Protocols in Molecular Biology*, Greene Publishing Associates and Wiley-Interscience, New York, NY (1993).
4. J. Chen and N. C. Seeman, *Synthesis from DNA of a molecule with the connectivity of a cube*, Nature **350** (1991), 631–633.
5. J. Chen, private communication.
6. J.-H. Chen, N. R. Kallenbach and N. C. Seeman, *A Specific Quadrilateral Synthesized from DNA Branched Junctions*, Journal of the American Chemical Society **111** (1989), 6402–6407.

7. N. R. Cozzarelli, *The structure and function of DNA supercoiling and catenanes*, The Harvey Lecture Series **87** (1993), 35–55.
8. M. Davis, R. Sigal and E. J. Weyuker, *Computability, Complexity, and Languages*, Academic Press, 1994 (sec. edition).
9. S. M. Du, H. Wang, Y. C. Tse-Dinh and N. C. Seeman, *Topological Transformations of Synthetic DNA Knots*, Biochemistry **34** (1995), 673–682.
10. T. Head, Gh. Paun and D. Pixton, *Language theory and molecular genetics*, in *Handbook of formal languages* (G. Rozenberg, A. Salomaa, editors) **2** (1997), 295–358.
11. N. Jonoska and S. A. Karl, *Ligation Experiments in DNA Computations* Proceedings of 1997 IEEE International Conference on Evolutionary Computation (ICEC'97), April 13–16, 1997, 261–265.
12. N. Jonoska, S. Karl and M. Saito, *Three dimensional DNA structures in computing*, in Ref. 26.
13. N. Jonoska, S. Karl and M. Saito, *Creating 3-dimensional graph structures with DNA*, in Ref. 25.
14. L. Kari, *DNA Computing: arrival of biological mathematics*, The Mathematical Intelligencer **19** No. 2 (1997), 9–22.
15. L. H. Kauffman and Y. B. Magarshak, *Vassiliev knot invariants and the structure of RNA Folding*, in *Knots and applications*, Ser. Knots Everything, **6**, World Sci. Publishing, River Edge, NJ (1995), 343–394.
16. L. H. Kauffman and Y. B. Magarshak, *Graph invariants and the topology of RNA folding*, Random knotting and linking (Vancouver, BC, 1993). J. Knot Theory Ramifications **3** No. 3 (1994), 233–245.
17. J. R. Koza, K. Deb, M. Dorigo, D.B. Fogel, M. Garzon, H. Iba and R. L. Riolo, eds. (1997). *Proc. 2nd Annual Genetic Programming Conference*, Morgan Kaufmann.
18. J. R. Koza, K. Deb, M. Dorigo, D. B. Fogel, M. Garzon, H. Iba and R. L. Riolo, eds. (1998). *Proc. 3rd Annual Genetic Programming Conference*, Morgan Kaufmann.
19. L. Landweber and L. Kari, *The evolution of cellular computing: nature's solution to a computational problem*, in Ref. 26.
20. R. Lipton, *DNA solution of hard computational problems*, Science **268** (1995), 542–545.
21. C. Mao, W. Sun and N. C. Seeman, *Construction of Borromean Rings from DNA*, Nature **386** (1997), 137–138.
22. Gh. Paun, G. Rozenberg and A. Salomaa, *Computing by splicing*, Theoretical Computer Science **168**, No. 2 (1996), 321–336.
23. Gh. Paun, *On the power of the splicing operation*, Intern. J. Computer Math. **59** (1995), 27–35.
24. Proceedings of the Second Annual Meeting on *DNA Based Computers*, DIMACS Workshop, (R. Lipton and L. Landweber editors) Princeton, NJ, June 10–12, 1996 (in press).
25. Proceedings of the Third Annual Meeting on *DNA Based Computers*, DIMACS Workshop, (D. Wood editor) University of Pennsylvania, June 23–25, 1997 (in press).

26. Proceedings of the Fourth Annual Meeting on *DNA Based Computers*, DIMACS Workshop, (L. Kari editor) University of Pennsylvania, June 15–19, 1998 (in press).
27. J. Reif, *Parallel molecular computation:models and simulation*, Seventh Annual ACM Symposium on Parallel Algorithms and Architectures (SPAA95), Santa Barbara, June 1995.
28. N. C. Seeman, private communication.
29. N. C. Seeman and N. R. Kallenbahc, *Nucleic Acid Junctions: A successful Experiment in Macromolecular Design*, in *Molecular Structure: Chamical Reactivity and Biological Activity* (J. J. Stezowski, J. L. Huang and M.C. Shao editors) Oxford Univ. Press, Oxford (1988), 189–194.
30. N. C. Seeman *et al. The Perils of Polynucleotides: The Experimental gap Between the Design and Assembly of Unusual DNA Structures* in Ref. 24.
31. N. C. Seeman *et al. Gel electrophoretic analysis of DNA branched junctions,* Electrophoresis, 10 (1989), 345–354.
32. N. C. Seeman, *Nucleic Acid Junctions and Lattices*' Journal of Theoretical Biology **99** (1982), 237–247.
33. N. C. Seeman, *The Design of Single-stranded Nucleic Acid Knots* Molecular Engineering **2** (1992), 197–307.
34. N. C. Seeman, Y. Zhang, S. M. Du and J. Chen, *Construction of DNA polyhedra and knots through symmetry minimization*, Supermolecular Stereochemistry, J. S. Siegel, ed. (1995), 27–32.
35. N. C. Seeman, F. Liu, C. Mao, X. Yang, L. Wenzler and E. Winfree, *DNA nanotechnology: Control of 1-D and 2-D arrays and the construction of a nanomechanical device* in Ref. 26.
36. D. W. Sumners, *Lifting the curtain: using topology to probe the hidden action of enzymes*, Notices of the AMS **42**, No. 5 (1995), 528–537.
37. S. Wasserman and N. Cozzarelli, *Biochemical topology: applications to DNA recombination and replication*, Science **232** (1986), 951–960.
38. E. Winfree, X. Yang and N. C. Seeman, *Universal Computation via Self-assembly of DNA: Some Theory and Experiments,* in Ref. 24.
39. E. Winfee, F. Liu, L. Wenzler and N. C. Seeman, *Design of self-assembly of two-dimensional crystals*, Nature, to appear, 1998.
40. E. Winfree, private communication.
41. E. Winfree, *Whiplash PCR for O*(1) *computing* in Ref. 26.

CIRCULAR SUGGESTIONS FOR DNA COMPUTING

TOM HEAD
Mathematical Sciences, Binghamton University
Binghamton, New York 13902-6000 USA
E-mail: tom@math.binghamton.edu

An introduction to DNA computing for non-specialists is given in a novel fashion. Wet lab solutions using circular double stranded DNA molecules are sketched for several classical algorithmic problems. Laboratory tests of these suggestions are in the initial stages.

1 Introduction

The solution of combinatorial problems using DNA molecules in aqueous solution was begun by Leonard Adleman with the report, in the journal *Science* [1], of his laboratory solution of an instance of the Hamiltonian path problem. Several scientific and computing magazines have now featured expository articles on the resulting birth of DNA computing. Surely the best of these introductions is Adleman's own recent recounting, for *Scientific American* [2], of the work he reported in [1]. In the present article we introduce DNA computing again, but in a manner that we hope is completely fresh. Solutions that use *circular double stranded* DNA molecules are suggested for several familiar algorithmic problems. Laboratory investigations of the procedures presented here are in progress at Binghamton University and at the Center for Natural Computing at Leiden University.

This article arose from a reading of the paper of Q.Ouyang, P.D.Kaplan, S.Liu and A.Libchaber [14] in which the authors reported their laboratory solution of an instance of the maximal clique problem. The first procedure suggested here is a variant of their solution in which their initial step, the construction of the exponential number of all potential solutions, is replaced by *initialization with one single circular molecular variety* which is seen to be capable of providing representations of all solutions as the computation progresses. Once the possibility of initializing DNA computations with a single circular molecular variety became clear, the remaining procedures suggested here were easily conceived. The technique is likely to be applicable to additional algorithmic problems.

Only toy instances of algorithmic problems have thus far been solved using DNA chemistry [1,14]. However, we believe that investigations and experiments currently in progress at institutions in several nations are of fundamental scientific significance and that they will contribute positively to the science and technology of the future. One should read carefully the considerations given in [1] for the potential of DNA computing in speed, energy consumption, and information storage. Adleman firmly established the following statement with which he concluded his [1] abstract: "This experiment demonstrates the feasibility of carrying out computations at the molecular level". Researchers in biomolecular computing are currently exploring this feasible new world of computations and being invigorated by the fresh new visions of computation that it provides. An antidote for excessive euphoria is available in [6].

Further valuable introductions and surveys are [11] and [17]. The books [16] and [15] incorporate a wide view of biomolecular computing and the formal representation of biomolecular processes. If additional background concerning molecules and enzymes is desired, help is available in [11,9] and [13]. Circular double stranded DNA molecules have been related to the theories of formal languages and computation previously in [7,8,21,18,22,9] and [19]. Early instances of the simulation of Turing machines using the cut and paste methods employed here appear in [3] and [20].

2 Naturally Occurring Plasmids

DNA molecules occur in nature in both linear and circular form. In bacteria the main genome is usually a circular double stranded DNA molecule consisting of a few million base pairs (bps). In addition to this large circular DNA there may occur small circular double stranded DNA molecules each consisting of a few thousand base pairs (bps). These smaller circular molecules are called *plasmids*. Plasmids may carry adequate information encoded in their sequences to provide for their replication and for the transcription of their genetic information into RNA within the bacterium in which they occur. Genetic engineers use this property of plasmids to convert bacteria into factories for the production of macromolecules of DNA, RNA, or protein, as desired. They do this using a cut and paste technique. The cutting is done using *restriction enzymes*. These enzymes are precision tools that cut double stranded DNA molecules only at segments having specific sequences called *sites*. The pasting is done with a general purpose *ligase* enzyme. Since these cut and paste operations are key steps in the procedures suggested here, we describe in detail a cut operation followed by a paste operation at the biomolecular level:

Suppose that a circular double stranded DNA molecule M has a unique segment of the following form:

```
...GAATTC...
...CTTAAG...
```

where:

(1) A, C, G, & T represent the deoxyribonucleotides having bases adenine, cytosine, guanine, & thymine, respectively;

(2) The '...' portion consists of an unspecified sequence of deoxyribonucleotides that wraps around to complete the circular form of M;

(3) the horizontal adjacencies represent strong (covalent) chemical bonds between successive deoxyribonucleotides;

(4) The vertical adjacencies represent weak (hydrogen) bonds; and

(5) we recall that, in well formed DNA, G bonds (vertically here) only with C, A only with T, T only with A, and C only with G.

If this molecule M is in an appropriate aqueous solution (buffer) and the restriction enzyme <u>Eco</u>RI is added, then the enzyme will find this segment of the DNA molecule M and cut one strong bond in the top strand and one strong bond in the bottom strand of M in just such a way that the segment shown above will be dissociated as follows:

```
...G        AATTC...
...CTTAA        G...
```

Since M was originally circular, after being cut as shown it remains a single DNA molecule although it is linear after this cut has been made. The protruding four base single stranded segments, each of which can be read, from the tips of the protrusions toward the double stranded portion, as '`AATT`', are called *sticky ends* (or, more technically: 5'-overhangs). Each of the pairs of `A`'s of one protrusion spends part of its time close to the pair of `T`'s of the other protrusion - allowing temporary weak (hydrogen) bonds to reform. However, since the firm (covalent) bonds in the backbones that originally connected the relevant `G`'s with the relevant `A`'s have been cut, each pair of `A`'s is bumped loose from the pair of `T`'s by random motion of the DNA and the surrounding water molecules.

If the restriction enzyme, EcoRI in this case, is removed from the solution and an enzyme called a *ligase* is added to the solution then the following can take place: At a moment at which the two sticky ends have formed temporary hydrogen bonds, the ligase will reconstruct the two strong bonds that were previously cut by EcoRI. Thus the original molecule M will be restored. We have now illustrated the operation CUT (with a restriction enzyme) and the operation PASTE (with a ligase). These two operations will be combined into one compound operation in the next section. It is this compound operation that is the basic computational operation for the procedures suggested here.

There are about 200 restriction enzymes commercially available at present. The sequential application of a set of several restriction enzymes acting at distinct non-overlapping sites in circular DNA molecules is fundamental for the procedures suggested here.

3 The Delphic Plasmid, DEL

The fundamental data structure for the computational work suggested here is an artificial plasmid DEL, which we call the *Delphic Plasmid.* This plasmid DEL is to be constructed such that, for a specified set of restriction enzymes $\{RE_1, RE_2, \ldots, RE_n\}$, the plasmid contains a segment of the form

$$\ldots C_1S_1C_1C_2S_2C_2\ldots C_iS_iC_i\ldots C_{(n-1)}S_{(n-1)}C_{(n-1)}C_nS_nC_n\ldots$$

for which:

(1) The subsegments $C_1, C_2, \ldots, C_n$ are sites at which the enzymes $RE_1, RE_2, \ldots, RE_n$, respectively, can cut the plasmid. Moreover, for each j, the two occurrences listed are the only occurrences of sites in DEL at which REj can cut DEL. [Mnemonic: 'C' is for Cut site.]

(2) The subsegments $S_1, S_2, \ldots, S_n$ are of fixed length (perhaps 21 base pairs). These subsegments will be called *stations.* Attention will be focused almost entirely on these stations. To allow more detailed analysis than will be discussed here, the sequence for each station may be chosen so that no other

subsegment of DEL has the same base pair sequence. [Mnemonic: 'S' is for Station.]

How many stations does the Delphic Plasmid have? As many as practicality allows. Preliminary testing is planned using a plasmid having six stations. This will allow a comparison with [14]. If results with six stations are encouraging then larger plasmids will be constructed and tested for use in computation.

The key computational operation to be used on plasmids, starting with the Delphic Plasmid, will be a compound operation that will consist of the removal of one station, say S_j, from a (circular) plasmid followed by the return of the resulting (linear) molecule to circular (plasmid) form. This will be a cut & paste process done as follows: (1) RE_j is added to an aqueous solution containing the plasmid. This cuts the plasmid into two linear pieces, a short piece and a long piece (each of which has the sticky ends determined by RE_j). The short piece is merely the *station* S_j with a fragment of C_j at each end. (2) RE_j is removed from the aqueous solution (and also the short fragment, if convenient or if necessary). (3) A ligase is added to the (dilute) aqueous solution containing the long fragment. The ligase then circularizes the long linear molecule into a plasmid.

The net effect of the compound operation just described on a plasmid of the sort we are concerned with here is to remove the station. In the case above, the net change in the plasmid operated on was that a new plasmid was produced that is identical with the original plasmid - except for the net omission of the station S_j and the equivalent of one of the adjacent RE_j sites C_j. Thus in the case that S_j has length 21 bps and RE_j is EcoRI, the resulting plasmid is $21 + 6 = 27$ bps shorter (i.e., less in circumference) than the initial plasmid. If each of the sites RE_j has the same length, then each deletion operation will shorten the plasmid by precisely the same number of base pairs. (If enzymes having sites of different length are used then the plasmid should be constructed to include additional bases between each S_j & C_j so that all deletions have the same length.) Thus after an unknown number of deletion operations has been made, the exact number of deletions can be immediately calculated from the length of the resulting plasmid. This fact is absolutely fundamental for the method of computation suggested here.

For efficient communication and algorithm presentation we provide a notation that will allow us to avoid reference to enzymes, and require only minimal reference to DNA molecules. The compound biochemical deletion process described above will be expressed as follows: DELETE(Station-Name). Since the station deleted above was named S_j, this specific instance would now be expressed: DELETE(S_j). This DELETE operation is the central computational step, but two more operations play significant roles. There are standard laboratory techniques that allow DNA molecules to be separated into sets that have the same length as measured in bps. One of these is gel separation, but more efficient methods are expected in the future. Such separation procedures allow the selection from a mixture of DNA molecules of those molecules having lengths that satisfy specifiable conditions. Such compound biochemical selection processes will be expressed here by focusing on the molecules that are discarded (or dropped). We express this process as follows: DROP(Condition-on-molecule-lengths). Thus the removal of all DNA molecules of length less than 50 bps from an aqueous solution is expressed as follows: DROP(length < 50). (In some of the procedures below optional DROP statements have been included and enclosed in braces.)

Our algorithms will frequently require that the contents of a single test tube be divided into a number of equal parts in different test tubes. The pattern of these statements is illustrated by:

DIVIDE(3)
  1: DELETE(S_1)
  2: DELETE(S_2)
  3: DELETE(S_3)
UNITE

The meaning here is that the content of one test tube is divided into three (always new) test tubes. In the first test tube the operation to be performed (in this specific case) is DELETE(S_1); in the second DELETE(S_2); and in the third DELETE(S_3). The contents of the three test tubes are then united into a single (always new) test tube. Steps 1,2,3 can be done in parallel or in any desired sequential order.

FOR ... END_FOR statements will be used in the standard manner that is familiar from many common programming languages. A special feature of the computational scheme described here is that every algorithm starts from the same initialization: *We always start with a single test tube containing copies of the Delphic Plasmid.* Reasonable initializations might consist of a few microliters of water (buffer) containing some number of billions of the Delphic Plasmids. Our algorithmic representations will not consider such details. Thus by INITIALIZE we mean begin with one test tube containing an appropriate number of the Delphic Plasmids.

4 Independent Subsets and Vertex Covers for Undirected Graphs

Let $G = (V, A)$ be an undirected graph, where V is a finite set of *vertices* and A is a subset of the set of all two element subsets of V. The elements of A will be called *arcs*. By an independent subset S for G we mean a subset of V for which, for each arc $\{p, q\}$ in A, it is not the case that both p and q are in S. An independent subset S is *a largest independent subset for* G if no independent subset for G contains more elements than S contains. The problem of finding the number of elements in a largest independent subset for a graph $G = (V, A)$ is known to be NP-complete [5]. Using the parallelism available through the action of enzymes the following procedure provides this largest number using a number of steps bounded linearly by the number of arcs of G:

Procedure 1. With each vertex p of G associate a station of DEL. Let each such p now denote both the original vertex and its associated station. Throughout the computation the *presence* in a DNA molecule, M, of the station p will have the interpretation that the vertex p is a member of the subset of V represented by M. Thus each (circular) DNA molecule will serve as an unordered list of the vertices that comprise the set represented by the molecule M.

```
INITIALIZE;
FOR each arc {p, q} in A DO
   DIVIDE(2)
      1: DELETE(p)
      2: DELETE(q)
   UNITE
END_FOR;
DROP(station count < maximum station count).
```

Observe that each of the remaining DNA molecules encodes a largest independent subset for G. Thus the number of stations (of those associated with elements of V) occurring in any one of these remaining DNA molecules is the answer to the problem proposed. The number of stations present is available immediately from the length of any remaining molecule. #

The cliques in a graph $G = (V, E)$ are precisely the independent subsets of $G' = (V, [V \times V] \backslash E)$. Consequently the number of vertices in a largest clique in a graph $G = (V, E)$ can be found by applying Procedure 1 to the graph $G' = (V, [V \times V] \backslash E)$.

Let $G = (V, A)$ be as above. By a vertex cover for G we mean a subset S of V for which, for each p in V, either p is in S or there is an arc $\{p, q\}$ in A for which q is in S. A vertex cover S is *a smallest vertex cover for* G if no vertex cover for G contains fewer elements than S contains. The problem of finding the number of elements in a smallest vertex cover for a graph $G = (V, A)$ is known to be NP-complete [5]. The following procedure provides this smallest number with a number of steps bounded linearly by the product of the number of vertices with the number of arcs of G:

Procedure 2. With each vertex p of G associate a station of DEL. Let each such p denote both the original vertex and its associated station. Throughout the computation the *absence* in a DNA molecule M of the station p will have the interpretation that the vertex p is a member of the subset of V represented by M. Thus each (circular) DNA molecule will serve as an unordered list of the vertices that comprise the *complement* of the set represented by the molecule M.

```
INITIALIZE;
FOR each isolated p in V DO
   DELETE(p)
END_FOR
FOR each non-isolated p in V DO
   Let q1 = p and let the vertices adjacent to p be q2, ..., qk(p);
   DIVIDE(k)
      1: DELETE(q1)
      2: DELETE(q2)
      ...
      k: DELETE(qk(p))
   UNITE
END_FOR;
DROP(station count < maximum station count).
```

Observe that each of the remaining DNA molecules encodes a smallest vertex cover for G. Thus the answer to the problem proposed is the number of missing stations of any one of these remaining DNA molecules. #

5 Partitions into Hamiltonian Subgraphs

Let $G = (V, E)$ be a directed graph where V is a finite set of n vertices and E is an *irreflexive* binary relation on V. The elements of E will be called (directed) edges. Can V be partitioned into disjoint subsets $V_1, V_2, \ldots, V_k$, for some positive integer k, such that, for each j, $(1 \leq j \leq k)$, V_j contains at least three vertices and the subgraph $G_j = (V_j, \{(p,q) \text{ in } E : p \text{ and } q \text{ in } V_j\})$ contains a Hamiltonian circuit? This problem is known to be NP-complete [5]. The following procedure provides the correct 'YES'/'NO' answer. If a uniform bound is placed on the degrees of the vertices occurring in the graphs to which the procedure is to be applied then the number of steps required is bounded linearly by the number of vertices. On the class of complete graphs the number of steps is cubic in the number of vertices.

Procedure 3. With each edge (p, q) of G associate a station of DEL. Let each such (p, q) denote both the original edge and its associated station. Throughout the computation the *presence* in a DNA molecule M of the station (p, q) will have the interpretation that the edge (p, q) is a member of the subset of E represented by M. Thus each (circular) DNA molecule will serve as an unordered list of the edges that comprise the set represented by the molecule.

```
INITIALIZE;
FOR each p in V DO
    FOR each unordered pair of distinct edges (p,q),(p,r) in E DO
        DIVIDE(2)
            1: DELETE(p,q)
            2: DELETE(p,r)
        UNITE
    END_FOR;
    FOR each unordered pair of distinct edges (s,p),(t,p) in E DO
        DIVIDE(2)
            1: DELETE(s,p)
            2: DELETE(t,p)
        UNITE
    END_FOR
END_FOR;
{DROP(station count < n);}
FOR each two element subset {p,q} of V with both (p,q) and (q,p) in E DO
    DIVIDE(2)
        1: DELETE(p,q)
        2: DELETE(q,p)
    UNITE
END_FOR;
DROP(station count ≠ n).
```

Observe that any DNA molecule that may remain encodes a set of n edges that constitute a collection of Hamiltonian circuits for which no two distinct circuits have a vertex in common. If k is the number of circuits, then V is partitioned into the k pairwise disjoint subsets $V_1, V_2, \ldots, V_k$ where each V_j $(1 \leq j \leq k)$ is the set of vertices occurring in the jth Hamiltonian circuit. The final FOR statement in the procedure guarantees that any such V_j will contain at least three vertices. Thus the answer to the question proposed is 'YES' if DNA molecules having n stations remain, and 'NO' otherwise. #

If in the original statement of the problem, each member of the partition is only required to contain at least two (instead of three) vertices then the problem is solvable in polynomial time [5,10,4]. If, from Procedure 3, one deletes the final FOR loop, then the resulting procedure decides this simpler algorithmic problem. We find it provocative that the procedure for solving the NP-complete problem differs from the procedure for solving the polynomial time problem by only this rather trivial final FOR loop.

6 Satisfiability of Boolean Formulas

Let $p_1, p_2, \ldots, p_n$ be atomic propositional variables. A Boolean formula in these variables is a well formed expression constructed from the three operations $\wedge$, $\vee$, $\neg$ (i.e., AND, OR, and NOT, respectively) and the variables $p_1, \ldots, p_n$. Thus the following three expressions are examples of such Boolean formulas:

(1) $p_1 \wedge p_2$

(2) $[\neg p_1] \vee (p_2 \wedge [\neg p_2])$

(3) $(p_1 \vee [\neg p_2] \vee [\neg p_4]) \wedge (p_2 \vee [\neg p_3] \vee p_4) \wedge ([\neg p_1] \vee [\neg p_2] \vee p_3)$.

A Boolean formula is *satisfiable* if there exist at least one choice of truth value settings of the atomic variables that provides the value TRUE for the formula. The three formulas just given are satisfiable: For (1) choose $p_1 = p_2 =$ TRUE. For (2) choose $p_1 =$ FALSE and choose p_2 at random. For (3) choose $p_2 =$ FALSE, $p_4 =$ TRUE, and choose values for the remaining variables at random.

The following formulas are not satisfiable since they evaluate to FALSE for every choice of truth values for the variables they contain.

(4) $p_1 \wedge [\neg p_1]$

(5) $(p_1 \vee p_2 \vee p_1) \wedge (p_1 \vee [\neg p_2] \vee p_1) \wedge ([\neg p_1] \vee p_2 \vee [\neg p_1]) \wedge ([\neg p_1] \vee [\neg p_2] \vee [\neg p_1])$.

A classical algorithmic problem of central significance in both the theory and the history of NP completeness is the problem of deciding, for a given Boolean formula, whether there exists at least one setting of the truth values of the atomic propositional variables for which the formula evaluates to TRUE. This problem is called the satisfiability problem and is known to be NP-complete [5]. The special case of satisfiability in which only those Boolean formulas considered are those {such as (3) and (5) above} that are given as conjunctions (ANDs) of three term disjunctions (ORs) is called 3SAT [5] and is known to be NP-complete.

The following procedure provides the correct 'YES'/'NO' answer for instances of 3SAT in a number of steps linearly bounded by the sum of the number of atomic propositional variables and the number of disjunctive triples that are conjoined.

Procedure 4. Let a conjunction of three term disjunctions be given. Let $p_1, p_2, \ldots, p_n$ be the pair-wise distinct atomic propositional variables that occur in the given formula. Let $\neg p_1, \neg p_2, \ldots, \neg p_n$ be their negations. With each of these $2n$ terms associate a station of DEL. For convenience let us redenote each p_i and also its associated station by the positive integer i. Let us redenote each $\neg p_i$ and also its associated station by the negative integer $-i$. We may use a letter j to denote either a positive or a negative integer and, in either case, $-j$ denotes the negation of j and $-(-j)$ and j have the same truth value. Throughout the computation, the *presence* in a DNA molecule of the station j (whether j is positive or negative) will mean that this molecule represents a truth assignment in which j has the value TRUE. Each (circular) DNA molecule will serve as an unordered list of truth values attributed to the $2n$ terms $p_1, p_2, \ldots, p_n, \neg p_1, \neg p_2, \ldots, \neg p_n$. The Delphic Plasmid encodes the n contradictory assertions: both p_j and $-p_j$ are TRUE. With the first FOR loop we remedy these contradictions:

```
INITIALIZE;
FOR each i in {1, 2, ..., n} DO
    DIVIDE(2)
        1: DELETE(i)
        2: DELETE(−i)
    UNITE
END_FOR;
{DROP(station count ≠ n);}
FOR each disjunction i ∨ j ∨ k in the compound proposition DO
    DIVIDE(3)
        1: DELETE(−i)
        2: DELETE(−j)
        3: DELETE(−k)
    UNITE
END_FOR;
DROP(station count ≠ n).
```

Observe that any DNA molecule that may remain encodes the specification of the truth values for the atomic propositions that provides the value 'TRUE' for the given proposition. Thus the answer to the question proposed is 'YES' if DNA molecules having n stations remain, and 'NO' otherwise. #

In Procedures 1, 2, and 3 we were able to avoid beginning by constructing an exponential number of distinct molecules. We constructed molecules only as they were needed. In Procedure 4 we have not avoided the initial construction of the exponential number of molecules that represent "all possible solutions". Consequently this procedure has much in common with the suggestion of R.Lipton [12] for approaching the satisfiability problem and with the initial stage of the laboratory solution of the maximum clique problem given in [14]. M.Yamamura has observed that, in some instances, a variant of Procedure 4, in which the order of the two FOR loops are interchanged, constructs fewer intermediate molecules in producing the solution.

A procedure for solving k-SAT (as defined in [5]) is easily obtained by altering the DIVIDE(3) statement in Procedure 4 to an appropriate DIVIDE(k) statement. A procedure for determining the k-colorability of a graph (as defined in [5]) can be obtained by altering the DIVIDE(2) statement in Procedure 4 to an appropriate DIVIDE(k) statement and also altering the DIVIDE(3) statement to an appropriate DIVIDE(2) statement. Both k-SAT and k-colorability are known to be NP-complete for $k > 2$ [5]. Polynomial time algorithms are available in both cases when $k = 2$. The property of being bipartite is equivalent to 2-colorability.

Acknowledgments

This article was written following valuable conversations with Peter Kaplan, Grzegorz Rozenberg, Herman Spaink and Nadrian Seeman. The basic approach was developed during the summer of 1998 which I spent at the newly established *Leiden Center for Natural Computing* (LCNC), headed by Prof. Dr. G. Rozenberg, at the *Leiden Institute for Advanced Computer Science* (LIACS) of Leiden University, the Netherlands. Procedure 2 was developed at the suggestion of Hudson Kronk. Procedure 3 reached its final NP-complete form following a valuable communication from David S. Johnson. I thank all these people for their stimulation and suggestions.

Partial support for this research through NSF CCR-9509831 and DARPA/NSF CCR-9725021 is gratefully acknowledged as is the support from LCNC and LIACS in Leiden.

References

1. L. Adleman, Molecular computation of solutions of combinatorial problems, *Science* **266** (1994) 1021–1024.
2. L. Adleman, Computing with DNA, *Scientific American* **297** No. 2 (August, 1998) 54–61.
3. K. L. Denninghoff and R. Gatterdam, On the undecidability of splicing systems, *Intern. J. Computer Math.* **27** (1989) 133–145.
4. H. Gabow, An efficient implementation of Edmund's algorithm for maximum matching on graphs, *J. Assoc. for Computing Machines* **23** (1976) 221–234.
5. M. R. Garey and D. S. Johnson, *Computers and Intractability — A Guide to the Theory of NP-Completeness*, W. H. Freeman, San Francisco, CA, (1979).
6. J. Hartmanis, On the weight of a computation, *Bull. European Assoc. for Theor. Computer Sci.* **55** (1995) 136–138.
7. T. Head, Formal language theory and DNA: an analysis of the generative capacity of specific recombinant behaviors, *Bulletin of Mathematical Biology* **49** (1987) 737–759.

8. T. Head, Splicing schemes and DNA, in: *Lindenmayer Systems — Impacts on Theoretical Computer Science, Computer Graphics, and Developmental Biology*, G. Rozenberg and A. Salomaa, eds., Springer Verlag, Berlin, pp. 371–383, 1992. Also in: *Nanobiology* **1** (1992) 335–342.
9. T. Head, Gh. Paun and D. Pixton, Language theory and molecular genetics: generative mechanisms suggested by DNA recombination, in: *Handbook of Formal Languages*, G. Rozenberg and A. Salomaa, eds., Springer Verlag, (1997).
10. J. E. Hopcroft and R. M. Karp, An $n^{5/2}$ algorithm for maximum matching in bipartite graphs, *SIAM J. Computing* **2** (1973) 225–231.
11. L. Kari, DNA computing: arrival of biological mathematics, *The Mathematical Intelligencer* **19** No. 2 (Spring 1997) 9–22.
12. R. Lipton, DNA solution of hard computational problems, *Science* **268** (1995) 542–545.
13. R. A. Meyers, Ed., *Molecular Biology and Biotechnology — A Comprehensive Desk Reference*, VCH Publishers Inc., New York (1995).
14. Q. Ouyang, P. D. Kaplan, S. Liu and A. Libchaber, DNA solution of the maximal clique problem, *Science* **278** (1997) 446–449.
15. Gh. Paun, Ed., *Biomolecular Computing. Theory and Experiment*, Springer Verlag, (1998).
16. Gh. Paun, G. Rozenberg and A. Salomaa, *DNA Computing. New Computing Paradigms*, Springer Verlag, (1998).
17. N. Pisanti, DNA computing: a survey, *Bull. European Assoc. for Theor. Computer Sci.* **64** (1998) 188–216.
18. D. Pixton, Linear and circular splicing systems, *Proc. Intelligence in Neural and Biological Systems*, IEEE Press, (1995) 181–188.
19. D. Pixton, Splicing in abstract families of languages, *Theor. Computer Sci.*, (to appear).
20. P. W. K. Rothemund, A DNA and restriction enzyme implementation of Turing machines, *DIMACS Series in Discrete Math. and Theor. Computer Sci.*, Vol. 27, Amer. Math. Soc., 1996, 75–119.
21. R. Siromoney, K. G. Subramanian and V. R. Dare, Circular DNA and splicing systems, in: *Parallel Image Analysis (LCNS 654)*, Springer Verlag (1992) 260–273.
22. T. Yokomori, S. Kobayashi and C.Ferretti, On the power of circular splicing systems and DNA computability, in: *Proc. ICEC'97*, IEEE Press (1997) 219–224.

DNA COMPUTING BY MATCHING: STICKER SYSTEMS AND WATSON-CRICK AUTOMATA

GHEORGHE PĂUN

Institute of Mathematics of the Romanian Academy
PO Box 1 – 764, 70700 Bucureşti, Romania
E-mail: gpaun@imar.ro

One of the most important properties of DNA molecules is their double stranded structure, based on the Watson–Crick complementarity of nucleotides. This new data structure and the operation of producing double stranded sequences by sticking together complementary single stranded strings can be used as basic ingredients in building two types of computability models, the so-called *sticker systems* and *Watson–Crick automata*. The former are language generating mechanisms, the latter are automata-like devices. Under certain conditions, both of them are proved to be computationally complete, that is, they have the same power as Turing machines. This is essentially due to the structure of DNA molecules, which is closely related to the *twin-shuffle language*, known to be a generator of the family of recursively enumerable languages.

1 Introduction

Speculations about the possibility of using bio-molecules, especially DNA, for computing were made already several decades ago, by visionary people like Ch. Bennett, R. Feynman, M. Conrad, T. Head, etc. We do not enter into details. It was L. Adleman's successful experiment of solving (a small instance of) the Hamiltonian Path Problem (known to be an NP-complete problem) by purely biochemical means, [1], which has shown that this is not only theoretically possible, but also practically feasible, at least at a small scale.

The flourishing researches which have followed L. Adleman report can be roughly clustered into two main classes: theoretically oriented studies and practically oriented attempts. The researches in the second class look for an immediate practical application, significant enough to compete with the usual electronic computers (the so-called, "killer app"). Up to now, the results can be estimated as disappointing (especially for those who stress on "immediate"); the main difficulties are the errors and the to-day impossibility to scale-up the procedures which work so nicely for toy problems... However, the efforts in this area are so intense and the matter is definitely so important that it is expected that the results will improve in the very next future.

The practical researches miss a central point, which is addressed by the theoretical researches: how powerful a DNA computer can be, in the broad sense, as compared with other computability models? The question should be carefully explained. We speak of a "DNA computer". This implicitly assumes that such a computer is of a type similar to the "usual" computers. Simply stated, such a computer should be *programmable*. The applicative researches look for *ad hoc* methods able to solve by biochemical means specific problems, they look for *computations*, not for *computers*. This is clearly of a practical importance, but from a theoretical point of view we have also to look for *computability models* based on DNA which, possibly/hopefully, can be implemented in the form of a *universal* computer. Thus, we have (1) to define computability models, (2) having universality properties, and (3) to study such models as abstract machines. However, we have to be prepared for the situation that the practical implementation of a "DNA computer" in DNA terms is too difficult, or error-prone, or inefficient, maybe simply impossible (for a long time). Remember that the nature "computes" by means of DNA processing since billions of years and the beautiful results we see around are obtained at the price of such a long time and of a similarly long row of "wrong" computations (species and individuals who were simply killed by evolution). Thus, a quite attractive idea raises here: to learn new computability models from the DNA area and to look for implementing them in different media, for instance in silicon, which is known enough in order to be thought to be reliable. Both these ideas make important the theoretical studies in DNA computing, in the sense of theoretical computer science: language theory, automata theory, complexity, process algebra, etc.

Anyway, it is a safe statement to say that at the moment DNA computing *in info* is much more advanced than DNA computing *in vitro*; the comparison with what happens *in vivo* is pointless, because the nature affords to "compute" without time and cost restrictions.

The present paper is a contribution to the theoretical direction of research in DNA computing and it deals with language theory and automata theory models based on the essential structural feature of DNA molecules: they are double stranded sequences, with the paired elements being in a specified complementarity relation. The basic operation we use is that also used by L. Adleman in his famous experiment: annealing, binding together single stranded sequences, according to the complementarity relation, in order to build complete, double stranded molecules. We give details in the next section. In fact, we survey here notions and results about two classes of mechanisms, *sticker systems* and *Watson–Crick automata*. The first mechanisms are language generating devices. They were introduced in [8] and then investigated in [5,12]. (It should be noted that what we call here sticker systems are mechanisms different from the sticker systems as introduced in [14], which are also based on the annealing operation, but are working in a quite different way.) The Watson–Crick automata are the automata counterparts of sticker systems and were introduced in [7]. Results about them can also be found in [6,9]. A comprehensive source of information is the monograph [13], where a chapter is devoted to each of the two devices.

We stress the fact that the basic ingredients we use in our models are characteristic to the DNA structure. We can say *bi-dimensional structure*. This is a very important aspect and we will elaborate about it when dealing with technical details. The practical researches in DNA computing try to exploit a series of features of DNA which are very attractive from a pragmatic point of view: huge parallelism (billions of molecules, hence of "processors", can be placed in the same test tube), informational compactness (at the level of a bit per nucleotide, much more efficient than electronic data storages), energy efficiency (a lot more elementary operations per one Joule than any electronic computer), many well-known tools for handling it (starting with enzymes, "one of the most intelligent gifts nature has made to us", as it was said somewhere, and ending with the PCR techniques able to produce 2^n copies of the same single molecule in n amplification steps). Implicitly, the practical researches also make use of the *structure* of DNA. In contrast, the theoretical studies — at least those of the sort we are going to deal with in this paper — are "forgetting" about all the pragmatic good features of DNA and they focus only on its syntactic/structural features, on the arrangements (patterns) of the constitutive units, the nucleotides. This is mainly because DNA brings to computer science a new data structure, the double stranded sequence with complementarity — let us call it a *Watson–Crick tape* — which, we will see, has a sort of intrinsic computational completeness. Only this, and it would be enough to make DNA an exciting media for computing. At the first sight, the two strands are nothing else than a redundant construction, because one strand can be deduced from the other one due to the precise complementarity. From a computational point of view, there is no redundancy here, but a surprising materialization of a very interesting "object" from formal language theory: the twin-shuffle language.

We anticipate a little, because the fact is worth emphasizing. Take a string over a given alphabet (a deck of cards); take the string which is obtained from the first one by marking in some way each symbol of it (take a second deck of cards, in the same order as the first one, but with the back of a different color). Shuffle the two strings, that is, interleave their symbols in such a way that the order of the symbols in the two strings remain the same (shuffle the two decks just once, interleaving in an arbitrary way the cards, without changing the order of cards in each deck). What is obtained is a string in the twin-shuffle language. All shufflings of two strings are permitted. For a given alphabet, all strings are taken and used as above for building strings in the twin-shuffle language.

Somewhat surprising (the result is proved about twenty years ago, [4]), the twin-shuffle language over the alphabet with only two letters (this means two letters and two twin letters of them, in total four) has universality properties: each recursively enumerable language (a language recognized by a Turing machine, according to the Church-Turing thesis, the most general notion of an algorithm) can be obtained from the twin-shuffle language over $\{0, 1\}$ by using a deterministic sequential transducer (any string in a given recursively enumerable language can be obtained by scanning a string in the mentioned twin-shuffle language from left to right, using a device with a finite memory, and producing an output which depends on the scanned symbol and the current state; thus, a sequential transducer is nothing else than a finite automaton with output).

Note that the twin-shuffle language is *unique*, one and the same twin-shuffle language over $\{0,1\}$; the sequential transducer depends, of course, on the recursively enumerable language we want to "recover" from this unique "blue print" of computability.

Moreover, the surprise is double: the twin-shuffle language over $\{0, 1\}$ can be obtained by reading in a certain way DNA molecules (details can be found in Section 4). This reading is made possible precisely by the double stranded structure of DNA molecules, based on complementarity. Thus, we can conclude that the DNA molecule has intrinsic universality properties from a computational point of view. (This fact has been first observed and discussed in [15].)

Our way of computing in a sticker system is a sort of a domino game with single strands or with incomplete molecules having hanging single strands (sticky ends). Just matching such single strands filled in with complementary nucleotides is not a very powerful operation: only regular languages can be obtained in this way (that is, languages recognized by finite automata, the weakest class of computing devices). The Adleman way of computing cannot lead to non-regular languages, whatever the complexity of the approached problem is. (A distinction should be made here between the *competence* of a class of algorithms — in particular, grammars or automata — and the *performances* of these algorithms. Saying that a device computes at the level of finite automata concerns the type of performed algorithms, not the — time and space — complexity of them and of the solved problems. Complex, hence difficult, hence interesting from a practical point of view, problems can be found even at low levels in a hierarchy dealing with the type of algorithms.) However, because there is no convenient universality result (and satisfactory definition of universality) at the level of finite automata (that is, no satisfactory programmable computer can be constructed at this level), it is necessary to supplement the model with features able to increase its power, if possibly at the level of Turing machines. Such features impose various types of controls on the operation of sticking. For many of the controls we consider we obtain the desired computational completeness, we reach the power of Turing machines (because we do not have in-built erasing possibilities, certain squeezing mechanisms are also necessary, in the form of projections, weak codings, etc). For Turing machines there were proved universality results since the beginning, see [18]. Thus, sticker systems with controls can be a model of a universal (programmable) DNA computer. The difficulty is that these controls are "non-natural", their implementation in purely biochemical terms does not look very feasible. But, this is a practical detail, of not too much interest here...

Building double stranded molecules out of single stranded strings and even building bi-dimensional patterns out of primary elements (in a way similar to computations with Wang dominoes) are also used in other models of DNA computing. The reader can find details and bibliographical information in [13] and in [19].

2 DNA: Structure and Annealing Operation

We stress the fact that we are interested here in the *formal* structure of DNA molecules, not in their biochemical structure. For what follows, a DNA molecule is a double stranded sequence composed of four letters, A, C, G, T, corresponding to the four nucleotides, adenine, cytosine, guanine, thymine, placed in the two strands in complementary pairs; A is paired with T and C with G. This pairing is called the *Watson–Crick complementarity*. Figure 1 presents a molecule.

$$5' - ATCGTTAGCTTACTCGCGCAAGCGCTCACG - 3'$$
$$3' - TAGCAATCGAATGAGCGCGTTCGCGAGTGC - 5'$$

Figure 1. A DNA molecule.

$$\begin{array}{l} ATCGTTAGCTTACTCGCGCAAG \\ \qquad\qquad\qquad\qquad + \\ \qquad\qquad TCGAATGAGCGCGTTCGCGAGTGC \end{array}$$

$$\Downarrow$$

$$\begin{array}{l} ATCGTTAGCTTACTCGCGCAAG \\ \qquad\qquad TCGAATGAGCGCGTTCGCGAGTGC \end{array}$$

Figure 2. An annealing operation.

Thus, we ignore everything else than the nucleotides present in the molecule (we also disregard the helicoidal structure, the possible folding, etc), but we keep the information about the directionality of the two strands, the so-called $5' - 3'$ orientation. Sometimes the orientation will be relevant also in our theoretical framework, but, because we suppose that always the upper strand is oriented $5' - 3'$ and the lower one $3' - 5'$, it is not necessary to specify the orientation. The reader interested in more precise biochemical details about the structure of DNA (and the operations we can apply to it) can consult the first chapter of [13] and the references given therein. We mention here only the operation of *annealing.* If in a test tube we have two single stranded sequences of nucleotides which are Watson–Crick complementary, in certain conditions (when the temperature of the solution decreases) the two strands will stick to each other. If the two strands are completely complementary, a double stranded molecule will be obtained. If the two strands are only partially complementary, then they will lead to an incomplete molecule; an example is given in Figure 2.

A molecule as in Figure 2 is said to have *sticky ends*, the overhanging single stranded sequences: *ATCGTT* in the left end and *GCGAGTGC* in the right end.

Of course, in reality, be it *in vivo* or *in vitro*, the annealing operation is never perfect, mismatches can appear, incomplete annealing, loops of one strand forced by a partial annealing of subsequences which match correctly and so on. We do not consider such cases here, always the matching is supposed complete and the annealing error-free.

In the subsequent sections we will work with an arbitrary alphabet, not with the four letter alphabet associated with the four nucleotides. This assumption is not reality losing, because using the four nucleotides we can encode the symbols of any larger alphabet. On such a general alphabet we will consider a *complementarity relation* in the form of a relation which is only supposed to be symmetric. (The Watson–Crick complementarity is also one-to-one, but we will not use this property here.)

Let us point out another aspect which will be implicitly used below: although there is no "up" and "down" in a solution placed in a test tube, we may safely speak about "upper strands" and "lower strands". First, the biochemists know many ways of marking nucleotides and sequences of nucleotides in order to distinguish them. Second, at a syntactic level, we can mark the strands by specific sequences of symbols A, C, G, T (for instance, we can insert an occurrence of A in between each two nucleotides in the upper strand — which means that in the corresponding positions of the lower strand we have T). At the limit, we can use only two nucleotides, for instance, A and C, in the upper strands and T, G in the lower strands (two symbols are still enough for encoding the symbols of a larger alphabet in a uniquely decypherable manner).

3 Formal Language Theory Prerequisites

We specify here some notions, notations and results from automata and formal language theory which will be used in the subsequent sections. A comprehensive source of further information is [16].

For an alphabet V, we denote by V^* the free monoid generated by V under the operation of concatenation; the empty string is denoted by λ. Moreover, $V^+ = V^* - \{\lambda\}$ and $|x|$ is the length of $x \in V^*$. A morphism $h : V^* \longrightarrow U^*$ such that $h(a) \neq \lambda$ for all $a \in V$ is said to be *λ-free.* A morphism $h : V^* \longrightarrow U^*$ is called a *coding* if $h(a) \in U$ for each $a \in V$, and a *weak coding* if $h(a) \in U \cup \{\lambda\}$, for each $a \in V$. If $U \subseteq V$, then the morphism $pr_U : V^* \longrightarrow U^*$ defined by $pr_U(a) = a$ if $a \in U$, and $pr_U(a) = \lambda$ if $a \in V - U$, is called *a projection* (on U).

By $mi(x)$ we denote the mirror image of $x \in V^*$.

The *shuffle* of two strings $x, y \in V^*$ is the language

$$x \sqcup\!\sqcup\, y = \{x_1y_1x_2y_2 \dots x_ny_n \mid x = x_1x_2 \dots x_n, y = y_1y_2 \dots y_n\,,$$
$$\text{for } n \geq 1, \text{ where } x_i, y_i \in V^*, 1 \leq i \leq n\}\,.$$

A Chomsky grammar is a system $G = (N, T, S, P)$, where N is the nonterminal alphabet, T is the terminal alphabet, S is the axiom, and P is the set of production rules, written in the form $u \to v$. The language generated by G is denoted by $L(G)$. We denote by *REG, LIN, CF, CS,* and *RE* the families of regular, linear, context-free, context-sensitive, and recursively enumerable languages, respectively. The sequence of inclusions $REG \subset LIN \subset CF \subset CS \subset RE$ is known under the name of *Chomsky hierarchy* and this is the usual test bed for any new type of language generating mechanisms.

The families *REG, LIN, CF, CS, RE* are characterized not only by classes of grammars but also by classes of automata: finite automata, one-turn push-down automata, push-down automata, linear bounded automata, and Turing machines, respectively.

A *finite automaton* is a system $M = (Q, V, s_0, F, \delta)$, where Q is the set of states, V is the alphabet, s_0 is the initial state, F is the set of final (accepting) states, and $\delta : Q \times V \longrightarrow \mathcal{P}(Q)$ is the transition mapping. ($\mathcal{P}(X)$ denotes the set of subsets of X; if X is an infinite set, then the set of all *finite* subsets of X is denoted by $\mathcal{P}_f(X)$.)

The transitions can be also presented as rewriting rules: $sa \to s'$ instead of $s' \in \delta(s, a)$, for $s, s' \in Q, a \in V$. Then the mapping δ can be replaced by a set P of rewriting rules of the form above.

A gsm (= generalized sequential machine) is a finite automaton with outputs: $g = (Q, V_I, V_O, s_0, F, \delta)$, where Q is the set of states, V_I is the input alphabet,

V_O is the output alphabet, s_0 is the initial state, F is the set of final states, and $\delta : Q \times V_I \longrightarrow \mathcal{P}_f(V_O^* \times Q)$; $(x, s') \in \delta(s, a)$ has the following meaning: reading the symbol $a \in V_I$ in state s, one produces the string $x \in V_O^*$ and goes to state s'. For $z, x \in V_O^*, s, s' \in Q, a \in V_I$, and $w \in V_I^*$, we write

$$zsaw \Longrightarrow zxs'w \text{ iff } (x, s') \in \delta(s, a).$$

Then, for $w \in V_I^*$ we define the result of the translation of w by g by

$$g(w) = \{y \in V_O^* \mid s_0 w \Longrightarrow^* y s_f,\ s_f \in F\},$$

where $\Longrightarrow^*$ is the reflexive and transitive closure of the relation $\Longrightarrow$. The mapping g is extended in the natural way to languages: for any $L \subseteq V_I^*$,

$$g(L) = \bigcup_{x \in L} g(x).$$

When for each pair (s, a) we have $card(\delta(s, a)) \leq 1$, then we say that the automaton M (the gsm g, respectively) is *deterministic*.

We are now going to introduce one of the basic theoretical tools of our investigation, the equality sets of morphisms.

For two morphisms $h_1, h_2 : V_1^* \longrightarrow V_2^*$, we define the equality set by

$$EQ(h_1, h_2) = \{x \in V_1^* \mid h_1(x) = h_2(x)\}.$$

(This is the set of strings in the common domain for which the two morphisms "agree".)

Proofs of the following representation of RE can be found in [2,17].

Theorem 1. *Each language $L \in RE, L \subseteq T^*$, can be written in the form $L = pr_T(EQ(h_1, h_2) \cap R)$, where $R \subseteq V_1^*$ is a regular language and $h_1, h_2 : V_1^* \longrightarrow V_2^*$ are two λ-free morphisms; $T \subseteq V_1$.*

A variant of this result is proved in [8] (Lemma 8 there).

Theorem 2. *Each language $L \in RE, L \subseteq T^*$, can be written in the form $L = pr_T(h_1(EQ(h_1, h_2)) \cap R)$, where $R \subseteq V_2^*$ is a regular language and $h_1, h_2 : V_1^* \longrightarrow V_2^*$ are two λ-free morphisms; $T \subseteq V_2$.*

Note the difference between these results: in the second case, the language L is obtained starting from the *image* of the equality set of the two morphisms by one of the morphisms, while in the first case we use the equality set itself.

4 The Twin-shuffle Language and its Relation with DNA

Consider an alphabet V and denote by $\bar{V}$ the alphabet $\{\bar{a} \mid a \in V\}$, of barred variants of symbols in V; $\bar{a}$ is the *twin* of a. We assume that $V \cap \bar{V} = \emptyset$. We also use "bar" as an operator: for $x = a_1 \dots a_n$, with $n \geq 1$ and $a_i \in V$ for $1 \leq i \leq n$, we write $\bar{x} = \bar{a}_1 \dots \bar{a}_n$.

The *twin-shuffle language over* V is defined by

$$TS_V = \bigcup_{x \in V^*} (x \sqcup\!\sqcup \bar{x}).$$

That is, we take all strings in V^*, we consider their barred twin strings, we shuffle each string with its twin in all possible ways and we collect all the resulting strings; the obtained language is the twin-shuffle language over V.

An alternative definition is the following. For $V, \bar{V}$ as above, consider the asociated projections, $pr_V, pr_{\bar{V}}$; then

$$TS_V = \{x \in (V \cup \bar{V})^* \mid \overline{pr_V(x)} = pr_{\bar{V}}(x)\}.$$

Because $\overline{pr_V(x)}$ is a morphism, if we denote it by $h : (V \cup \bar{V})^* \longrightarrow \bar{V}^*$, the above definition can be read as $TS_V = EQ(h, pr_{\bar{V}})$. In view of Theorem 1, this makes the following result plausible.

Theorem 3. ([4]) *Each language $L \in RE, L \subseteq T^*$, can be written in the form $L = pr_T(TS_V \cap R)$, where R is a regular language and V an alphabet.*

Now, because the projection pr_T, the intersection with a regular language, and the encoding/decoding of symbols in V using the symbols of $\{0, \bar{0}, 1, \bar{1}\}$ can be simultaneously performed by a gsm, we get:

Theorem 4. ([4]) *Each language $L \in RE$ can be written in the form $L = g_L(TS_{\{0,1\}})$, where g_L is a deterministic gsm.*

This result is important for DNA computing, because of the relation we have mentioned in the Introduction between the twin-shuffle language over two letters and the DNA molecules. Indeed, let us consider the encoding of nucleotides present in the two strands of a DNA molecule as suggested in the table below:

	upper strand	lower strand
A, T	0	$\bar{0}$
C, G	1	$\bar{1}$

That is, both nucleotides A and T are identified with 0, without a bar when appearing in the upper strand and barred when appearing in the lower strand; the nucleotides C, G are identified with 1 in the upper strand and with $\bar{1}$ in the lower strand.

Assume now that we have two "intelligent-nucleotide-readers" which are placed at the beginning of the two strands of a DNA molecule, and that these guys read the nucleotides in their strands, from left to right, with arbitrary speeds each (uncorrelated), and telling us what they read modulo the codification in the previous table. What we get is a string in $TS_{\{0,1\}}$. The reader might try with the molecule in Figure 1.

Conversely, we can obtain *all* strings in $TS_{\{0,1\}}$ if we consider *all* molecules (complete double stranded sequences) and *all* possibilities to read them as specified above. The same result is obtained if we use molecules containing in the upper strand only nucleotides in any of the pairs

$$(A, C),\ (A, G),\ (T, C),\ (T, G).$$

Thus, in the sense of the above reading, in Theorem 4 we can put "DNA" instead of $TS_{\{0,1\}}$ (the reader should see and accept the methaphorical flavour of this assertion). Thus, all computations are "hidden" in (the readings of) DNA molecules; in order to actualize them, it is enough to use a sequential transducer, even a deterministic one.

There is a sort of a biochemical flaw here: the two "intelligent-nucleotide-readers" were both placed at the left end of a molecule, somewhat contradicting the fact that the two strands of a DNA molecule have opposite orientations. On the one hand, this will not be important for our model, on the other hand, there is no difficulty to take care of the orientation, because the twin-shuffle representations of recursively enumerable languages prove to be robust from this point of view.

Consider the *reverse twin-shuffle* language over an alphabet V, defined by

$$RTS_V = \bigcup_{x \in V^*} (x \sqcup\!\sqcup\, mi(\bar{x}))$$

(or, if you prefer, by $RTS_V = \{x \in (V \cup \bar{V})^* \mid \overline{pr_V(x)} = mi(pr_{\bar{V}}(x))\}$). That is, a string $y \in RTS_V$ is obtained by shuffling a string $x \in V^*$ with the mirror image of its twin $\bar{x}$.

The following counterpart of Theorem 4 was proved in [3]:

Theorem 5. *Each language $L \in RE$ can be written in the form $L = g_L(RTS_{\{0,1\}})$, where g_L is a deterministic gsm.*

In a speculative manner, Theorems 4 and 5 can be seen as a mathematical justification of the fact that four nucleotides are needed in order to have computational completenes: we use four letters, $0, \bar{0}, 1, \bar{1}$. Actually, this is not true, three letters are enough; we mention this result as a curiosity, without having a clear biochemical interpretation.

Consider the coding $c : \{0,1\}^* \longrightarrow \{\bar{0}, 1\}^*$ defined by $c(0) = \bar{0}$ and $c(1) = 1$. The *semi-twin-shuffle* language over $\{0,1\}$, denoted by $STS_{\{0,1\}}$, is defined by

$$STS_{\{0,1\}} = \bigcup_{x \in \{0,1\}^*} (x \sqcup\!\sqcup\, c(x)).$$

Theorem 6. ([10]) *For each recursively enumerable language L there is a gsm g_L such that $L = g_L(STS_{\{0,1\}})$.*

In order to have computational completeness, three nucleotides are enough...

5 The Operation of Sticking

The basic objects we use in this paper are double stranded sequences whose corresponding symbols are paired according to a complementarity relation. First, let us fix some notation and terminology concerning such objects.

Consider an alphabet V and a symmetric relation ρ on V, $\rho \subseteq V \times V$ (of *complementarity*).

Besides the monoid V^*, of strings over V, we associate to V also the monoid of pairs, $V^* \times V^*$. In accordance with the way of representing DNA molecules, where one considers the two strands as placed one over the other, we write the elements (x_1, x_2) of $V^* \times V^*$ in the form $\binom{x_1}{x_2}$. Therefore, the concatenation of two pairs $\binom{x_1}{x_2}, \binom{y_1}{y_2}$ is $\binom{x_1y_1}{x_2y_2}$. We also write $\binom{L_1}{L_2}$ instead of $L_1 \times L_2$, for $L_1, L_2 \subseteq V^*$.

We denote

$$\begin{bmatrix} V \\ V \end{bmatrix}_\rho = \left\{ \begin{bmatrix} a \\ b \end{bmatrix} \mid a, b \in V, (a,b) \in \rho \right\},$$

$$WK_\rho(V) = \begin{bmatrix} V \\ V \end{bmatrix}_\rho^*.$$

The set $WK_\rho(V)$ is called the *Watson–Crick domain* associated to V and ρ. The elements $\begin{bmatrix} a_1 \\ b_1 \end{bmatrix} \begin{bmatrix} a_2 \\ b_2 \end{bmatrix} \cdots \begin{bmatrix} a_n \\ b_n \end{bmatrix} \in WK_\rho(V)$ are also written in the form $\begin{bmatrix} w_1 \\ w_2 \end{bmatrix}$, for $w_1 = a_1a_2 \ldots a_n$, and $w_2 = b_1b_2 \ldots b_n$. We call such elements $\begin{bmatrix} w_1 \\ w_2 \end{bmatrix} \in WK_\rho(V)$ well-formed double stranded sequences, or simply *molecules*, in order to remind the reality they are modeling. When used as input or output sequences for automata, such objects are also called *Watson–Crick tapes.* The two component strings, w_1, w_2, are called strands, the *upper strand* and the *lower strand*, respectively.

We emphasize the two properties characterizing the elements in $WK_\rho(V)$: (1) the two strands are of the same length, (2) the corresponding symbols in the two strands are complementary in the sense of the relation ρ.

We shall also use below "incomplete molecules," that is elements in the set

$$W_\rho(V) = L_\rho(V) \cup R_\rho(V) \cup LR_\rho(V),$$

where

$$L_\rho(V) = \left(\binom{\{\lambda\}}{V^*} \cup \binom{V^*}{\{\lambda\}} \right) \begin{bmatrix} V \\ V \end{bmatrix}_\rho^*,$$

$$R_\rho(V) = \begin{bmatrix} V \\ V \end{bmatrix}_\rho^* \left(\binom{\{\lambda\}}{V^*} \cup \binom{V^*}{\{\lambda\}} \right),$$

$$LR_\rho(V) = \left(\binom{\{\lambda\}}{V^*} \cup \binom{V^*}{\{\lambda\}} \right) \begin{bmatrix} V \\ V \end{bmatrix}_\rho^+ \left(\binom{\{\lambda\}}{V^*} \cup \binom{V^*}{\{\lambda\}} \right).$$

The possible shapes of elements in $W_\rho(V)$ are illustrated in Figure 3. In all cases, we have a well-formed double stranded sequence x and overhangs y, z in one or two sides of x. These overhangs (sticky ends) can be placed in the upper strand or in the lower one. Note that the block x may be empty. Also the overhangs can be empty; what remains in the latter case is an element of $WK_\rho(V)$.

Among the elements of $W_\rho(V)$ we can define a partial operation, modeling the annealing operation: a well-started molecule (hence a sequence having at least a column filled in both of the two strands) can be prolonged to the right or to the left with an element of $W_\rho(V)$, providing that the sticky ends match, that is they are complementary in the corresponding positions. The result should always be a well-started molecule, hence a sequence which does not have empty places surrounded by symbols from V.

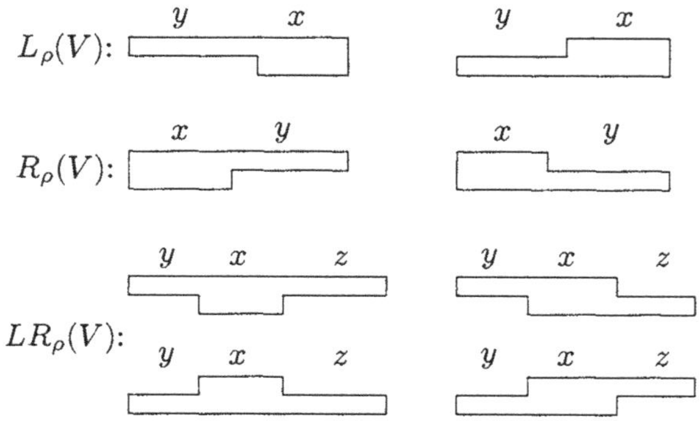

Figure 3. Possible shapes of incomplete molecules.

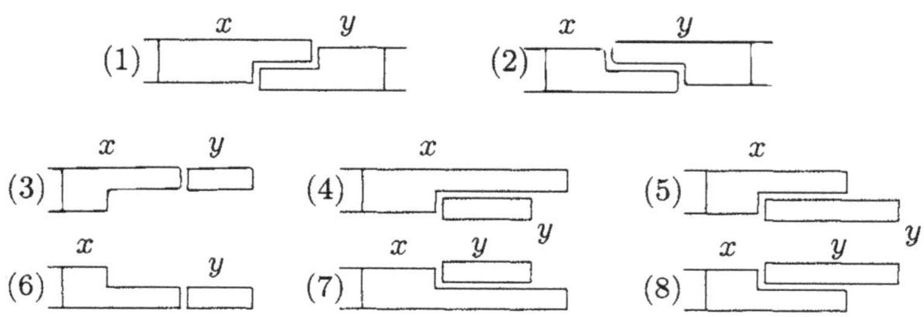

Figure 4. The sticking operation.

If the starting molecules are x, y, the resulting molecule is denoted by $\mu(x, y)$.

We do not define formally here this operation (this can be found in Chapter 4 of [13]); the previous explanations and the prolongations to right presented in Figure 4 should be enough for what follows.

Note that we do not need to distinguish the "left prolongation" from the "right prolongation" by denoting them in different ways: in any case at least one of the terms of the operation must be a well-started molecule and the result — it is a well-started molecule, too — entirely depends on the order of the two sequences and on their sticky ends.

The maximal length of an overhang in a sequence $z \in W_\rho(V)$ is also called the *delay* of z and it is denoted by $d(z)$; it represents the delay in completing the two strands with symbols in V. (Hence, in cases 1, 2 in Figure 4, the "right delay" of x and the "left delay" of y should coincide when y is also a well-started double stranded sequence.)

6 Sticker Systems

We introduce now one of the two basic devices we discuss in this paper, the sticker systems. They are based on the sticker operation defined above in the same manner as the Chomsky grammars (and other devices considered in formal language theory) are based on the rewriting operation. We will mainly look for characterizations of recursively enumerable languages, hence we do not present here all variants investigated in the literature. Moreover, we give no proof; technical details can be found in [8,5,12], and in Chapter 4 of [13].

A *sticker system* is a construct $\gamma = (V, \rho, A, D)$, where V is an alphabet, $\rho \subseteq V \times V$ is a symmetric relation, A is a finite subset of $LR_\rho(V)$, and D is a finite subset of $W_\rho(V) \times W_\rho(V)$.

The relation ρ is the complementarity relation on V, the elements of A are called *axioms*. Starting from these axioms and using the pairs (u, v) from D, we can obtain a set of sequences in $WK_\rho(V)$, hence complete molecules, by using the operation μ of sticking.

Formally, for a given sticker system $\gamma = (V, \rho, A, D)$ and two sequences $x, y \in LR_\rho(V)$, we write

$$x \Longrightarrow y \text{ iff } y = \mu(u, \mu(x, v)), \text{ for some } (u, v) \in D\,.$$

A sequence $x_1 \Longrightarrow x_2 \Longrightarrow \cdots \Longrightarrow x_k$, with $x_1 \in A$, is called a *computation* in γ. A computation $\sigma : x_1 \Longrightarrow^* x_k$ is *complete* when $x_k \in WK_\rho(V)$ (no sticky end is present in the last step).

The set of all molecules over V produced at the end of complete computations in γ is denoted by $LM_n(\gamma)$ (LM stands for "language of molecules" and the subscript n stands for "non-restricted": there is no restriction on the computations except that they are complete):

$$LM_n(\gamma) = \{w \in WK_\rho(V) \mid x \Longrightarrow^* w, x \in A\}\,.$$

In what follows we consider the sticker systems as generating languages of strings. To this aim, we associate with $LM_n(\gamma)$ the language

$$L_n(\gamma) = \left\{w \in V^* \middle| \begin{bmatrix} w \\ w' \end{bmatrix} \in LM_n(\gamma) \text{ for some } w' \in V^*\right\}.$$

We say that $L_n(\gamma)$ is the language *generated* by γ (at the end of non-restricted complete computations).

Several types of restricted computations in γ can be considered, but we mention only one: a computation is said to be *of delay d* when the delay of each intermediate molecule in the computation is smaller than or equal to d. The language generated by γ by computations of delay at most d is denoted by $L_d(\gamma)$. If $L_n(\gamma) = L_d(\gamma)$, for some integer d, then we say that γ is a sticker system *of a bounded delay*.

Several restricted variants of sticker systems are of interest. A system $\gamma = (V, \rho, A, D)$ is said to be:

- *one-sided*, if for each pair $(u, v) \in D$ we have either $u = \lambda$ or $v = \lambda$,
- *regular*, if for each pair $(u, v) \in D$ we have $u = \lambda$,
- *simple*, if for each $(u, v) \in D$ we have either $u, v \in \binom{V^*}{\{\lambda\}}$ or $u, v \in \binom{\{\lambda\}}{V^*}$.

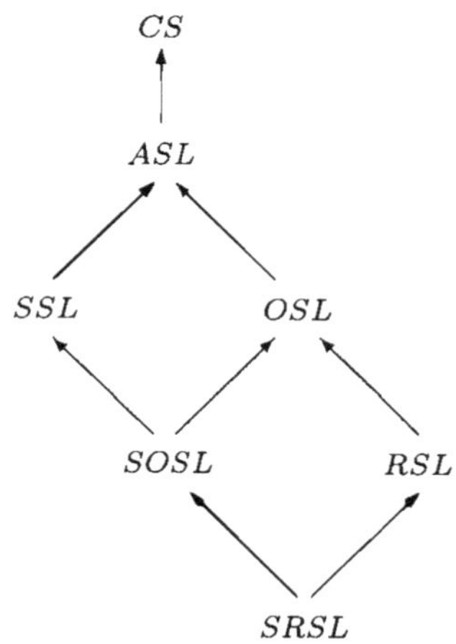

Figure 5. Relationships between families of languages generated by sticker systems (preliminary results).

In one-sided systems, the prolongation to the left is independent of the prolongation to the right; in regular systems we only prolong the sequences to the right (hence the axioms must be of the form x_1x_2, with $x_1 \in WK_\rho(V)$ and $x_2 \in \binom{V^*}{\{\lambda\}} \cup \binom{\{\lambda\}}{V^*}$). In a computation in a simple sticker system we add symbols only to one of the two strands.

We denote by ASL the family of languages of the form $L_n(\gamma)$, for γ a sticker system of an arbitrary form (SL stands for "sticker language" and A indicates the use of sticker systems of "arbitrary forms"). When only sticker systems which are one-sided, regular, simple, simple and one-sided, or simple and regular are used, we replace A in front of SL by O, R, S, SO, SR, respectively. When we consider only sticker systems of a bounded delay, we add the subscript b, thus obtaining families ASL_b, OSL_b, etc.

We stress the fact that these families contain *string languages*, not languages of molecules, hence we can discuss their relationships with families in the Chomsky hierarchy without further precautions.

From the definitions (note that we do not have erasing possibilities, hence the inclusion in CS easily follows), we obtain:

Lemma 1. *The relations in the diagram in Figure 5 hold, where the arrows indicate inclusions which are not necessarily proper; an identical diagram holds when adding the subscript b to each family XSL.*

Before starting to investigate the size of the families XSL, XSL_b, let us examine an example. Consider the sticker system $\gamma = (V, \rho, A, D)$, with:

$$V = U \cup \bar{U} \cup U', \text{ for some alphabet } U; U' = \{a' \mid a \in U\},$$

$$\rho = \{(a, a), (\bar{a}, \bar{a}), (a', a') \mid a \in U\},$$

$$A = \left\{ \begin{bmatrix} a'_0 \\ a'_0 \end{bmatrix} \right\}, \text{ for some fixed } a_0 \in U,$$

$$D = \left\{ \left(\begin{pmatrix} a' \\ \lambda \end{pmatrix}, \begin{bmatrix} a \\ a \end{bmatrix} \right), \left(\begin{pmatrix} \lambda \\ a' \end{pmatrix}, \begin{bmatrix} \bar{a} \\ \bar{a} \end{bmatrix} \right) \middle| a \in U - \{a_0\} \right\}.$$

We start from $\begin{bmatrix} a'_0 \\ a'_0 \end{bmatrix}$ and we build a molecule by adding columns $\begin{bmatrix} a \\ a \end{bmatrix}, \begin{bmatrix} \bar{a} \\ \bar{a} \end{bmatrix}$ to the right hand side of it and, simultaneously with $\begin{bmatrix} a \\ a \end{bmatrix}$ and $\begin{bmatrix} \bar{a} \\ \bar{a} \end{bmatrix}$, respectively, we add symbols a' in the upper strand and symbols a' in the lower strand to the left of $\begin{bmatrix} a'_0 \\ a'_0 \end{bmatrix}$. This means that, modulo the primes and the bars, the sequence of columns $\begin{bmatrix} a' \\ a' \end{bmatrix}$ to the left of $\begin{bmatrix} a'_0 \\ a'_0 \end{bmatrix}$ is identical to the reversed sequence of columns $\begin{bmatrix} a \\ a \end{bmatrix}$ to the right of $\begin{bmatrix} a'_0 \\ a'_0 \end{bmatrix}$, and the same sequence $\begin{bmatrix} a' \\ a' \end{bmatrix}$ to the left of $\begin{bmatrix} a'_0 \\ a'_0 \end{bmatrix}$ is identical to the reversed sequence of columns $\begin{bmatrix} \bar{a} \\ \bar{a} \end{bmatrix}$ to the right of $\begin{bmatrix} a'_0 \\ a'_0 \end{bmatrix}$. Consequently,

$$LM_n(\gamma) = \left\{ \begin{bmatrix} x' \\ x' \end{bmatrix} \begin{bmatrix} a'_0 \\ a'_0 \end{bmatrix} \begin{bmatrix} w \\ w \end{bmatrix} \middle| x \in U^*, mi(w) \in x \sqcup\!\sqcup \bar{x} \right\},$$

$$L_n(\gamma) = \{x' a'_0 w \mid x \in U^*, mi(w) \in x \sqcup\!\sqcup \bar{x}\},$$

where $x', \bar{x}$ are the primed and the barred versions of $x \in U^*$, respectively.

Therefore, the twin-shuffle language over U is obtained to the right of $\begin{bmatrix} a'_0 \\ a'_0 \end{bmatrix}$, on both strands, together with the copy of the shuffled strings which is present in a primed version in both strands to the left of $\begin{bmatrix} a'_0 \\ a'_0 \end{bmatrix}$. (Because $mi(U^*) = U^*$, the mirror image operation can be ignored.)

Using Theorem 4, we get in this way a representation of recursively enumerable languages as gsm images of languages in the family ASL.

7 The Power of Sticker Systems

Let us first recall some results concerning the relationships between the families XSL, XSL_b, where $X \in \{A, O, R, S, SO, SR\}$, and between them and the families in the Chomsky hierarchy.

Theorem 7. *The relations in the diagram from Figure 6 hold; the arrows indicate inclusions which are not necessarily proper.*

Because the inclusion $SOSL \subset REG$ is proper (probably also the inclusion $SSL_b \subseteq LIN$ is proper), it is of interest to supplement the sticker systems with a squeezing mechanism in order to obtain representations of regular and of linear languages starting from languages in families $SOSL, SSL_b$, respectively. This is possible, even using weak squeezing mechanisms, like codings and weak codings.

Theorem 8.. (i) *Each regular language is the coding of a language in the family* $SRSL_b$. (ii) *Each linear language is the weak coding of a language in the family* SSL_b.

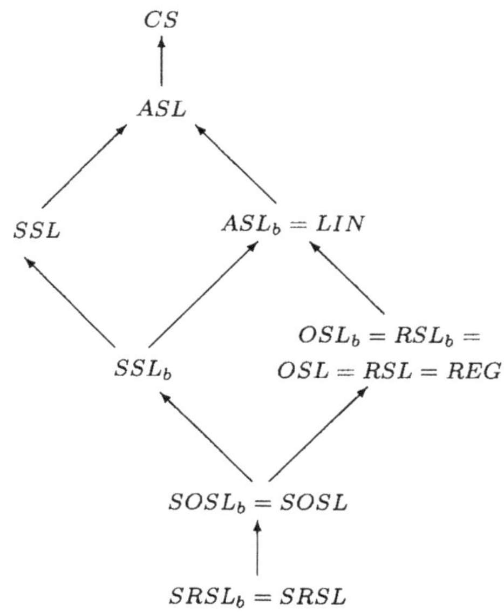

Figure 6. Relationships between families of languages generated by sticker systems.

From the point of view of DNA computing, of more interest is the possibility of representing (hence characterizing) the recursively enumerable languages by means of sticker languages. We have already presented such a possibility at the end of the previous section. Because of its importance, we state this result as a theorem:

Theorem 9. *Every language* $L \in RE$ *can be written in the form* $L = g(L')$, *where* $L' \in ASL$ *and* g *is a deterministic gsm mapping.*

In view of the relations summarized in the diagram in Figure 6, such a representation cannot be obtained for L' in any other family than ASL, SSL (all other families contain only linear languages, and the family LIN is closed under arbitrary gsm mappings). However, SSL can be used in order to obtain a representation of recursively enumerable languages, thus strengthening the result in Theorem 9.

Theorem 10. *Every language* $L \in RE$ *can be written in the form* $L = h(L')$, *where* h *is a weak coding and* $L' \in SSL$.

The regular sticker systems generate only regular languages, hence they cannot characterize RE by using usual operations as squeezing mechanisms. On the other hand, mainly in the simple variant, such devices are attractive from a mathematical and a biochemical point of view. This makes important the question of modifying the definition of simple sticker systems or of their language in such a way as to obtain characterizations of recursively enumerable languages for these sticker systems.

We consider here one further restriction on the language generated by a simple regular sticker system.

As we work only with right-sided pairs of "dominoes", we shall ignore the left hand member, the empty one. Moreover, we write separately the "upper dominoes" and the "lower dominoes". Thus, we write a simple regular sticker system in the form $\gamma = (V, \rho, A, D_l, D_u)$, where V, ρ, A are as above, and $D_l \subseteq \binom{\{\lambda\}}{V^*}, D_u \subseteq \binom{V^*}{\{\lambda\}}$, with D_l and D_u finite sets. The language $L_n(\gamma)$ is defined as in the previous section.

Consider two given labelings of elements in D_l, D_u with elements in a set Lab, $e_\pi : D_\pi \longrightarrow Lab, \pi \in \{l, u\}$. For a complete computation $\sigma : x_1 \Longrightarrow x_2 \Longrightarrow \cdots \Longrightarrow x_k$ in γ we define the control words $e_\pi(\sigma), \pi \in \{l, u\}$, consisting of the labels of elements of D_l, D_u, respectively, used in σ. Formally, we denote

$$\delta(\pi, x \Longrightarrow y) = \begin{cases} e_\pi\left(\begin{pmatrix} w_1 \\ w_2 \end{pmatrix}\right), & \text{if } x \Longrightarrow y \text{ uses } \begin{pmatrix} w_1 \\ w_2 \end{pmatrix} \in D_\pi, \\ \lambda, & \text{otherwise,} \end{cases}$$

for $\pi \in \{l, u\}$. Then,

$$e_\pi(\sigma) = \delta(\pi, x_1 \Longrightarrow x_2)\delta(\pi, x_2 \Longrightarrow x_3) \cdots \delta(\pi, x_{k-1} \Longrightarrow x_k), \ \pi \in \{l, u\}.$$

A complete computation $\sigma : x_1 \Longrightarrow^* x_k$, $x_1 \in A$, is said to be *coherent*, if $e_l(\sigma) = e_u(\sigma)$.

Thus, in a coherent computation we require that the sequence of labels associated to the upper blocks used in the computation is equal to the sequence of labels associated to the lower blocks used in the computation.

We denote by $L_c(\gamma)$ the language (of strings in the upper strand of molecules in $LM_n(\gamma)$) generated by γ using only coherent computations. The obtained family is denoted by $SRSL_c$.

The coherence condition leads again to a representation of recursively enumerable languages.

Theorem 11. *Each recursively enumerable language is the weak coding of a language in the family $SRSL_c$.*

The proofs of these characterizations of recursively enumerable languages make an essential use of Theorems 1 and 2.

We conclude by emphasizing the fact that we get computational completeness in many cases, but all of them involve certain restrictions on the sticker operation: prolonging the molecules in both directions, synchronously, or using only coherent computations. Because all the proofs are constructive, universality results known for Turing machines and for type-0 Chomsky grammars can be algorithmically transferred to sticker systems. In this way, universal (hence programmable) sticker systems are obtained (modulo squeezing mechanisms as specified above). They can serve as theoretical models of universal programmable DNA-like computers based on the sticker operation. (We say "DNA-like" because of the non-biochemical ingredients involved in the model, the control on the operations and the squeezing mechanisms, not to speak about the idealization of our operation itself, which is supposed to run perfectly, that is error-free.)

8 Watson–Crick Finite Automata

We consider now the automata counterpart of the sticker systems. Because they work on Watson–Crick tapes, they are called Watson–Crick automata. First, we introduce the basic variant, corresponding to finite automata.

A *Watson–Crick finite automaton* is a construct $M = (Q, V, \rho, s_0, F, \delta)$, where Q and V are disjoint alphabets, $\rho \subseteq V \times V$ is a symmetric relation, $s_0 \in Q$, $F \subseteq Q$, and $\delta : Q \times \binom{V^*}{V^*} \longrightarrow \mathcal{P}(Q)$ is a mapping such that $\delta\left(s, \binom{x}{y}\right) \neq \emptyset$ only for finitely many triples $(s, x, y) \in Q \times V^* \times V^*$.

The elements of Q are called *states*, V is the (input) alphabet, ρ is a complementarity relation on V, s_0 is the initial state, F is the set of final states, and δ is the transition mapping. The interpretation of $s' \in \delta\left(s, \binom{x_1}{x_2}\right)$ is: in state s, the automaton passes over x_1 in the upper strand and over x_2 in the lower strand of a double stranded sequence, and enters the state s'.

As in the case of finite automata, we can also write the transitions of M as rewriting rules of the form $s\binom{x_1}{x_2} \rightarrow \binom{x_1}{x_2}s'$; such a rule has the same meaning as $s' \in \delta(s, \binom{x_1}{x_2})$.

A transition in a Watson–Crick finite automaton can be defined as follows: For $\binom{x_1}{x_2}, \binom{u_1}{u_2}, \binom{w_1}{w_2} \in \binom{V^*}{V^*}$ such that $\begin{bmatrix} x_1u_1w_1 \\ x_2u_2w_2 \end{bmatrix} \in WK_\rho(V)$ and $s, s' \in Q$, we write

$$\binom{x_1}{x_2} s \binom{u_1}{u_2} \binom{w_1}{w_2} \Longrightarrow \binom{x_1}{x_2} \binom{u_1}{u_2} s' \binom{w_1}{w_2} \text{ iff } s' \in \delta\left(s, \binom{u_1}{u_2}\right).$$

We denote by $\Longrightarrow^*$ the reflexive and transitive closure of the relation $\Longrightarrow$.

As in the case of sticker systems, we investigate here the language of strings appearing in the upper strands of Watson–Crick tapes recognized by our automata, that is the language

$$L_u(M) = \left\{ w_1 \in V^* \,\middle|\, s_0 \begin{bmatrix} w_1 \\ w_2 \end{bmatrix} \Longrightarrow^* \begin{bmatrix} w_1 \\ w_2 \end{bmatrix} s_f, \text{ for } s_f \in F, \right.$$
$$\left. \text{and } w_2 \in V^*, \begin{bmatrix} w_1 \\ w_2 \end{bmatrix} \in WK_\rho(V) \right\}.$$

Another important language associated with a Watson–Crick automaton can be defined taking into account the transitions, not the recognized sequence.

For a Watson–Crick finite automaton $M = (Q, V, \rho, s_0, F, P)$ (hence with the transition given as rewriting rules) consider a labeling $e : P \longrightarrow Lab$, of rules in P with elements in a set Lab. For a computation $\sigma : s_0w \Longrightarrow^* ws_f$, $w \in WK_\rho(V)$, $s_f \in F$, we denote by $e(\sigma)$ the control word of σ, that is the sequence of labels of transition rules used in σ. In this way we obtain the language

$$L_{ctr}(M) = \{e(\sigma) \mid \sigma : s_0w \Longrightarrow^* ws_f, w \in WK_\rho(V), s_f \in F\}.$$

We say that the languages $L_\alpha(M), \alpha \in \{u, ctr\}$, are *recognized* by the Watson–Crick finite automaton M.

We note again that the work of Watson–Crick automata is defined for elements of $WK_\rho(V)$ only, that is, for double stranded sequences of elements in V paired

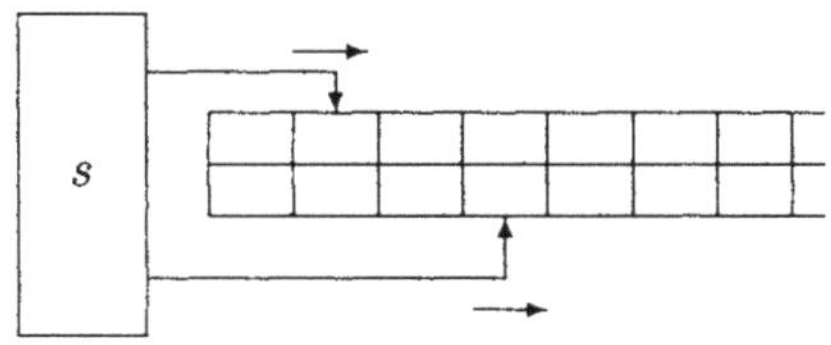

Figure 7. A Watson–Crick finite automaton.

according to the complementarity relation ρ. We can represent such a machine as consisting of a double tape on which an element of $WK_\rho(V)$ is written, a finite memory, able to store a state from a finite set of states, and two read only heads, one of them scanning the upper level and the other one scanning the lower level of the tape. Start with the two heads placed before the first symbol of each level, in state s_0. The two heads are moved to the right, according to the current state of the machine, as indicated by the transition mapping (the transition rules). A transition step means to move the two heads across blocks defined by a specific transition rule. Stop and accept the starting sequence when both heads reach the right hand end of the sequence written on the tape, entering a final state. Figure 7 illustrates this representation.

We consider several variants of Watson–Crick finite automata. We say that $M = (Q, V, \rho, s_0, F, P)$ is:

- *stateless*, if $Q = F = \{s_0\}$;
- *all-final*, if $F = Q$,
- *simple*, if for all $s\binom{x_1}{x_2} \rightarrow \binom{x_1}{x_2}s' \in P$ we have either $x_1 = \lambda$ or $x_2 = \lambda$,
- 1-*limited*, if for all $s\binom{x_1}{x_2} \rightarrow \binom{x_1}{x_2}s' \in P$ we have $|x_1x_2| = 1$.

In the stateless automata, the components Q, s_0, F can be omitted, and the transition rules can be written in the form $\binom{x_1}{x_2}$. Then, the automaton is given as $M = (V, \rho, P)$.

We denote by AWK_α, NWK_α, FWK_α, SWK_α, $1WK_\alpha$, $NSWK_\alpha$, $N1WK_\alpha$, $FSWK_\alpha$, $F1WK_\alpha$, the families of languages of the form $L_\alpha(M)$, $\alpha \in \{u, ctr\}$, recognized by Watson–Crick finite automata which are arbitrary (A), stateless (N, from "no state"), all-final (F), simple (S), 1-limited (1), stateless and simple (NS), stateless and 1-limited (N1), all-final and simple (FS), and all-final and 1-limited (F1), respectively.

9 The Power of Watson–Crick Finite Automata

As for sticker systems, we first investigate the relationships between the families of languages recognized by the variants of our automata, as well as their relationships with families in the Chomsky hierarchy.

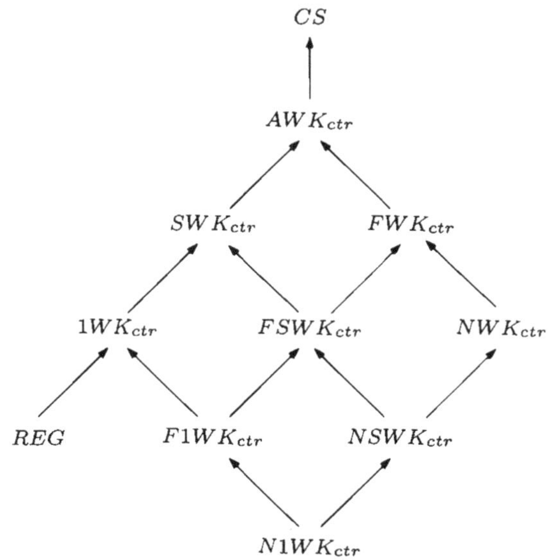

Figure 8. The hierarchy of *ctr* families.

Without entering into proof details (they can be found in [7,6], and in Chapter 5 of [13]) we summarize the known results in diagrams as in the case of sticker systems.

Theorem 12. *The relationships between families XWK_{ctr} are shown in the diagram in Figure 8, while those between families XWK_u are shown in the diagram in Figure 9; the arrows indicate inclusions which are not necessarily proper.*

In some sense, the families above are "small", their languages satisfy very strong conditions.

Lemma 2. (i) *If $L \in NWK_\alpha$, for $\alpha \in \{u, ctr\}$, then $L = L^+$.* (ii) *If $L \in N1WK_u$, then there is an alphabet V such that $L = V^+$.* (iii) *Every one-letter language in AWK_u is regular.*

However, the families discussed above (with the exception of $NSWK_u$) contain languages of a very high complexity. As we shall see immediately, characterizations of recursively enumerable languages can be obtained starting from them and using weak codings and deterministic gsm's.

Such results can be obtained as consequences of the following assertions, which establish links between Watson–Crick automata and notions considered above: twin-shuffle languages, equality sets of morphisms, sticker systems.

Lemma 3. (i) *For every alphabet V, we have $TS_V \in N1WK_{ctr}$.* (ii) *$SRSL(c) \subseteq FWK_u$.* (iii) *If $h_1, h_2 : V_1^* \longrightarrow V_2^*$ are two morphisms, then $h_1(EQ(h_1, h_2)) \in NWK_u$.*

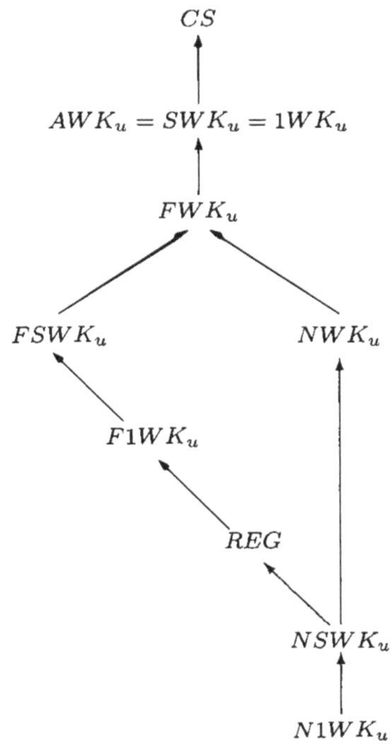

Figure 9. The hierarchy of u families.

Theorem 13. (i) *Each recursively enumerable language is the image of a deterministic gsm mapping of a language in any family* $XWK_{ctr}, X \in \{A, N, F, S, 1, NS, N1, FS, F1\}$.

(ii) *Each recursively enumerable language is the weak coding of a language in any family* $XWK_u, X \in \{A, F, S, 1\}$.

(iii) *Each language* $L \in RE$ *can be written in the form* $L = h(L' \cap R)$, *where* $L' \in NWK_u$ *or* $L' \in FSWK_u$, $R \in REG$, *and* h *is a projection.*

The power of Watson–Crick finite automata is rather surprising, especially if we consider their finite memory and the resemblance with usual finite automata. This power stands in the tape the automata use, the double stranded tape based on the complementarity relation. This is worth emphasizing, because of our starting point, the DNA computing: using DNA molecules as "automata tapes" adds a significant power to the device as compared to the usual one straded tapes.

As we know from Section 2, the two strands of a DNA molecule have opposite $5' - 3'$ orientations. This suggests considering a variant of Watson–Crick finite automata that parse the two strands of a Watson–Crick tape in opposite directions. Figure 10 illustrates the initial configuration of such an automaton.

Formally, a *reverse Watson–Crick finite automaton* is a construct $M = (Q, V, \rho, s_0, F, \delta)$, with the components defined exactly as for Watson–Crick finite automata, but with the relation $\Longrightarrow$ defined as follows:

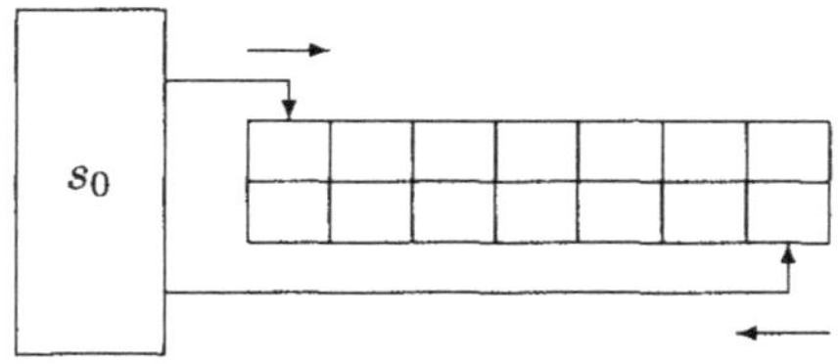

Figure 10. A reverse Watson–Crick finite automaton.

For $w_1, w_2, w_1', w_2', x, y \in V^*$, $s, s' \in Q$, we write

$$\binom{w_1}{w_2 y} s \binom{x w_1'}{w_2'} \Longrightarrow \binom{w_1 x}{w_2} s' \binom{w_1'}{y w_2'} \text{ iff}$$

$$\begin{bmatrix} w_1 x w_1' \\ w_2 y w_2' \end{bmatrix} \in WK_\rho(V),\ s' \in \delta\left(s, \binom{x}{y}\right).$$

Then $\begin{bmatrix} w_1 \\ w_2 \end{bmatrix} \in WK_\rho(V)$ is recognized by M if and only if

$$\binom{\lambda}{w_2} s_0 \binom{w_1}{\lambda} \Longrightarrow^* \binom{w_1}{\lambda} s_f \binom{\lambda}{w_2}, \text{ for } s_f \in F.$$

As above, we can associate two languages $L_\alpha(M), \alpha \in \{u, ctr\}$, with a reverse Watson–Crick finite automaton M. The families of languages recognized by reverse Watson–Crick finite automata corresponding to families XWK_α are denoted by $XRWK_\alpha, X \in \{A, N, F, S, 1, NS, N1, FS, F1\}$.

Clearly, a diagram like that in Figure 8 also holds true for these families. Moreover, because of the following lemma, again we get characterizations of recursively enumerable languages.

Lemma 4. *For every alphabet V, we have $RTS_V \in N1RWK_{ctr}$.*

Theorem 14. *Each language in RE is the image by a deterministic gsm of a language in any family $XRWK_{ctr}, X \in \{A, N, F, S, 1, NS, N1, FS, F1\}$.*

10 Variants of Watson–Crick Automata

First, let us associate an output to a Watson–Crick finite automaton in the same way as an output is associated to a finite automaton to form a gsm. This output is written on a normal tape rather than on a Watson–Crick double stranded tape (the other possibility will be considered later). Figure 11 illustrates this idea.

We do not give formal definitions of a Watson–Crick gsm and of its mapping; the reader can easily fill in the details, following the model of Watson–Crick finite automata.

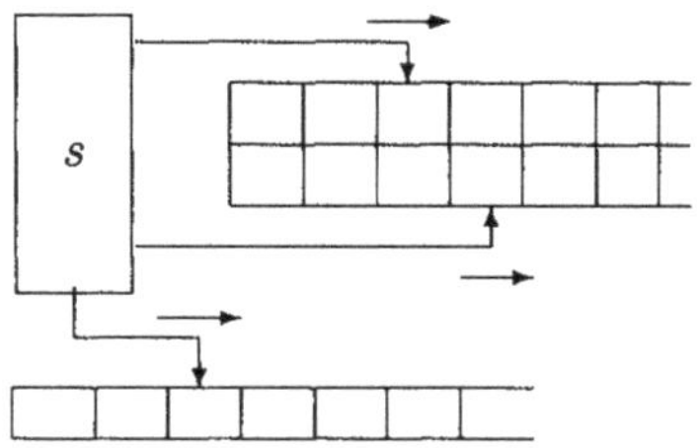

Figure 11. A Watson–Crick transducer.

As for Watson–Crick finite automata, we can consider stateless, simple, 1-limited, all-final, stateless simple, stateless 1-limited, all-final simple, and all-final 1-limited Watson–Crick gsm's.

Because the labeling of transitions in a stateless Watson–Crick finite automaton can be expressed as the output function of a Watson–Crick gsm, from Lemma 3 we get that for every alphabet V, there is a stateless 1-limited Watson-Crick gsm g_V such that we have $TS_V = g_V(WK_\rho(V))$, for $\rho = \{(a,a) \mid a \in V\}$. This implies a result like that in Theorem 13(i): each language $L \in RE$ can be written in the form $L = g(g_V(WK_\rho(V)))$, where g is a usual gsm (deterministic) and g_V is a stateless 1-limited Watson–Crick gsm. A stronger version of this result can be proved:

Theorem 15. *For each language $L \in RE, L \subseteq V^*$, there is a simple all-final Watson–Crick gsm g_L such that $L = g_L(WK_\rho(V'))$, for some Watson–Crick domain $WK_\rho(V')$.*

Hence, Watson–Crick finite transducers not only do not preserve the families in the Chomsky hierarchy, but they map (very simple) regular languages in such a way that they cover the whole family RE.

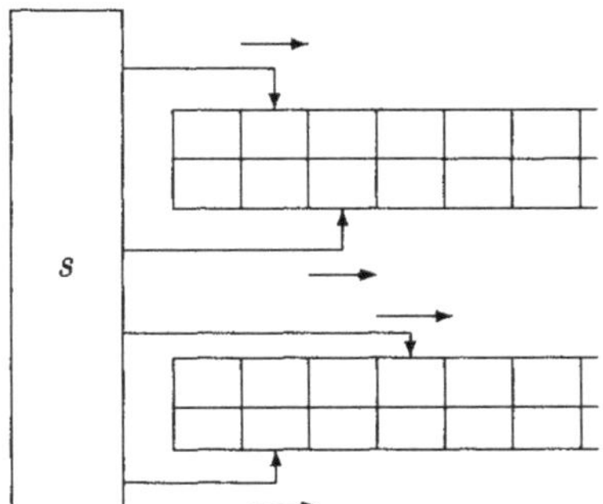

Figure 12. A Watson–Crick automaton with memory.

The Watson–Crick finite transducers are somewhat hybrid devices, as they use an input Watson–Crick tape and a usual output tape (single stranded). This suggests considering output tapes to be also Watson–Crick tapes, leading to a device as that in Figure 12. We call it a *Watson–Crick prefix automaton* because of the way it works: one starts in state s_0, with the four heads placed at the left hand end of the four strands of the two tapes, and stops in a final state, with all the four heads placed at the right ends of the four strands.

We can interpret and use such a machinery in two ways: as a recognizing device, and as a transducer.

In the first interpretation, as a recognizer, the second tape is used as a control of the automaton, as a memory. This can be done in two modes: (1) start with the second tape empty and at each transition $\left(\binom{y_1}{y_2}, s'\right) \in \delta\left(s, \binom{x_1}{x_2}\right)$ the strings y_1, y_2 are written on the strands of the second tape, starting with the leftmost empty positions, or (2) start with the second tape containing a molecule $\begin{bmatrix} w_1 \\ w_2 \end{bmatrix}$ and at each transition $\left(\binom{y_1}{y_2}, s'\right) \in \delta\left(s, \binom{x_1}{x_2}\right)$ move the heads reading the second tape over y_1, y_2, respectively. In the first mode, the automaton stops correctly only if it reaches a molecule $\begin{bmatrix} w_1 \\ w_2 \end{bmatrix}$ in the last configuration. Clearly, the two modes of using the second tape when M is a recognizing machine are equivalent.

In either case, we can also define accepting configurations to be ones with the second tape empty: in case (1) any column $\begin{bmatrix} a \\ b \end{bmatrix}$ with $(a, b) \in \rho_2$, appearing in the second tape is deleted immediately after being produced; in case (2) we delete any column $\begin{bmatrix} a \\ b \end{bmatrix}$ with $(a, b) \in \rho_2$, immediately after moving both the two heads to the right of it. Thus, in the first case we have a sort of FIFO (first-in-first-out) memory, which "melts" from the left as soon as completing the columns. This is why we call these automata Watson–Crick prefix automata.

Again, we avoid defining formally the transitions with respect to these devices. We only mention that a computation (recognition or translation) is correctly terminated if the contents of the first tape — which was a Watson–Crick tape — is exhausted and on the second tape one produces another Watson–Crick tape, a complete molecule observing the complementarity relation on the output alphabet.

It is easy to see that the Watson–Crick gsm's can be simulated by Watson–Crick prefix transducers: take the identity relation as a complementarity relation on the output alphabet and $y_1 = y_2$ for each transition $\left(\binom{y_1}{y_2}, s'\right) \in \delta\left(s, \binom{x_1}{x_2}\right)$, etc. Thus, the assertion in Theorem 15 holds true also for (simple and all-final) Watson–Crick prefix transducers.

As for Watson–Crick finite automata, we can associate two languages also to a Watson–Crick prefix automaton, $L_u(M)$ and $L_{ctr}(M)$. Moreover, we can consider Watson–Crick prefix automata which are simple, 1-limited, stateless, all-final, simple and stateless, 1-limited and stateless, simple and all-final, 1-limited and all-final. In the simple case, for each transition $\left(\binom{y_1}{y_2}, s'\right) \in \delta\left(s, \binom{x_1}{x_2}\right)$ we have one of x_1, x_2 and one of y_1, y_2 empty, but not necessarily on the same strand in the two tapes; similarly, in the 1-limited case we have at the same time one of x_1, x_2 and one of y_1, y_2 empty and the others equal to symbols. We can also consider initial stateless Watson–Crick prefix automata or reverse automata.

We denote by $XPWK_\alpha$ the families obtained in this way, for $X \in \{A, N, F, S, 1, NS, N1, FS, F1\}$ and $\alpha \in \{u, ctr\}$.

A Watson–Crick finite automaton can be simulated by a Watson–Crick prefix automaton with the two tapes identical, and the transitions operating in the same way on both of them (hence no further control is provided by the second tape). In this way, all the representations of RE obtained for Watson–Crick finite automata hold true also for Watson–Crick prefix automata.

Also the relations between families $XPWK_{ctr}$ corresponding to the diagram from Figure 8 are true.

The use of the second tape increases the power of Watson–Crick automata, hence of prefix automata compared to finite automata.

Lemma 5. (i) *If $h_1, h_2 : V^* \longrightarrow U^*$ are two morphisms, then $h_1(EQ(h_1, h_2)) \in NSPWK_u$.* (ii)*For every alphabet V we have $TS_V \in N1PWK_u$.*

As consequences of these results, we get:

Theorem 16. (i) *Each language $L \in RE$ can be written in the form $L = h(L' \cap R)$, where $L' \in NSPWK_u$, $R \in REG$, and h is a projection.*

(ii) *Each language in RE is the image by a deterministic gsm mapping of a language in $N1PWK_u$.*

11 Conclusions

The main goal of this paper was to explore the computational power of the double stranded structure of the DNA molecule, building both language generating devices and automata which essentially work with such double stranded data structures (called Watson–Crick tapes, because of the complementarity relation between the symbols appearing in pairs in the two strands). We hope to have convinced the reader that "there is universality in complementarity". Passing from strings to double sequences is unexpectedly useful, the power of our devices are almost equal (that means: equal modulo certain squeezing mechanisms, such as weak codings, projections, gsm mappings) to the power of Turing machines (and type-0 Chomsky grammars), which, according to Church-Turing thesis, are the most powerful representations of the notion of an algorithm.

From a meta-scientific point of view, we can claim that formal language theory and automata theory have many unexplored resources, both tools developed decades ago for their sake and relevant now to DNA computing (such as equality sets and twin-shuffle characterizations of recursively enumerable languages), and possibilities of development motivated by DNA computing. On the other hand, to the extent to which dealing with DNA is a syntactic matter, a well investigated syntactic framework as that provided by language and automata theory seems to be not only useful, but also unavoidable. The hope is that, besides the esthetical value, the theoretical results (of DNA computing *in info*) will also have a practical relevance — which is out of the scope of this paper...

Acknowledgments

Work supported by the Academy of Finland, Project 137358. Useful remarks by N. Seeman on a previous version of the paper are gratefully acknowledged.

References

1. L. M. Adleman, Molecular computation of solutions to combinatorial problems, *Science* **226** (1994), 1021–1024.
2. K. Culik II, A purely homomorphic characterization of recursively enumerable sets, *Journal of the ACM* **26** (1979), 345–350.
3. J. Engelfriet, Reverse twin-shuffles, *Bulletin of the EATCS* **60** (1996), 144.
4. J. Engelfriet and G. Rozenberg, Fixed point languages, equality languages, and representations of recursively enumerable languages. *Journal of the ACM* **27** (1980), 499–518.
5. R. Freund, Gh. Păun, G. Rozenberg and A. Salomaa, Bidirectional sticker systems, *Third Annual Pacific Conf. on Biocomputing*, Hawaii, 1998 (R. B. Altman, A. K. Dunker, L. Hunter, T. E. Klein, eds.), World Scientific, Singapore, 1998, 535–546.
6. R. Freund, Gh. Păun, G. Rozenberg and A. Salomaa, Watson–Crick finite automata, *Proc of the Third Annual DIMACS Symp. on DNA Based Computers*, Philadelphia, 1997, 305–317.
7. R. Freund, Gh. Păun, G. Rozenberg and A. Salomaa, Watson–Crick automata, *Techn. Report 97-13*, Dept. of Computer Sci., Leiden Univ., 1997.
8. L. Kari, Gh. Păun, G. Rozenberg, A. Salomaa and S. Yu, DNA computing, sticker systems, and universality, *Acta Informatica* **35** (1998), 401–420.
9. C. Martin-Vide, Gh. Păun, G. Rozenberg and A. Salomaa, Universality results for finite H systems and for Watson–Crick finite automata, in [11], 200–220.
10. A. Mateescu, Gh. Păun and A. Salomaa, Variants of twin-shuffle languages, submitted, 1998.
11. Gh. Păun, ed., *Computing with Bio-Molecules. Theory and Experiments*, Springer-Verlag, Singapore, 1998.
12. Gh. Păun and G. Rozenberg, Sticker systems, *Theoretical Computer Sci.* **204** (1998), 183–203.
13. Gh. Păun, G. Rozenberg and A. Salomaa, *DNA Computing. New Computing Paradigms*, Springer-Verlag, Heidelberg, 1998.
14. S. Roweis, E. Winfree, R. Burgoyne, N. Chelyapov, M. Goodman, P. Rothemund and L. Adleman, A sticker based architecture for DNA computing, in *DNA Based Computers*, Proc. of the Second Annual Meeting, Princeton, 1996 (E. Baum, D. Boneh, P. Kaplan, R. Lipton, J. Reif and N. Seeman, eds.), 1–27.
15. G. Rozenberg and A. Salomaa, Watson–Crick complementarity, universal computations and genetic engineering, *Techn. Report* 96-28, Dept. of Computer Sci., Leiden Univ., 1996.
16. G. Rozenberg and A. Salomaa, eds., *Handbook of Formal Languages*, 3 volumes, Springer-Verlag, Heidelberg, 1997.

17. A. Salomaa, *Jewels of Formal Language Theory*, Computer Science Press, Rockville, Maryland, 1981.
18. A. M. Turing, On computable numbers, with an application to the Entscheidungsproblem, *Proc. London Math. Soc.*, Ser. 2 **42** (1936), 230–265; a correction **43** (1936), 544–546.
19. E. Winfree, *Algorithmic Self-Assembly of DNA*, PhD Thesis, California Institute of Technology, Pasadena, CA, 1998.

Images and Perception

ASPECTS OF HUMAN SHAPE PERCEPTION

JACQUES NINIO
Laboratoire de Physique Statistique[1]
Ecole Normale Supérieure
24, rue Lhomond
75231 Paris cedex 05, France
E-mail: jacques.ninio@physique.ens.fr

The first insightful approach to visual shape perception was provided in the 1910's by the Gestalt school, with the formulation of rules according to which local visual features are perceptually grouped into larger units forming separate shapes (proximity, closeness, good continuation, etc.). Similarly oriented textural elements are perceptually grouped, and used to support 3D interpretations, suggesting that some of the tools used in the perception of static shapes are derived from tools in motion perception and optic flow analysis.

Another approach, rooted in differential geometry, was developed in the 1970's. Here, shape perception is conceived as mainly concerned with the computation of local surface curvature. Features like shading gradients, surface reflection spots, internal contour terminators would be clues from which local surface curvature would be inferred, in a consistent manner, over the whole surface.

The nature of the shape-generators used by the brain to make sense of the visual inputs is still mysterious. It is speculated here that in 3D, the shape generators would be rather crude ones, and would make use of rules such as "the height associated to a point inside a contour (assumed to be horizontal) increases with its remoteness from the contour". The 2D shape generators seem to be economical and accurate. As a metaphor, one might suggest the equipotential curves that are produced in electrostatics by judiciously positioning a few charges.

1 Shape from motion

Primitive eyes, such as those of clams or barnacles, merely detect abrupt changes in the incoming flux of light, a change being taken as a signal for the arrival of a predator. At a higher evolutionary stage, the visual systems detect moving targets and analyze their speed and direction. In addition, through the motion of the eyes carried by the head, an insect retrieves a crude description of the environment. When an animal moves or moves its eyes, the visual system must sort out, from the changes in the distribution of the incoming light intensity, the information pertaining to its eye or body motion, from that on the speed, direction and shape of moving objects in the environment. This theme has generated several lines of experimental or theoretical work.

There are chiefly three ways in which the observer's eyes move with respect to the environment. When the observer moves forward in a still environment, keeping his eyes in a fixed direction, the various points of the retinal image of the scene are subject to radial expansion. The eyes may also slightly rotate around the visual axes, or be subject to saccadic, translational movements. The vectors describing the transformations of the retinal image form the so-called "optic flow" [1]. The three main cases are expansion, translation and rotation (Figure 1, top). The

[1]Research laboratory associated to the CNRS and to Paris 6 and Paris 7 Universities

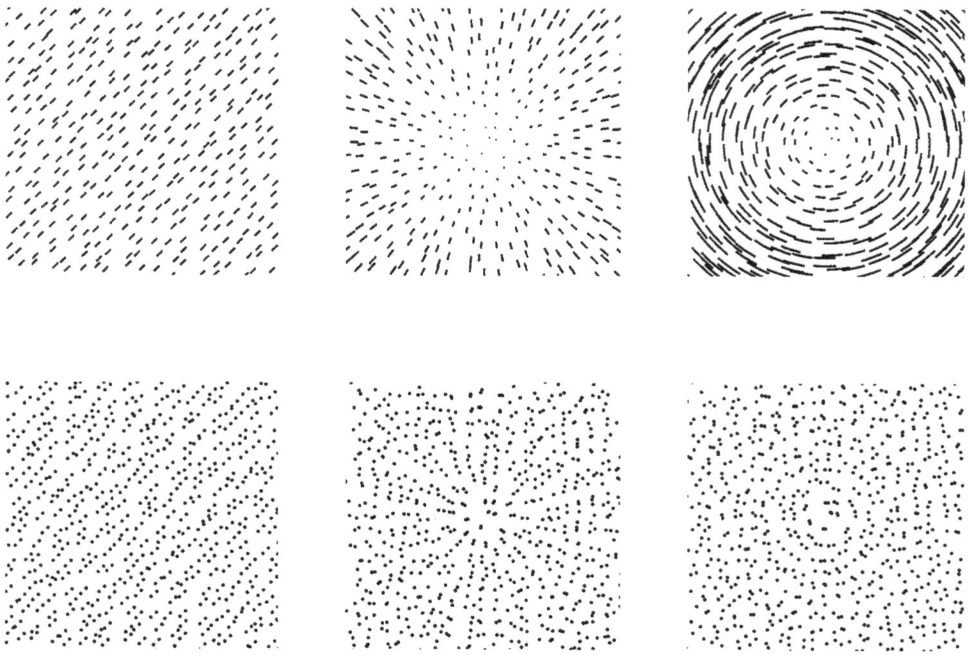

Figure 1. (Top) *Optic flow.* The small arcs or segments describe the apparent motion of points in the visual field, when the gaze shifts by translation (left), when the observer moves forward (center) or when the eye rotates around the visual axis (right). (Bottom) *Glass patterns.* When the arcs or segments are replaced with their two endpoints, characteristic patterns are obtained [13].

motion parameters of the observer (his "egomotion") can be extracted from the flow field. More interestingly, the shapes of the objects in the environment can also be extracted from the flow field.

One striking phenomenon, showing the human brain's ability at extracting shape from motion is the kinetic depth effect [2]. If one looks at an image on a screen (for instance, the shadow of an object on a silver screen or a computer graphic representation of a complex protein structure on a computer monitor) and the image becomes animated, showing the object in slow motion, then the structure of the object, confusing as it might have been initially, reveals itself. The represented object takes on a vivid 3D appearance. In a modern variant, the object — usually a cylinder or a hemisphere — is represented with a camouflaging random-dot texture, such that every single image is a random dot pattern devoid of structural information. However, as soon as the image is animated, the observer perceives the 3D shape that is represented [3].

In principle, if 9 points belonging to the surface of a rigid object are followed, and one has 3 images giving their successive 2D coordinates, one can deduce both the parameters of the motion in space of the object, and the 3D configuration of the points [4]. However, shape is easily derived from random-dot representations, in which the dots describing the object are frequently refreshed, and do not last more than 100 msec [5]. Rather than making geometric inferences from the comparison of static images, the brain would directly use the vector field describing the velocities of the various points in the retinal image.

To-day, most theoretical approaches focus, first on how surface boundaries are detected, then on how surface curvature informations are extracted (e.g. [6,7]). The underlying neurophysiological model is that of a 2D array of neurons arranged in a square lattice. Each set of four neurons at the corner of a square would act as an instrument that compares the light intensity changes at the four corners and would, by various combinations of the results, approximate various operators of differential geometry through which the main radii of curvature may be computed.

Experimentally, several problems have been tackled: In what circumstances does the brain interpret the animated images as representing a rigid object, and when is this assumption relaxed? Can the brain interpret as belonging to a deformating object the successive images of a moving rigid object? There are known such cases, e.g. [8]. In random-dot displays representing two surfaces that move independently, the two surfaces can be interpreted in 3D, as, e.g., one surface seen by transparency through the other, although the representative dots are intermingled in the images. This finding is a challenge to current neuron network models. Last, if 3D interpretation is not metrically perfect, is there at least a precise appreciation of affine or projective properties (see e.g. [9–12])?

2 From moving to static shapes

The vector fields corresponding to expansion, rotation and translation are immediately recognizable (Figure 1, top). When vectors are replaced with their two endpoints, these points give rise to the characteristic "Glass patterns" (Figure 1, bottom) [13].

Arrays of correlated oriented segments are readily interpreted as flows of moving matter, or as relief maps, showing valleys and mountains (Figure 2, top) [14]. Here, replacing the segments by their endpoints gives rise, with difficulty, to similar patterns, but the third dimension is completely lost (Figure 2, bottom).

When a simple planar geometric figure is moved in its plane, it is often seen to change shape — therefore, the rigidity assumption is not too powerful. One well-studied case is that of an ellipse rotating around its center. It may be seen as an ellipse flipping back and forth, out of the plane, or as a gelatinous ellipse moving in the plane like a caterpillar [8]. In the "hulla-hoop illusion" (Figure 3, left, see also Figure 3,right), it seems that positional information concerning the median circle and shape extraction are lagging behind global motion analysis, hence the strange behavior of the median circle.

3 Static shapes: the gestalt approach

At the beginning of the 20th century, a very imaginative series of studies on shape perception was initiated in Germany by the "Gestalt" school of psychology. This school was mainly concerned with the issues now discussed under the names of "segregation" and "fusion". What makes us see different elementary parts in a

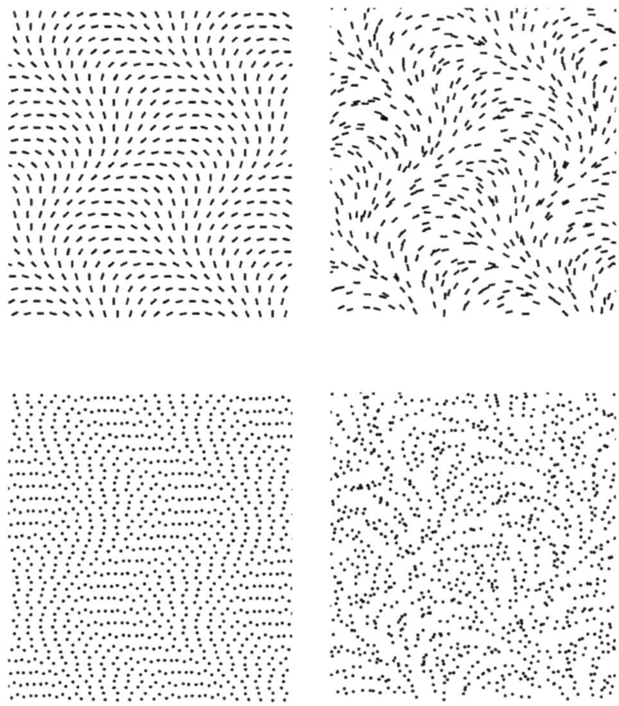

Figure 2. (Top left) *Field stimulus*. This image is made from a unique stripe of segments, the orientations of which vary regularly and periodically from left to right. The stripes are reproduced from top to bottom with small shifts to the left or to the right. The generated patterns convey a sensation of flowing matter, and also of a 3D landscape with hills and valleys. According to Field *et al.*, these interpretations are due to the alignments formed by sets of segments in close proximity [14]. (Top right) However, when the segments are randomly shifted, thus breaking their alignments, the 3-dimensionality is still striking. The organizing principle is therefore more likely to be a conjunction of spatial and orientational proximity.
(Bottom) When the segments are replaced with their endpoints, 2D patterns may still be discerned, mainly in the left image, but they do not give rise to 3D interpretations.

visual stimulus, and what are the rules according to which the elementary parts are combined into visually distinct wholes?

According to the gestalt school, the main factors favoring perceptual grouping are those of proximity, resemblance, continuity in direction, closure (elements are assigned to a same unit if they lie on a common, virtual closed curve or surface) and "pregnance" (the structural coherence which may be brought by simplicity, order, regularity, symmetry). Although these definitions may appear a bit vague, they capture a number of striking effects in shape perception (see, e.g. [14]) and have equally striking counterparts in the auditory field [16].

Here, the Glass patterns (Figure 1, bottom) can be taken as examples of groupings by proximity. The Field image (Figure 2, top left) was initially thought to reflect grouping by collinearity [14], but more probably reflects grouping by "prox-

Figure 3. *Lagging shapes.* (Left) The "hulla-hoop illusion". When a motion along a circular path is communicated to the page, without changing its orientation (the way one shakes a glass of water, to dissolve granules) the intermediate circle wobbles around the center. (Right) When this pattern, taken from [42] is gently shaken along a horizontal axis, the parabolic separation between the two striped halves undulates with respect to the border of the image.

Figure 4. *Fraser's spiral* [17]. The tiling pattern in the background seems to be formed of curved stripes, comprising either alternating black and grey quadrangles, or alternating white and grey quadrangles. However, the straight alignments of alternating black and white quadrangles emanating from the center, are rather difficult to perceive. This illustrates a principle of association according to proximity in grey level (or dissociation due to excessive contrast). The essence of Fraser's illusion is that black and white arcs appear to form a spiral, although, they belong to concentric circles. Actually, if one considers the set of black or white arcs separately, the arcs are indeed typical of "spiral motion" - a combination of the two top flow patterns of Figure 1.

imity in the orientation space". Figure 4, which represents the famous Fraser's spiral [17] offers a good illustration of the gestalt principles. The apparent background organization is explained by the fact that black and grey tiles on one hand, white and grey tiles on the other form coherent wholes, while black and white tiles do not. Similarly, the spiral aspect of the arcs indicates that white and black arcs are treated separately (see the legend to Figure 4).

4 Static shapes: the importance of orientations

Rather than analyzing shapes on a point by point basis, the brain seems to be particularly sensitive to the presence of oriented linear elements, and makes a great use of the orientational information. As an example, consider the stimulus in Figure 5. There, a section of planar sinusoid is translated repetitively. The overall pattern is interpreted as the orthographic projection of a 3D surface having the shape of a staircase. The observers can describe accurately how they perceive the surface orientation at each point [18]. According to Stevens, the observer makes the following implicit assumptions [19]:

(i) the parallel aspect of the lines is "generic", i.e., had the same surface been observed under a slightly different viewpoint, the same lines, assumed to be engraved on the surface, would have maintained their parallel appearance. From there, it follows that the lines do indeed follow parallel paths in 3D, and must be on a cylindrical surface. The genericity principle [20] is emerging as one of the main guiding principle in image understanding.

(ii) The represented lines are lines of curvature on the surface.

I suggested elsewhere that parallelism is one of the very first features the visual system learns to detect [21]. When exploring a scene, the eye accomplishes small saccadic movements. The retinal images of most of the points are then displaced by small translations (as in Figure 1, top left). Hence, during the early development of a child, the brain can learn through experience to recognize parallel displacements. Having mastered parallelism, it becomes able to detect straight lines : such lines

Figure 5. *Stevens' staircase* [18,19]. This figure is formed by sections of parallel, equidistant sinusoids, and is interpreted in 3D as a cylindrical surface.

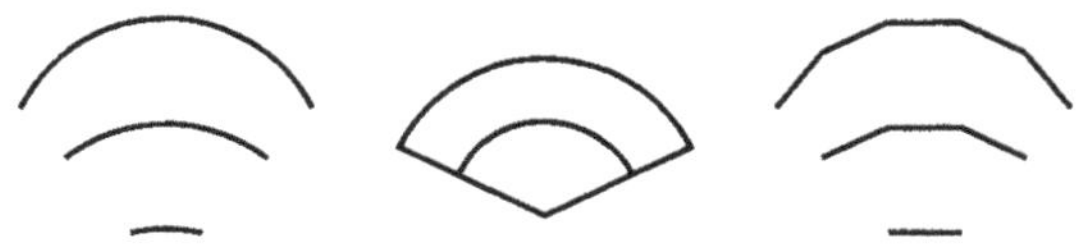

Figure 6. *Flattening of short arcs.* In the left figure the three arcs are sections of a same circle. They look increasingly flatter from top to bottom, and this was considered to be an illusion. However, what we call shape is invariant by homothecy. The two arcs in the central figure have different radii of curvature, they are homothetical and look about equally curved. The arcs on the left are different sections of a same figure, and need not have the same appearance, in the same way that the polygonal shapes in the right are really increasingly flatter from top to bottom [27]. Note that the top arcs in the left and central figures are equal, although the latter looks flatter. Therefore, there is some illusory component in the left figure, but it is due to the relative placing of the arcs, When the smaller arcs are drawn above the larger one, the flattening effect is reduced.

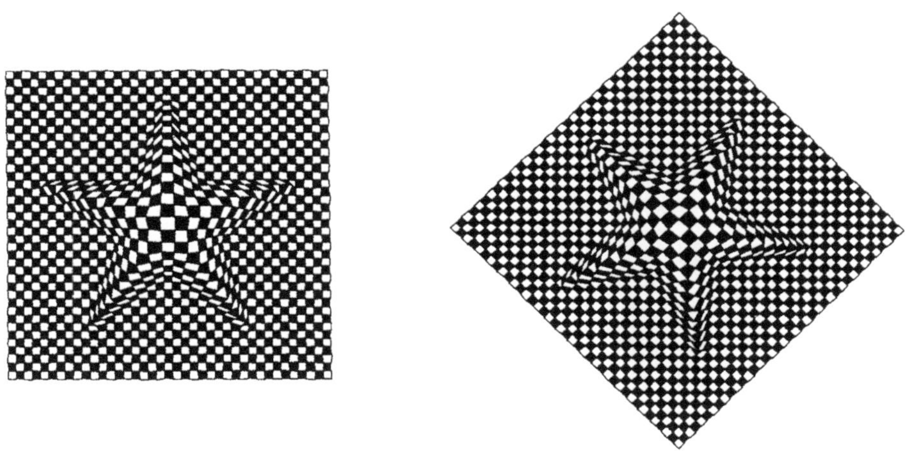

Figure 7. *Anisotropies.* The two figures are equal, yet the figure on the left seems smaller than the right one. In the latter, the quadrangles are organized according to their grey level. One sees threads of all-black or all-white quadrangles along the horizontal and vertical directions. In the left image, one sees horizontal and vertical stripes, but this time, they are composed of alternating black and white squares (an exception to the gestalt principles).

would be characterized by the property that their local orientation is uniform all along the line [22]. For earlier proposals, see [23].

Parallelism being readily appreciated, it becomes in principle possible to recognize that two figures are homothetical — through a triangulation procedure, involving constructions, i.e., addition of lines to the figures. However, the brain does not add lines to appreciate the properties of a figure (except in the case of "subjective contours" — see Section 7). In any event homothecy recognition is very powerful in shape perception — it is known under the name of "shape constancy", i.e., a 2D shape looks the same at various distances. Shape constancy explains for instance the irksome effect of flattening of short arcs (Figure 6). It does not apply in 3D, because of the foreshortening effect of distance. Depth is overestimated

Figure 8. *Vertical symmetry.* This figure is made of three juxtaposed symmetrical vertical stripes. Motifs are seen along the symmetry axis, and acquire complexity as they are being inspected. When the page is turned by 90 degrees, the symmetry axes become horizontal, and the patterns regress, to the benefit of the original aniseed texture.

with respect to width at short distances (think of a nose seen in a close-up picture) and underestimated at large distances (consider the squeezing effects in pictures taken with a telephoto lens). Consistent with this trend, the slope of a hill or that of a pyramid, when appreciated frontally, looks steeper and steeper at increasing distances [24].

While we are very sensitive to orientations, and see that two lines at slightly different orientations are not strictly parallel, there are marked anisotropies in our perception of orientations (Figure 7). The terrestrial vertical direction plays a particular role. Symmetries with respect to the vertical are much better appreciated than symmetries with respect to any other direction (Figure 8, reviews in [25]). When a picture is turned upside down, faces are hard to recognize, bumps are sometimes transformed into hollows, and drawn material may organize itself differently.

5 2D perceptual geometry

We are extremely sensitive to shapes in 2D : slight defects in the alignments of points, in the parallelism of lines drawn on a plane (seen at all orientations of the plane), slight departures from regularity in a smooth curve are readily noticed. On the other hand, we are subject to many "illusions" in the appreciation of the metric proportions in a figure. The two aspects are not contradictory. In cartography, when a planar map of a portion of the earth is constructed, according to a rigorous algorithm, it looks distorted. The pattern of distortions tells us something about the procedure used to construct the map. In vision, the retinas are slightly spherical, but I do not think this feature plays a role in geometrical illusions.

I hypothesize that geometrical illusions arise from the necessity to put the informations pertaining to shape, acquired through various channels, into a uniform format [22,24]. When the magnitude of a geometrical illusion is measured, it amounts, typically, to about 5% error on the ratio between the lengths of two segments. Now, the illusion for the right and left eyes agree to about 10%, which means that the two eyes provide geometrical estimates that agree within 5 parts in a thousand. (Actually, the agreement must be quite good, otherwise stereoscopic vision cannot work).

When a scene is viewed with a fixed eye, the input is structured according to the rules of linear perspective. When the eye moves and explores a scene, the input is structured according to curvilinear perspective. Imagine being in front of a wall of uniform height which extends very far to the left and to the right. According to linear perspective, it should be represented as a rectangle. But when it is inspected with moving eyes, its apparent (angular) height is maximum in front of the observer, and decreases as the line of sight moves to the extremes. The results from the two viewing modes must be harmonized, perhaps by retaining the alignments and linearities from the static viewing mode, and the metric proportions derived from the dynamic viewing mode. Note also that shapes determined from motion must agree with statically determined shapes.

When we perceive a shape, we are not aware of the channel through which this shape was retrieved, and this is an indication that the outputs of all the channels are made to agree. This probably requires systematic corrections to be made at the output of each channel, and it is because we do not understand culturally the rationale for these corrections, that we are puzzled by the geometric distortions in the so-called illusions [24].

I have started mapping systematically the distortions in orientation or alignment illusions such as the Zöllner illusion [26] and the Poggendorff illusion (unpublished) and believe that they are related to the problem of correcting for perspective effects. A rectangular array, presented laterally, is squeezed by perspective. Yet, when we look at the horizontal stripes corresponding to the successive floors in buildings, viewed laterally in a street, the horizontals seem nearly parallel, and do not show the strong convergence one usually finds in perspective representations. Therefore, our perceptual apparatus has exerted an "anti-squeeze" correction, making elongated parallelograms look rectangular. However, I have not yet been able to establish a more precise link between the postulated correction and the experimental results.

The distortions in metric illusions are perhaps easier to characterize. I find that there is a major theme behind these illusions : If x is the size of a segment and $m(x)$ its perceived size, then $m(x)$ is a convex function of x — something like $x^{1.2}$. Such a convex relationship preserves shape-constancy, since ratios of perceived sizes do not vary with distance, under this law, and it enhances the contrast between small and large sections in a figure. To describe correctly most of the illusions, this basic law must be completed by two laws on how to normalize the measurements, so as to compensate for undesirable side-effects of the first law [27].

6 From 2D to 3D

The brain has a natural tendency to interpret 2D images as representing 3D objects, even when 3D clues are minimal, or non-existent. Even the simple drawing of a trapezium, which one expects to be interpreted as a planar shape, gives rise to a compelling twisted-ribbon 3D interpretation (Figure 9).

As pointed out by Stevens [19], there is a tendency, in line drawings, to represent a surface using its lines of curvature (Figure 5), but one may also use unrelated lines. For instance, in Figure 10 initially parallel and equidistant lines were distorted by vertical displacements. This representation is obtained in two steps. In a first step, a grid of black equidistant parallel lines is projected orthogonally onto the surface to be represented. The surface is thus defined by the textural markings of the projected grid. In the second step, these markings are projected conically on the picture plane. This procedure is computationally easy, but its ecological validity is questionable. The remarkable fact is that it is perceptually adequate. Sudden orientation changes in the lines signal sudden orientation changes on the surface. On the other hand, an increase in the spacing between two lines does not necessarily signal a greater proximity. (For further discussions on the role of compression cues, see e.g. [28].) In the Stevens type images (Figure 5), in which the lines are all parallel, there is no compression cue, and the brain must make the additional restrictive hypothesis that the lines are lines of curvature on the surface.

Beyond the intellectual appreciation of depth, it is possible to experience depth sensations from single images, with the same acuity as stereoscopic depth. When an image is explored with a single eye through a narrow tube, so as to eliminate the influences of the border and the environment (which normally contribute to

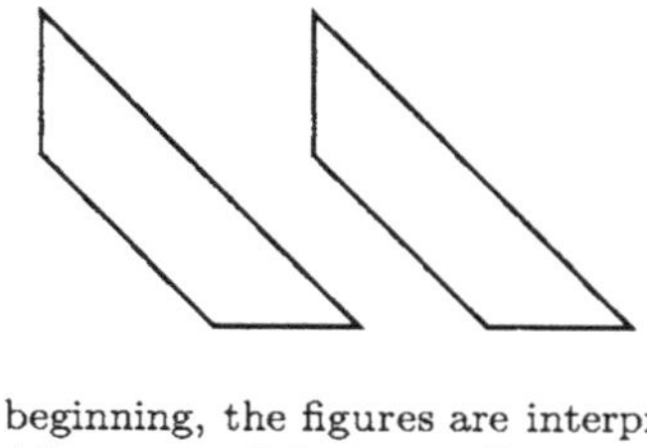

Figure 9. *3D salience.* At the beginning, the figures are interpreted as trapeziums in the plane of the sheet, or trapeziums in oblique, parallel planes. However, they can be also interpreted as non-planar shapes, and seen as portions of twisted ribbons, the lower sides of the trapeziums being in the forefront, and the upper sides being at the back. Thereafter, it may become difficult to see the trapeziums as planar again.

Figure 10. *Shape from perspective cues.* A surface is represented here with a set of lines which could have been produced as shadows from a grid of parallel, equidistant lines. The same principle of representation (cylindrical projection of a grid on a surface, followed by conical projection on a plane) applies to Figure 7.

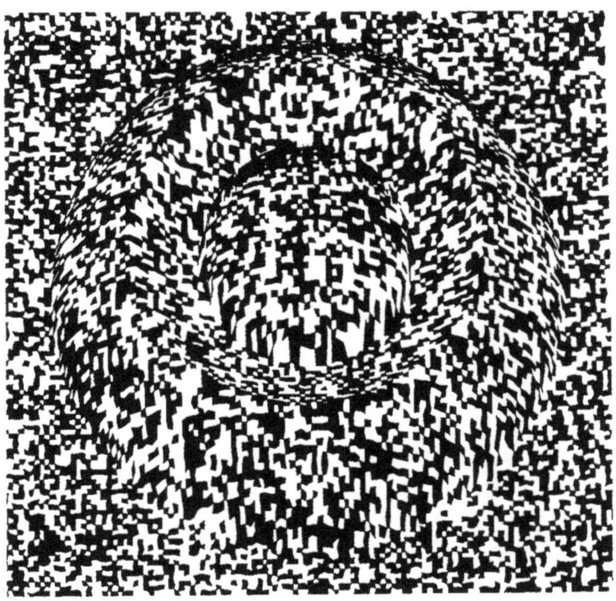

Figure 11. *Paradoxical monocular stereoscopy.* Observe this picture through a narrow cylinder (a folded sheet, or a folded hand), following the regions of changes in textural density. The picture acquires a realistic relief, as compelling as the stereoscopic relief.

inform us that we are dealing with a planar stimulus) one sees the image as a real 3D surface. The phenomenon can be observed even in the absence of shading cues, with perspective cues (Figure 10) [29–31].

Contemporary info-graphic methods in which 3D surfaces are represented with distorted lattices are much less efficient than the traditional methods of the artists who know how to represent a rather complex 3D surface with a few judiciously selected lines. According to which criteria are these lines selected? The first is that of the external contour : the locus of the points at which rays emanating from the eye meet tangentially the surface. When a surface has bumps, one must associate to the bumps additional lines of contours. Since the bumps are not independent closed surfaces, their contour lines must be open. They must start or end at some place, but where? The logic seems to be as follows [32]: if a ray emanating from the eye is tangent to the bump and intersects the surface further away, then the point of tangency must be included in the contour. But as soon as, after the point of tangency, there is no additional path to meet the surface, the contour must end. Both the external contour and the additional open contours depend upon the viewing position, and how the topology of the contour lines vary with viewpoint is being actively investigated (e.g. [33]).

The additional open contours should in general terminate with convexity reversals, but artists have a tendency to represent protruding shapes (women's breasts, for instance) by exclusively convex contour lines [32]. More generally, it seems that artists tend to represent shapes by the accretion of convex units. Concavities would be conceived as depressions created in otherwise convex units (as for instance, when pushing a finger into a ball [32]).

The connecting regions between concave and convex portions have a saddle shape (hyperbolic), which is not well conceptualized, even by artists [32]. The curves that delimit the frontiers between hyperbolic regions on one side, convex or concave regions on the other, are called parabolic curves. Those that separate hyperbolic from convex regions have an important visual role: it is there that new reflections, new shadows or new open contours arise when an object rotates [34], although these parabolic lines are very rarely lines of physical separation between different parts of an object. The "back" of a baker's croissant is convex, its "belly" is hyperbolic, and the subjects locate quite accurately the delimiting frontier, even though they may not have the concept of parabolic lines [35].

In a series of systematic experiments, Koenderink and collaborators [36] asked subjects to evaluate the 3D surface orientation at different points of the photographic picture of a human torso (Figure 12). The surface orientation was defined by two parameters (the slant and the tilt) which the subjects had to evaluate consecutively at arbitrary points of the surface, presented in a random order. The pictorial cues, in this picture, were the lines of contour and the shading gradients. The responses formed a remarkably consistent set. From the subjects' evaluations of the 3D orientation of the tangent planes, the authors were able to calculate, with a good approximation, a mathematical surface compatible with the set, and this surface was in good agreement with the represented object (Figure 12, right).

Among the various pathways through which 3D shapes may be interpreted, stereopsis has received considerable attention. In descriptive geometry, the 3D shape of an object is inferred from a couple of cylindrical projections on two orthogonal planes. Stereopsis is similar in principle [21], but uses the conical projections

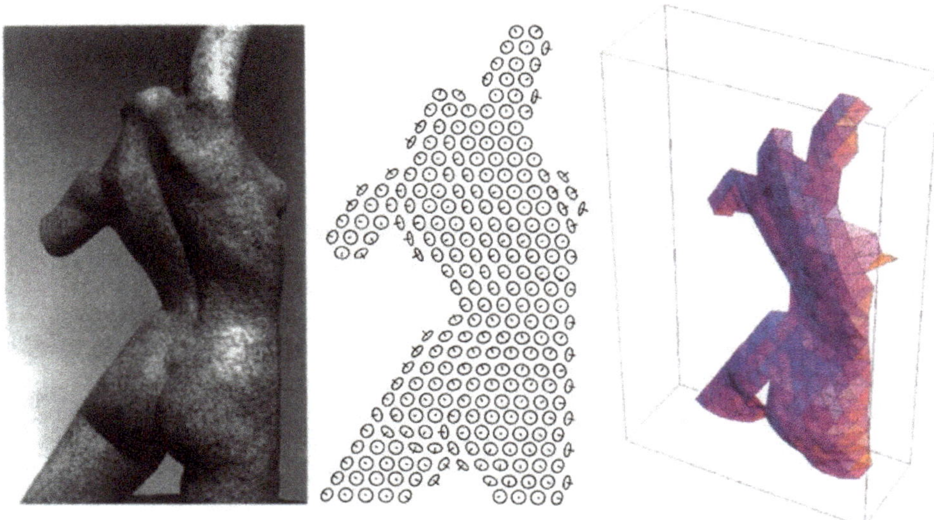

Figure 12. *Surface orientation perception in pictures.* Observers were asked to appreciate the orientation of the tangent plane at various points on the torso in the left photographic picture. Their judgments relative to the slant and the tilt of the tangent planes are represented by the gauges in the central figure. From these results, a continuous mathematical surface could be computed (right), which agreed with the original shape [36]. I thank J.J. Koenderink for permission to reproduce these documents.

of a scene on the slightly curved surfaces of the retinas, these surfaces forming an angle of 0 to about 45 degrees.

There is a current, very influential misconception, in which the task of stereopsis is assimilated to that of a range-finder. According to this school, stereopsis must produce a depth map, from the comparison of the two retinal images. Consider four points in 3D space, and assume you know all their coordinates. From these, you may compute 3D shape indexes and, in particular, deduce whether the quadrangle is planar or not. However, and very strikingly, when looking at stereoscopic couples, it is as though stereopsis was extracting directly the information on coplanarity or on surface curvature, from local comparisons, without prior determination of the depth map.

In most theoretical treatments of stereopsis (e.g. [37,38]), the two views to compare are distributions of points, known by their x and y coordinates in the projection planes, and all discussions are about extracting information from the horizontal disparities (the differences between the x coordinates of corresponding points in the two views) or from the vertical disparities (the differences in y coordinates), or both. The fact that linear elements may have different orientations in the two views, and that these differences in orientations may be used to assign 3D orientations in space has not yet attracted much interest, although it seems that stereo vision is apt at extracting this orientational disparity information [39]. The problem is that most stereoscopic stimuli used in laboratory experiments ignore linear elements.

7 Shape synthesis

Once a number of clues, relative to shape, have been derived from the visual input, a global shape is perceived and this perception entails, in my opinion, the synthesis of an internal model of the shape. One supporting argument is provided by the phenomenon of subjective contours (Figure 13). The rings and the disks seen in this figure are not defined by their external contours. Such contours are illusory, and cannot even be generated by interpolation between the cuts in the textures. It rather seems that the disks and the rings were synthesized as shapes that could best explain the patterns of cuts in the background textures.

Subjective 3D surfaces are also produced in stereoscopic vision, as shown in the classical example of Figure 14, in which one perceives a saddle-shaped surface [40]. The subjective perceived surface cannot have been interpolated, or computed point by point since the projections of the surface are uniformly black, and therefore do not contain any salient mark that might be used for disparity calculations.

Figure 13. *Subjective contours.* In both figures, one sees a very regular annulus, over a background of spaced hexagons. Contours drawn by interpolation from one terminator to the next would have produced less regular shapes. Usually, much thinner background lines are used, and they are chosen black. Here, black and white are treated more symmetrically, making more striking the fact that the subjective shapes must be "cut away" from lines of their own color.

Figure 14. *Illusory surface in 3D.* When viewed stereoscopically, this couple of images gives rise to the perception of a saddle-shaped surface [40]. Such a shape could have been obtained by inserting rigid sticks at the four corners of a flexible rectangular sheet, holding the sticks rigidly in 3D (in the orientations expected from the real disparities at the corners) and letting the surface adjust to the constraints.

In order to define a particular curve or surface, the mathematician selects the most economical procedure, on a case by case basis. I suspect that the brain also uses many techniques to represent shapes, and in my view, the representations of curves and those of surfaces follow different paths. The representation of curves is very precise, in the sense that minute differences between two curves are immediately noticed, even though the shapes are perceived with the systematic distortions called illusions. The representation of surfaces is more approximate; hyperbolic regions are poorly conceptualized, we do not see that a surface is ruled unless the straight lines are explicitly figured on it, we appreciate very crudely the ratio of depth to the two frontal dimensions, as we discover if we wish to sculpt a head.

The shape of perceived surfaces can be understood from just a few lines of contour, and this suggests methods for producing 3D surfaces from planar contours. It seems to me that we tend to attribute to a closed planar contour, a relief given by the simple rule that the height associated to a given point inside a contour increases with its remoteness from the contour. In other words, narrow 2D portions are thin in the orthogonal direction, large 2D sections are thick. One, among the many ways to achieve this goal, would be to take a variable sphere the center of which would remain in the horizontal plane, and make it move inside the contour with the constraint that the sphere be tangent to the contour on two sides simultaneously [41]. This process has the advantage of generating a backbone from which an economical description of the initial shape and many variants could be derived.

The representation of planar curves appears more delicate and mysterious to me. I think first of physical models involving propagating fields created by punctual sources. A stone thrown in the water produces a family of concentric circles. An isolated electric charge in a plane produces an electrostatic field of which the equipotentials in this plane are concentric circles. With two charges of the same or of different signs, it is possible to generate other more complex families of curves. A neuronal implementation of such a model is conceivable. If emitting neurons sending impulses in every direction impersonate the charges, a family of contour lines could be generated, at loci receiving a same total number (or some more pertinent function) of impulses. The risk, with most propagative models from punctual sources is to produce scalloped curves. It seems to me that visual perception produces curves having maximum regularity (see Figure 13), and this brings to my mind the analogy of an ovoid elastic surface, which expands as one keeps insufflating air inside, and is then remodeled with a few localized external pushes.

References

1. J. J. Gibson (1950). *The perception of the visual world.* Houghton Mifflin, Boston, Mass.
2. H. Wallach and D. N. O'Connell. The kinetic depth effect (1953). *J. Exp. Psychol.* **45**: 205–217.
3. J. S. Lappin, J. F. Doner and B. L. Kottas (1980). Minimal conditions for the visual detection of structure and motion in three dimensions. *Science* **209**: 717–719.

4. S. Ullman (1979). *The interpretation of visual motion.* The M.I.T. Press, Cambridge, Mass.
5. R. M. Siegel and R. A. Andersen (1988). Perception of three-dimensional structure from motion in monkey and man. *Nature* **331**: 259–261.
6. K. Nakayama and J. M. Loomis (1974). Optical velocity patterns, velocity sensitive neurons, and space perception: a hypothesis. *Perception* **3**: 63–80.
7. J. Droulez and V. Cornilleau-Pérès (1990). Visual perception of surface curvature. The spin variation and its physiological implications. *Biol. Cybernetics* **6**: 211–224.
8. C. L. Musatti (1924). Sui fenomeni stereocinetici. *Arch. Ital. Psicol.* **3**: 105–120.
9. J. M. Foley (1964). Desarguesian property in visual space. *J. Opt. Soc. Am.* **54**: 684–692.
10. J. T. Todd and P. Bressan (1990). The perception of 3-dimensional affine structure from minimal apparent motion sequences. *Perception and Psychophysics* **48**: 419–430.
11. K. N. Niall (1992). Projective invariance and the kinetic depth effect. *Acta Psychologica* **81**: 127–168.
12. H. T. Kukkonen, D. H. Foster, J. R. Wood, J. Wagemans and L. van Gool (1996). Qualitative cues in the discrimination of affine-transformed minimal patterns. *Perception* **25**: 195–206.
13. L. Glass (1969). Moiré effects from random dots. *Nature* **223**: 578–580 (1969).
14. D. J. Field, A. Hayes and R. F. Hess (1993). Contour integration by the human visual system: evidence for a local "association field". *Vision Research* **33**: 173–193.
15. G. Kanizsa (1979). *Organization in vision.* Praeger, New-York.
16. A. S. Bregman (1990). *Auditory scene analysis: the perceptual organization of sound.* MIT Press, Cambridge, USA.
17. J. Fraser (1908). A new visual illusion of direction. *Brit. J. Psychol.* **2**: 307–320.
18. K. A. Stevens and A. Brookes (1987). Probing depth in monocular images. *Biological Cybernetics* **56**: 355–366.
19. K. A. Stevens. Inferring shape from contours across surfaces. In *From pixels to predicates : recent advances in computational vision*, ed. A.P. Pentland (Ablex, Norwood, USA, 1986), pp. 93–110.
20. J. J. Koenderink and A. J. van Doorn (1979). The internal representation of solid shape with respect to vision. *Biological Cybernetics* **32**: 211–216.
21. J. Ninio (1977). The geometry of the correspondence between two retinal projections. *Perception* **6**: 627–643.
22. J. Ninio (1996). *L'empreinte des sens. Perception, mémoire, langage.* 3rd ed. Odile Jacob, Paris.
23. J. R. Platt (June 1960). How we see straight lines. *Scientific American* **202**^{6}: 121–129.
24. J. Ninio (1998). *La science des illusions.* Odile Jacob, Paris.
25. C. W. Tyler (ed.) (1996). *Human symmetry perception and its computational analysis.* VSP, Zeist, Netherlands.

26. J. Ninio and J. K. O'Regan (1996). The half-Zöllner illusion. *Perception* **25**: 77–94.
27. J. Ninio (1979). An algorithm that generates a large number of geometric visual illusions. *J. Theoret. Biol.* **79**: 167–201.
28. A. Blake, H. H. Bülthoff and D. Sheinberg (1993). Shape from texture: ideal observers and human psychophysics. *Vision Research* **33**: 1723–1737.
29. A. Jr. Ames (1925). The illusion of depth from single pictures. *J. Opt. Soc. Am.* **10**: 137–148.
30. J. J. Koenderink, A. J. van Doorn and A. M. L. Kappers (1994). On so-called paradoxical monocular stereoscopy. *Perception* **23**: 583–594.
31. J. Ninio (1994). *Stéréomagie.* Seuil, Paris.
32. J. J. Koenderink and A. J. van Doorn (1982). The shape of smooth objects and the way contours end. *Perception* **11**: 129–137.
33. T. van Effelterre (1994). Aspect graphs for visual recognition of three-dimensional objects. *Perception* **23**: 563–582.
34. M. S. Longuet-Higgins (1960). Reflection and refraction at a random moving surface. I. Pattern and paths of specular points. *J. Opt. Soc. Am.* **50**: 838–844.
35. P. Mamassian, D. Kersten and D. C. Knill (1996). Categorical local-shape perception. *Perception* **25**: 95–107.
36. J. J. Koenderink, A. J van Doorn and A. M. L. Kappers (1996). Pictorial surface attitude and local depth comparisons. *Perception & Psychophysics* **58**: 163–173.
37. B. Julesz (1971). *Foundations of cyclopean perception.* Chicago University Press, Chicago Illinois.
38. I. P. Howards and B. J. Rogers (1995). *Binocular vision and stereopsis.* Oxford University Press, Oxford.
39. J. Ninio (1985). Orientational versus horizontal disparity in the stereoscopic perception of slant. *Perception* **14**: 305–314.
40. J. P. Harris and R. L. Gregory (1973). Fusion and rivalry of illusory contours. *Perception* **2**: 235–247.
41. H. Blum (1973). Biological shape and visual science. *J. Theoret. Biol.* **38**: 205–287.
42. H. R. Wilson and W. A. Richards (1992). Curvature and separation discrimination at texture boundaries. *J. Opt. Soc. Am. A* **9**: 1653–1662.

PATTERN RECOGNITION IN THE VISUAL SYSTEM AND THE NATURE OF NEURAL CODING

SIMON THORPE
Centre de Recherche Cerveau & Cognition
133 route de Narbonne
31062, Toulouse, France
E-mail: thorpe@cerco.ups-tlse.fr

The ability of the human visual system to rapidly categorize complex natural scenes presents a major challenge to our understanding of pattern recognition mechanisms. In particular, recent experiments that have measured the speed with which the visual system processes information impose very severe constraints on the coding mechanisms that are used. Indeed, the classical view, which holds that the information is transmitted in the form of a firing rate code, appears to be too slow and inefficient to be compatible. We argue that alternative coding schemes that make use of the pattern of spikes across a population of cells, and more specifically the order in which they fire could be more plausible hypothesis.

1 Introduction

The most sophisticated pattern recognition system we know of is the one that we have in our own heads. We are all able, in a fraction of a second, to recognize events and objects in the visual world. This remarkable ability is something that is orders of magnitude more impressive than even the most sophisticated artificial vision systems. In this chapter, I will first review some recent experiments that have quantified the remarkable accuracy of the human and primate visual system in scene categorization tasks. I will then argue that the temporal constraints imposed by such data oblige us to rethink currently accepted views of how information is encoded in the nervous system. The vast majority of brain scientists believe that neurons transmit information in the form of a firing rate code, but the evidence suggests that some forms of sensory processing are performed so quickly that it would be impossible to determine firing rates with sufficient accuracy because very few cells would have time to generate more than one spike. I will then propose an alternative approach to neural coding which makes use of the temporal pattern of pulses across a population of neurons. Specifically, I suggest that one could use the order of firing across a set of cells in order to send information rapidly and efficiently. The task for the visual system then becomes one of being able to detect particular temporal patterns of firing across neurons.

2 The efficiency of pattern recognition by the human visual system

In a recent series of experiments we have explored the abilities of human subjects in a task where they have to decide whether a briefly flashed color photograph contains an animal [14]. The task was a go/no-go categorization task — the subjects maintained their finger on a button and released the button as quickly as possible on target trials, which included photographs of mammals, birds, reptiles, fish

Figure 1. Examples of the images used in the experiments by Thorpe *et al.* (1996).

and insects in their natural environments. No response was required to distractor stimuli, which included nature scenes such as trees, flowers, rocks, waterfalls and so forth. Figure 1 illustrates the sorts of images used in this task.

Performance was remarkably good (around 94% correct) despite the very brief presentation duration (20 ms) which prevents any possibility of eye movements, and the fact that none of the images had ever been seen before by the subjects, which prevents the possibility of learning to use particular image features. Note that the subjects have no advance information concerning the type of animal to look for, its size, shape, an orientation or even the number of animals present. This is clearly a task that would present insurmountable problems for even the most sophisticated artificial pattern recognition systems. And yet human subjects find this task almost trivially easy.

The ease with which subjects can perform this task is also reflected in the behavioral reaction times of the subjects. The mean reaction time tends to be about 400–450 ms in this task, but this value varies substantially from subject to subject — some subjects have median reaction times well under 400 ms. It is particularly interesting to examine the reaction time distributions in detail because we can see that some of the behavioral responses can occur with reaction times as short as 260–280 ms. It might be thought that these very short reaction times could be due to subjects "guessing" or making early errors. However, this is clearly

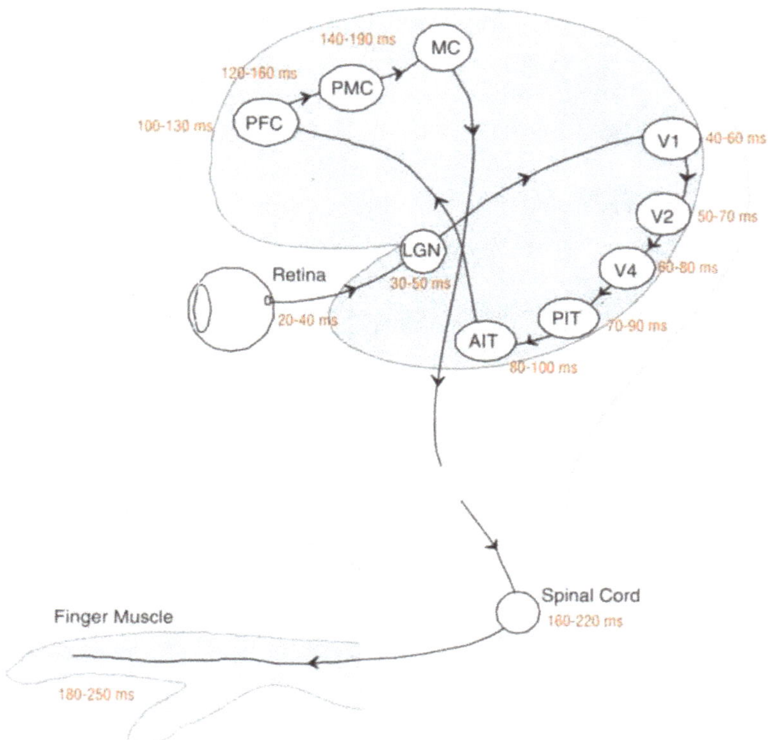

Figure 2. Diagram showing the pathway by which visual information could reach the muscles in the hand. After a relay in the Lateral Geniculate Nucleus (LGN), the image is processed by at least 5 cortical areas including V1, V2, V4, the Posterior and Anterior Inferotemporal Cortex (PIT and AIT), after which signals are sent via the Prefrontal Cortex (PFC), Premotor Cortex (PMC) and Motor Cortex (MC) to motoneurones in the Spinal Cord which activate the muscles in the hand. The numbers refer to estimates of the response latencies of neurons in each area, the first number being the earliest latencies, whereas the second gives a typical average latency.

not the case because when one looks at the reaction time distribution for errors on distractor trials it is clear that, even if we just take the interval from 260–300 ms, the probability of making a response on target trials is roughly 7 times higher than for distractors. Given that targets and distractors were equally probable, guessing would lead to equal proportions of responses to targets and distractors — clearly the 260 ms is long enough to execute the entire processing sequence between the arrival of the image on the retina and the movement of the muscles in the hand necessary to make the behavioral response.

Recently, we succeeded in training monkeys to do the same task [3]. Interestingly, not only were the monkeys remarkably successful (achieving roughly 90% correct), but they were also substantially faster than even the fastest human subjects, with reaction times starting at around 180 ms. In this time, information has to pass through a very large number of intermediate processing stages, which means that the amount of time available for processing at each stage will be very limited. Given the complexity of the visual task we used, it seems very likely that information has to go through all the brain areas illustrated in Figure 2 on its way from the eye to the hand. The figure also gives rough indications of the latencies of the neural responses in the different areas in the monkey, and from this it can be seen that in general, response latency increases by only about 10 ms between each processing stage. Note that each cortical processing stage involves not just one synaptic relay, but at least two (since the neurons that receive the inputs from the previous layer are not the output neurons). Furthermore, one also has to take into account the fact that quite a lot of time is taken up by conduction delays, since the conduction velocity of nerve fibers within the cortex is probably quite slow — about 1–2 m/s [10].

3 The Coding Problem

Such data emphasize the remarkable speed with which the brain can do computations, but more specifically, they raise an important issue concerning the nature of the code used for transmitting information. The conventional view is that neurons use a firing rate code to send information, and it is indeed true that when the intensity of a sensory stimulus is increased, the firing rate of sensory neurons increases. However, the firing rates of neurons in the visual system only rarely go over about 100 spikes/second, which means that in the 10 ms or so that is available for processing in each stage, the vast majority of neurons will generate at most one spike. How can one reasonably expect to code information accurately under such conditions?

In fact, the problem is made even worse by the fact that when a neuron is stimulated by a constant sensory stimulus, the spike generation process is highly stochastic, corresponding roughly to a Poisson process. Jacques Gautrais recently looked at just how much information could be transmitted using such neurons on the basis of such a firing rate [4,5]. The conclusions were clear: using a firing rate code with stochastic neurons is extremely inefficient with short observation periods.

To illustrate this point, consider a neuron that fires one spike in a 10 ms time window — what can we say about its firing rate? Gautrais showed that in such a case the true frequency of the neuron lies somewhere in the range 5 Hz to 474 Hz, with only a 90% chance of being correct. Of course, one could improve accuracy by increasing the number of neurons, but it turns out that this is very costly, since in order to get an estimate as precise as 100 Hz ± 10 Hz in 10 ms, you would need no less than 281 redundant and independent neurons!

It could be argued that neurons are not designed to code precise analog values, but rather to allow comparisons to be made. But even such simple problems are difficult to solve using stochastic neurons and firing rate codes. Consider two pools of neurons A and B, one of which generates spikes according to a Poisson process at 100 spikes/second, the other at 75 spikes/second. Suppose that we need to decide within 10 ms which is the more active pool. How many independent neurons would be needed in each pool in order to obtain an error rate of less than 5%? Gautrais showed that even this trivial problem in which less than one bit of information is transmitted would need no less than 76 neurons in each pool!

It seems clear that such a poor level of efficiency is incompatible with the anatomical constraints such as the number of fibers in the optic nerve. The primate optic nerve contains roughly 1 million fibers, and it is clear that within at most a few tens of milliseconds, this number of fibers is sufficient to send detailed information concerning the entire visual field. If each bit of information required tens or hundreds of redundant neurons, there simply would not be enough capacity in the system.

4 An alternative — Using patterns of spikes

Is there some other way in which information could be sent more rapidly than with the classic rate coding scheme? In recent years, an increasing number of researchers have become interested in the possibility of using temporal codes [1,6,8,11,13,17]. These are codes in which information is contained not in the overall firing rate of neurons, but rather in the precise timing of spikes either within a single neuron, or across a population of cells. A recently published book entitled "Spikes" [12] took a detailed look at the power of temporal spike codes at the single cell level. Here, we will briefly examine another temporal coding scheme that uses the relative timing of spikes across a population of cells, and more specifically the order in which they fire.

Recent data indicates that the initial transient response of a neuron may in fact be temporally quite precise and reliable [9]. This opens up possibilities for alternative coding schemes that depend on the timing of spikes. One such alternative considers a neuron not so much as an analog-to-frequency converter, but rather as an analog-to-delay converter. The idea is very simple and is based on the fact that the time taken for an integrate-and-fire neuron to reach threshold depends on the strength of the input [see Figure 3(a)]. This would lead to the sort of latency-intensity function illustrated in Figure 3(b). Consider now what would happen with several receptor cells and an intensity profile as illustrated in Figure 3(c). Because of the intensity-delay transformation, the neurons will tend to generate spikes in an order that reflects the input distribution.

Given this asynchrony in spike generation, there are actually various ways that one could develop a coding scheme. One would be to use the relative timings of spikes across the different cells [15]. This certainly allows a large amount of information to be transmitted within a very short time. Suppose that we have N neurons, and that precise arrival times of spikes can be measured to the nearest millisecond. Within a time window of T milliseconds one can transmit a maximum of $\log_2(T^N)$ bits of information. Thus with 6 neurons, one can transmit up to $\log_2(10^6) \approx 20$ bits of information. Of course, in order to be useable, this information has to be decoded, and although it is possible to conceive of biological circuits capable of responding selectively to particular temporal patterns across a set of inputs, it is likely that this would require a great deal of specialized hardware. Although sensory systems often rely on timing differences between spikes coming from different sources [2], in general, these tend to involve only two sources as in the case of calculating interaural delay times or the timing differences between electrical signals arriving on the left and right hand sides of the body in electric fish [7]. Even these

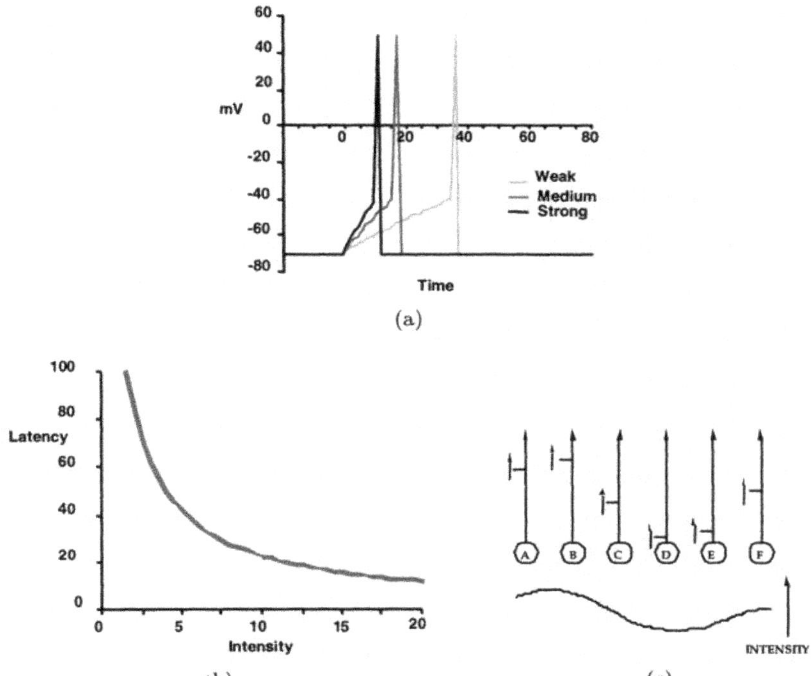

Figure 3. (a) The response of a leaky integrate-and-fire neuron to inputs of different strengths. (b) A typical intensity-latency response function. (c) Generation of an asynchronous spike wave in a population of receptors.

require very large amounts of specialized neural hardware. As far as we are aware, there is little or no direct evidence for neural circuits sensitive to delays arising from several sources at once.

5 Rank Order Coding

However, it appears that a simpler coding scheme can be used which also uses the relative timing of spikes across a population of cells, but which is considerably easier to compute [16]. The idea is to throw away the precise timing information, and use a code that depends only on the order in which the spikes arrive. Returning to the six neurons in Figure 3, we can now code the stimulus by the ordering $B > A > F > C > E > D$. Since there are factorial N possible orderings, a rank order code can transmit up to $\log_2(N!)$ bits of information under conditions where each input neuron can only emit one spike.

Figure 4(a) compares the maximum amount of information that can be transmitted by a range of candidate codes as a function of the number of input neurons. We have included a simple binary code in which the state of each input is either 0 or 1. Here the maximum information transfer is $\log_2(2^N) = N$ bits. We have also

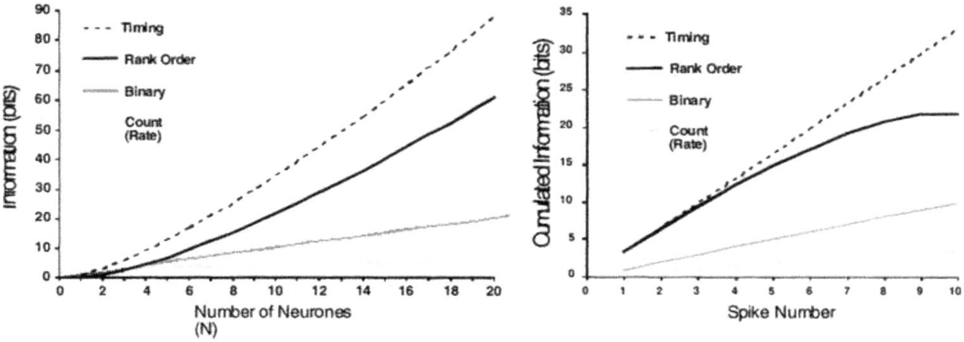

(a) (b)

Figure 4. (a) Maximum information transfer for four types of coding schemes as a function of the number of inputs (N) and on the assumption that each input can generate a maximum of one spike. (b) Information accumulation over time for four different codes assuming 10 inputs, a temporal window of 10 ms, and a temporal precision of 1 ms.

included values for a rate code in which the only information that is available comes from the number of neurons that have fired. In that case, the maximum transfer is $\log_2(N+1)$ bits ($N+1$ because we include the case where none of the units fires). From this graph it is clear that the two codes that rely on spike timing have much greater potential, and that although the timing code is clearly the most powerful, rank order coding is nevertheless a good second best. Figure 4(b) compares the way in which information accumulates over time for each of the four codes, under the assumption that we have 10 input neurons, a 10 ms time window, and a temporal resolution of 1 ms. After one spike has arrived, the rate (count) code can provide only a tiny amount of information, because only 1 of the $N+1$ possibilities has been eliminated, namely, the situation where the final count would have been zero. In contrast, the rank order code can already eliminate 90% of the 10! (i.e. 3628800) possible orderings, which means that the very first spike can transmit up to $\log_2(N)$ bits of information. It can be seen that with rank order coding the amount of information provided by each spike drops progressively, until at the end, the final spike provides no information at all. Thus although the absolute timing code can ultimately provide more information ($\log_2(10^{10}) \approx 33$ bits) than a rank order code ($\log_2(10!) \approx 22$ bits), this difference only becomes apparent towards the end of the propagation sequence. Thus, when decisions need to be made rapidly, a rank order code can be almost as good as a code in which the precise spike timings are preserved.

6 Rank Order Decoding

Rank order coding would be of no use if there were no practical way of decoding the information. There is, however, a very simple way of making neurons sensitive to the order in which their inputs fire and which does not depend on complicated delay lines. The idea takes a simple integrate and fire neuron with inputs that have differing synaptic weights and adds a mechanism which makes the neuron increasingly less sensitive over time as a function of the number of spikes that have already arrived. This could be done by a variety of mechanisms, including some form of feed-forward shunting inhibition. But whatever the precise mechanism, rank order decoding requires that the effect of activating a synapse depends not just on its synaptic weight, but also on a general sensitivity term which gets progressively smaller over time. For a neuron with m afferents with weights $\{w_1, w_2, w_3, \ldots, w_m\}$ and an arbitrary modulation factor *mod* between 0 and 1, the activation level at time t is given by

$$Activation(t) = \sum_{i \in [1,m]} mod^{order(a_i)} w_i$$

where $order(a_i)$ is the firing rank of afferent i. By convention, $order(a_i) = +\infty$ if the afferent a_j has not fired at time t, setting the corresponding term in the above sum to zero.

Figure 5 shows how the activation level of such a neuron would change as a function of the order in which its 16 inputs (A to P) are activated. The synaptic weights are distributed linearly with input A having a weight of 16 arbitrary units, input B a weight of 15 units and so on down to input P which has a weight of 1. The value for *mod* was set to 0.85 so that the first input to fire produces an activation equal to the synaptic weight, the second to the weight*0.85, the third equal to the weight*0.852 and so on. The upper curve illustrates how the neuron would

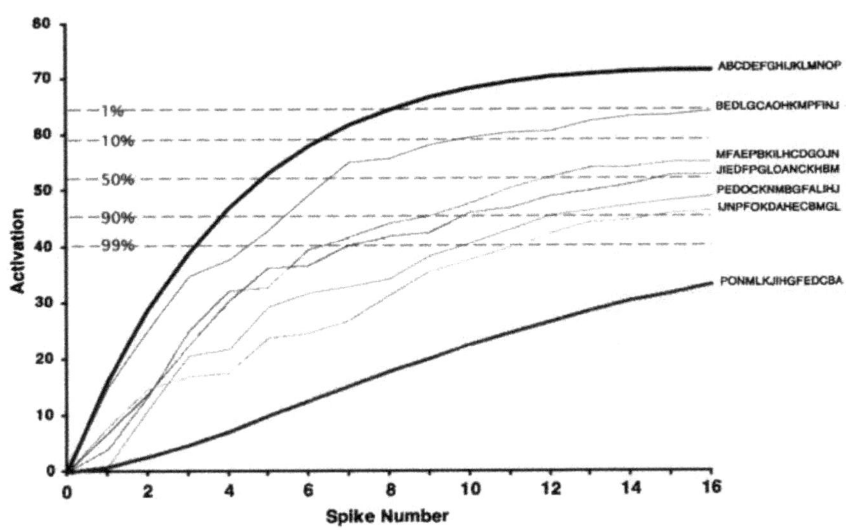

Figure 5. Response of a rank order-sensitive neuron with 16 inputs to variations in the order in which its inputs fire. The upper solid line shows how its activation increases as spikes arrive when the order of arrival exactly matches the weights of the inputs. The lower solid line shows the response to the exact opposite order. Intermediate lines give examples of the response to 5 randomly selected input patterns. The five dotted lines specify the proportion of random patterns that will exceed a given final activation value.

react to the inputs firing in alphabetical order. In that case, the final activation value is maximal (71.7 units). The bottom curve gives the worst possible result, i.e. when the order of activation is *PONMLKJIHGFEDCBA*. The other lines correspond to a variety of random sequences taken from the 16! (i.e. over 2×1013) possible orderings of 16 inputs, none of which produce activation values close to the optimal. In fact, both mathematical analysis and Monte Carlo simulations can be used to demonstrate that the distribution of final activation values is very close to a normal distribution, with the mean value situated exactly halfway between the minimum and maximum values. This is illustrated by the dotted lines that indicate the percentage of random sequences that will be expected to exceed a certain final activation value. It can be seen that only 10% of all possible orderings will produce values above 59.0 units, and only 1% will go beyond 64.0 units. The low variance of this distribution means that such a neuron is actually extremely sensitive to the order of its inputs, so much so that if one was to set a threshold for firing at 69.0, it would only respond to a random input about 0.001% of the time!

Another interesting feature of the Rank Order decoding mechanism proposed here is that by adjusting the threshold level, the output cell will fire at different times depending on how well the input pattern matches the weights. If we set the threshold at around the 50% level in Figure 4, the neuron would fire after only five inputs have been activated if the order was perfect (i.e. *ABCDE*). With random inputs, firing when only 5 inputs have been activated would only occur roughly 0.01% of the time, and the few sequences that do work would be very close indeed to optimal. As more and more spikes arrive, the proportion of patterns capable of resulting in a spike increase, from 4.4% after 6 inputs have fired, 14% after 8 have fired and 50% when all 16 inputs have fired. Note that varying the threshold provides a simple way of adjusting the speed-accuracy trade-off of such a mechanism. With a high threshold (e.g. corresponding to the 1% line), the mechanism is very selective, but will take a relatively long time to respond. In contrast, with a low threshold (e.g. the 99% line), the neuron will respond to virtually all possible inputs but it will do so much more quickly in the case of a "good" pattern (e.g. "*ABCD*..."). Note that this ordering information could be reused again by the next stage of processing. Imagine what would happen with a whole array of neurons each receiving inputs from the same 16 inputs, but where each cell has a different pattern of weights. Even with the threshold set so low that essentially all the cells fire to virtually all the input patterns, a cell in the next layer could still recover information from the order in which the cells fire. To put it another way, the rank order coding scheme that we propose is cascadable.

7 Conclusions and Perspectives

In this chapter we have seen two different issues related to the importance of patterns in biology. In the first, we saw how the pattern recognition capacities of the primate visual system impose very serious constraints on our models of how information is processed within the nervous system. In the second we saw how another type of pattern may be critical for understanding the remarkable speed of information processing, namely patterns that involve the order in which different neurons discharge. Coding schemes that make use of these temporal patterns are exceptionally information rich and it seems that simple biological mechanisms can be used to make neurons sensitive to the order in which their inputs fire.

References

1. Abeles, M., *Corticotonics: Neural circuits of the cerebral cortex*, Cambridge University Press, Cambridge, 1991.
2. Carr, C. E., Processing of Temporal Information in the Brain, *Annu. Rev. Neurosci.* **16** (1993) 223–243.
3. Fabre-Thorpe, M., Richard, G. and Thorpe, S. J., Rapid categorization of natural images by rhesus monkeys, *NeuroReport* **9** (1998) 303–308.
4. Gautrais, J., *Théorie et simulations d'un nouveau type de codage impulsionnel pour le traitement visuel rapide: Le codage par ordre d'activation*, Doctoral Thesis, E.H.E.S.S. Paris, 1997.
5. Gautrais, J. and Thorpe, S. J., Rate coding vs temporal order coding: A theoretical approach *Biosystems* In press (1998).
6. Hopfield, J. J., Pattern recognition computation using action potential timing for stimulus representation, *Nature* **376** (1995) 33–36.
7. Kawasaki, M., Rose, G. and Heiligenberg, W., Temporal hyperacuity in single neurons of electric fish. *Nature* **336** (1988) 173–176.
8. Lestienne, R., Gary-Bobo, E., Imbert, M. and Przybyslawski, J., Temporal coding in the visual cortex of the cat. *C.R.A.S.* **307** (1988) 511-000.
9. Mainen, Z. F. and Sejnowski, T. J., Reliability of spike timing in neocortical neurons, *Science* **268** (1995) 1503–1506.
10. Nowak, L. G. and Bullier, J., The timing of information transfer in the visual system, In *Extrastriate cortex in primates*, eds. Kaas, J., Rocklund, K. and Peters, A. (Plenum Press, New York, 1997), Vol. 13, pp. 205–241.
11. Parodi, O., Combe, P. and Ducom, J. C., Temporal coding in vision: coding by the spike arrival times leads to oscillations in the case of moving targets, *Biol. Cybern.* **74** (1996) 497–509.
12. Rieke, F., Warland, D., Ruyter van Steveninck, R. and Bialek, W., *Spikes: Exploring the neural code*, MIT Press, Cambridge, MA, 1996.
13. Singer, W. and Gray, C. M., Visual Feature Integration and the Temporal Correlation Hypothesis, *Annu. Rev. Neurosci.* **18** (1995) 555–586.
14. Thorpe, S., Fize, D. and Marlot, C., Speed of processing in the human visual system, *Nature* **381** (1996) 520–522.
15. Thorpe, S. J., Spike arrival times: A highly efficient coding scheme for neural networks. In *Parallel processing in neural systems*, eds. Eckmiller, R., Hartman, G. and Hauske, G. (Elsevier, North-Holland, 1990), pp. 91–94.
16. Thorpe, S. J. and Gautrais, J., Rank Order Coding: A new coding scheme for rapid processing in neural networks, In *Computational Neuroscience: Trends in Research*, ed. Bower, J. (Plenum Press, New York, 1998), pp. 113–118.
17. von der Malsburg, C. and Schneider, A neural cocktail-party processor, *Biological Cybernetics* (1986) pp. 29–40.

HOW CAN SINGULARITY THEORY HELP IN IMAGE PROCESSING?

M. BRISKIN
Ort Academic College for Teachers in Technology
Givat Ram, Jerusalem 95435, Israel

Y. ELICHAI
Grid Point Ltd., 1, Hamalaha Street, Ashdod 70700, Israel

Y. YOMDIN
Department of Theoretical Mathematics
The Weizmann Institute of Science
Rehovot 76100, Israel

A method for compressed representation of photo realistic images by means of geometric models (vectorization) is described. This method is based on a general approach, developed in the framework of Singularity Theory. Applications to compressed representation and processing of three-dimensional scenes and animations are given. No specific knowledge in Singularity Theory or in Image Processing is assumed.

1 Introduction

In this paper we present an approach to compressed representation and fast processing of photorealistic two and three-dimensional images and animations. This approach is based, on one side, on some ideas, developed in mathematical Singularity Theory, and, on the other side, on an extensive experimental work, involving direct estimation of psychovisual significance of various geometric parameters.

Following a creative and stimulating attitude of most of the talks given at this conference, where specialists in quite different fields, like mathematics, biology, physics and image processing succeeded to find a common language, we'll try to make a presentation in this paper accessible for this wide auditory.

Consequently, a part of the paper can possibly look somewhat speculative. Our only justification is that the methods and the algorithms, presented here, do work quite efficiently in practice and we just tried to stress main ideas, avoiding numerous not very instructive implementation details.

Let us describe shortly the problems in Image Processing, which we deal with.

Compression Problem. *What is the minimal amount of information, necessary to represent (with full visual quality) any digital image?*

Clearly, this is a question of major practical importance, since storing and transmitting digital images is one of the most difficult and important parts of modern communication systems.

We believe that this is also a major scientific problem, penetrating and relating several quite different fields, from mathematics through the theory of color to the psychology of human visual perception.

Notice, that, stated as above, this problem escapes any possibility of a "brute force" solution: from the very beginning we do not restrict processing time or computing power applied.

In Section 3, we discuss this problem in much more detail, in particular, relating it to mathematical notion of ε-entropy and suggesting (in an admittedly disputable way) certain lower and upper bounds for a possible compression.

In Sections 5, 6 and 7 we describe a compressed representation of photo realistic 2D-images, which we call "Vectorization" or "Normal Forms representation" and which is basic in our approach.

Another problem we discuss in this paper, is the following:

Depth Problem. *How much additional information do we need to make a 3D-scene from one or several 2D-images of this scene?*

Also here practical importance is immediate: existing representations of three-dimensional scenes require a lot of information to be stored and processed (and are very difficult in preparation). As a result, photo realistic 3D-imaging is bounded at present to "high-end" applications.

Common experience with stereo-photographs and stereo-movies shows, that twice the same 2D-image, slightly shifted according to the distance between our eyes, is enough (if only very small changes of a viewer position are allowed).

In Section 8, we show that much less than this suffices: by adding a negligeable amount of "depth" information to a 2D still image (properly represented, i.e., vectorized according to our scheme), it can be transformed into a "vectorized 3D-image", which creates a complete 3D-illusion, for relatively big changes of a viewer position.

The last problem, we discuss in this paper, concerns a possibility to "animate" photorealistic images:

Animation Problem. *How much additional information do we need to describe a photorealistic motion of objects and characters, represented by one or several 2D-images of a three-dimensional scene?*

We discuss this problem in Section 8. The main conclusion is similar to that for 3D-information: if the 2D-image is represented in a "proper way" (vectorized), then an addition of a negligible amount of a "motion information" allows us to create a complete illusion of a photorealistic motion of objects and characters on the scene.

Let us stress that in all the three problems above we appeal to our visual perception, and not to any objective measurement of the image quality or of the 3D or the motion geometry. In simple words, we want high quality visual illusions!

Let us conclude this introduction by explaining shortly how Singularity Theory can help us to create these illusions.

Mathematical objects (functions, surfaces, differential equations) are considered in Singularity Theory up to a certain equivalence relation (provided by a certain group of allowed transformations).

Normal Form is "the simplest" representative in each equivalence class of objects (for example, containing the minimal possible number of free parameters).

In no way were normal forms invented in Singularity Theory. The idea to bring a mathematical problem to its simplest possible form by changes of variables and other permitted transformations is certainly one of the most powerful and oldest in Analysis and in mathematics in general.

Singularity Theory provided a general approach and techniques for finding normal forms of singularities of smooth mappings and other related objects. It stressed the importance of this notion in many theoretic and applied fields, extending far beyond the classical examples.

In what follows, we apply the approach (or, better, the philosophy) of Singularity Theory in the domain, where its usual machinery does not work.

Indeed, our objects are digital images (which can be considered as functions of two variables — see Section 2). The first main difficulty is that these functions are usually very far from being smooth in any reasonable way. A good portion of noise is also usually present.

We have a natural equivalence relation between images. From all the discussion above it is clear, that this equivalence cannot be anything else, but "visual indistinguishability": two images are equivalent, if any normal viewer cannot distinguish between them. The second (and the most important) difficulty is that this equivalence turns out to be very complicated. In particular, it does not allow any straightforward mathematical description. Neither norms on the space of images nor transformation groups suffice (although both provide important partial information).

Nevertheless, we do have a short list of Normal Forms (or models, or vectors), sufficient for a faithfull representation of any image, but these Normal Forms have been found and "tuned" in experiments, involving viewer evaluation of the results (and not deduced mathematically, as one could hope, applying Singularity Theory).

Stressing these differences with the usual applications of Singularity Theory, we nevertheless insist that essentially we remain in its framework. *Digital Images do have an hierarchy of singularities. Normal Forms from our list do capture these singularities. On a global scale, the vectorized image is a Normal Form of the initial one. The vectorized image plus a depth distribution along vectors is a Normal Form of a 3D-scene. The vectorized object plus its "skeleton" with the motion scenario is a Normal Form of an animated scene.*

In a detailed presentation, we explain how our "Normal Forms" look like, how they can be extracted from a given image in a simple automatic way, and how they can be used in a 3D and motion representation. We also try to explain why do we believe that all the experimental work, involved in construction of visual Normal Forms, is indeed necessary.

The paper is organized as follows: in Section 2 a short introduction to digital images is given;

in Section 3 we discuss in some detail the problem of images compression;

in Section 4 we give a simple example of an application of Singularity Theory, which can be instructive for our main purpose;

in Section 5 typical singularities of digital images are described, as well as our list of Normal Forms;

in Section 6 we present the algorithm used for detection of image singularities and for their representation by Normal Forms;

in Section 7 we describe the structure of our vectorized images and various operations on them, including quantization and coding;

Section 8 is devoted to Normal Form representation of 3D-scenes, and to animation of vectorized images.

Finally, in Section 9 we discuss some possible implications of our results for human visual perception models.

We do not try to give an extensive bibliography, refereeing the reader to a couple of basics books in Image Processing and Singularities.

2 Digital images

Color or black and white images are digitally represented by rectangular arrays of points, called pixels (picture elements). At each pixel the brightness (or the color) of the image at this point is stored.

There exist various standards, concerning space and brightness (color) resolution of digital images. They strongly depend on reproduction means: TV or computer screen, printers, projectors, etc. Typical printers space resolution varies at present between 300 to 1200 "dots per inch" (or "dpi"), while a typical computer screen resolution is 80–100 dpi.

Thus an image of the size 12×18 sm on a computer screen is represented by a pixel array of about the size $500 \times 700 = 350,000$ pixels.

As far as brightness is concerned, a typical resolution of 256 grey levels (or 8 bits of information) for black and white images is fairly enough for most of applications, although in medical imaging 12 bits and higher resolutions are used.

Color representation in digital images is a complicated problem, which is far away from a complete solution. It is enough to mention that color calibration of printers or TV-screens remains generally unavailable: any two randomly picked TV-sets will show the same movie in quite different colors.

We do not discuss this problem here, referring the reader to [1], where a good introduction to image processing, and in particular, to color representations is given. In computer imaging the high quality standard is to represent a color at each pixel by three components R (red), G (green) and B (blue), each one with 256 brightness levels, or 8 bits. Altogether this gives around 16,000,000 different colors. (Notice, that in computer games one normally uses only 256 different colors, although this palette is adapted to the images to be shown).

Thus a typical image on a computer screen is represented by an array of, say, 480×700 pixels, with $8 \times 3 = 24$ bits of a color information at each pixel.

This amounts roughly to 8,000,000 bits or 1 MByte of information per image.

The following picture shows a $480 \times 700 = 336,000$ pixels 24 bits RGB digital image with a total data size of 1 MByte. The printing resolution is 600 dpi.

In what follows, to simplify the presentation, we'll mostly refer to digital images of this size and resolution.

3 Compression problem

A data volume of 1 MB per a moderate size and resolution image is prohibitive for most applications in modern digital data communications (in particular, for the Internet). Consequently, various methods for reducing this volume (or for image compression) have been developed.

We'll not discuss here numerous achievements and questions in practical implementation of various compression methods, concentrating instead on the principal aspects of the compression problem, as stated in the introduction: *what is the minimal amount of information necessary to represent (with full visual quality) any digital image?*

Let us try to reformulate this problem in a more accurate way. First of all, we restrict ourselves to 500×700 pixels 24 bits RGB images as above. (And thus we get an upper bound for information required: 1 MB is enough). A compression method is a computer program, which obtains as an imput an image I, and produces its compressed representation — a file $C(I)$. There is also a decompression program, which "opens" the file $C(I)$ and produces a new digital image $\tilde{I}$. We want $\tilde{I}$ to preserve the quality of I and the size of $C(I)$ to be as small as possible. The ratio of the bit-sizes of $C(I)$ and I is called the compression ratio $CR(I)$.

The following picture shows the image of Fig. 1, compressed to the size of 12 Kbytes, or with the compression ratio of 1:80. (In fact, this size counts also the depth information, which has been inserted into the image. The viewer position has been slightly shifted. Also some changes have been interactively inserted. See Sections 7 and 8 for the details).

Quality requirements strongly depend on the application we have in mind. They are decisively important for the compression. In fact, each compression method is

Figure 1. An example of 480 by 700 pixels RGB digital image

Figure 2. Image of Fig. 1 compressed 1:80, after some interactive processing

characterized not by a number (average, maximal, etc. compression ratio over a set of images), but rather by a "compression curve", giving compression ratios, achieved by the method for different quality requirements.

In this paper we are interested in the part of the compression curve, corresponding to a fairly high image quality. So let us adopt a quality assessment procedure used, for example in MPEG1 competitions: the original image I and its version $\tilde{I}$, obtained after decompression, are shown on the same computer screen, one aside another. An average viewer looking at the screen from the distance of 40 sm (and allowed to compare the images for an unlimited time) should be unable to distinguish between I and $\tilde{I}$.

(It may be instructive to notice, that the size of the images, as reproduced in this paper, is too small for a compression assessment: distorsions, easily detectable under the MPEG1 assessment procedure, will dissappear after the size reduction).

Setting these conditions, we can now produce in a quasi-rigorous way a nontrivial upper bound for possible compression. Consider images as above, subdivided in cells of the size 10×10-pixels (about 2.5×2.5 mm on the screen) and assume that the pixels of each cell have the same color, chosen from a very restricted palette of 8 colors (or 3 bits). Easy experiments convince us that taking enough time, from the distance 40 sm, we can distinguish between any two such images, which have at least one cell, colored differently.

To represent all these "block" images we need exactly $3 \times 50 \times 70 = 10{,}500$ bits or 1.3 KB per image. Hence under the above conditions we cannot achieve a compression ratio higher than 1.3 KB/1,000 KB $\sim$ 1:800.

Let us remind a mathematical notion of ε-entropy. Assume X is a compact metric space. A subset $\{x_1, \ldots, x_\ell\} \subseteq X$ is called an ε-separated set, if the distance $d(x_i, x_j) \geq \varepsilon$ for any $i \neq j$. Let $M(\varepsilon, X)$ be the maximal number of elements in an

ε-separated set. $H(\varepsilon, X) = log_2 M(\varepsilon, X)$ is called an ε-entropy of X. This is one of the most basic notions in Information and Complexity Theories.

$H(\varepsilon, X)$ is the minimal amount of information (measured in bits) necessary (and essentially sufficient) to specify any element of X *up to accuracy* ε.

In our case X is the space of all 500×700 pixels 24-bits RGB digital images. The distance $d(I_1, I_2)$ is the "visual distinction" between the images I_1 and I_2. As it was explained above (and will be illustrated by many examples below) it is very difficult to express this distance mathematically. Except for this problem, the computation above *is exactly* the computation of the ε_0-entropy of the image space, where ε_0 is the distance, corresponding to "visual indistinguishability". The subset of "block images", constructed above, is an ε_0-separated set, and hence the ε_0-entropy of the space of images is at least 10,500 bits.

Of course, both theoretical and practical value of the above type of reasoning is very restricted. The main (and very serious) objection is that most of the "block images" above do not appear in practical applications (although the situation with aerophotographs is approximated by such images fairly well).

As far as the lower bound for possible compression is concerned, we have to appeal to existing compression methods. First of all, the engineers, who determined the standard representation of digital images, did a good job! "Naive" compression methods fail: any serious reduction of the density of pixels or of the color resolution *is easily distin guishable.*

Now a "lossless" statistical compression, which leaves the digital image unchanged, and only utilized nonuniformity of color distribution, usually produces a quite low compression ratio of around 1:1.5.

All the "lossy" compression methods, which change the image, but in a "visually acceptable" way, utilize specifics of human visual perception.

Let us describe shortly the so called JPEG (Joint Picture Experts Group) compression. First of all, JPEG utilizes a correlation between the color components R, G and B. We describe this step in some detail, since in one or another way it is used in most compression methods, including the one, given in this paper.

Experiments show that the color components R, G, B (considered in the L_2-space of functions on the pixel grid) are statistically strongly correlated. They can be "uncoupled" by the linear transformation

$$Y = 0.29900\,R + 0.58700\,G + 0.11400\,B$$

$$I = -0.16874\,R - 0.33126\,G + 0.50000\,B$$

$$Q = 0.50000\,R - 0.4186\,G - 0.08131\,B\,.$$

In this new color representation the component Y statistically comprises 90% of the total energy (the L_2-norm) of the image (although one can easily produce images with $Y \equiv 0$).

Now each component Y, I, Q is represented in the JPEG standard by piece-wise trigonometric polynomials, but with a different space and brightness resolution.

The representation of the Y-component is performed as follows: the image is subdivided into 8×8-pixel blocks. On each block a kind of a Fourier transform — a discrete cosine transform (DCT) is performed. As a result, the brightness function

$z(x, y)$ of the Y-component on the block (where x and y are the pixel coordinates on the block, assuming integer values from 0 to 7) is represented as follows:

$$z(x,y) = \sum_{i,j=0}^{7} a_{ij} \cos \frac{\pi i}{7} x \cos \frac{\pi j}{7} y\,.$$

Till now the transformation is essentially lossless. However, on the next step the coefficients a_{ij} are quantized, according to the prescribed reproduction quality. The quantization step depends on this quality and on the frequencies i, j of the quantized component. This dependence is expressed by the so-called "quantization tables", which have been tuned in a long experimental process. They express different psychovisual significance of different frequencies in the human visual perception, and their proper choice is the most important factor in successful compression. (In the ε-entropy framework above, the quantization tables describe the form of the ball in the "visual distinction" distance on the space of images). Let us stress that DCT is a *piecewise-linear* approximation of the images, and hence the quantization tables provide, in a sense, a piecewise-linear approximation to the "visual indistinguishability" domain. However, we believe (and to some extent demonstrate below) that the form of this domain is also determined by *nonlinear geometric effects, which cannot be seen in the frequency domain.*

The last step of the JPEG compression method (of the Y-color component) is a lossless statistical encoding of the quantized coefficients a_{ij} for all the 8×8-cells of the image. Usually after a quantization most of a_{ij} vanish, and the JPEG encoding takes this into account (so-called run-length encoding).

Now I and Q color components are compressed according to the same scheme, but on a subsampled grid (only each second pixel in each direction is taken into account). Also color quantization tables are usually coarser than those for the Y-component.

JPEG provides a visually lossless compression (in the above setting) with the compression ratio between 1:5 to 1:10 essentially for any reasonable image and 1:20 for most of images. Usually 30% to 50% of this compression is due to the "colour decoupling", as described above. (In the best case the colour decoupling can reduce three components RGB to one — Y, and thus to provide 66% of the total compression. In our method, presented below, a typical figure is about 55%).

There are other successful compression schemes, especially wavelets compression and (to some extent) fractal compression. Although wavelets methods have several serious advantages, as compared to JPEG, our experience shows, that their compression-to-quality curves rise significantly higher than that of JPEG only at unexceptably low qualities. Thus we believe that the compression ratio between 1:5 to 1:10 for a visually lossless compression is shared by all the conventional lossy compression methods.

Thus we've firmly established the lower and the upper bounds 1:5 and 1:800, respectively, for a possible compression. There is a lot of space to work!

The method, described in this paper, at present produces compression ratios of 1:50 and higher for images, not containing much of "texture" areas. (See Figures 1 and 2). For highly textured images its compression at present is roughly the same, as for JPEG. Ultimately, we expect it to produce a visually lossless compression of

1:15 for any image, and 1:100 for most of images. As far as the upper bound for a possible compression is concerned, we believe it to be around 1:50 for any image and around 1:200 for most of images. However, it will be, probably, quite difficult either to approach these bounds with a working compression algorithm, or to prove their sharpness.

4 An example of data representation, based on normal forms of singularities

Before we proceed with images vectorization, it could be instructive to consider a much simpler example of data representation, based on hierarchy of singularities and on their normal forms. (For a general introduction to Singularity Theory and its applications see [2,3,4]).

Assume, our problem is to memorize and to represent on the computer screen graphs of singular functions of the form $f(x) = P(x)/Q(x)$, with P and Q — two regular functions. We want these graphs to give an accurate visual description of the functions under consideration. [See Fig. 3(a)].

Clearly, the singularities of $f(x)$ are poles, corresponding to the zeros of the denominator $Q(x)$. Main characteristic of the pole is its order, which is the multiplicity of the corresponding zero of Q.

Thus we get an "hierarchy of singularities" of $f(x)$ — most of the points x are regular ($Q(x) \neq 0$), then there are poles of the first, the second and higher orders.

Now at regular points splines of the third order represent in a fairly accurate way all the visually important features of the graph of $f(x)$: its extrema, curvature, inflection points, etc. Therefore we can take *a polynomial of degree three as the normal form of f at any regular point.*

However, if we try to push this representation to the poles, we see that the nodes of the spline must go denser and denser as we approach the pole. [See Fig. 3(b)]. Thus ignoring the special nature of singularities we shall need more and more parameters to store and process.

To get a normal form of f at the poles, assume that $Q(x) = (x - x_0)\tilde{Q}(x)$, with $\tilde{Q}(x_0) \neq 0$ (the first order pole at $x = x_0$).

Then $f(x) = \frac{1}{x-x_0}(P(x)/\tilde{Q}(x))$, where $P(x)/\tilde{Q}(x)$ is a regular function at x_0. Approximating it by a polynomial of degree four in $(x - x_0)$, we get

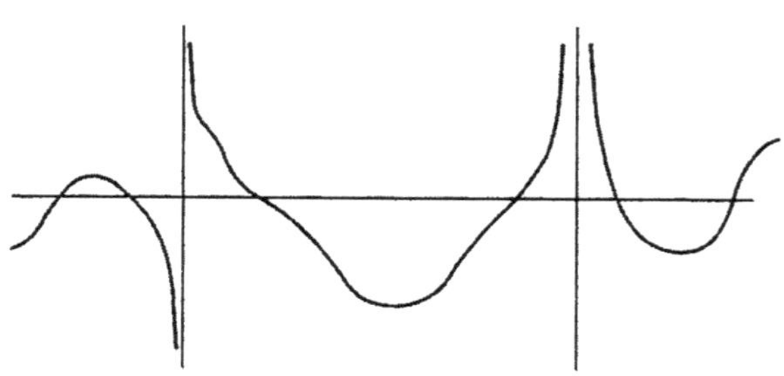

Figure 3. Function to be represented

$f(x) \approx \frac{1}{x-x_0}(C_0 + C_1(x-x_0) + C_2(x-x_0)^2 + C_3(x-x_0)^3 + C_4(x-x_0)^4) = \frac{C_0}{x-x_0} + C_1 + C_2(x-x_0) + C_3(x-x_0)^2 + C_4(x-x_0)^3 = \frac{C_0}{x-x_0} + P_3(x)$, with P_3 — a polynomial of degree 3.

We can take *an expression* $\frac{C_0}{x-x_0} + P_3(x)$ *as a normal form of* $f(x)$ *at the pole* x_0. Indeed, it contains only 6 free parameters: x_0, C_0 and the coefficients of $P_3(x)$, and, as the experiments show, it faithfully represents the graph of $f(x)$ near poles of the first order.

Respectively, *the normal form for the second order pole is* $\frac{C_0}{(x-x_0)^2} + \frac{C_1}{x-x_0} + P_3(x)$. It contains 7 free parameters.

Figure 4 shows the graphs of these normal forms.

Now we memorize $f(x)$ as follows: first of all, we find and memorize all the locations of the singularities $x_0, x_1, \ldots, x_\ell$, and their order. Then at each singularity we find and memorize its normal form as above. We take a regular grid at each interval of regular points between subsequent singularities, and find and memorize a third order spline, approximating $f(x)$ on this interval. Finally, we use free parameters of the third degree polynomials to glue together the different pieces of this representation (see Fig. 5).

We do not give further details here: it would require much more accurate specification of the origin of the functions $f(x)$ and of the tolerances of the representation. We believe that in many important situations the scheme above gives serious ad-

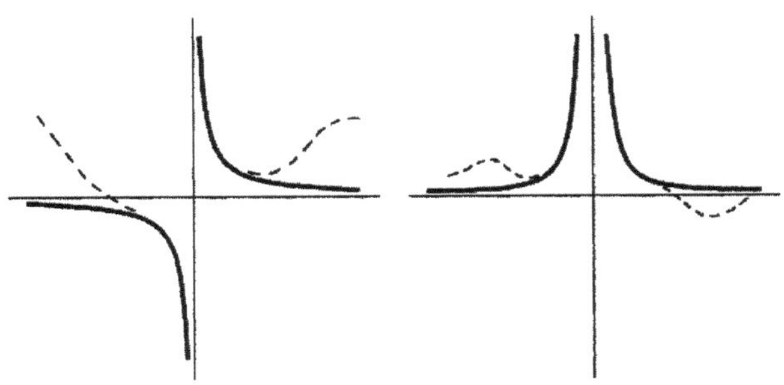

Figure 4. Normal forms

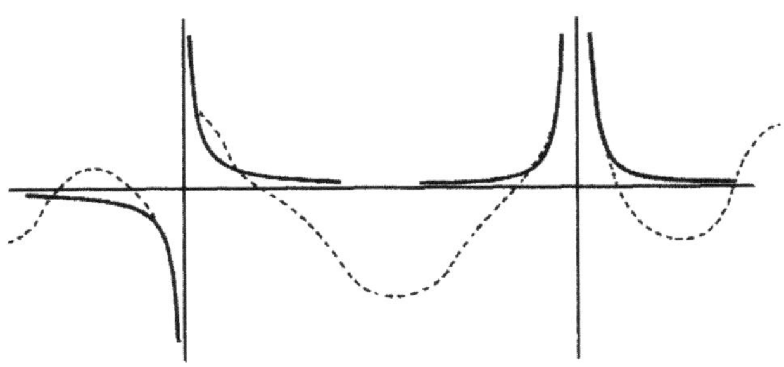

Figure 5. Normal Form representation of the orignal function

vantages both in the storage space, in the quality of the representation and in the processing time (see [5-8]), where a detailed description of our general approach and of its applications in solving elliptic PDE's and in motion planning in robotics is given). Notice that from a computational point of view normal forms can be very convenient: their main "singular" elements can be computed once forever and then in actual computations only simple transformations must be performed. For example, in the normal form above we can use precomputed "look-up-tables" for $\frac{1}{x}$, $\frac{1}{x^2}$ etc.

Remark. In a more rigorous approach, one can easily show, that $f = P/Q$ can be reduced, by change of the free variable x, to $c/\tilde{x}$, $c/\tilde{x}^2$, etc. Thus we can store alternatively at each singularity its location, the coefficient c and the polynomial approximation of the change of variables from x to $\tilde{x}$.

5 Singularities of digital images and their normal forms

In what follows, we shall consider only black and white images (in particular, a Y-component of a color image), represented by 256 grey levels, or 8 bits at each pixel. We return to color at the end of Section 6.

The brightness function $z(x, y)$ of such an image is a discrete function of two discrete variables x, y, running over the coordinates of all the pixels of the image. We'll ignore for a while this discreteness and consider z as a function of continuous arguments.

Considering real world images from this point of view, we find immediately that their brightness function has a lot of discontinuities of different types. Visually the most prominent are "edges" — the lines, separating areas of different brightness. Another apparent type of singularities we call "ridges" — these are thin dark lines on a bright background (or bright lines on a dark background). The last singularity type, which is immediately apparent, corresponds to small bright patches on a dark background or vice versa. We call them "hills".

In real world images edges usually correspond to occlusions of some objects by others, ridges represent thin objects or lines of various origin (like hair or branches of trees) and hills mostly appear as isolated bright or dark patches or as elements of "textured" areas.

In most images relatively large "background" areas exist, where the brightness function changes slowly. The background points are regular points of the brightness function.

Clearly, the above classification is very fussy. While some parts of a typical image can unambiguously be described in this way, others show various kinds of a mixture of edges, ridges and hills, and some others seem not to be representable in their terms at all.

Nevertheless, experiments prove the following basic fact:

Any image can be represented (with a full visual quality and with a major data reduction) in terms of background areas, edges, ridges and hills and their adjacency relations.

Of course, we have to be much more accurate to give the above statement any "positive" meaning: for example, if we allow hills of one pixel size, any image can be reconstructed exactly (but without any data reduction!).

We will now proceed as follows: first we describe Normal Forms (or models) for each of the singularities types above. Then we discuss their typical scales and discretization ranges, and finally return to the main statement above and present in more detail experimental and theoretical results behind it.

Our constructions are motivated and justified by various experimental observations, concerning human visual perception, which we formulate below. Some of them are well-known in the field, others seem to be new. We would like to stress that *in no way do we pretend to put any of these results as an established scientific fact in the field of psychology of human vision. None of them has been formulated or verified according to the criteria and the methodologies in this field.*

The only purpose of our experiments was to construct a good compression algorithm (exactly as in the experimental tuning of the JPEG quantization tables). However, below we discuss some possible consequences of our results for our knowledge about human visual perception, which hopefully could be verified in additional experiments.

According to the general approach of Singularity Theory, explained above, Normal Forms for the background, edges, ridges and hills are mathematical models, providing a visually faithful approximation of these image structures, and containing a minimal possible number of free parameters. Visual quality of the approximation is assessed according to the procedure, described in Section 3 above.

The following picture shows a 48 by 48 pixels square, marked on the image of Fig. 1 (zoomed 1:5) and its Normal Form representation.

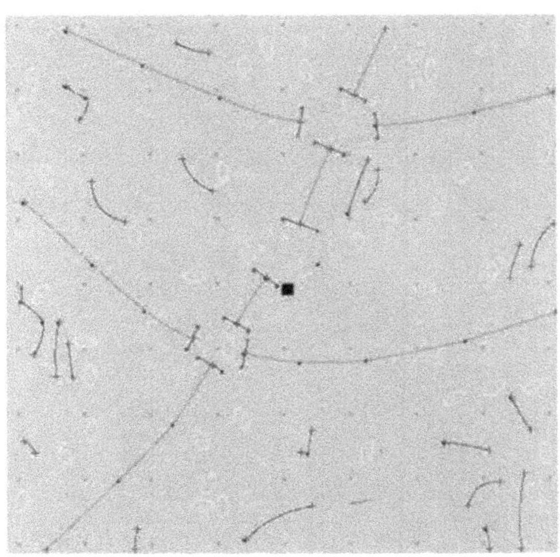

Figure 6. Normal Form, representating the 48 by 48 block, marked on Fig. 1

Background. It is approximated (separately on each component, contoured by edges and ridges) with the first or the second degree splines on the fixed nodes. For the first degree splines, one node per 6×6-pixels cell usually provide a faithful approximation of the background. On Fig. 6 the background nodes are shown as the 6×6 grid of black dots.

Edge. We approximate the central lines of the edges by spline curves (of degree 2 or 3, with nonuniform nodes). On Fig. 6 they are shown as yellow lines. At each node we keep a "brightness profile", which is a S-shaped curve, given in Fig. 8(a), (b), (c). The profile values are interpolated between the nodes, and the margin profile values, shown by punctured lines, coincide with the corresponding background values. The profile parameters are the width h and the two brightness values b_1 and b_2, shown on Fig. 8.

There are several experimental facts behind this choice of the profiles. First of all, the width h usually does not exceed 6 pixels (wider edges visually become a part of the background). Secondly, although the actual brightness profiles can be rather complicated, the three parameters, shown on Fig. 8 are usually sufficient for a faithful representation of the edge, if we adopt one of the typical shapes, three of which are shown on Fig. 8(a), (b) and (c). These shapes depend on the scanner or digital camera used in image preparation. In particular, the shape a corresponds to a normal scanner, the shape c — to a high resolution scanner, and the shape b — to synthetic computer-generated images.

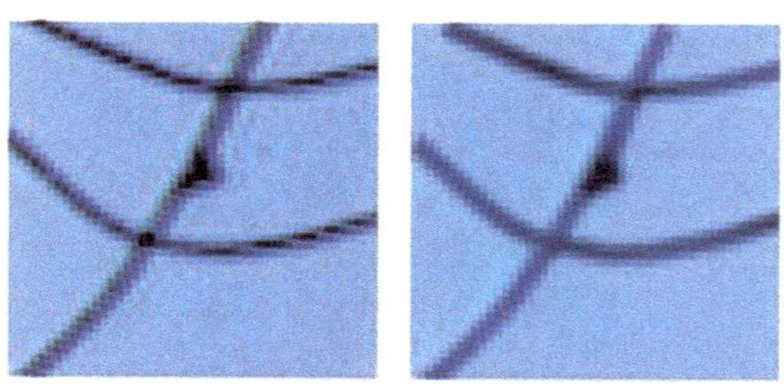

Figure 7. The same block, original and represented by normal form (after 1:5 zoom)

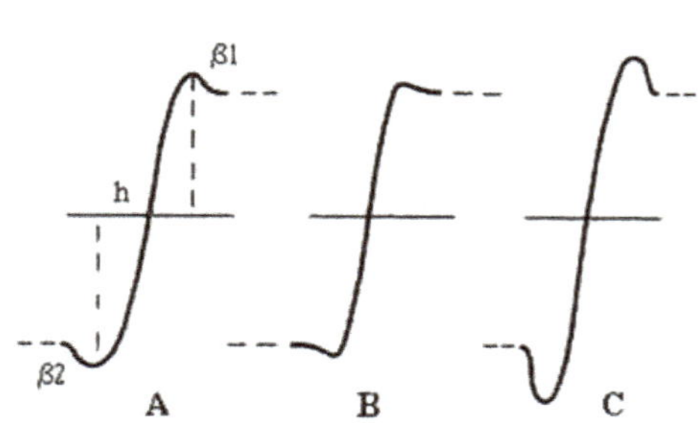

Figure 8. Edge profiles

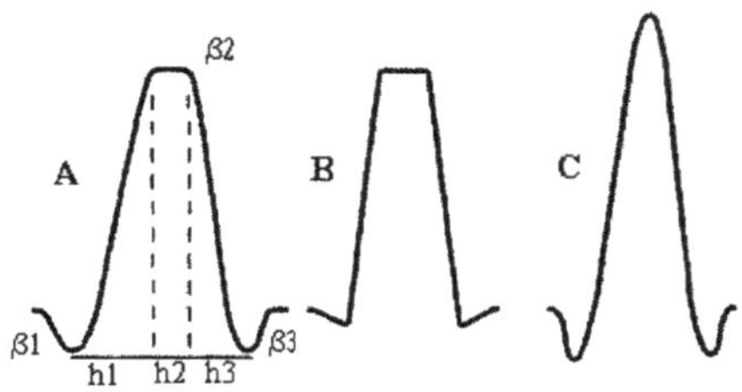

Figure 9. Ridge Profiles

Ridge. (Green lines on Fig. 6). Its representation is completely similar to that of the edge, but the profiles are replaced by those shown on Fig. 9(a), (b), (c). There are 6 free parameters: the widths h_1, h_2, h_3 and the brightness values b_1, b_2, b_3.

Usually the side widths h_1 and h_3 of the ridge do not exceed 6 pixels (wider margins make a ridge to be visually a part of the background). The middle width h_2 can be rather big (long and relatively wide strips are usually percepted as "line structures"), but the locality requirement in detection (see Section 6) restrict h_2 by 6–7 pixels. Wider strips are treated as two edges; however, their geometry can be utilized in coding.

Let us stress that our visual sensitivity to the shapes of the ridge and the edge profiles is extremely high. Their inaccurate tuning destroys the image. The shapes shown on Figures 7 and 8 and other shapes in our library have been tuned in a long experimental process. (Of course, if we allow more parameters, we can approximate any shape with a sufficiently high accuracy. The point was to keep their number at minimum).

Hills are represented by the bell-like shapes, shown on Fig. 10. In the basis they have an ellipse (characterized by the center coordinates, the size of the two semi-axes and by their direction). The only bright ness parameter is the maximal brightness at the center. As above, the shapes a, b, c correspond to a regular scanner or camera, to a high-resolution one and to synthetic computer images.

The size of the longer axis of a hill usually does not exceed 6–7 pixels (for larger sizes a hill is visually percepted as a ridge, if the second axis is small, or as a background area, separated by an edge, or as a part of the background).

Adjacency relations. Experiments show that any situation, where two or more edges or ridges come closer to one another than 3–4 pixels, is visually percepted as an "adjacency". We are very sensitive to the presence and to the geometric shape

Figure 10. Hill

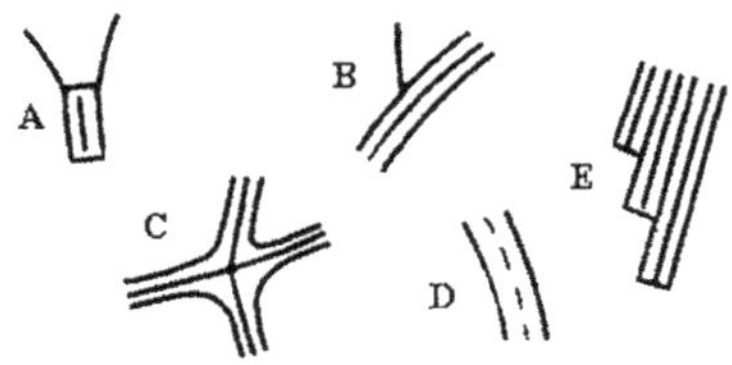

Figure 11. Adjacency relations

of these adjacencies, and they must be represented as additional types of singularities. (In the language of Singularity Theory, edges and ridges are "codimension one" singularities, while these adjacencies correspond to "higher codimension" singularities).

The normal forms for adjacencies we call *joints.* They are shown on Fig. 11. We keep the coordinates of the point (or the interval) of the contact, and the types and the parameters of the edges and the ridges, joined at this point (or along the interval).

Figure 11(a) shows the splitting of a ridge into two edges. The joint, shown on Fig. 11(b), represents three ridges, emanating from one center. Both these joints present on the part of a vectorized photorealistic image, shown on Fig. 6. Figure 11(c), shows an edge, adjacent to a ridge. Two edges, adjacent to one another along a certain interval, are shown on Fig. 11(d). In fact, joining, these two edges form a ridge, and this is the form, in which this adjacency is captured and stored. Adjacencies of two or several ridges, shown on Fig. 11(e), commonly appear in representation of "wave-like" image patterns.

This completes the list of singularities and of their normal forms we use in representation (vectorization) of photorealistic images.

Let us stress once more that in no way is this representation unique or "canonical". Many image patterns can be approximated equally well by different models. In fact, the profile parameters, given above, are intendedly somewhat redundant: as shows Fig. 12, ridge profile for a special choice of parameters can represent a typical edge-shape and vice-versa.

The goal we try to achieve is *not to classify any part of a given image as an "edge", a "ridge" or a "hill", but to represent this image in a visually faithful way by a whatever combination of these models.*

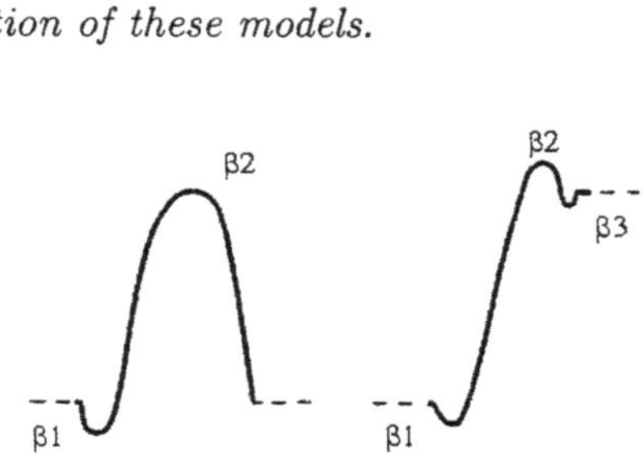

Figure 12. redundancy in the edge and the ridge profiles

(Nevertheless, as a byproduct, in most of image regions we usually obtain quite a convincing structural description of image patterns, which helps very much in further interactive high-level processing. See Section 7).

Scales of normal forms. Let us summarize the range of image scales. covered by the normal forms.

Background. Covers large scale patterns. Finest resolution (for the first degree splines) — 5–6 pixels.

Edge. In the spline representation of the edge central line the minimal distance we allow between the nodes is 2–3 pixels. So this is the finest geometric scale we can capture. The profile width is between 1 to 8 pixel, and a typical width is 1.5–3 pixels.

Ridge. The finest geometric scale is about 3–4 pixels, the profile width is between 18 to 1.5 pixels, typically 2 to 4 pixels.

Hills. A typical size of a small semiaxes is about 1.5–2 pixels, and of a big one — 1.5–3 pixels.

The fact that the scales of our normal forms and their "ranges of representability" overlap, is very important for a faithful image reconstruction: approximation of image patterns by each of the normal forms becomes poor on the limits of its scale range.

Discretization of the normal forms parameters. An *ultimate* accuracy of the parameters discretization is determined by the requirement that the approximation of the brightness function of the image by normal forms be accurate up to a couple of grey levels.

For different parameters of our normal forms this requirement is translated into different discretization steps. For various brightness parameters about one grey level accuracy is required. The geometric parameters are mostly very sensitive: since the brightness jumps in singular regions of the image, an error can be as big as 256 grey levels for one pixel replacement of a normal form. This calls for at least 0.01 pixel accuracy for geometric parameters. However, their range is typically 2–3 pixels. As a result, 8 bits or 256 levels discretization is fairly sufficient for an initial representation of both the brightness and the geometric parameters. (In this discussion we did not yet touch a *psychovisual* significance of our parameters. This is one of the most important issues in all the compression process, and we discuss it in more detail in Section 7).

Now we are ready to return to our main "representability claim", formulated in the beginning of this section, and to discuss it in more detail. In fact, we reformulate it in a "local" setting:

Each 6 by 6-pixels cell of any image can be faithfully represented by the above normal forms, with a total number of free parameters at most 12.

Allocating 8 bits per parameter, we get an upper bound of 12 bytes of information required to represent in a visually indistinguishable way any 6×6-pixels cell of a black and white image. Since initially such a cell takes 36 bytes, we get an *initial* compression ratio 1:3 in Normal Form representation of black and white images.

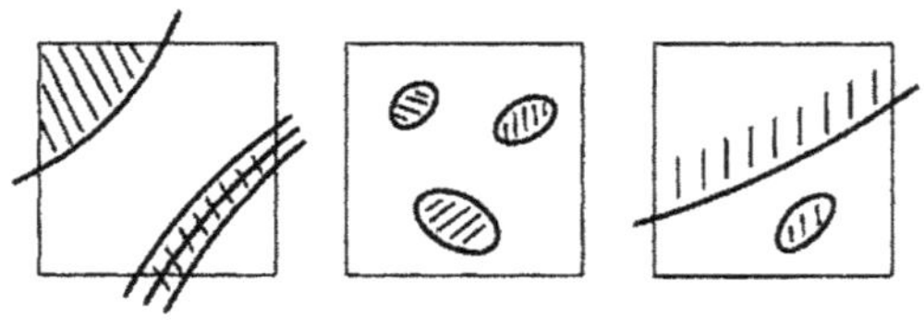

Figure 13. Normal Forms on 6 by 6 pixels cell

Local representability, as a fact concerning our visual perception, is much more difficult to check experimentally than its "global" counterpart. Indeed, it is very difficult to visually estimate a "quality" of a 6×6 pixels image (a block of 1.5 mm size, seen from 40 sm distance). In fact, we've deduced it, analyzing a visual content of each of 6×6-cells in full quality normal form representations of standard size images.

On the other hand, 6×6 pixels is essentially the minimal block size, for which all our normal forms and their visual interaction can still be traced. As a result some very important "low scale" visual effects become more apparent. In particular, the following simple dyhotomia, confirmed by out experiments, has no chance of being traced on larger scales (where possible image complexity becomes immense):

Either the image at the 6 by 6 pixels scale presents some "curvilinear patterns" (edges or ridges), or it is appreciated as a "texture" and then our sensitivity to the distorsions in this area drops drastically.

In the first case, our sensitivity to geometric distorsions is very high. But "by definition" edge and ridge normal forms provide a required accuracy in approximation of "curvilinear structures".

In the second case, the "patches" forming the image, can be captured by "hills". Of course, usually these patches have complicated shapes, only very roughly approximated by elliptic hill normal forms. But this is visually enough!

Now the local representability statement above can be "mathematically deduced": we represent "curvilinear patterns" by edges and ridges normal forms, and the remaining "texture" by several "hills". Experiments show that in various combinations of normal forms, occurring in 6×6-cell, the total number of free parameters very rarely exceeds 12. (See Fig. 13). This bound is in good agreement with the typical scales of our normal forms. (The total number of free parameters of the normal forms, representing a certain image part, can be naturally considered as a *complexity measure* of this part.

6 Detection of singularities and their approximation by normal forms

A typical image contains between several hundreds to a couple of thousands of separate edge and ridge components and up to 7–8 thousands of hills. Thus a normal form representation (vectorization) can be practical only if a fast, completely automatic and computationally stable defection and approximation algorithm is available.

Our vectorization algorithm consists of several stages. In the first stage image singularities are analyzed locally, on the scale between 3×3 to 5×5-pixels. The analysis is based on the second and the third order approximation of the brightness function.

In the second stage chains of locally detected elements of edges and ridges are traced and adjacencies between them are marked. Also here the analysis is local with the scale of 5 to 10 pixels.

In the third stage central lines of edges and ridges are approximated by splines (usually of the second order), the detected adjacencies are approximated by the corresponding joints, and the profile parameters are estimated. Although this stage has a global nature, it is usually subdivided into two steps.

On the first step we allow the distance between the spline nodes to be at most 8 to 16 pixels, and hence preserve the local nature of the processing. Only on the second step "long parabolas" enter (which can greatly reduce the volume of geometric data).

We stress the locality of the image processing in our vectorization algorithm. It is this property that makes the algorithm straightforward and robust, excludes complicated ramifications and provides almost unlimited parallelizability.

Below we describe each of the stages in the algorithm in some detail. Figures 14–16 and 6 represent them schematically, for a 48×48 pixels part, indicated on the image of Fig. 1. We shall constantly refer to this example.

Let us forget for a while a discrete nature of digital images, and explain, what we are looking for, assuming the brightness function $z(x, y)$ to be a continuous (and even twice differentiable) function of two continuous coordinates x, y on the image. We start with ridges (see Fig. 6). An intuitive definition of a ridge, as a bright thin

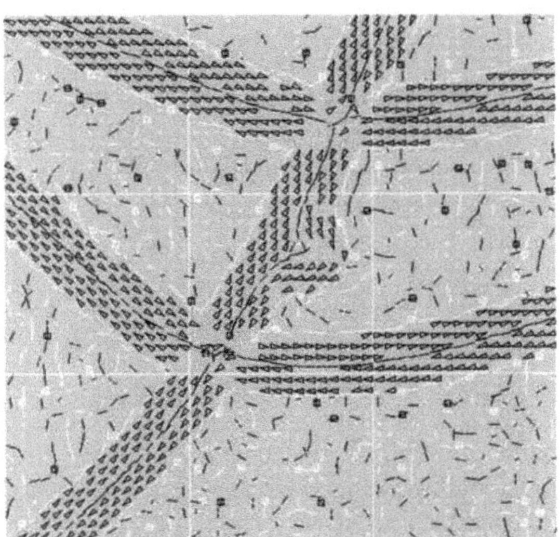

Figure 14. Second order detection

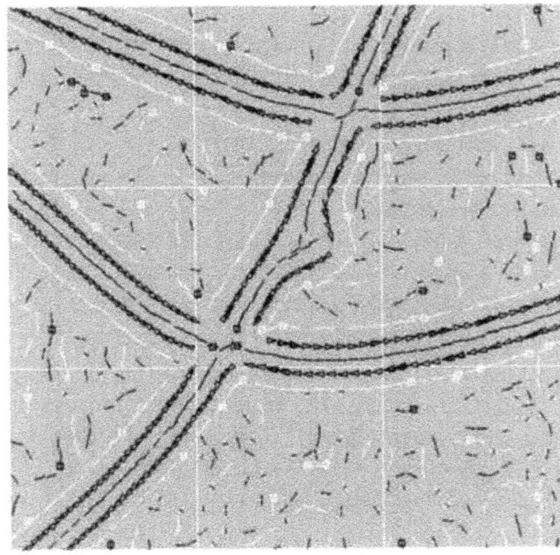

Figure 15. Third order detection

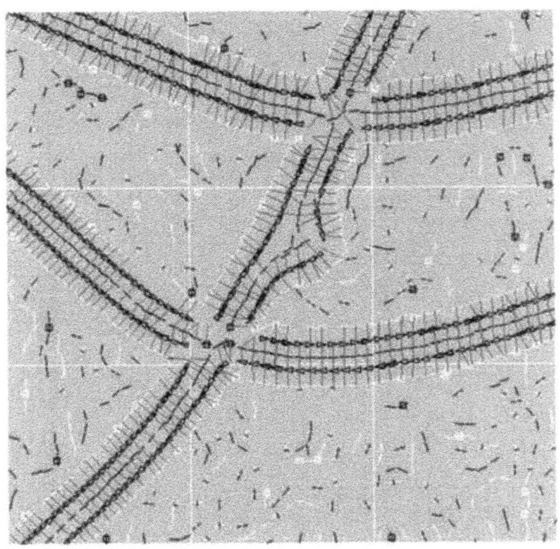

Figure 16. Some adjacencies

strip on a dark background leads to several mathematical consequences: first of all, the brightness on the central line of the ridge is maximal. Hence we can assume that the gradient field *grad* $z(x, y)$ is tangent to the ridge central line. On the other hand, since our ridge is "thin", the second derivative of z in the orthogonal direction must be negative and "big" in absolute value, while the second derivative along the ridge must be small. Thus the ridge central line must be tangent to the field of eigenvectors of the second derivative of z. Mathematically these two properties are incompatible: the first leads to gradient lines, and the second — to the curvature lines. (Of course, practically both the properties are roughly satisfied on what we would like intuitively to be ridges).

Mostly by the considerations of numerical stability, we use a "compromise" mathematical definition: ridges are defined by a condition that one eigenvalue λ_1 of d^2z is "big" in absolute value, the second λ_2 is "small", and the derivative of z in the direction of λ_1 is zero.

Practically this definition works quite well (and the ridges mostly *are curvature lines*, since $grad\ z$ is usually very small on their central lines). However, we believe that further experiments and a resulting better mathematical understanding of what we call "ridge" can improve our detection range and accuracy.

Let us now turn to edges. Mathematically edges present some additional problems. Indeed, the above definitions for the ridge are *local*: the point (x, y) of the image belongs to the central line of a certain ridge if and only if certain equations and inequalities between the partial derivatives of $z(x, y)$ at this point are satisfied.

Intuitively, edges are lines, where the brightness $z(x, y)$ jumps. Mathematically this is expressed by the fact that $grad\ z(x, y)$ is "big" at the edge points. However, this last condition usually defines not lines, but some *strips* on the image (see Fig. 14, where small black triangles denote points with the norm of $grad\ z(x, y)$ bigger than a certain parameter G. If we put a higher value of G, we get instead of a strip, a set of separated points, which hardly form a smooth line, because of the noise in the image brightness values). This problem is well known in edge detection.

We would like the center line of our edge to be the central line of the strip, given by $\|grad\ z(x, y)\| \geq G$. But this property cannot be expressed in terms of the derivates at the points of the central line itself.

Once more, we believe that further experiments can improve our mathematical understanding of what we call "edges". At present we've adopted a "semilocal" detection scheme, based on experimental brightness shapes of the edge, shown on Fig. 8. Indeed, these shapes are well approximated by cubic parabolas, and the center of the edge can be taken as the center of the cubic parabola, approximating the brightness profile. (See Fig. 17.)

Detection of "hills" is mathematically quite straightforward: these are sharp extrema of the brightness function, i.e. the points (x, y), where $grad\ z(x, y) = 0$,

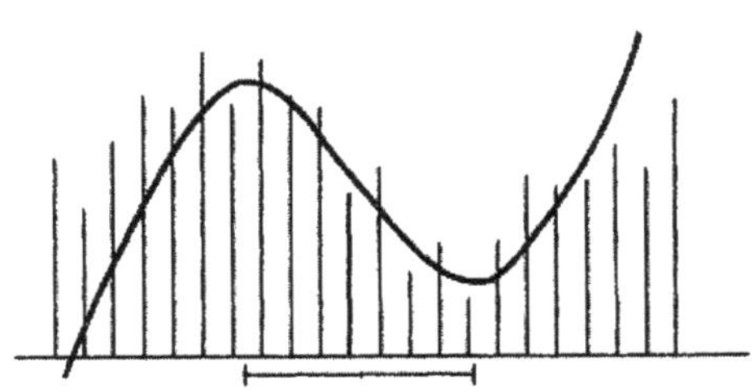

Figure 17. Third order approximation of the edge profile

and the eigenvalues λ_1, λ_2 of $d^2z(x,y)$ are of the same sign and "big" in absolute value.

What is not covered by edges, ridges and hills, is the background.

Let us now describe a realization of the mathematical definitions above in a numerically stable algorithm, working with discrete and noised image data.

First of all, for each pixel on the image we take a 3×3-pixel box, centered at this pixel, and compute a second degree polynomial P, providing the best (mean square) approximation of the brightness $z(x,y)$ on this box. If the gradient of P at the center is bigger than a certain threshold G, we mark this central pixel by a black triangle (see Fig. 14. We call such marks "gradients") and pass on to the next pixel.

(In the coordinates scaling, where the distance between two neighbor pixels is one and the brightness is measured in the grey levels, and so varies between 0 and 255, a typical value of G is 10).

If *grad* P is small, we compute the eigenvalues λ_1, λ_2 of d^2P, $|\lambda_1| \geq |\lambda_2|$. If both λ_1, λ_2 have the same sign, and $|\lambda_1| \geq |\lambda_2| \geq \Lambda_1$, we find the point, where *grad* $P = 0$. If the distance of this point to the central pixel is less than d_1 (roughly 0.5 in the same coordinates scaling), we mark this point by the "hill" sign (small black and white crosses on Fig. 14, corresponding to the positive and the negative hills). If $|\lambda_1| \geq \Lambda_2$ and $|\lambda_2| \leq \Lambda_3$ with $\Lambda_1, \Lambda_2, \Lambda_3$-certain parameters (roughly 10, 20 and 5, respectively), we find a straight line L, defined by the condition that the derivative of P in the direction of λ_1 be zero. Finally, if the distance of the line L to the central pixel is less than d_2 (roughly, 0.5), we mark the part of L inside the box as a short segment (see Fig. 14. We call such marks "segments"). If one of the above conditions is not satisfied, we pass on to the next pixel.

As the second degree analysis of the image is completed, we pass on to the third degree approximation. At each pixel, marked by the gradient sign, a third degree polynomial Q, providing the best approximation of $z(x,y)$ on the 5×5-pixels box, is computed. Now we restrict Q to the line, orthogonal to the gradient and passing through the central pixel, and analyze the shape of the resulting cubic parabola. If it has two extrema, as on Fig. 17, and if the distance of these extrema to the pixel does not exceed d_3 (roughly 6), we find the center of the parabola, and shift the gradient mark to this center. Otherwise the gradient mark is deleted.

The result of this correction procedure is shown on Fig. 15. The "gradient strips" of Fig. 14 become now coherent gradient "central lines".

At this stage the background is represented just by pixels, not marked with segment, gradient or hill marks (and not neighboring to the marked ones).

Notice that till now the analysis has been performed separately for each pixel. Although the segments and the gradients on Fig. 15 present apparent ridge and edge patterns, the computer still does not know it. Another important point is that at each segment, gradient or hill mark we memorize the brightness and geometric data detected, which will be later used in normal forms construction.

For a segment we memorize (besides its coordinates and the direction) the brightness value of the approximating polynomial P at the central points, and the maximal eigenvalue λ_1 of d^2P.

For a gradient its coordinates and direction, and the brightness values at the extrema of the approximating cubic parabola (see Fig. 17) are memorized as well as the distance between these extrema.

For a hill, its coordinates, the brightness at its center, the eigenvalues λ_1 and λ_2 of d^2P and their directions are memorized.

In the next stage we build chains of segments and gradients. This is done by connecting neighboring segments (or gradients), if their centers and directions are coherent. At this stage we also mark adjacency relations between segments and gradients, not belonging to the same chain. These relations are shown on Fig. 16.

Now we analyze these adjacencies along each of the edge and ridge chains and finally classify them as edges or ridges. At this moment also joints between them are marked. (See Fig. 6, where the green parts have been finally classified as ridges, splitting at the joints into two edges).

Let us stress once more, that this classification is not canonical, and in many cases different interpretations of the initial detection results in terms of edges and ridges are possible, providing equally faithful image reconstruction.

In the third stage of the algorithm we construct the normal forms, representing the edges, ridges, hills and adjacency relations, detected in the first two stages. First of all we approximate the chains of segments and gradients, classified as ridges and edges, by the second degree splines with free nodes (only the nodes, corresponding to the endpoints of the chains and to the joints, are fixed). We also restrict the minimal and the maximal possible distance between the neighboring nodes, thus prescribing the geometric resolution and the locality scale.

Then the information, stored at each segment and gradient is translated (taking into account the adjacent segments or gradients) into the profile parameters. The resulting values are usually quite noisy, so a certain averaging procedure along the chains is performed.

As for hills, the information, stored at each hill mark, directly provides the parameters of the approximating elliptic paraboloid, which is the normal form of the hill.

As it was explained above, we represent the background by the first order splines with fixed nodes of a density roughly 1 node per 6×6-pixels block. This choice is the result of a tradeoff between the requirements of a faithful image representation, a high compression ratio and a fast decompression. In fact, our basic background representation is much more flexible: we memorize background values at all the nodes of a certain nonuniform grid Z (which is also memorized). Background values are also stored along each side of edges and ridges (exactly in the same way as the profile parameters are stored). Figure 18 gives an example of the background representation. As this example suggests, edges and ridges subdivide the background into connected components. While the brightness function $z(x, y)$ is relatively smooth on each component, its values on different components can jump.

These discontinuities present a serious problem in all the stages: in detection, approximation, quantization, coding and decompression. First of all, we must capture accurately the topology of the background partition. In each connected component of the background we are forced to use low-resolution splines, and hence if there are gaps in the detected edges, the background values will "escape" through them

Figure 18. A typical Background pattern

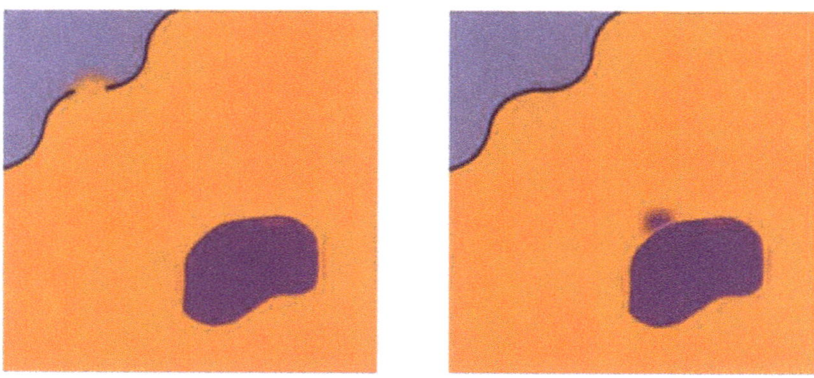

Figure 19. Results of "topological" errors in detecion

(see Fig. 19, left). Also if one of the background nodes passes from one component to another (as a result of a quantization, etc.), a serious image distorsion occurs (see Fig. 19, right).

Consequently, the main requirement to the grid Z is that its nodes do not come closer to the edges and to the ridges than 2–3 pixels. This property guarantees that as we quantize the geometric parameters, the nodes of Z cannot cross the boundaries of the background components.

The brightness values of the background at the nodes of Z are obtained by averaging over the pixels, not marked as segments or gradients on the first stage of the detection, and not neighboring to the marked ones. (In fact, a little bit more sophisticated "expansion over free pixels" is used in this averaging, to guaranty that we never cross the boundaries of the background components).

The brightness values of the background on the margins of the edges and the ridges are obtained by averaging over the corresponding segments (or the gradient ends).

This completes the construction of the Normal Form representation of the image (black and white).

Representation of color images is quite similar. In principle, we can represent each color component R, G and B (or Y, I and Q) separately and independently, using the above procedure.

However, experiments show that for most of images (except some synthetic, specially constructed examples) *singularities of the R, G and B (or Y, I and Q) color components are geometrically the same. They differ only in brightness profiles.*

This fact has very important consequences for the color representation, and our scheme fits perfectly to utilize it. Indeed, *we compute and memorize the types of normal forms (edges, ridges, hills), and their geometric parameters (spline approximation of the central lines, widths, coordinates of the hill centers and of the joints, etc.), only for one color component (usually Y). For the two remaining components (I and Q) we compute and memorize only the brightness parameters of the normal forms.*

Consequently, we reduce drastically the amount of information we need to represent color components I and Q. Together with certain effects in human color visual perception, which are also very easy to utilize in Normal Form representation (see Section 7), this allows us to reduce most of theoretically possible 66% of the data in color image representation.

7 Vectorized images

The algorithm, described in Section 6, produces for each image its Normal Form, or vectorized representation. This representation preserves the image quality and is usually at least 5 times more compact. Indeed, since to represent any 6×6-pixels cell of a black and white image we need normal forms with a total of at most 12 free parameters, each discretized to 8 bits (see Section 5 above), we get a compression ratio of 36 bytes/12 bytes = 1:3 for black and white images, which gives (without additional quantization) at least 1:5 for color ones.

However, the most important fact is that from the moment the vectorization is obtained, *we do not need to return to pixels anymore. All the image processing operations, including interactive creation of 3D-images and animations, can be performed on our vectorized images in a much easier way than on the original pixel ones, and with a better visual quality.*

We call our vectorized images *v-images.*

In what follows, we describe shortly some of the operations on v-images, concentrating on those, essential for further compression: quantization and coding. We also discuss shortly creation and structure of 3D-images and animations.

The role of v-images in our approach can be illustrated by the diagram, shown on Fig. 20:

Let us stress once more the main feature of our Normal Form (or vectorized) representation: as the original digital image has been vectorized, we do not need pixels any more (until we want to see the image again).

Now we explain in somewhat more detail, how the operations, shown on Fig. 20 are performed.

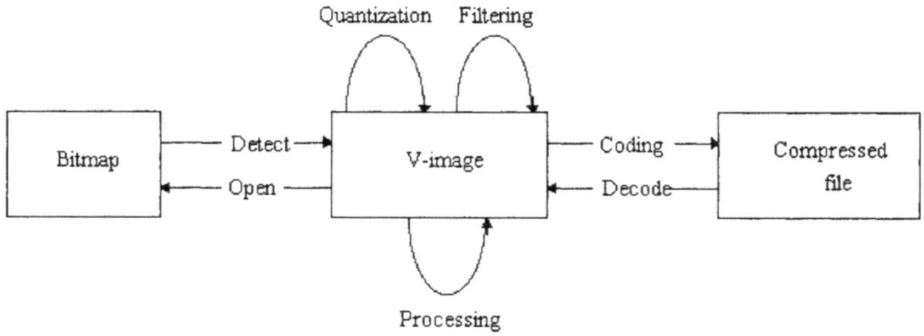

Figure 20. V-images and operations with them.

Quantization. Originally the parameters of the v-image are represented by 8 bits (or quantized to 256 levels). However, psychovisual significance of these parameters is different. As a result, some of them can usually be quantized to a much smaller number of levels without any visual degradation.

First of all, our representation preserves the structure of the main image patterns (continuity of the brightness on the background components, the geometry of the edges and the ridges and of their adjacencies, etc.), for rather big variations of parameters. Consequently, we can hope that a much stronger quantization than the original 8-bits one will generally preserve the image quality. Experiments in this direction give the following "initial quantization table":

Without any visual distortion,

All the brightness parameters can be quantized up to a 2 grey level step.

The spline parameters of the central lines of edges and ridges (and the center coordinates of hills) can be quantized up to a geometric accuracy of quarter a pixel.

The width parameters of the profiles can be quantized up to 1/20 of a pixel.

Already this initial quantization provides a rather high compression in many cases. However, many additional features of our visual perception can be incorporated into the table. We'll give just a couple of examples.

It is well known that our color sensitivity drops for image patterns of a small angular size. As a result, I and Q brightness values for edges, ridges and hills can be quantized up to 2–3 bits or even less.

Also the Y-brightness for the "small scale" normal forms can be quantized much stronger than for the background.

We are much less sensitive to the "absolute" geometry of edges, ridges and hills (especially, if the curvature patterns of edges and ridges are preserved) than to their mutual position.

This list can be continued. We have altogether about 40 different types of parameters in our v-images, and their initial tuning included experimental determination of their quantization scales and of their mutual dependence.

Exactly as for the JPEG quantization tables, a proper tuning of our quantization threcholds and scales is decisevely important for the overall compression. However, a "naive" optimization of 40 independent parameters does not help much. We have to build an hierarchical structure in the representation of v-images, wich would take into account the experimentally found mutual dependences between the parameters.

Not very surprizingly, constructing this hierarchy reveals some additional mathematical patterns, which one can naturally call "Quantization Normal Forms". We plan to present this part of our vectorization approach separately.

Filtering. Closely related to quantization is a filtering operation, which allows us to drop visually negligeable normal forms. For example, we can drop all the hills, whose brightness differs from the background less than to 5 grey levels. We can drop all the ridges, shorter than 10 pixels, whose brightness differs from the background one by less than 10 grey levels, etc. Once more, a correct tuning of filtering parameters is important for the overall compression and it is achieved in experiments.

Encoding. As the parameters of the image are quantized, they can be finally encoded by lossless statistical methods, like Huffmann encoding. This step is important in overall compression, as it can give a serious additional data reduction. However, its relative importance in our method is, probably, smaller than that in JPEG. This is because we utilize correlations between parameters in the earlier stage of "Quantization Normal Forms". Moreover, these Normal Forms not only decorrelate different parameters, but also make their distributions relatively uniform.

Processing of v-images. Any image processing operation (whose desired effect is expressed in "visual" terms) can be easily translated to the structure of v-images.

For example, sharpness and contrast changes are translated into corres ponding corrections of the brightness profiles. Global or local color corrections are interpreted in a similar way.

Filtering, as described above, can be a powerful image processing operation. For example, filtering out all the "texture" on the image (this texture is usually represented by hills), we get from a photorealistic image a "cartoon" with the same characters.

Colors of any desired image patterns (or "objects") can be easily corrected or changed. (Compare Figures 21 and 22).

Various feature extraction operations become very easy: our vector representation by itself is a highly structured one. As a result, we can ask questions concerning v-images on a certain rather reach language, for example:

Show all the closed edges of the size between 30 and 40 pixels, of the width at most 3 pixels and of the curvature radius at least 5 pixels.

Show all the texture areas, where the hills have a relative density at least 2 per 6×6-pixels cell, are directed between 45° and 60° and are not longer than 4 pixels.

Especially convenient v-images are for geometric processing: various normal forms can be deformed, moved, etc. New ones can be added, colored and joined with the existing ones (Compare Figures 1 and 2).

In the next section we describe two very important operations with v-images: insertion of depth and animation.

Figure 21. Image of Fig. 22 after processing

Figure 22. The original image (some result of processing are shown on Figures 21 and 23)

8 Creation of virtual worlds

Availability of v-images allows us to arrange some important experiments, concerning human visual perception of the 3D-scenes and of the motion.

Figure 23. Results of animation and change of a viewpoint in the image of Fig. 22

Vectorized 3D-images. Assume we have a two dimensional image of a 3D-scene. A variant of the "depth problem", as stated in the introduction, is: *how much additional information do we need to create an illusion of a photorealistic 3D-scene, for small changes of the viewer position?*

Assume also that the depth distribution on the 2D-image is available: we know the distance from the viewer to each pixel on the image.

It is quite easy to construct a "rendering" algorithm, which will show each pixel according to a new viewer position. However, experiments of this sort, stress immediately some serious difficulties. The first one is that the depth information, stored for each pixel, at least doubles the total amount of data. Another one concerns occlusions: as the viewer position changes, some parts on the image should occlude the other. This asks for a kind of a "z-buffering" algorithm, where only the pixels wich are closer to the viewer, are shown, and to rather complicated additional structures, necessary to preserve a visual quality. The last problem is that from a new position some parts of the image can be visible, that were originally occluded, and so did not existed on the initial image.

Our vectorization solves all these problems. First of all, we associate the depth information with our normal forms, as an additional parameter of the "brightness type". Consequently, it adds only 10–20% of the total (already highly compressed) information.

Secondly, we can easily perform within a rendering process, a "z-buffering" procedure on the vectorized data, and show only non-occluded parts of the normal forms. A high visual quality of the occlusion lines is provided by the profiles of the "separating" edges and ridges, which in this case are specially tuned. (Notice that these operations are performed on v-images, which are at least 5 times more compact that the pixel ones, and whose data is arranged following geometric patterns. Consequently, our rendering and "z-buffering" are very fast).

Finally, if as a result of a change in the viewer position, new image areas appear, which did not exist on the original image, they can be filled in by an "analytic continuation" of the existing normal forms. This "analytic continuation" can be performed completely automatically, and for small viewer position changes it does not create serious visual distortions.

Now, the first basic experimental fact is that *the extension of a 2D v-image to vectorized 3D-image, as described above, creates a completely photorealistic illusion of a 3D-scene, as the viewer position changes within 5°.*

However, there are additional features of our visual perception of 3D-scenes, which allow us to extend significantly the area of possible viewer position changes.

The first one we have found in our experiments: *visual sensitivity to the content of the "occluded strips" near the boundaries of the occluding objects, as the viewer moves, is very low.*

Consequently, we can shift the viewer position to much more than 5°, to use the "analytic continuation" as above, and most likely we will not notice any serious visual distortions.

The second fact has been presented in the talk of Jacques Ninio at this conference (see [9]): *we appreciate the depth of the objects only very roughly, and essentially replace the actual depth inside the object contours by its rough approximation.*

(We hope, Professor Ninio will agree with this interpretation of his results).

Our experiments strongly support this observation. In fact, as we only very roughly approximate the objects depth (by planes and ellipsoids), *a 3D-visual illusion remains practically complete, for the viewer position changes as big as 20–30 degrees.*

Figure 19 presents several views of a vectorized 3D-image, created from a single 2D picture by first vectorizing it and then interactively inserting a (plane approximation of a) depth function.

In conclusion, we believe that it is quite natural to consider *a vectorized 3D-image as a "normal form" of a three-dimensional scene: it represents the scene with the minimal number of parameters, up to a natural equivalence — creation of a photorealistic visual illusion.*

Animation of v-images. In real world what moves are usually objects. Our vectorization does not detect "objects" in the usual sense (objects detection is a difficult high level global image processing problem, while all our techniques are "low level" and local).

However, as a byproduct of the vectorization, we do detect a good portion of what we would like to call "objects": the computer simply does not know about it.

Consequently, before animating a v-image, we explain to the computer, what objects we would like to move. This is a simple interactive procedure, which consists in marking the edges and the ridges, contouring the desired object (and, if necessary, closing their gaps).

Then we insert to each object a "skeleton" — a simple polygonal line, and prescribe the way, in which all the normal forms, representing the object, follow the motions of the skeleton.

Now to animate the object it remains to define interactively the motions of the skeleton.

Figures 24–26 represent a couple of typical objects with their skeletons, and one of the results of their animation.

Figure 24. An original image and a vectorized object with its skleton

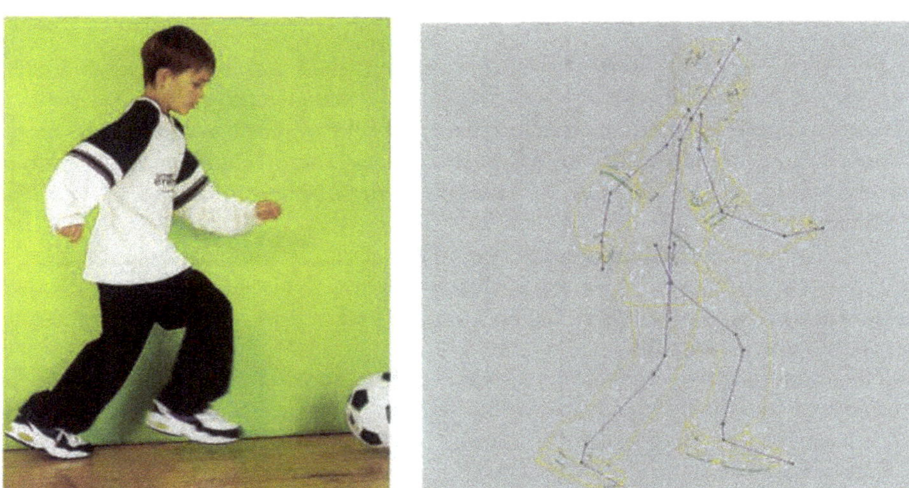

Figure 25. Another example of an object and of its vectorized represntation (with the skleton)

Figure 26. An animated virtual world, combining the objects shown on Figures 25 and 26

An important experimental fact here is that *this very simple animation scheme produces an illusion of a photorealistic motion not only for cartoon-like objects, as on Fig. 24, but also for fairly photorealistic ones, as on Fig. 25.*

Remember that we control only a very few degrees of freedom of the skeleton, while the number of the degrees of freedom of, say, a human body (and of the clothing!) is very big.

This allows us to call *a vectorized image of an object, together with its skeleton (and a motion scenario) the normal form of the object motion: it contains very few parameters and represents the motion up to our main equivalence — an optical illusion.*

Besides the main problem of motion representation, animation presents roughly the same additional difficulties, as the depth insertion: occlusions and completion of hidden areas. We solve them, creating (from a single image or from a combination of several ones) a vectorized 3D-virtual world, some objects of which can be animated independently. Figure 23 gives an example of such a world, created from a single still image.

Although our current implementation of depth insertion and animation is completely tailored for interactive work, this restriction is not too principal. Both the 3D-structure and the motion can be captured and represented by our v-images completely automatically.

The main point here is that the element of "objects detection" which we perform interactively (and which probably will be difficult to leave to the computer, at least in the near future), is not essential for our approach.

Exactly as for the initial normal form construction, we insist on a *faithful visual representation of the 3D-world and of its motions*, and not on a correct capturing of "objects". The structure of 3D animated vectorized images is sufficient for this purpose.

9 Possible implications for human visual perception

We've described above various experiments, concerning psychovisual significance of our normal forms and their parameters. Some of the possible conclusions for human visual perception are well-known, some other can be possibly new. Let us repeat that *in no way do we pretend to put any of these results as an established scientific fact in the field of psychology of human vision. None of them have been formulated or verified according to the criteria and the methodologies in this field.*

The only purpose of our experiments was to construct a good compression algorithm (exactly as in the experimental tuning of JPEG quantization tables).

On the other hand, we hope very much, that the vectorization technique, described above, can help in arranging rigorous experiments in this direction. Indeed, our v-images on one side give a faithful representation of any image, and, on the other side, provide a plenty of easily controlable image parameters, each one with a simple geometric or brightness meaning. Moreover, if v-images indeed provide a much higher compression and a much easier processing than other representation schemes, one can hope, that their models and parameters can really help in understanding of the human visual system, which still strongly overperforms any of our creations.

References

1. Weidong Kou, *Digital Image Compression: Algorithms and Standards*, Kluwer Academic Publishers, Boston, 1995.
2. V. I. Arnold, S. M. Gusein-Zade and A. N. Varchenko, *Singularities of differentiable maps, I.* Birkhäuser, Boston, 1985.
3. M. Golubitsky and V. Guillemin, *Stable mappings and their singularities. Graduate Texts in Mathematics* **14**, Springer-Verlag, New York, Heidelberg, Berlin, 1973.
4. R. Thom, *Structural Stability and Morphogenesis.* Benjamin/Cummings Publ., 1976.
5. Y. Elihai and Y. Yomdin, Flexible high order discretization, Preprint, 1989.
6. Y. Elihai and Y. Yomdin, Normal forms representation: A technology for image compression, *SPIE* Vol.1903, Image and Video Processing, 1993, 204–214.
7. Y. Elihai and Y. Yomdin, Flexible high order discretization of geometric data for global motion planning. *Theoretical Computer Science A,* **157**, Elsevier Science B.V. (1996), 53–77.
8. Z. Wiener and Y. Yomdin, From formal numerical solutions of elliptic PDE's to the true ones, *Mathematics of Computations*, to appear, 1998.
9. J. Ninio, Aspects of human shape perception, this volume.

www.ingramcontent.com/pod-product-compliance
Lightning Source LLC
Chambersburg PA
CBHW040806230926
51851CB00009B/11
* 9 7 8 9 8 1 0 2 3 7 9 2 9 *